Selected Titles in This Series

77 **Fumio Hiai and Dénes Petz,** The semicircle law, free random variables and entropy, 2000

76 **Frederick P. Gardiner and Nikola Lakic,** Quasiconformal Teichmüller theory, 2000

75 **Greg Hjorth,** Classification and orbit equivalence relations, 2000

74 **Daniel W. Stroock,** An introduction to the analysis of paths on a Riemannian manifold, 2000

73 **John Locker,** Spectral theory of non-self-adjoint two-point differential operators, 2000

72 **Gerald Teschl,** Jacobi operators and completely integrable nonlinear lattices, 1999

71 **Lajos Pukánszky,** Characters of connected Lie groups, 1999

70 **Carmen Chicone and Yuri Latushkin,** Evolution semigroups in dynamical systems and differential equations, 1999

69 **C. T. C. Wall (A. A. Ranicki, Editor),** Surgery on compact manifolds, second edition, 1999

68 **David A. Cox and Sheldon Katz,** Mirror symmetry and algebraic geometry, 1999

67 **A. Borel and N. Wallach,** Continuous cohomology, discrete subgroups, and representations of reductive groups, second edition, 2000

66 **Yu. Ilyashenko and Weigu Li,** Nonlocal bifurcations, 1999

65 **Carl Faith,** Rings and things and a fine array of twentieth century associative algebra, 1999

64 **Rene A. Carmona and Boris Rozovskii, Editors,** Stochastic partial differential equations: Six perspectives, 1999

63 **Mark Hovey,** Model categories, 1999

62 **Vladimir I. Bogachev,** Gaussian measures, 1998

61 **W. Norrie Everitt and Lawrence Markus,** Boundary value problems and symplectic algebra for ordinary differential and quasi-differential operators, 1999

60 **Iain Raeburn and Dana P. Williams,** Morita equivalence and continuous-trace C^*-algebras, 1998

59 **Paul Howard and Jean E. Rubin,** Consequences of the axiom of choice, 1998

58 **Pavel I. Etingof, Igor B. Frenkel, and Alexander A. Kirillov, Jr.,** Lectures on representation theory and Knizhnik-Zamolodchikov equations, 1998

57 **Marc Levine,** Mixed motives, 1998

56 **Leonid I. Korogodski and Yan S. Soibelman,** Algebras of functions on quantum groups: Part I, 1998

55 **J. Scott Carter and Masahico Saito,** Knotted surfaces and their diagrams, 1998

54 **Casper Goffman, Togo Nishiura, and Daniel Waterman,** Homeomorphisms in analysis, 1997

53 **Andreas Kriegl and Peter W. Michor,** The convenient setting of global analysis, 1997

52 **V. A. Kozlov, V. G. Maz′ya, and J. Rossmann,** Elliptic boundary value problems in domains with point singularities, 1997

51 **Jan Malý and William P. Ziemer,** Fine regularity of solutions of elliptic partial differential equations, 1997

50 **Jon Aaronson,** An introduction to infinite ergodic theory, 1997

49 **R. E. Showalter,** Monotone operators in Banach space and nonlinear partial differential equations, 1997

48 **Paul-Jean Cahen and Jean-Luc Chabert,** Integer-valued polynomials, 1997

47 **A. D. Elmendorf, I. Kriz, M. A. Mandell, and J. P. May (with an appendix by M. Cole),** Rings, modules, and algebras in stable homotopy theory, 1997

46 **Stephen Lipscomb,** Symmetric inverse semigroups, 1996

(Continued in the back of this publication)

The Semicircle Law, Free Random Variables and Entropy

Mathematical
Surveys
and
Monographs

Volume 77

The Semicircle Law, Free Random Variables and Entropy

Fumio Hiai
Dénes Petz

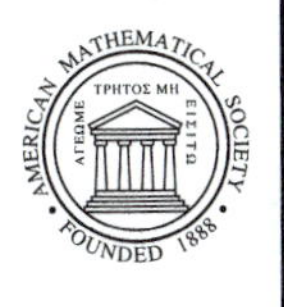

American Mathematical Society

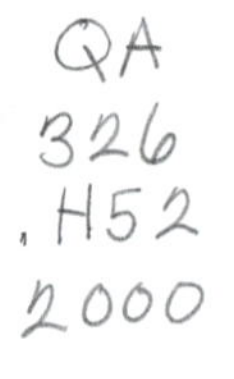

2000 *Mathematics Subject Classification.* Primary 46L54; Secondary 15A52, 60F10, 94A17, 46N50, 60J65, 81S25, 05A17.

ABSTRACT. This is an expository monograph on free probability theory. The emphasis is put on entropy and random matrix models. The highlight is the very far-reaching interrelation of free probability and random matrix theories. Wigner's theorem and its broad generalizations, such as asymptotic freeness of independent matrices, are expounded in detail. The parallelism between the normal and semicircle laws runs through the book. Many examples are included to illustrate the results. The frequent random matrix ensembles are characterized by maximization of their Boltzmann-Gibbs entropy under certain constraints, and the asymptotic eigenvalue distribution is treated in the almost everywhere sense and in the form of large deviation. Voiculescu's multivariate free entropy is presented with full proofs and extended to unitary operators. Some ideas about applications to operator algebras are also given.

Library of Congress Cataloging-in-Publication Data

Hiai, Fumio, 1948–
The semicircle law, free random variables, and entropy / Fumio Hiai, Dénes Petz.
p. cm. — (Mathematical surveys and monographs, ISSN 0076-5376 ; v. 77)
Includes bibliographical references and index.
ISBN 0-8218-2081-8 (alk. paper)
1. Free probability theory. 2. Random matrices. 3. Entropy. I. Petz, Dénes, 1953– II. Mathematical surveys and monographs ; no. 77.
QA326.H52 2000
512′.55–dc21 99-088288

∞ The paper used in this book is acid-free and falls within the guidelines
established to ensure permanence and durability.
Visit the AMS home page at URL: `http://www.ams.org/`

10 9 8 7 6 5 4 3 2 1 05 04 03 02 01 00

Contents

Preface

This book is based on the recent brilliant discoveries of Dan Voiculescu, which started from free products of operator algebras, but grew rapidly to include all sorts of other interesting topics. Although we both were fascinated by Voiculescu's beautiful new world from the very beginning, our attitude changed and our interest became more intensive when we got an insight into its interrelations with random matrices, entropy (or large deviations) and the logarithmic energy of classical potential theory.

There are many ways to present these ideas. In this book the emphasis is not put on operator algebras (Voiculescu's original motivation), but on entropy and random matrix models. It is not our aim to make a complete survey of all aspects of free probability theory. Several important recent developments are completely missing from this book. Our emphasis is on the role of random matrices. However, we do our best to make the presentation accessible for readers of different backgrounds.

The basis of this monograph was provided by lectures delivered by the authors at Eötvös Loránd University in Budapest, at Hokkaido University in Sapporo, and at Ibaraki University in Mito.

The structure of the monograph is as follows. Chapter 1 makes the connection between the concepts of probability theory and linear operators in Hilbert spaces. A sort of ideological foundation of noncommutative probability theory is presented here in the form of many examples. Chapter 2 treats the fundamental free relation. Again several examples are included, and the algebraic and combinatorial aspects of free single and multivariate random variables are discussed. This chapter is a relatively concise, elementary and self-contained introduction to free probability. The analytic aspects come in the next chapter. The infinitely divisible laws show an analogy with classical probability theory. This chapter is not much required to follow the rest of the monograph. Chapter 4 introduces the basic random matrix models and the limit of their eigenvalue distribution. Voiculescu's concept of asymptotic freeness originated from independent Gaussian random matrices. Since its birth, asymptotic freeness has been a very important bridge between free probability and random matrix theory. The strong analogy between the free relation and statistical independence is manifested in the asymptotic free relation of some independent matrix models. Entropy appears on the stage in Chapter 5—first the Boltzmann-Gibbs entropy, which is considered here as the rate function in some large deviation theorems. The frequent random matrix ensembles are characterized by maximization of the Boltzmann-Gibbs entropy under certain constraints. Several large deviation results are given following the pioneering work of Ben Arous and Guionnet on symmetric Gaussian random matrices. The main ingredient of the rate functional is the logarithmic energy, familiar from potential theory. For an n-tuple of noncommutative random variables, the probabilistic–measure theoretic model is lacking; hence Chapter 6 is technically in the field of functional analysis. Properties of Voiculescu's multivariate free entropy are discussed in the setting of operator algebras, and we introduce an analogous concept for n-tuples of unitaries. Chapters 3–6 comprise the main part of the monograph. The last chapter is mostly on free group factors, and gives ideas on applications to operator algebras.

Since rather different areas in mathematics are often combined, it was our intention to make the material nearly self-contained for the sake of convenience. This was a heavy task, and we had to cope with the combination of probabilistic, analytic, algebraic and combinatorial arguments. Each chapter concludes with some notes giving information on our sources and hints on further developments. Furthermore, we supply standard references for the reader who is not familiar with the general background of the chapter. The "Overview" is an attempt to show the place of the subject and to give orientation. It replaces an introduction, and the reader is invited to consult this part either before or after studying the much more technical chapters.

We thank many colleagues for helping us to finish this enterprise. Imre Csiszár, Roland Speicher, and Masaki Izumi can be named specifically. We are also grateful to several institutions for supporting our collaboration: funds of the Hungarian Academy of Sciences (OTKA F023447 and AKP 96/2-6782), of the Canon Foundation, of the Grant-in-Aid for Scientific Research (C)09640152, and of the Erdős Center are acknowledged.

Fumio Hiai and Dénes Petz

Overview

0.1 The isomorphism problem of free group factors

John von Neumann established the theory of so-called *von Neumann algebras* in the 1930's. The comprehensive study of "rings of operators" (as von Neumann algebras were called at that time) was motivated by the spectral theorem of selfadjoint Hilbert space operators and by the needs of the mathematical foundation of quantum mechanics. A von Neumann algebra is an algebra of bounded linear operators acting on a Hilbert space which is closed with respect to the topology of pointwise convergence. *Factors* are in a sense the building blocks of general von Neumann algebras; they are von Neumann algebras with trivial center. In a joint paper with F.J. Murray, a classification of the factors was given. Von Neumann was fond of the type II_1 factors, which are continuous analogues of the finite-dimensional matrix algebras. The normalized trace functional on the algebra of $n \times n$ matrices is invariant under unitary conjugation, and it takes the values k/n $(k = 0, 1, \dots, n)$ on projections. A type II_1 factor admits an abstract trace functional τ which is invariant under unitary conjugation, and it can take any value in $[0, 1]$ on projections. The N-fold tensor product of 2×2 matrix algebras is nothing else than the matrix algebra of $2^N \times 2^N$ matrices on which the normalized trace of projections is in the set $\{k/2^N : k = 0, 1, 2, \dots, 2^N\}$. In the limit as $N \to \infty$, the dyadic rationals fill the interval $[0, 1]$ and we arrive at a type II_1 factor. What we are constructing in this way is the infinite tensor product of 2×2 matrices, and the construction yields the *hyperfinite* type II_1 factors. ("Hyperfinite" means a good approximation by finite-dimensional subalgebras; in the above case approximation is by the finite tensor products with growing size.) This was the first example of a type II_1 factor. Murray and von Neumann showed that any two hyperfinite type II_1 factors are isomorphic, and they were looking for a non-hyperfinite factor.

Countable discrete groups give rise to von Neumann algebras; in fact one can associate to a discrete group G a von Neumann algebra $\mathcal{L}(G)$ in a canonical way. On the Hilbert space $\ell_2(G)$ the group G has a natural unitary representation $g \mapsto L_g$, the so-called left regular one, which is given by

$$(L_g\xi)(h) := \xi(g^{-1}h) \qquad (\xi \in \ell^2(G),\ g, h \in G).$$

The group ring $R(G)$ is the linear hull of the set $\{L_g : g \in G\}$ of unitaries. The *group von Neumann algebra* $\mathcal{L}(G)$ associated to G is by definition the closure of $R(G)$ in the topology of pointwise convergence. If the group under consideration is ICC (i.e. all its non-trivial conjugacy classes contain infinitely many elements), then the von Neumann algebra $\mathcal{L}(G)$ is a factor. When the closure of $R(G)$ is taken with respect to the norm topology of $B(\ell^2(G))$, we arrive at another important object, that is, the *reduced group C^*-algebra $C^*_r(G)$*. There exists a *canonical trace* τ on $\mathcal{L}(G)$, which is given by the unit element e of G. Let $\delta_e \in \ell_2(G)$ stand for the characteristic function of $\{e\}$ and define

$$\tau(\,\cdot\,) := \langle\,\cdot\,\delta_e, \delta_e\rangle\,.$$

Then it is easy to check that τ is a trace, i.e. it satisfies

$$\tau(ab) = \tau(ba) \quad \text{for all } a, b \in \mathcal{L}(G).$$

Von Neumann started from the free group with two generators and proved that the corresponding factor is not hyperfinite. Historically this led to the first example of two non-isomorphic type II_1 factors. Much later it was discovered that the group factor of an ICC group is hyperfinite if and only if the group itself is amenable, and free groups are the simplest non-amenable groups. (The concept of *amenable groups* also goes back to von Neumann.) Actually, von Neumann showed that the free product of groups leads to a non-hyperfinite factor. It seems that Richard Kadison was the person who explicitly posed the question of whether free groups with different numbers of generators could produce the same factor. This question is still open, and it was the main motivation for Dan Voiculescu to study the free relation and to develop free probability theory.

Let $\mathbf{F}_n$ denote the *free group with n generators*. If $n \neq m$ then $\mathbf{F}_n$ and $\mathbf{F}_m$ are not isomorphic. This can be seen by considering the group homomorphisms from $\mathbf{F}_n$ to $\mathbb{Z}_2$, the two element group, which is actually a field. Consider the space X_n of all homomorphisms from $\mathbf{F}_n$ to $\mathbb{Z}_2$. This is a bimodule over $\mathbb{Z}_2$ of dimension exactly 2^n. Since a group isomorphism between $\mathbf{F}_n$ and $\mathbf{F}_m$ induces a module isomorphism between X_n and X_m, the groups $\mathbf{F}_n$ and $\mathbf{F}_m$ for $n \neq m$ cannot be isomorphic. Although the free group $\mathbf{F}_n$ is contained in $\mathcal{L}(\mathbf{F}_n)$ in the form of the group ring, $\mathcal{L}(\mathbf{F}_n)$ is much larger than $R(\mathbf{F}_n)$ due to the closure procedure in the definition of the group factor. Hence simple proofs, for example the proof that $\mathbf{F}_2$ is not isomorphic to $\mathbf{F}_3$, do not extend to the topological closures, to the group C^*-algebras, or to the group von Neumann algebras.

If one expects $\mathcal{L}(\mathbf{F}_n)$ and $\mathcal{L}(\mathbf{F}_m)$ to be non-isomorphic, a possible strategy to prove this is to read out n intrinsically from $\mathcal{L}(\mathbf{F}_n)$. Each von Neumann factor of type II_1 has a unique canonical (faithful normal) tracial state τ, so this is intrinsically at our disposal. A.N. Kolmogorov in 1958 and Ya.G. Sinai in 1959 introduced the so-called Kolmogorov-Sinai (or dynamical) entropy of a measure-preserving transformation on a probability space, which has been a very successful tool in the isomorphism problem of Bernoulli shifts. Bernoulli shifts are simple, but they are the most important probabilistic dynamical systems. It turned out that the Kolmogorov-Sinai entropy is a complete invariant for Bernoulli shifts; in 1970

Donald Ornstein proved that two Bernoulli shifts are conjugate if and only if they have the same dynamical entropy. It seems to the authors that Dan Voiculescu was deeply influenced by these ideas. Independently of the success of his approach toward the isomorphism problem, his analysis created a new world in which the free relation and the free entropy of noncommutative random variables play the leading roles. Random matrices give the simplest examples of noncommutative random variables; they can model the free relation and are present in the definition of free entropy.

0.2 From the relation of free generators to free probability

The free group factor $\mathcal{L}(\mathbf{F}_n)$ is an important example in von Neumann algebra theory, since it is the first-found and simplest non-hyperfinite type II_1 factor. The free group itself is an important object in harmonic analysis; in fact, there is a strong and intimate relation between harmonic analysis on the free group and the structure of the free group factor.

The fundamental relation of the generators of a free group is the algebraic "free relation"; namely, there are no identities expressed in operations of the group which are satisfied by the generators. Eventually, this algebraic free relation is present in non-free groups as well. For instance, if G is the free product of G_1 and G_2, then the elements of G_1 are in free relation to the elements of G_2. The study of harmonic analysis on free product groups is the obvious continuation of its study on free groups.

Next we outline an idea which starts from a random walk on free groups and leads to the key concept of free relation; on top of that, the role of the semicircle law becomes clear as well. Let $g_1, g_2, \dots, g_n$ be generators of $\mathbf{F}_n$ and consider a random walk on this group which starts from the group unit and goes from the group element g to hg with probability $1/2n$ if $h \in \{g_1, g_2, \dots, g_n, g_1^{-1}, g_2^{-1}, \dots, g_n^{-1}\}$. Then the probability of return to the unit e in m steps is of the form

$$P(n,m) := \frac{1}{(2n)^m} \left\langle \left(L_{g_1} + L_{g_1^{-1}} + L_{g_2} + L_{g_2^{-1}} + \ldots + L_{g_n} + L_{g_n^{-1}}\right)^m \delta_e, \delta_e \right\rangle .$$

(This vanishes if m is not even.) In the probabilistic interpretation of quantum mechanics, it is standard to interpret the number $\langle A\xi, \xi \rangle$ as the expectation of the operator A in the state vector ξ. Adopting this view, we have the expectation

$$\left\langle \Big(\sum_{i=1}^{n} A_i^{(n)} \Big)^m \delta_e, \delta_e \right\rangle$$

of selfadjoint operators, $A_i^{(n)}$ standing for $(L_{g_i} + L_{g_i^{-1}})/\sqrt{2}$. The operators $A_i^{(n)}$

might be called random variables in the sense that they have expectations, and to distinguish this concept from classical probability theory, we speak of *noncommutative random variables.* The asymptotic behavior of the return probability $P(n, 2m)$ is given as follows:

$$P(n, 2m) \approx \frac{1}{(2n)^m} \frac{1}{m+1} \binom{2m}{m} \quad \text{as} \quad n \to \infty.$$

What we have is a sort of *central limit theorem* for the array

$$A_1^{(n)}, A_2^{(n)}, \dots, A_n^{(n)}$$

of noncommutative random variables, because

$$\frac{A_1^{(n)} + A_2^{(n)} + \dots + A_n^{(n)}}{\sqrt{n}}$$

converges in moments (or in distribution) to the even classical probability density whose $2m$th moment is the so-called *Catalan number*

$$\frac{1}{m+1} \binom{2m}{m} \qquad (m \in \mathbb{N}).$$

This limit density is the *semicircle* or *Wigner law* $\frac{1}{2\pi}\sqrt{4-x^2}$ with support on $[-2, 2]$.

The different generators g_i are free in the algebraic sense, and this led Voiculescu to call the relation of the selfadjoint operators $A_i^{(n)}$ (for a fixed n) free as well. When one aims to formulate the concept in the spirit of probability, the plausible free relation of these noncommutative random variables must be formulated in terms of expectation. The noncommutative random variables $A_1, A_2, \dots, A_n$ are called *free* with respect to the expectation φ if

$$\varphi\big(P_1(A_{i(1)})P_2(A_{i(2)}) \dots P_k(A_{i(k)})\big) = 0$$

whenever $P_1, P_2, \dots, P_n$ are polynomials, $P_j(A_{i(j)}) = 0$ and $i(1) \neq i(2) \neq \dots \neq i(k)$. In the motivating example we had $\varphi(\,\cdot\,) = \langle \,\cdot\, \delta_e, \delta_e \rangle$, and of course the above defined operators $A_1^{(n)}, A_2^{(n)}, \dots, A_n^{(n)}$ are free in the sense of Voiculescu's definition. It is not immediately obvious from the definition that the free relation of the noncommutative random variables $A_1, A_2, \dots, A_n$ is a particular rule to calculate the joint moments

$$\varphi\big(A_{i(1)}^{m(1)} A_{i(2)}^{m(2)} \dots A_{i(k)}^{m(k)}\big)$$

(for positive integers $m(1), m(2), \dots, m(k)$) from the moments $\varphi(A_i^m)$ of the given variables.

It is always tempting to compare Voiculescu's free relation with the independence of classical random variables. The comparison cannot be formal, since the free relation and independence do not take place at the same time. What we have in mind is only the analogy of the free relation in noncommutative probability theory with independence in classical probability theory. The part of noncommutative probability in which the free relation plays a decisive role is called *free probability theory.*

Free probability theory has its celebrated distributions. The semicircle law is one of those celebrities; it comes from the free central limit theorem and its moment sequence is the Catalan numbers. The free analogue of the Poisson limit theorem can be established, and it gives the free analogue of the Poisson distribution. Given measures μ and ν, we can find noncommutative random variables a and b with these distributions. When a and b are in free relation, the distribution measure of $a+b$ (or ab) can be called the *additive* (or *multiplicative*) *free convolution* of μ and ν. The notations $\mu \boxplus \nu$ and $\mu \boxtimes \nu$ are used to denote the two kinds of free convolution. The class of semicircle laws is closed under additive free convolution. The distributions $\mu_0 \boxplus w_{\sqrt{4t}}$ form a convolution semigroup when $w_{\sqrt{4t}}$ stands for the semicircle law with variance t (which corresponds to radius $\sqrt{4t}$). The analytic machinery to handle additive free convolution is based on the Cauchy transform of measures. If μ_t is a freely additive convolution semigroup and $G(z,t)$ is the Cauchy transform of μ_t, then the complex Burger equation

$$\frac{\partial G(z,t)}{\partial t} + R(G(z,t)) = \frac{\partial G(z,t)}{\partial z}$$

is satisfied with initial condition $G(z,0) = G_{\mu_0}(z)$, and R is the so-called R-transform of μ_1. If μ_t is the semigroup of semicircle laws, then $R(z) = z$ and we have the analogue of the heat equation.

There is an obvious way to associate a measure to a noncommuting random variable via the spectral theorem if it is a normal operator. If the operator is not normal, then a measure can still be constructed by using type II_1 von Neumann algebra techniques. (What we have in mind is the Brown measure of an element of a type II_1 von Neumann algebra.) However, if we deal with several noncommuting random variables which are really noncommuting operators, then there is no way to reduce the discussion to measures. At this level of generality free probability theory has to work with joint moments, power series on noncommuting indeterminates, and a new kind of combinatorial arguments. The picture becomes very different from classical probability theory. Computation of the joint moments of free noncommutative random variables is a rather combinatorial procedure.

0.3 Random matrices

The joint eigenvalue density for certain symmetric random matrices has been known for a long time. J. Wishart found it for the so-called (real) Wishart matrix back in 1928. In quantum physics the energy is represented by the Hamiltonian operator, and one is interested in the point spectrum of the Hamiltonian, which is the set of eigenvalues. The problem is difficult; we do not know the exact Hamiltonian, and even if we did, it would be too complicated to find the eigenvalues. An approach to this problem is based on the following statistical hypothesis: The statistical behavior of the energy levels is identical with the behavior of the eigenvalues of a large random matrix. Reasoning along this line, in 1955 E.P. Wigner obtained the semicircle law for the limiting eigenvalue density of random matrices with independent Bernoulli entries. Wigner's work initiated a huge and deep interest in random matrices from theoretical and nuclear physicists. From the point of view of physical applications the most interesting question is to treat the correlation functions of the eigenvalues and the so-called "level spacing".

Random matrices are noncommutative random variables with respect to the expectation

$$\tau_n(H) := \frac{1}{n}\sum_{i=1}^{n} E(H_{ii})$$

for an $n \times n$ random matrix H, where E stands for the expectation of a classical random variable. It is a form of the *Wigner theorem* that

$$\tau_n(H(n)^{2m}) \to \frac{1}{m+1}\binom{2m}{m} \quad \text{as} \quad n \to \infty$$

if the $n \times n$ real symmetric random matrix $H(n)$ has independent identical Gaussian entries $N(0, 1/n)$ so that $\tau_n(H(n)^2) = 1$. The semicirle law is the limiting eigenvalue density of $H(n)$'s as well as the limiting law of the free central limit theorem in the previous section. The reason why this is so was made clear by Voiculescu. Let $X_1(n), X_2(n), \ldots$ be independent random matrices with the same distribution as $H(n)$. It follows from the properties of Gaussians that the distribution of the random matrix

$$\frac{X_1(n) + X_2(n) + \cdots + X_n(n)}{\sqrt{n}}$$

is the same as that of $H(n)$. Hence the convergence in moments to the semicircle law would be understandable if $X_1(n), X_2(n), \ldots, X_n(n)$ were in free relation. Their free relation with respect to the expectaion τ_n would include the condition

$$\tau_n\Big(\big[X_1(n)^k - \tau_n(X_1(n)^k)\big]\big[X_2(n)^l - \tau_n(X_2(n)^l)\big]\Big) = 0\,,$$

which is equivalently written as

$$\tau_n\big(X_1(n)^k X_2(n)^l\big) = \tau_n\big(X_1(n)^k\big)\tau_n\big(X_2(n)^l\big)\,. \tag{1}$$

For notational simplicity we write A and B for $X_1(n)$ and $X_2(n)$, respectively.

Then what we have on the left hand side is

$$\begin{aligned}
&\frac{1}{n^2}\sum E\big(A_{i(1)i(2)}A_{i(2)i(3)}\cdots A_{i(k)i(k+1)}B_{i(k+1)i(k+2)}\cdots B_{i(k+l)i(1)}\big)\\
&\quad=\frac{1}{n^2}\sum E\big(A_{i(1)i(2)}A_{i(2)i(3)}\cdots A_{i(k)i(k+1)}\big)E\big(B_{i(k+1)i(k+2)}\cdots B_{i(k+l)i(1)}\big)
\end{aligned}$$

with summation for all indices. The matrix elements are independent and have zero expectation. Hence a term in which a matrix element appears only once among the factors must vanish. On the right hand side of (1) we have

$$\begin{aligned}
&\frac{1}{n}\sum E\big(A_{i(1)i(2)}A_{i(2)i(3)}\cdots A_{i(k)i(1)}\big)\\
&\quad\times\frac{1}{n}\sum E\big(B_{i(k+1)i(k+2)}\cdots B_{i(k+l)i(k+1)}\big)\,.
\end{aligned}$$

The two expressions are equal in many cases, in particular when k or l is odd or when both are 0. However the summation is over slightly different sequences of indices. The difference goes to 0 as $n \to \infty$. In this way, instead of the equality in (1) for a finite n, we have identical limits as $n \to \infty$. The free relation appears only in the limit; this is called the *asymptotic freeness* of $X_1(n)$ and $X_2(n)$. More generally, the asymptotic freeness of the sequence $X_1(n), X_2(n), \dots$ is formulated. The free relation is present in the random matrix context asymptotically, and this fact explains why the semicircle law is the limiting eigenvalue distribution of the random matrix $H(n)$.

Independent symmetric Gaussian matrices are asymptotically free, but there are several other interesting examples too. For instance, independent Haar distributed unitary matrices are asymptotically free (as the matrix size tends to infinity). The asymptotic freeness may serve as a bridge connecting random matrix theory with free probability theory.

In a very abstract sense, the distribution of a not neccessarily selfadjoint non-commutative random variable A is the collection of all joint moments

$$\varphi(A^{m(1)}A^{*m(2)}A^{m(3)}A^{*m(4)}\dots)$$

of A and its adjoint A^*. A random matrix model of A is a sequence of random matrices $X(n)$ such that $X(n)$ is $n \times n$ and the joint moments of these matrices reproduce those of A as $n \to \infty$, that is,

$$\tau_n(X(n)^{m(1)}X(n)^{*m(2)}X(n)^{m(3)}\dots) \to \varphi(A^{m(1)}A^{*m(2)}A^{m(3)}\dots)$$

for all finite sequences $m(1), \dots, m(k)$ of nonnegative integers. It is really amazing that many important distributions appearing in free probability theory admit a suitable random matrix model. The selfadjoint Gaussian matrices with independent entries were mentioned above in connection with the Wigner theorem. The free analogue of the Poisson distribution is related to the Wishart matrix, and the circular and elliptic laws come from non-selfadjoint Gaussian matrices. These facts provide room for the interaction between random matrices and free probability. On the one hand, approximation by random matrices gives a powerful method to study free probability theory, and on the other hand techniques of free probability can be used to determine the limiting eigenvalue distribution of some random matrices, for example. It seems that bi-unitarily invariant matrix ensembles form an important class; they give the random matrix model of R-diagonal distributions. One way to define an R-diagonal noncommutative random variable is to consider its polar decomposition uh. If u is a Haar unitary and it is free from h, then uh is called R-diagonal. A bi-unitarily invariant random matrix has a polar decomposition UH in which U is a Haar distributed unitary matrix independent of H. As the matrix size grows, independence is converted into freeness according to some asymptotic freeness result.

The *empirical eigenvalue distribution* of an $n\times n$ random matrix H is the random atomic measure

$$R_H := \frac{1}{n}\big(\delta(\lambda_1) + \delta(\lambda_2) + \cdots + \delta(\lambda_n)\big),$$

where $\lambda_1, \lambda_2, \ldots, \lambda_n$ are the eigenvalues of H. It is a stronger form of the Wigner theorem that $R_{H(n)}$ goes to the semicircle law almost everywhere when $H(n)$ is symmetric with independent identically distributed entries. The almost sure limit of the empirical eigenvalue distribution is known to be a non-random measure in many examples, and it is often called the *density of states*. Sometimes it cannot be given explicitly, but is determined by a functional equation for the Cauchy transform or by a variational formula. Results are available for non-selfadjoint random matrices as well.

The best worked out example of symmetric random matrices is the case of independent identically distributed Gaussian entries. If the entries are Gaussian $N(0,1/n)$ and if G is an open subset of the space of probability measures on $\mathbb{R}$ such that the semicircle law (the density of states) w is not in the closure, then we have

$$\mathbf{Prob}\big(R_{H(n)} \in G\big) \approx \exp(-n^2 C(w,G))$$

with a positive constant $C(w,G)$ depending on the distance of w from G. The bigger the distance of w from G, the larger the constant $C(w,G)$ is. Large deviation theory expresses this constant as the infimum of a so-called *rate function* I defined on the space of measures:

$$C(w,G) = \inf\{I(\mu) : \mu \in G\}.$$

In our example, I is a stricly convex function which is stricly positive for $\mu \neq w$. Ingredients of $I(\mu)$ are the logarithmic energy and the second moment of μ. What we are sketching now is the pioneering large deviation theorem of Ben Arous and Guionnet for symmetric Gaussian matrices. More details can be found in the next section.

0.4 Entropy and large deviations

Originally entropy was a quantity from physics. Entropy as a mathematical concept is deeply related to large deviations, although the two had independent lives for a long time. A typical large deviation result was discovered by I.N. Sanov in 1957; however, the general abstract framework of large deviations was given by S.R.S. Varadhan in 1966.

Let $\xi_1, \xi_2, \ldots$ be independent standard Gaussian random variables and let G be an open set in the space $\mathcal{M}(\mathbb{R})$ of probability measures on $\mathbb{R}$ (with the weak topology). The *Sanov theorem* says that if the standard Gaussian measure ν is not in the closure of G, then

$$\mathbf{Prob}\Big(\frac{\delta(\xi_1)+\delta(\xi_2)+\cdots+\delta(\xi_n)}{n} \in G\Big) \approx \exp(-nC(\nu, G))$$

and

$$C(\nu, G) = \inf\{I(\mu) : \mu \in G\}\,.$$

In the above case, the rate function $I(\mu)$ is the *relative entropy* (or the *Kullback-Leibler divergence*) $S(\mu, \nu)$ of μ with respect to ν. So it is also written as

$$I(\mu) = -S(\mu) + \frac{1}{2}\int x^2\, d\mu(x) + \frac{1}{2}\log(2\pi) \tag{2}$$

with the *Boltzmann-Gibbs entropy*

$$S(\mu) := -\int p(x)\log p(x)\, dx\,, \tag{3}$$

whenever μ has the density $p(x)$ and the logarithmic integral is meaningful. This rate function I is a strictly convex function such that $I(\mu) > 0$ if $\mu \neq \nu$.

The rate functions in some large deviation results are called entropy functionals. Eventually, this could be the definition of entropy. The logarithmic integral (3) of a probability distribution $p(x)$ has been used for a long time, but it was identified much later as a component of the rate function in the Sanov theorem.

The Boltzmann-Gibbs entropy $S(\mu)$ can be recovered from the asymptotics of probabilities. Let ν^n be the n-fold product of the standard Gaussian measure on $\mathbb{R}$. For each $x \in \mathbb{R}^n$ we have the discrete measure

$$\kappa_x := \frac{\delta(x_1)+\delta(x_2)+\ldots+\delta(x_n)}{n},$$

which can be used to approximate the given measure μ. The asymptotic volume of the approximating $\mathbb{R}^n$-vectors up to the first r moments is given by

$$\frac{1}{n}\log\nu^n\big(\{x \in \mathbb{R}^n : |m_k(\kappa_x) - m_k(\mu)| \le \varepsilon,\, k \le r\}\big). \tag{4}$$

A suitable limit of this as $n \to \infty, r \to \infty$ and $\varepsilon \to 0$ is exactly the above described rate function (2). Furthermore, from (4) the entropy $S(\mu)$ can be recovered. This crucial argument deduces the entropy from the asymptotics of probabilities. The extension of this argument works for multivariables, but first we consider the analogous situation in which the atomic measures κ_x are replaced by symmetric matrices.

In the large deviation result of Ben Arous and Guionnet the explicit form of the rate function I on $\mathcal{M}(\mathbb{R})$ is

$$I(\mu) = -\frac{1}{2} \iint \log|x-y| \, d\mu(x)\, d\mu(y) + \frac{1}{4} \int x^2 \, d\mu(x) + \text{const.} \tag{5}$$

The above double integral is the (negative) logarithmic energy of μ, which is very familiar from potential theory.

Here we give an outline of how the rate function in (5) arises in the large deviation theorem of Ben Arous and Guionnet. In an abstract setting, a large deviation is considered for a sequence (P_n) of probability distributions, usually on a Polish space $\mathcal{X}$ in the scale (L_n); in our example $\mathcal{X} = \mathcal{M}(\mathbb{R})$ and $P_n(G) = \mathbf{Prob}(R_{H(n)} \in G)$. A standard way of proving the large deviation in this setting is to show the following equality:

$$I(x) = \sup\Big[-\limsup_{n\to\infty} L_n \log P_n(G)\Big] = \sup\Big[-\liminf_{n\to\infty} L_n \log P_n(G)\Big],$$

where the supremum is over neighborhoods G of $x \in \mathcal{X}$. This equality gives the large deviation of (P_n) with the rate function I if (P_n) satisfies an additional property of a stronger form of tightness. The scale in the Sanov large deviation is $L_n = n^{-1}$, but the scale in large deviations related to random matrices is $L_n = n^{-2}$, corresponding to the number of entries of an $n \times n$ matrix. The joint eigenvalue density of the relevant random matrix $H(n)$ is known to be

$$\frac{1}{Z_n} \exp\Big(-\frac{n+1}{4} \sum_{i=1}^n x_i^2\Big) \prod_{i<j} |x_i - x_j| \tag{6}$$

on $\mathbb{R}^n$ with the normalizing constant Z_n. This means that for a neighborhood G of $\mu \in \mathcal{M}(\mathbb{R})$ one has

$$\begin{aligned}\mathbf{Prob}&(R_{H(n)} \in G) \\ &= \frac{1}{Z_n} \int \cdots \int_{\tilde G} \exp\Big(\sum_{i<j} \log|x_i - x_j| - \frac{n+1}{4} \sum_{i=1}^n x_i^2\Big) dx_1 \cdots dx_n \,,\end{aligned}$$

where $\tilde G \subset \mathbb{R}^n$ is defined by

$$\tilde G := \Big\{ x \in \mathbb{R}^n : \frac{1}{n} \sum_{i=1}^n \delta(x_i) \in G \Big\}.$$

Very roughly speaking, when G goes to a point μ, the approximation

$$\sum_{i<j} \log|x_i - x_j| - \frac{n+1}{4}\sum_{i=1}^{n} x_i^2 \approx n^2\left(\frac{1}{2}\iint \log|x-y|\,d\mu(x)\,d\mu(y) - \frac{1}{4}\int x^2\,d\mu(x)\right)$$

holds for $x \in \tilde{G}$, and one gets

$$-\frac{1}{n^2}\log \mathbf{Prob}\big(R_{H(n)} \in G\big) \approx -\frac{1}{2}\iint \log|x-y|\,d\mu(x)\,d\mu(y) + \frac{1}{4}\int x^2\,d\mu(x) + \frac{1}{n^2}\log Z_n\,.$$

This gives rise to the rate function (5), and the constant term there comes from $\lim_{n\to\infty} n^{-2}\log Z_n$.

Besides the symmetric Gaussian matrix, we know the exact form of the joint eigenvalue density for several other random matrices, such as the selfadjoint or non-selfadjoint Gaussian matrix, the Wishart matrix, the Haar distributed unitary matrix, and so on. The joint densities are distributed on $\mathbb{R}^n, \mathbb{C}^n, (\mathbb{R}^+)^n, \mathbb{T}^n$ depending on the type of matrices, but they have a common form which is a product of two kernels as in (6); one is the kernel of Vandermonde determinant type

$$\prod_{i<j} |x_i - x_j|^{2\beta}$$

with some constant $\beta > 0$, and the other is of the form

$$\exp\left(-\sum_{i=1}^{n} Q_n(x_i)\right)$$

with some function Q_n depending on n. Applying the method outlined above to this form of joint density, we can show the large deviations for the empirical eigenvalue distribution of random matrices as above. Corresponding to the form of joint density, the rate function is a weighted logarithmic potential as in (5) and its main term is always the logarithmic energy.

What is the free probabilistic analogue of the Boltzmann-Gibbs entropy (3) of a probability distribution μ on $\mathbb{R}$? Voiculescu answered this question by looking at the asymptotic behavior of the Bolzmann-Gibbs entropy of random matrices: The free entropy of μ should be

$$\Sigma(\mu) := \iint \log|x-y|\,d\mu(x)\,d\mu(y)\,, \tag{7}$$

that is, minus the logarithmic energy. Below we shall explain that the above double integral is the real free analogue of the Boltzmann-Gibbs entropy.

Let $\mathcal{M}$ be a von Neumann algebra with a faithful normal tracial state τ; so τ gives the expectation of elements of $\mathcal{M}$. Moreover, $m_k(a) := \tau(a^k)$ is viewed as the kth moment of a noncommutative random variable $a \in \mathcal{M}$. In order to approximate a selfadjoint a with $\|a\| \le R$ in distribution, Voiculescu suggested using symmetric matrices $A \in M_n(\mathbb{R})$; approximation means that $|\mathrm{tr}_n(A^k) - \tau(a^k)|$ is small. Hence the analogue of (4) is

$$\frac{1}{n^2} \log \nu_n\bigl(\{A \in M_n(\mathbb{R})^{sa} : \|A\| \le R,\ |\mathrm{tr}_n(A^k) - \tau(a^k)| \le \varepsilon,\ k \le r\}\bigr).$$

The scaling is changed into n^2 corresponding to the higher degree of freedom, and the measure ν_n must be a measure on the space of real symmetric matrices; again the standard Gaussian measure will do. The limit as $n \to \infty$ and then $r \to \infty$, $\varepsilon \to 0$ is

$$\frac{1}{2}\iint \log|x-y|\,d\mu(x)\,d\mu(y) - \frac{1}{4}\tau(a^2) + \text{const.}, \tag{8}$$

where μ is the probability measure on $\mathbb{R}$ which has the same moments as a: $\int x^k\,d\mu(x) = \tau(a^k)$. This limit is minus the rate function (5), and the first term gives the free entropy $\Sigma(\mu)$. Instead of $M_n(\mathbb{R})^{sa}$ one may use the space $M_n(\mathbb{C})^{sa}$ of selfadjont matrices together with the standard Gaussian measure on it.

Another analogy between the two entropies $S(\mu)$ and $\Sigma(\mu)$ is clarified by their maximization results. The entropy $S(\mu)$ can take any value in $[-\infty, +\infty]$. Instead of the value itself, rather important is the difference of $S(\mu)$ from the maximum under some constraint. For instance, under the constraint of the second moment $m_2(\mu) \le 1$, the Boltzmann-Gibbs entropy has the upper bound $S(\mu) \le \frac{1}{2}\log(2\pi e)$, and equality is attained here if and only if μ has the normal distribution $N(0,1)$. This fact is readily verified from the positivity of the relative entorpy $S(\mu,\nu)$ with $\nu = N(0,1)$. On the other hand, under the constraint of the second moment $m_2(\mu) \le 1$, the free entropy has the upper bound $\Sigma(\mu) \le -1/4$, and equality is attained if and only if μ is the semicircle law of radius 2. This maximization of $\Sigma(\mu)$ resembles that of $S(\mu)$; their maximizers are the normal law and the semicircle law, and the latter is the free analogue of the former. The Gaussian and semicircle maximizations are linked by random matrices. The symmetric random matrix with maximal Boltzmann-Gibb entropy under the constraint $\tau_n(H^2) \le 1$ is the standard Gaussian matrix, which is a random matrix model of the semicircle law.

0.5 Voiculescu's free entropy for multivariables

The free entropy as well as the Boltzmann-Gibbs entropy can be extended to multivariables. The multivariate case is slightly more complicated, but conceptually it is exactly the same. First we consider the Boltzmann-Gibbs entropy of multi-random variables (i.e. random vectors). For a random vector $(X_1, X_2, \dots, X_N)$ one has the joint distribution μ on $\mathbb{R}^N$, and the logarithmic integral (3) is meaningful and functions well whenever μ has the density $p(x)$ on $\mathbb{R}^N$. For $x = (x_1, x_2, \dots, x_n) \in (\mathbb{R}^N)^n$ let κ_x be the atomic measure on $\mathbb{R}^N$ defined analogously to the above single variable case. Now $k = (k(1), k(2), \dots, k(p))$ must be a multi-index of length $|k| := p$. For a measure μ on $\mathbb{R}^N$ whose support is in $[-R, R]^N$, we define

$$m_k(\mu) := \int x_{k(1)} x_{k(2)} \cdots x_{k(p)} \, d\mu(x)$$

and consider

$$\frac{1}{n} \log \nu^n \big(\{x \in ([-R, R]^N)^n : |m_k(\kappa_x) - m_k(\mu)| \le \varepsilon, \ |k| \le r\}\big), \tag{9}$$

where ν^n is the n-fold product of Gaussian measures on $\mathbb{R}^N$. The usual limit as $n \to \infty$ and then $r \to \infty$, $\varepsilon \to 0$ is

$$-S(\mu) + \frac{1}{2} \int \left(x_1^2 + x_2^2 + \cdots + x_N^2\right) d\mu(x) + \text{const.}$$

Now let $(a_1, a_2, \dots, a_N)$ be an N-tuple of selfadjoint noncommutative random variables. Due to the noncommutativity we cannot have the joint distribution (as a measure); however the mixed joint moments of $(a_1, a_2, \dots, a_N)$ with respect to the tracial state τ are available and we can consider the analogue of (9). For a multi-index k we set $m_k(a_1, a_2, \dots, a_N) := \tau(a_{k(1)} a_{k(2)} \cdots a_{k(p)})$, and similarly $m_k(A_1, A_2, \dots, A_N) := \mathrm{tr}_n(A_{k(1)} A_{k(2)} \cdots A_{k(p)})$ for an N-tuple $(A_1, A_2, \dots, A_N)$ of $n \times n$ matrices. To deal with the quantity

$$\begin{aligned} \frac{1}{n^2} \log \nu_n \big(\{(A_1, A_2, \dots, A_N) \in (M_n(\mathbb{C})^{sa})^N : \|A_i\| \le R,& \\ |m_k(A_1, A_2, \dots, A_N) - m_k(a_1, a_2, \dots, a_N)| \le \varepsilon, \ |k| \le r\}\big),& \end{aligned} \tag{10}$$

we again put a product measure ν_n on the set of selfadjoint matrices. Since it is not known whether the limit as $n \to \infty$ of (10) exists, the lim sup as $n \to \infty$ may be considered. The limit as $r \to \infty$, $\varepsilon \to 0$ of $\limsup_{n \to \infty}$ of the quantity (10) is of the form

$$\chi(a_1, a_2, \dots, a_N) + \frac{1}{2} \tau(a_1^2 + a_2^2 + \cdots + a_N^2) + \text{const.},$$

independently of the choice of $R > \|a_i\|$. This defines Voiculescu's free entropy $\chi(a_1, a_2, \dots, a_N)$. The multivariate free entropy generalizes the above free entropy $\Sigma(\mu)$ (up to an additive constant); to be more precise, the equality

$$\chi(a) = \Sigma(\mu) + \frac{1}{2} \log(2\pi) + \frac{3}{4}$$

is valid with the distribution μ of a. The term "free" has nothing to do with thermodynamics; it comes from the additivity property:

$$\chi(a_1, a_2, \ldots, a_N) = \chi(a_1) + \chi(a_2) + \cdots + \chi(a_N) \tag{11}$$

when $a_1, a_2, \ldots, a_N$ are in free relation with respect to the expectation τ. In this spirit the Boltzmann-Gibbs entropy must be called "independent" since it is additive if and only if $X_1, X_2, \ldots, X_N$ are independent; that is, the joint distribution μ is a product measure. When the noncommutative random variables are far away from freeness (in particular, when an algebraic relation holds for $a_1, a_2, \ldots, a_N$), their free entropy becomes $-\infty$. This is another reason for the terminology "free entropy". The additivity (11) is equivalent to the free relation of the a_i's when $\chi(a_i) > -\infty$, but the subadditivity

$$\chi(a_1, a_2, \ldots, a_N) \le \chi(a_1) + \chi(a_2) + \cdots + \chi(a_N)$$

always holds. The free entropy $\chi(a_1, a_2, \ldots, a_N)$ is upper semicontinuous in the convergence in joint moments. Furthermore, certain kinds of change of variable formulas are available. Under the constraint for $a_i = a_i^*$ that $\sum_i \tau(a_i^2)$ is fixed, $\chi(a_1, a_2, \ldots, a_N)$ is maximal when (and only when) all a_i's are free and semicircular. There are possibilities to extend $\chi(a_1, a_2, \ldots, a_N)$ to the case of non-selfadjoint noncommutative random variables. One possibility is to allow non-selfadjoint matrices in the definition (10), and another is to split the non-selfadjoint operators into their real and imaginary parts. The two approaches give the same result, say $\hat{\chi}(a_1, a_2, \ldots, a_N)$, where the N-tuple is arbitrary and not necessarily selfadjoint. The subadditivity is still true, and $\hat{\chi}(a_1, a_2, \ldots, a_N) = -\infty$ when one of the a_i's is normal, in paricular when all of them are unitaries.

For an N-tuple of unitaries $(u_1, u_2, \ldots, u_N)$ the appropriate way leading to a good concept of entropy is to use unitary matrices in a definition similar to (10), and to measure the volume of the approximating unitary matrices by the Haar measure. In this way we arrive at $\chi_u(u_1, u_2, \ldots, u_N)$. The free entropy of unitary variables has properties similar to the free entropy of selfadjoint ones; namely, the subadditivity and the upper semicontinuity hold, and additivity is equivalent to freeness. The three kinds of free entropies are connected under the polar decompositions $a_i = u_i h_i$ in the following way:

$$\hat{\chi}(a_1, a_2, \ldots, a_N) \le \chi_u(u_1, u_2, \ldots, u_N) + \chi(h_1^2, h_2^2, \ldots, h_N^2) + \frac{N}{2}\left(\log\frac{\pi}{2} + \frac{3}{2}\right),$$

and, furthermore. equality is valid under a freeness assumption.

0.6 Operator algebras

The study of (selfadjoint) operator algebras is divided into two major categories, C^*-algebras and von Neumann algebras (i.e. W^*-algebras). *C^*-algebras* are usually introduced in an axiomatic way: A C^*-algebra is an involutive Banach algebra satisfying the C^*-norm condition $\|a^*a\| = \|a\|^2$. But any C^*-algebra is represented as a norm-closed *-algebra of bounded operators on a Hilbert space (Gelfand-Naimark representation theorem). Von Neumann algebras are included in the class of C^*-algebras; however, the ideas and methods in the two categories are very different. A commutative C^*-algebra with unit is isomorphic to $C(\Omega)$, the C^*-algebra of continuous complex functions on a compact Hausdorff space Ω with sup-norm (another Gelfand-Naimark theorem). So a general C^*-algebra is sometimes viewed as a "noncommutative topological space". On the other hand, a von Neumann algebra is a noncommutative analogue of (probability) measure spaces. In fact, a commutative von Neumann algebra with a faithful normal state is isomorphic to the space $L^\infty(\Omega, \mu)$ over a standard Borel space (Ω, μ).

According to von Neumann's reduction theory, a von Neumann algebra $\mathcal{M}$ on a separable Hilbert space is a sort of direct integral of factors:

$$\mathcal{M} = \int_\Gamma^\oplus \mathcal{M}(\gamma)\, d\nu(\gamma)\,.$$

Therefore factors are building blocks of general von Neumann algebras. Factors are classified into the types I_n $(n = 1, 2, 3, \ldots, \infty)$, II_1, II_∞, and III. The I_n $(n < \infty)$ factor is the matrix algebra $M_n(\mathbb{C})$ and the I_∞ factor is $B(\mathcal{H})$ with $\dim \mathcal{H} = \infty$. A II_1 factor is sometimes said to have continuous dimensions because, as already mentioned in the first section, it has a normal tracial state whose values of projections in $\mathcal{M}$ are all reals in $[0, 1]$. A type II_∞ factor is written as the tensor product of a II_1 factor and the I_∞ factor, and it has a normal semifinite trace. The type I factors are trivial from the operator algebra point of view. Infinite tensor products of matrix algebras with normalized traces and the group von Neumann algebras of ICC discrete groups are typical examples of type II_1 factors; the simplest construction is $\mathcal{R} := \bigotimes_{n=1}^\infty (M_2(\mathbb{C}), \mathrm{tr}_2)$, as described in the first section. All factors except type I or II are said to be of type III, and they are further classified into the types III_λ $(0 \le \lambda \le 1)$; the latter subclasses were introduced by A. Connes. For $0 < \lambda < 1$ a typical example of a III_λ factor is the *Powers factor*

$$\mathcal{R}_\lambda := \bigotimes_{n=1}^\infty (M_2(\mathbb{C}), \omega_\lambda), \quad \text{where} \quad \omega_\lambda(\,\cdot\,) := \mathrm{tr}_2\left(\begin{bmatrix} \frac{1}{1+\lambda} & 0 \\ 0 & \frac{\lambda}{1+\lambda} \end{bmatrix} \cdot\right).$$

Furthermore, the tensor product $\mathcal{R}_\lambda \otimes \mathcal{R}_\mu$ of two Powers factors with $\log \lambda / \log \mu \notin \mathbb{Q}$ becomes a III_1 factor. The Tomita-Takesaki theory is fundamental in the structure analysis of type III factors.

A von Neumann algebra $\mathcal{M}$ (on a separable Hilbert space) is said to be *approximately finite dimensional* (AFD), or sometimes *hyperfinite*, if it is generated by an increasing sequence of finite-dimensional subalgebras. On the other hand, $\mathcal{M} \subset B(\mathcal{H})$ is said to be *injective* if there exists a conditional expectation (i.e. norm one projection) from $B(\mathcal{H})$ onto $\mathcal{M}$. The epoch-making result of Connes in 1976 shows that the injectivity of $\mathcal{M}$ is equivalent to the AFD of $\mathcal{M}$, and there is a unique injective factor for each type II_1, II_∞, III_λ $(0 < \lambda < 1)$. The fact that the above $\mathcal{R}$ is a unique hyperfinite type II_1 factor had been proved long ago by Murray and von Neumann, and the uniqueness of the injective III_1 factor was later proved by U. Haagerup in 1987. Furthermore, all AFD factors of type III are completely classified in terms of the flow of weights introduced by Connes and Takesaki. In this way, $\mathcal{R}\otimes B(\mathcal{H})$, the Powers factor $\mathcal{R}_\lambda$ and the above $\mathcal{R}_\lambda\otimes\mathcal{R}_\mu$ are unique AFD factors of type II_∞, III_λ $(0<\lambda<1)$ and III_1, respectively. There are many AFD III_0 factors; all of them are constructed as *Krieger factors* $L^\infty(\Omega,\mu)\rtimes_T\mathbb{Z}$, where T is an ergodic transformation on a standard Borel space (Ω,μ). The so-called measure space-group construction is to make the crossed product von Neumann algebra $L^\infty(\Omega,\mu)\rtimes_\alpha G$, where α is an action of a group G on (Ω,μ). According to J. Feldman and C.C. Moore, there is a more general construction of von Neumann algebras from a measurable equivalence relation on (Ω,μ) with countable orbits, and $L^\infty(\Omega,\mu)$ is a *Cartan subalgebra* of the constructed von Neumann algebra. Moreover, the existence of a Cartan subalgebra is sufficient for a von Neumann algebra to enjoy such a measure space construction. Any injective factor has a Cartan subalgebra as a consequence of the above classification result. The notion of amenability can be defined for a measurable relation, and a profound result of Connes, Feldman and Weiss in 1982 says that a measurable relation is amenable if and only if it is generated by a single measurable transformation. A consequence of this is that any two Cartan subalgebras of an injective factor $\mathcal{M}$ are conjugate by an automorphism of $\mathcal{M}$.

After the AFD factors were classified as above, a huge class of non-AFD factors remained unclassified. Since any type III factor can be canonically decomposed into the crossed product $\mathcal{N}\rtimes_\theta\mathbb{R}$ with a II_∞ von Neumann algebra $\mathcal{N}$ according to the Takesaki duality, and since type II_∞ can be somehow reduced to type II_1, we may say that the problem returns to the type II_1 theory in some sense. A type II_1 von Neumann algebra with a faithful normal tracial state is a noncommutative probability space most appropriate to free probability theory. A typical example of non-AFD factors is the free group factor $\mathcal{L}(\mathbf{F}_n)$, and it is generated by a free family of noncommutative random variables. This is the reason why some techniques from free probability and free entropy are so useful in analyzing free group factors. There has been much progress in the theory of free group factors; for instance, the non-existence of a Cartan subalgebra in $\mathcal{L}(\mathbf{F}_n)$, proved by Voiculescu, is remarkable because it means the impossibility of the measure space construction as above for $\mathcal{L}(\mathbf{F}_n)$.

The K-theory of C^*-algebras yields algebraic invariants to study isomorphism problems of C^*-algebras. Two abelian groups $K_0(\mathcal{A})$ and $K_1(\mathcal{A})$ are associated with each C^*-algebra $\mathcal{A}$. The construction of $K_0(\mathcal{A})$ follows the old idea of Murray and von Neumann for the classification of von Neumann factors. Since a C^*-algebra may not have enough projections, we pass to the algebra $\bigcup_{n=1}^\infty M_n(\mathcal{A})$. An equivalence relation is defined on the collection of all projections in $\bigcup_{n=1}^\infty M_n(\mathcal{A})$, whose equivalence classes form an abelian semigroup $K_0(\mathcal{A})^+$ with zero element. Then the Grothendieck group of $K_0(\mathcal{A})^+$ is the K_0-group $K_0(\mathcal{A})$. On the other hand, the K_1-group $K_1(\mathcal{A})$ is defined as the inductive limit of the quotient groups $\mathcal{U}(M_n(\mathcal{A}))/\mathcal{U}_0(M_n(\mathcal{A}))$, where $\mathcal{U}(M_n(\mathcal{A}))$ is the group of unitaries in $M_n(\mathcal{A})$ and $\mathcal{U}_0(M_n(\mathcal{A}))$ is the connected component of the identity.

The first major contribution toward classification of C^*-algebras was made by

G.A. Elliott in 1976. He showed that the AF C^*-algebras are completely classified by the ordered group $(K_0(\mathcal{A}), K_0(\mathcal{A})^+)$, the *dimension group.* A C^*-algebra $\mathcal{A}$ is said to be *nuclear* if the minimal C^*-norm is a unique C^*-norm on the algebraic tensor product of $\mathcal{A}$ with any C^*-algebra $\mathcal{B}$. The class of nuclear C^*-algebras is in some sense a C^*-counterpart of the class of injective von Neumann algebras. Many characterizations of nuclear C^*-algebras are known; for example, $\mathcal{A}$ is nuclear if and only if $\pi(\mathcal{A})''$ is injective for any representation of $\mathcal{A}$. The nuclear C^*-algebras are closed under basic operations such as inductive limits, quotients by closed ideals, tensor products and crossed products by actions of amenable groups. In particular, AF algebras are nuclear, and the amenability of a discrete group G is equivalent to the nuclearity of the reduced C^*-algebra $C_r^*(G)$ (similarly to the hyperfiniteness of the group von Neumann algebra $\mathcal{L}(G)$).

Exact C^*-algebras form an important class. A C^*-algebra is said to be *exact* if the sequence of the minimal C^*-tensor products

$$0 \to \mathcal{A} \otimes \mathcal{J} \to \mathcal{A} \otimes \mathcal{B} \to \mathcal{A} \otimes (\mathcal{B}/\mathcal{J}) \to 0$$

is exact whenever $\mathcal{J}$ is a closed ideal of an arbitrary C^*-algebra $\mathcal{B}$. This class includes the nuclear C^*-algebras. For example, the reduced C^*-algebra $C_r^*(\mathbf{F}_n)$ of the free group $\mathbf{F}_n$ is not nuclear but exact. Indeed, it is open whether $C_r^*(G)$ is exact for every countable discrete group G. A C^*-subalgebra of an exact C^*-algebra is exact, and the exact C^*-algebras are closed under the operations of inductive limits, minimal tensor poducts and quotients. It seems that the role of exact C^*-algebras has increased in recent development of C^*-algebra theory since the appearance of the work of Kirchberg.

The group C^*-algebra $C_r^*(\mathbf{F}_n)$ is out of the scope of well-established algebraic invariants; nevertheless, K-theory can detect n from the reduced C^*-algebra of $\mathbf{F}_n$. In 1982 Pimsner and Voiculescu computed the K-groups

$$K_0(C_r^*(\mathbf{F}_n)) = \mathbb{Z} \quad \text{and} \quad K_1(C_r^*(\mathbf{F}_n)) = \mathbb{Z}^n ,$$

and this computation proves that $C_r^*(\mathbf{F}_n)$ is not isomorphic to $C_r^*(\mathbf{F}_m)$ for $n \neq m$. The isomorphism question of whether $\mathcal{L}(\mathbf{F}_n) \not\cong \mathcal{L}(\mathbf{F}_m)$ if $n \neq m$ is still open, and a possible approach uses the free entropy dimension, which is a candidate for a reasonable entropic invariant.

Let $a_1, \dots, a_N \in \mathcal{M}^{sa}$ and assume that $S_1, \dots, S_N \in \mathcal{M}^{sa}$ is a free family of semicircular elements which are in free relation to $\{a_1, \dots, a_N\}$. Then the *free entropy dimension* $\delta(a_1, \dots, a_N)$ is defined in terms of the multivariate free entropy by

$$\delta(a_1, \dots, a_N) := N + \limsup_{\varepsilon \to +0} \frac{\chi(a_1 + \varepsilon S_1, \dots, a_N + \varepsilon S_N)}{|\log \varepsilon|} .$$

Note that the above $S_1, \dots, S_N$ always exist when we enlarge $\mathcal{M}$ by taking a free product with another von Neumann algebra, and that the joint distribution of

$a_1 + \varepsilon S_1, \ldots, a_N + \varepsilon S_N$ is independent of the choice of $S_1, \ldots, S_N$, so $\delta(a_1, \ldots, a_N)$ is well-defined.

Let $g_1, g_2, \ldots, g_N$ be generators of $\mathbf{F}_N$. Then $a_i = L_{g_i} + L_{g_i^{-1}}$ form a free family of semicircular noncommutative random variables, and $\delta(a_1, \ldots, a_k) = k$ for $k \leq N$. Proving the lower semicontinuity of the free entropy dimension would be very exciting, for the following reason. If $a_1, a_2, \ldots, a_N$ are in free relation and $b_i \in \{a_1, a_2, \ldots, a_N\}''$, then

$$\delta(a_1, a_2, \ldots, a_N) \leq \delta(a_1, a_2, \ldots, a_N, b_1, \ldots, b_M) \tag{12}$$

with equality when the b_i's are polynomials of the a_j's. This is a proven fact. If the lower semicontinuity held, we would have equality in (12) without any further hypothesis on the b_i's. The isomorphism $\mathcal{L}(\mathbf{F}_N) \cong \mathcal{L}(\mathbf{F}_M)$ would imply that in this von Neumann algebra $\mathcal{M}$ there exist two systems $a_1, \ldots, a_N$ and $b_1, \ldots, b_M$ of generators, each of which consists of free semicircular variables. Hence, the above equality in (12) gives

$$N = \delta(a_1, \ldots, a_N) = \delta(a_1, \ldots, a_N, b_1, \ldots, b_M) = \delta(b_1, \ldots, b_M) = M\,.$$

In this way, the lower semicontinuity of $\delta(a_1, \ldots, a_N)$ would imply the solution of the isomorphism problem.

In the parametrization of the von Neumann algebras $\mathcal{L}(\mathbf{F}_n)$ the integer n is a discrete parameter when free group factors are considered. However, in the work of K. Dykema and F. Rădulescu a continuous interpolation $\mathcal{L}(\mathbf{F}_r)$ appears, where r is real and $r > 1$. Those are the so-called *interpolated free group factors*. It turned out that they are either isomorphic for all parameter values or non-isomorphic for any two different values of r. So one of the two extreme cases holds true. However, the stable isomorphism $\mathcal{L}(\mathbf{F}_r) \otimes B(\mathcal{H}) \cong \mathcal{L}(\mathbf{F}_s) \otimes B(\mathcal{H})$ is known.

The existence of the interpolation of the free group factors may suggest that the $\mathcal{L}(\mathbf{F}_n)$ are all isomorphic, contrary to the indication from the free entropy dimension.

Chapter 1

Probability Laws and Noncommutative Random Variables

Random variables are functions defined on a measure space, and they are often identified by their distributions in probability theory. In the simplest case when the random variable is real valued, the distribution is a probability measure on the real line. In this chapter it is first demonstrated that probability distributions can be represented by means of linear Hilbert space operators as well. This observation is as old as quantum mechanics; the standard probabilistic interpretation of the quantum mechanical formalism is strongly related. Examples will be given to show how some basic probability distributions, such as the Poisson distribution and the normal, arcsine and semicircle laws, may arise in the context of a Hilbert space. There is a parallelism between the normal and semicircle laws which runs through this book. The similar combinatorial meaning of the moments is the first appearance of this parallelism. Another emergence is in the Fock space. The normal law is the distribution of field operators in the symmetric Fock space. When the symmetry condition is dropped and the full Fock space is considered, the analogous operators are semicircularly distributed. Creation operators on the full Fock space are of central importance in our later considerations.

In an algebraic generalization, elements of a typically noncommutative algebra together with a linear functional on the algebra are regarded as noncommutative random variables. The linear functional evaluated on an element is the expectation value, and application to powers of the selected element yields the moments of the noncommutative random variable. One does not distinguish between two random variables when they have the same moments. A very new feature of this theory appears when several noncommutative random variables are really noncommuting with each other. Then one cannot have a joint distribution in the sense of classical probability theory, but a functional of the algebra of polynomials of noncommuting indeterminates may work as an abstract concept of joint distribution. Random matrices with respect to the expectation of their trace are natural "noncommuting" noncommutative random variables.

This chapter is mostly a collection of examples. Many of them are strongly related to our main stream, but some of them are presented on the basis of curiosity; they merely indicate possible other directions, or a sort of interpolation between the normal and semicircle laws.

1.1 Distribution measure of normal operators

Let N be a bounded normal operator acting on a Hilbert space $\mathcal{H}$, $N^*N = NN^*$. According to the *spectral theorem* ([163], Sec. 111), there exists a projection-valued measure E on the complex plane $\mathbb{C}$ such that

$$N = \int_{\mathbb{C}} z\, dE(z), \tag{1.1.1}$$

and more generally, if q is a polynomial of two (commuting) variables, then

$$q(N, N^*) = \int_{\mathbb{C}} q(z, \bar{z})\, dE(z)\,.$$

Here the integration with respect to an operator-valued measure may be understood as

$$\langle q(N, N^*)\xi, \eta\rangle = \int_{\mathbb{C}} q(z, \bar{z})\, d\langle E(z)\xi, \eta\rangle$$

for every $\xi, \eta \in \mathcal{H}$, and $\langle E(\,\cdot\,)\xi, \eta\rangle$ is a complex-valued measure depending on the vectors ξ and η. It is a part of the spectral theorem that all these measures are concentrated on the spectrum of the operator N.

For a normal operator N and a vector $\xi \in \mathcal{H}$, we define the *distribution measure* of N at ξ as the Borel measure

$$H \mapsto \langle E(H)\xi, \xi\rangle \qquad (H \subset \mathbb{C}), \tag{1.1.2}$$

where E is the projection-valued measure from the spectral decomposition (1.1.1) of N. Since $\mu(\mathbb{C}) = \|\xi\|^2$, μ is a probability measure if and only if ξ is a unit vector.

Example 1.1.1 If N is a normal $n \times n$ matrix with n different eigenvalues and with (pairwise orthogonal) eigenvectors $\eta_1, \eta_2, \ldots, \eta_n$, then the distribution measure of N at ξ is the discrete measure which gives the weight $|\langle \eta_i, \xi\rangle|^2$ to the eigenvalue corresponding to η_i.

□

Example 1.1.2 Let G be a countable group, and consider the unitary operator U on $l^2(G)$, defined as

$$(Uf)(h) := f(g^{-1}h) \qquad (f \in l^2(G),\, h \in G),$$

where g is a fixed element of G. (U is the left-translation by g.) If g^n is never the group unit e for $n \in \mathbb{N}$ and

$$\delta(h) := \begin{cases} 1 & \text{if } h = e, \\ 0 & \text{otherwise,} \end{cases}$$

then the distribution measure of U at the vector $\delta \in l^2(G)$ is the normalized Lebesgue measure on the unit circle.

The spectrum of a unitary is in the unit circle $\mathbb{T}$. Any measure μ on $\mathbb{T}$ is determined by the integrals

$$\int_{\mathbb{T}} z^n \, d\mu(z) \qquad (n \in \mathbb{Z}).$$

This follows from the fact that every continuous function on $\mathbb{T}$ can be approximated by linear combinations of positive and negative powers of z; they are the trigonometric polynomials. If μ is the distribution measure of U at δ, then the equality

$$\int_{\mathbb{T}} z^n \, d\mu(z) = \langle U^n \delta, \delta \rangle \qquad (n \in \mathbb{Z})$$

must hold. Since $\langle U^n \delta, \delta \rangle = 0$ for $n \neq 0$, the normalized Lebesgue measure of $\mathbb{T}$ satisfies this condition.

□

We shall speak of the distribution measure of an unbounded selfadjoint operator at a vector in the domain of the operator. This is defined by the spectral theorem and by (1.1.2) similarly to the bounded case. The distribution measure of a selfadjoint operator is a measure on the real line. The next example is essential in quantum mechanics.

Example 1.1.3 Let $\mathcal{H}$ be the Hilbert space $L^2(\mathbb{R})$ of square integrable functions on $\mathbb{R}$. The *position operator* is the selfadjoint operator Q with domain $\mathcal{D}(Q) := \{f \in L^2(\mathbb{R}) : xf(x) \in L^2(\mathbb{R})\}$, and it is defined by $(Qf)(x) := xf(x)$. A unit vector $g \in L^2(\mathbb{R})$ is usually called a wave function in quantum mechanics. The distribution measure of Q at the vector g gives the probability that a quantum particle of one degree of freedom is confined to a subset $H \subset \mathbb{R}$. Since the spectral projection of Q corresponding to H is the operator of multiplication by the characteristic function of H, we have that $\int_H |g(x)|^2 \, dx$ is the probability that a particle of quantum state g is in H.

□

Let $\mathcal{H}$ be a Hilbert space and denote by $\mathcal{H}^{\otimes n}$ the n-fold tensor product $\mathcal{H} \otimes \mathcal{H} \otimes \cdots \otimes \mathcal{H}$. A permutation π of the set $\{1, 2, \ldots, n\}$ gives rise to a unitary U_π on $\mathcal{H}^{\otimes n}$ which is determined by

$$U_\pi : \eta_1 \otimes \eta_2 \otimes \cdots \otimes \eta_n \mapsto \eta_{\pi(1)} \otimes \eta_{\pi(2)} \otimes \cdots \otimes \eta_{\pi(n)} \,.$$

(In this way one gets an action of the symmetric group $\mathbf{S}_n$.) The subspace of all common fixed vectors of all U_π's is called the nth symmetric tensor power of $\mathcal{H}$, and we denote it by $\mathcal{H}^{\odot n}$. The *full Fock space* $\mathcal{F}(\mathcal{H})$ and the *symmetric Fock space* $\mathcal{F}_s(\mathcal{H})$ over $\mathcal{H}$ are defined as the orthogonal sum of all tensor powers and, respectively, symmetric tensor powers of $\mathcal{H}$. Namely,

$$\mathcal{F}(\mathcal{H}) := \bigoplus_{n=0}^{\infty} \mathcal{H}^{\otimes n}\,, \qquad \mathcal{F}_s(\mathcal{H}) := \bigoplus_{n=0}^{\infty} \mathcal{H}^{\odot n}\,.$$

If $\mathcal{H} = \mathbb{C}$ then both kinds of Fock space reduce to $l^2(\mathbb{Z}^+)$. It is customary to set $\mathcal{H}^{\otimes 0} = \mathcal{H}^{\odot 0} = \mathbb{C}\Phi$ and to call the unit vector Φ the *vacuum vector.* Another name motivated by physics is the *particle number operator.* This is a positive selfadjoint operator on $\mathcal{F}_s(\mathcal{H})$ such that $\mathcal{H}^{\odot n}$ is an eigensubspace corresponding to the eigenvalue n ($n \in \mathbb{Z}^+$).

For $\eta \in \mathcal{H}$ the vector

$$\psi(\eta) := \bigoplus_{n=0}^{\infty} \frac{1}{\sqrt{n!}} \eta^{(1)} \otimes \eta^{(2)} \otimes \cdots \otimes \eta^{(n)} \qquad (\eta^{(i)} = \eta) \tag{1.1.3}$$

belongs to $\mathcal{F}_s(\mathcal{H})$ and is called an *exponential vector* or *coherent vector.* It is easy to check that

$$\langle \psi(\eta_1), \psi(\eta_2) \rangle = \exp\langle \eta_1, \eta_2 \rangle \,,$$

and it is useful to know that the linear hull of all exponential vectors is dense in $\mathcal{F}_s(\mathcal{H})$.

Example 1.1.4 The distribution measure of the number operator at the normalized exponential vector

$$\xi := \exp\bigl(-\|\eta\|^2/2\bigr)\psi(\eta)$$

is the *Poisson distribution* with parameter $\lambda = \|\eta\|^2$.

One has to compute $\langle P_n \xi, \xi \rangle$, where P_n is the projection of $\mathcal{F}_s(\mathcal{H})$ onto $\mathcal{H}^{\odot n}$:

$$\langle P_n \xi, \xi \rangle = \exp\bigl(-\|\eta\|^2\bigr) \frac{\|\eta\|^{2n}}{n!}\,,$$

and this is really a Poisson distribution.

□

For each $h \in \mathcal{H}$ a bounded operator $\ell(h)$ is defined on the full Fock space $\mathcal{F}(\mathcal{H})$ by the formula

$$\ell(h)\eta := \begin{cases} h & \text{if } \eta = \Phi, \\ h \otimes \eta & \text{if } \langle \eta, \Phi \rangle = 0. \end{cases} \tag{1.1.4}$$

$\ell(h)$ is called the *left creation operator*, or for short just the *creation operator*. For its adjoint $\ell(h)^*$, called the *annihilation operator*, we have $\ell(h)^*\Phi = 0$ and

$$\ell(h)^*\eta_1 \otimes \eta_2 \otimes \cdots \otimes \eta_k = \langle \eta_1, h \rangle \eta_2 \otimes \cdots \otimes \eta_k \,.$$

Evidently $\ell(h_1)^*\ell(h_2) = \langle h_2, h_1 \rangle \mathbf{1}$ ($\mathbf{1}$ denotes the identity operator). In particular, $\ell(h)$ is an isometry for every unit vector $h \in \mathcal{H}$.

The *semicircle law* is a probability measure on $\mathbb{R}$ whose density is

$$w_{m,r}(x) := \begin{cases} \dfrac{2}{\pi r^2}\sqrt{r^2 - (x-m)^2} & \text{if } m - r \le x \le m + r, \\ 0 & \text{otherwise,} \end{cases} \tag{1.1.5}$$

where m and $r > 0$ are real numbers. Instead of $w_{0,r}$ we write simply w_r. The graph of the semicircle law is a semiellipse. The expectation value of the semicircle law $w_{m,r}$ is m due to the symmetry, and the variance is $r^2/4$.

Theorem 1.1.5 *The distribution measure of $\ell(h)^* + \ell(h)$ at the vacuum vector $\Phi \in \mathcal{F}(\mathcal{H})$ is the semicircle law w_r with $r = 2\|h\|$.*

Proof: We may assume that h is a unit vector, and we write ℓ for $\ell(h)$. Since $\|\ell\| = 1$, the distribution measure of $\ell^* + \ell$ is concentrated on $[-2, 2]$. A compactly supported measure is determined by its moments, due to the Weierstrass approximation theorem. The distribution measure μ of $\ell^* + \ell$ at Φ is determined by the conditions

$$\int x^n \, d\mu(x) = \langle (\ell^* + \ell)^n \Phi, \Phi \rangle \qquad (n \in \mathbb{Z}^+).$$

We compute $m_n := \langle (\ell^* + \ell)^n \Phi, \Phi \rangle$ for every n. We have

$$m_n = \sum \langle a_{i(n)} a_{i(n-1)} \cdots a_{i(1)} \Phi, \Phi \rangle \,,$$

where $a_{i(k)} \in \{\ell, \ell^*\}$ and the summation is over all such possibilities.

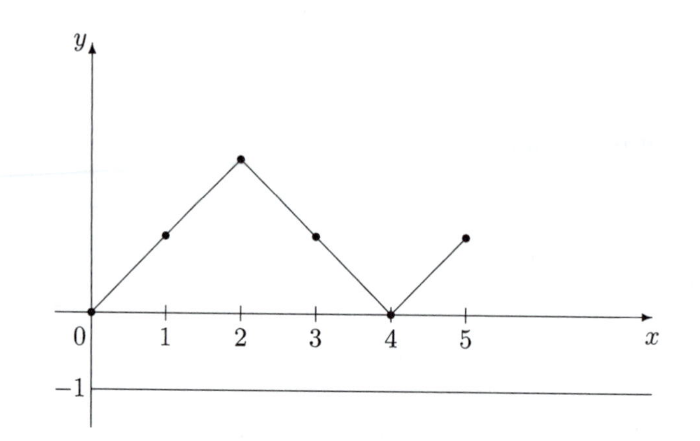

Fig.1.1. The polygonal line associated to the product $\ell\ell^*\ell^*\ell\ell$.

We can associate to an n-fold product $a_{i(n)}a_{i(n-1)}\cdots a_{i(1)}$ a polygonal line on the x-y-plane, where the x-axis is horizontal. The associated polygonal line starts from the origin $(0,0)$ and possesses slope 1 on the interval $(k-1,k)$ if $a_{i(k)}=\ell$ and slope -1 if $a_{i(k)}=\ell^*$. If the polygonal line touches the horizontal line $y=-1$, or its end point is not on the x-axis, then $\langle a_{i(n)}a_{i(n-1)}\cdots a_{i(1)}\Phi,\Phi\rangle=0$. So, for an odd n we have $m_n=0$, because the corresponding polygonal line cannot end on the x-axis. If $n=2k$, then we have to count the number of polygonal lines that contribute to m_{2k} (see [76], III.1 or [166], Sec. I.10). So

$$m_n=\begin{cases} 0 & \text{if } n=2k+1,\\ \binom{2k}{k}-\binom{2k}{k-1}=\frac{1}{k+1}\binom{2k}{k} & \text{if } n=2k.\end{cases} \tag{1.1.6}$$

Now we have to show that the semicircle law has these moments.

We integrate by parts, and get

$$\begin{aligned}\alpha_{2k} &:= \frac{1}{2\pi}\int_{-2}^{2}x^{2k}\sqrt{4-x^2}\,dx=-\frac{1}{2\pi}\int_{-2}^{2}\frac{-x}{\sqrt{4-x^2}}x^{2k-1}(4-x^2)\,dx\\ &= \frac{1}{2\pi}\int_{-2}^{2}\sqrt{4-x^2}\big(x^{2k-1}(4-x^2)\big)'\,dx\\ &= 4(2k-1)\alpha_{2k-2}-(2k+1)\alpha_{2k}.\end{aligned}$$

In this way the recursion

$$\alpha_{2k}=\frac{2(2k-1)}{k+1}\alpha_{2k-2}\qquad (k\in\mathbb{N})$$

is obtained, and the moment sequence m_{2k} in (1.1.6) satisfies this recursion.

□

We shall use the notation

$$c_k := \frac{1}{k+1}\binom{2k}{k} \tag{1.1.7}$$

for the so-called *Catalan numbers.*

Before we turn to the distribution of the field operators in the symmetric Fock space, we give a second proof for the previous theorem. The starting point will be the observation that computation of the nth moment of $\ell(h)^* + \ell(h)$ is essentially an n-dimensional problem.

Second proof: Let $e_1, e_2, \dots, e_n$ be an orthonormal basis in an n-dimensional space and consider the tridiagonal matrix

$$\mathbf{A}_n := \begin{bmatrix} 0 & 1 & & & & \\ 1 & 0 & 1 & & 0 & \\ & 1 & 0 & 1 & & \\ & & \ddots & \ddots & \ddots & \\ & & & \ddots & \ddots & 1 \\ & 0 & & & 1 & 0 \end{bmatrix}.$$

Then it is easy to see that

$$\int x^m \, d\mu(x) = \langle \mathbf{A}_n^m e_1, e_1 \rangle$$

must hold for the distribution measure μ of $\ell^* + \ell$ when $m < 2n$. The matrix $\mathbf{A}_n$ is rather well-known; its eigenvalues were computed in 1759 by Lagrange. The eigenvalues are $2\cos j\pi/(n+1)$, and the eigenvectors are

$$\sqrt{\frac{2}{n+1}} \sum_{k=1}^{n} \left(\sin \frac{jk\pi}{n+1} \right) e_k \qquad (j = 1, 2, \dots, n).$$

Therefore,

$$\int x^m \, d\mu(x) = \frac{2^{m+1}}{n+1} \sum_{j=1}^{n} \cos^m \frac{j\pi}{n+1} \sin^2 \frac{j\pi}{n+1}.$$

For finite m this seems to be complicated, but it is easy to take the limit as $n \to \infty$, since the sum is an approximation of an integral. In this way we get

$$\int x^m \, d\mu(x) = \frac{2^{m+1}}{\pi} \int_0^{\pi} \cos^m t \sin^2 t \, dt = \frac{1}{2\pi} \int_{-2}^{2} x^m \sqrt{4 - x^2} \, dx.$$

□

Let $\mathcal{H}$ be a Hilbert space. For each $f \in \mathcal{H}$ we introduce three operators on the symmetric Fock space $\mathcal{F}_s(\mathcal{H})$. All of them are defined on the linear hull of the exponential vectors:

$$\begin{aligned} W(f)\psi(h) &:= \exp\bigl(-\|f\|^2/2 - \langle h, f\rangle\bigr)\psi(f+h)\,, \\ a(f)\psi(h) &:= \langle h, f\rangle \psi(h)\,, \\ a^*(f)\psi(h) &:= \frac{d}{dt}\psi(h+tf)\Big|_{t=0}\,. \end{aligned} \tag{1.1.8}$$

It is immediate from the definition that the $W(f)$'s satisfy the so-called *Weyl commutation relation*

$$W(f)W(g) = \exp(\mathrm{i}\,\mathrm{Im}\,\langle f, g\rangle)W(f+g)$$

on exponential vectors. (More about the *canonical commutation relation* is in [148].) This relation implies that $W(f)$ can be extended to $\mathcal{F}_s(\mathcal{H})$, and a unitary operator arises. It is worthwhile to note that the normalized exponential vectors form the orbit of the vacuum vector under the *Weyl unitaries.* Namely,

$$W(f)\Phi = \exp\bigl(-\|f\|^2/2\bigr)\psi(f)\,.$$

The correspondence $t \mapsto W(tf)$ gives a strongly continuous group of unitaries acting on $\mathcal{F}_s(\mathcal{H})$. The *Stone theorem* ([163], Sec. 137) tells us that there exists a selfadjoint operator $B(f)$ such that

$$W(tf) = \exp(\mathrm{i}\,tB(f)) \qquad (t \in \mathbb{R})$$

and

$$-\mathrm{i}\frac{d}{dt}W(tf)\xi\Big|_{t=0} = B(f)\xi \tag{1.1.9}$$

whenever ξ is in the domain of $B(f)$. Performing this derivation at exponential vectors, we obtain

$$\begin{aligned} -\mathrm{i}\frac{d}{dt}W(tf)\psi(g)\Big|_{t=0} &= \mathrm{i}\,\langle g, f\rangle\psi(g) - \mathrm{i}\,\frac{d}{dt}\psi(g+tf)\Big|_{t=0} \\ &= \mathrm{i}\,a(f)\psi(g) - \mathrm{i}\,a^*(f)\psi(g) \\ &= \bigl(a(-\mathrm{i}\,f) + a^*(-\mathrm{i}\,f)\bigr)\psi(g)\,. \end{aligned}$$

Therefore the operator $a(-\mathrm{i}\,f) + a^*(-\mathrm{i}\,f)$ defined on exponential vectors has a selfadjoint extension $B(f)$. In fact, $B(f)$ is the closure of the densely defined operator $a(-\mathrm{i}\,f) + a^*(-\mathrm{i}\,f)$. This can be seen as follows. The linear hull $\mathcal{D}$ of exponential vectors is invariant under $\{W(tf) : t \in \mathbb{R}\}$; therefore $\mathcal{D}$ is a core of the generator $B(f)$ (cf. [162], Theorem VIII.10).

We recall that the *normal distribution* $N(m,\sigma^2)$ with mean m and varance σ^2 has the density function

$$\frac{1}{\sqrt{2\pi}\sigma}\exp\Big(-\frac{(x-m)^2}{2\sigma^2}\Big).$$

Theorem 1.1.6 *For $f \in \mathcal{H}$ the selfadjoint closure of the operator $a(f)+a^*(f)$ is normally distributed at the vacuum vector $\Phi \in \mathcal{F}_s(\mathcal{H})$.*

Proof: For the distribution measure μ of $B := \overline{a(f)+a^*(f)}$ the equality

$$\langle \exp(\mathrm{i}\,tB)\Phi, \Phi\rangle = \int e^{\mathrm{i}\,tx}\,d\mu(x) \qquad (t \in \mathbb{R}) \tag{1.1.10}$$

holds. Since $\exp(\mathrm{i}\,tB)$ is the Weyl unitary $W(\mathrm{i}\,tf)$, the left-hand side is easily computed, and we have

$$\langle W(\mathrm{i}\,tf)\psi(0), \psi(0)\rangle = \exp(-t^2\|f\|^2/2)\,.$$

This must be the characteristic function (or Fourier transform) of μ according to (1.1.10), and we conclude that μ is $N(0,\|f\|^2)$ (cf. [77], Part XV).

□

The notation for creation operators in the symmetric and full Fock spaces is not compatible. The left creation operator $\ell(h)$ on the full Fock space is unstarred while the creation operator $a^*(h)$ on the symmetric space is starred. The difference fades away when the selfadjoint field operators are considered. The distribution of $\ell(h)^*+\ell(h)$ was identified by means of the moments. In the case of $a(f)+a^*(f)$ the Fourier transform offered a faster way. However, moments of the normal law can be easily obtained either from the Fourier transform or by direct integration. The $2k$th moment is

$$(2k-1)!! := \frac{(2k-1)!}{2^{k-1}(k-1)!}\,,$$

which is the number of all pair partitions of a set of $2k$ elements. In comparison, the $2k$th moment c_k of the semicircle law is the number of syntactically correct orders of k copies of "(" and k copies of ")". It will be seen in the next chapter that c_k is the number of pair partitions of a certain kind.

Example 1.1.7 Let U be a Haar distributed unitary from Example 1.1.2. We want to compute the distribution measure of the selfadjoint operator $U+U^*$ at the vector δ. The nth moment is

$$\langle (U+U^*)^n\delta, \delta\rangle,$$

which is 0 for $n \neq 2k$; otherwise it is the number of orders of k copies of U and k copies of U^*. We have

$$\int_{-2}^{2} \frac{4x^{2k-2} - x^{2k}}{\sqrt{4-x^2}}\, dx = \int_{-2}^{2} x^{2k-2}\sqrt{4-x^2}\, dx = \frac{1}{2k-1}\int_{-2}^{2} \frac{x^{2k}}{\sqrt{4-x^2}}\, dx$$

integrating by parts, and this gives the recursion $\alpha_k = 2k^{-1}(2k-1)\alpha_{k-1}$ for $\alpha_k := \pi^{-1}\int_{-2}^{2} x^{2k}/\sqrt{4-x^2}\, dx$. Thus the *arcsine law*

$$\frac{1}{\pi\sqrt{4-x^2}}\chi_{(-2,2)}(x)\, dx$$

has exactly the desired even moments $\binom{2k}{k}$, so it is the distribution of $U + U^*$.

□

Beyond the classical probability laws, rather unusual probability distributions may come up in the Hilbert space probability theory. Next we want to exhibit such an example. Given a real number $q \in (-1, 1)$, we construct a selfadjoint operator on $l^2(\mathbb{Z}^+)$ from a certain weighted shift operator. Let δ_n be the standard orthonormal basis in $l^2(\mathbb{Z}^+)$. Set

$$[n]_q := 1 + q + \cdots + q^{n-1} = \frac{1-q^n}{1-q} \qquad (n \in \mathbb{N}).$$

The *weighted shift* S_q, defined as

$$S_q\delta_n := \sqrt{[n+1]_q}\,\delta_{n+1} \qquad (n \in \mathbb{Z}^+),$$

is a bounded operator such that

$$\|S_q\| = \sup\left\{\sqrt{[n]_q} : n \in \mathbb{N}\right\} = \begin{cases} \dfrac{1}{\sqrt{1-q}} & \text{if } q \in [0,1), \\ 1 & \text{if } q \in (-1,0]. \end{cases}$$

For $q = 0$ the ordinary shift is recovered. It is worthwhile to note that $S_q^* S_q = \mathbf{1} + qS_qS_q^*$ is a generalization of the canonical commutation relation, called the *q-commutation relation.* Thus S_q seems to be the "q-creation operator", and the spectrum of the "q-number operator" $S_qS_q^*$ consists of the "q-natural numbers" $[n]_q$.

Example 1.1.8 The distribution measure of the selfadjoint operator $S_q + S_q^*$ at the vector δ_0 is a measure supported on the interval $\left[-2/\sqrt{1-q}, 2/\sqrt{1-q}\right]$. Its density is known to be

$$C_q\sqrt{4-(1-q)x^2}\prod_{k=1}^{\infty}\left(1 - \frac{(1-q)q^k x^2}{(1+q^k)^2}\right),$$

where

$$C_q := \frac{\sqrt{1-q}}{2\pi} \prod_{k=1}^{\infty}(1-q^{2k}) \prod_{k=1}^{\infty}(1+q^k).$$

This probability distribution may be called the q-deformed Gaussian distribution, or simply the *q-Gaussian distribution*. The value $q = 0$ gives the semicircle law, and the limit distribution as $q \to 1$ is the normal law. Indeed,

$$\lim_{q\to 1} \sum_{k=1}^{\infty} \log\left(1 - \frac{(1-q)q^k x^2}{(1+q^k)^2}\right) = \lim_{q\to 1} \sum_{k=1}^{\infty}\left(-\frac{(1-q)q^k x^2}{(1+q^k)^2}\right) = -\frac{x^2}{2}$$

because $\sum_{k=1}^{\infty}(q^k - q^{k+1})/(1+q^k)^2$ is an infinite Riemannian sum approximation to $\int_0^1 dt/(1+t)^2$.

□

Above, the distribution measure of a weighted shift operator at the vacuum led to the q-normal distribution. However, many more distributions can be obtained from weighted shifts. Now we review this subject and show its relation to the theory of orthogonal polynomials.

Let $(b_n)_{n=1}^{\infty}$ be a sequence of positive real numbers, and define a *weighted shift operator*

$$S\delta_n := b_{n+1}\,\delta_{n+1} \qquad (n \in \mathbb{Z}^+)$$

on the space $l^2(\mathbb{Z}^+)$. When (b_n) is bounded, so is S, and we mainly consider this case. The truncation of the selfadjoint operator $S + S^*$ is the tridiagonal matrix

$$\mathbf{A}_n := \begin{bmatrix} 0 & b_1 & & & & \\ b_1 & 0 & b_2 & & & \mathbf{0} \\ & b_2 & 0 & b_3 & & \\ & & \ddots & \ddots & \ddots & \\ & & & \ddots & \ddots & b_{n-1} \\ & \mathbf{0} & & & b_{n-1} & 0 \end{bmatrix}.$$

The eigenvalues of $\mathbf{A}_n$ are the roots of the monic polynomial $P_n(x)$ defined by the recursion formula

$$P_{n+1}(x) = xP_n(x) - \lambda_n P_{n-1}(x) \qquad (n \in \mathbb{Z}^+) \tag{1.1.11}$$

$(P_{-1}(x) \equiv 0,\ P_0(x) \equiv 1)$, where $\lambda_n := b_n^2$. In fact, this is nothing else but the recursion for the characteristic polynomials of the $\mathbf{A}_n$'s. The very basics of the theory of orthogonal polynomials say (see the book [49], Chap. I, II, for example) that $(P_n(x))_{n=0}^{\infty}$ is an *orthogonal polynomial sequence* with respect to a symmetric positive-definite moment functional represented by a probability measure μ (of compact support), a solution of the Hamburger moment problem.

The monic polynomial $P_n(x)$ has the simple real zeros $x_{n1}, x_{n2}, \ldots, x_{nn}$, which are the eigenvalues of $\mathbf{A}_n$. The famous Gauss quadrature formula says that there are positive numbers $c_{n1}, c_{n2}, \ldots, c_{nn}$ (the so-called Christoffel numbers) with $\sum_{k=1}^{n} c_{nk} = 1$ such that

$$\int q(x)\, d\mu(x) = \sum_{k=1}^{n} c_{nk} q(x_{nk}) \tag{1.1.12}$$

for every polynomial $q(x)$ of degree less than $2n$. From this we know that μ is the limit of the sequence of atomic measures $\mu_n := \sum_{k=1}^{n} c_{nk}\delta(x_{nk})$ supported on the zeros of $P_n(x)$, where $\delta(x_0)$ denotes the point measure at x_0. Moreover, it is known ([49], Chap. III, Theorem 4.3) that the *Cauchy transform* of μ_n,

$$\int \frac{d\mu_n(x)}{z - x} = \sum_{k=1}^{n} \frac{c_{nk}}{z - x_{nk}},$$

is equal to the nth partial approximant of the following *continued fraction*:

$$\cfrac{1}{z - \cfrac{\lambda_1}{z - \cfrac{\lambda_2}{z - \ddots}}}\,. \tag{1.1.13}$$

In this way, we can conclude that the Cauchy transform of μ has the above continued fraction expansion. (See Sec. 3.1 for more about the Cauchy transform for measures.)

Proposition 1.1.9 *The distribution measure of $S + S^*$ at the vector δ_0 is μ, given above. More generally,*

$$\langle (S + S^*)^m \delta_i, \delta_j \rangle = \int x^m p_i(x) p_j(x)\, d\mu(x) \qquad (m, i, j \in \mathbb{Z}^+),$$

where the $p_n(x)$ are the normalized orthogonal polynomials for μ.

Proof: We need the following facts from the theory of orthogonal polynomials (see [49]):

$$\int P_n(x)^2\, d\mu(x) = \lambda_1 \lambda_2 \cdots \lambda_n \qquad (n \in \mathbb{N})$$

and

$$\sum_{i=0}^{n-1} p_i(x_{nk})^2 = \frac{1}{c_{nk}} \qquad (1 \le k \le n,\ n \in \mathbb{N}).$$

From the recursion (1.1.11) together with these facts, one can easily check that

$$\sqrt{c_{nk}}\big(p_0(x_{nk}), p_1(x_{nk}), \dots, p_{n-1}(x_{nk})\big)^t$$

is the normalized eigenvector of $\mathbf{A}_n$ corresponding to the eigenvalue x_{nk}. From the spectral decomposition of $\mathbf{A}_n$ we have

$$\langle \mathbf{A}_n^m \delta_i, \delta_j \rangle = \sum_{k=1}^{n} c_{nk} x_{nk}^m p_i(x_{nk}) p_j(x_{nk}) .$$

Now we can use (1.1.12) with $q(x) = x^m p_i(x) p_j(x)$ when $m+i+j < 2n$, to conclude that

$$\langle (S+S^*)^m \delta_i, \delta_j \rangle = \lim_{n\to\infty} \langle \mathbf{A}_n^m \delta_i, \delta_j \rangle = \int x^m p_i(x) p_j(x)\, d\mu(x) .$$

□

It is a consequence of the proposition that the distribution measure of $S + S^*$ is determined by (1.1.13). (Every symmetric measure can arise in this way.)

In the example of a creation operator in Theorem 1.1.5 we have $b_n = 1$ for every $n \in \mathbb{N}$. Then the measure μ is the semicircle law, and the corresponding orthogonal polynomials are the *Jacobi polynomials* $P_n^{(\alpha,\beta)}$ with $\alpha = \beta = 1/2$. In this case, the number $\langle (S+S^*)^m \delta_i, \delta_j \rangle$ is an integer, and is interpreted as the number of certain polygonal lines. In the spirit of the proof of Theorem 1.1.5 we count the polygonal lines from $(0, i)$ to (m, j) which go through lattice points, have slopes ± 1 and do not touch the horizontal line $y = -1$ on the x-y-plane. (It is not difficult to write the explicit formula, which can be the difference of two binomial coefficients.)

The weighted shift S_q in Example 1.1.8 arises when $b_n = \sqrt{[n]_q}$ ($n \in \mathbb{N}$). In this case, we have to treat the recursion formula

$$P_{n+1}(x) = xP_n(x) - \frac{1-q^n}{1-q} P_{n-1}(x) \qquad (P_{-1} \equiv 0,\ P_0 \equiv 1);$$

but these orthogonal polynomials are known as "continuous q-Hermite polynomials" and the associated measure is contained in Example 1.1.8.

1.2 Noncommutative random variables

It was shown in the previous section how a Hilbert space operator together with a unit vector of the Hilbert space can induce a probability distribution. Now we continue to generalize the usual setting of classical probability theory in an algebraic way. Random variables over a probability space form an algebra. Indeed, they are measurable functions defined on a set Ω, and so are the product and sum of two of them. The *expectation value* is a linear functional on this algebra. The algebraic approach to probability stresses that point. If $\mathcal{A}$ is a unital algebra over the complex numbers and φ is a linear functional of $\mathcal{A}$ such that $\varphi(\mathbf{1}) = 1$, then $(\mathcal{A}, \varphi)$ will be called a *noncommutative probability space* and an element a of $\mathcal{A}$ will be called a *noncommutative random variable.* The number $\varphi(a^n)$ is called the *nth moment* of a.

Example 1.2.1 Let $B(\mathcal{H})$ denote the algebra of all bounded operators acting on a Hilbert space $\mathcal{H}$. If the linear functional $\varphi : B(\mathcal{H}) \to \mathbb{C}$ is defined by means of a unit vector $\xi \in \mathcal{H}$ as $\varphi(A) = \langle A\xi, \xi\rangle$, then any element of $B(\mathcal{H})$ is a noncommutative random variable.

□

If $A \in B(\mathcal{H})$ is selfadjoint, then a probability measure is associated to A and φ, as discussed in the previous section. The algebra used in the definition of a noncommutative random variable is often replaced by a *-algebra. Actually, $B(\mathcal{H})$ is a *-algebra if the operation A^* stands for the adjoint of A.

A **-algebra* is a unital algebra over the complex numbers which is endowed with an *involution* *. The involution recalls the adjoint operation of Hilbert space operators as follows:

(1) $a \mapsto a^*$ is conjugate linear,

(2) $(ab)^* = b^*a^*$,

(3) $a^{**} = a$.

When $(\mathcal{A}, \varphi)$ is a noncommutative probability space over a *-algebra $\mathcal{A}$, φ is always assumed to be a *state* on $\mathcal{A}$, that is, a linear functional such that

(4) $\varphi(\mathbf{1}) = 1$, where $\mathbf{1}$ denotes the unit of $\mathcal{A}$,

(5) $\varphi(a^*) = \overline{\varphi(a)}$ and $\varphi(a^*a) \geq 0$ for every $a \in \mathcal{A}$.

A matrix whose elements are (classical) random variables on a (classical) probability space is called a *random matrix.* Random matrices form a *-algebra. For instance, let $a_{11}, a_{12}, a_{21}, a_{22}$ be four bounded (classical) random variables on a probability space. Then

$$T = \begin{bmatrix} a_{11} & a_{12} \\ a_{21} & a_{22} \end{bmatrix}$$

is a bounded 2×2 random matrix. The set $\mathcal{A}$ of all such matrices has a *-algebra structure when the usual matrix operations are considered, and $(\mathcal{A}, \varphi)$ is a non-commutative probability space when $\varphi(T) = E(a_{11})$, for example.

A C^**-algebra* is a *-algebra $\mathcal{A}$ which is endowed with a norm such that

$$(6) \quad \|a^*a\| = \|a\|^2, \ \|ab\| \leq \|a\| \, \|b\| \text{ for every } a, b \in \mathcal{A}, \text{ and } \|\mathbf{1}\| = 1,$$

and furthermore $\mathcal{A}$ is a Banach space with respect to this norm. Two important theorems (due to Gelfand and Naimark) concern the representation of C^*-algebras. A commutative unital C^*-algebra is isometrically isomorphic to the algebra of all continuous complex functions on a certain compact Hausdorff space if the function space is endowed with the supremum norm and the involution of pointwise conjugation. A general C^*-algebra is isometrically isomorphic to an algebra of operators acting on a Hilbert space if this algebra is endowed with the operator norm and the involution of adjoint operation. (Combination of the two Gelfand-Naimark theorems yields a form of the spectral theorem.) For a linear functional φ of a C^*-algebra, $\|\varphi\| = \varphi(1)$ is equivalent to $\varphi(a^*a) \geq 0$ $(a \in \mathcal{A})$.

A noncommutative probability space $(\mathcal{A}, \varphi)$ will be called a C^**-probability space* when $\mathcal{A}$ is a C^*-algebra and φ is a state on $\mathcal{A}$.

Example 1.2.2 All real bounded classical random variables may be considered as noncommutative random variables.

Let ξ be a bounded real random variable with distribution μ and let $K \subset \mathbb{R}$ be the compact support of μ. The continuous complex functions defined on K form a C^*-algebra $\mathcal{A}$ with the supremum norm. A state φ is defined on $\mathcal{A}$ as $\varphi(f) := \int f(x)\, d\mu(x)$. So $(\mathcal{A}, \varphi)$ is a degenerate noncommutative probability space because $\mathcal{A}$ is actually commutative. The identity function f_0 of K (viewed as an element of $\mathcal{A}$) corresponds to the classical random variable ξ. In particular, ξ and f_0 have the same moments: $\varphi(f_0^n) = E(\xi^n)$ $(n \in \mathbb{Z}^+)$. The C^*-algebra $\mathcal{A}$ has a representation on the Hilbert space $L^2(\mu)$ by multiplication. If M_0 is the multiplication operator by the above f_0, then the distribution measure of M_0 at the unit vector 1 (the identically 1 function) is exactly μ.

□

The representation of a classical random variable as a noncommutative one is far from unique. Let $\mathbb{C}\langle X\rangle$ denote the algebra of polynomials in an indeterminate X over $\mathbb{C}$, and let ξ be a classical random variable whose moments $E(\xi^n)$ exist for all $n \in \mathbb{N}$. A linear functional $\varphi : \mathbb{C}\langle X\rangle \to \mathbb{C}$ is produced as $\varphi(X^k) = E(\xi^k)$. In this way the indeterminate X is a noncommutative random variable with the same moment sequence as ξ. When A is an operator on a Hilbert space $\mathcal{H}$ and (A, Φ) is a noncommutative random variable with $\Phi \in \mathcal{H}$, one is tempted to call the linear functional $\varphi : \mathbb{C}\langle X\rangle \to \mathbb{C}$ given as $\varphi(X^k) = \langle A^k\Phi, \Phi\rangle$ the *distribution* of (A, Φ). For our purposes two noncommutative random variables are equivalent if they have the same distributions (that is, the same moments).

Example 1.2.3 Let $\mathcal{A}$ be the *-algebra generated by the creation and annihilation operators ($a^*(f)$ and $a(f)$) acting on the symmetric Fock space $\mathcal{F}_s(\mathcal{H})$. Then $\mathcal{A}$

with the functional $\varphi(b) = \langle b\Phi, \Phi\rangle$ is a noncommutative probability space.

For the sake of simplicity, we assume that $\mathcal{H}$ is one-dimensional. Then $\mathcal{F}_s(\mathcal{H}) = l^2(\mathbb{Z}^+)$, and up to a constant factor we have only one creation operator and one annihilation operator. In (1.1.8) they were defined on exponential vectors. Any polynomial of the annihilation operator a leaves invariant the linear hull of exponential vectors. However, the creation operator a^* does not behave so. In order to have the algebra generated by a and a^* on a commom (large) domain, we need something else—the linear hull of exponential vectors will not do. Let δ_n be the canonical basis vectors of $l^2(\mathbb{Z}^+)$. The operators a and a^* act on these vectors as

$$a\delta_n = \begin{cases} \sqrt{n}\,\delta_{n-1} & \text{if } n > 0, \\ 0 & \text{if } n = 0, \end{cases} \qquad a^*\delta_n = \sqrt{n+1}\,\delta_{n+1}\,. \tag{1.2.1}$$

Hence any polynomial of a and a^* leaves invariant the (finite) linear combinations of basis vectors, and a^* is really the adjoint of a. The algebra generated by a and a^* makes sense, and $(\mathcal{A}, \varphi)$ is a noncommutative probability space.

We note that the functional φ is called the expectation value at the vacuum, and $\mathcal{A}$ is not a C^*-algebra because it consists of unbounded operators. The latter statement is obvious from (1.2.1). According to Theorem 1.1.6, the noncommutative random variable $a + a^*$ has the normal distribution.

□

From our point of view the distribution of the operator $\ell(h)^* + \ell(h)$ from Theorem 1.1.5 is rather important. It is an example of what we call a semicircular noncommutative random veriable. Let a be a random variable in the noncommutative probability space $(\mathcal{A}, \varphi)$. We say that a is (selfadjoint) *standard semicircular* if its distribution is the same as that of $\ell(h)^* + \ell(h)$ with $\|h\| = 1$ at the vacuum vector Φ. In other words, a standard semicircular variable is characterized by its moment sequence (1.1.6).

Let a_1 and a_2 be two random variables in the same noncommutative probability space $(\mathcal{A}, \varphi)$. Their *joint distribution* will be defined as a linear functional of a noncommutative algebra of polynomials. Let $\mathbb{C}\langle X_1, X_2\rangle$ be the algebra of polynomials in two noncommuting indeterminates X_1 and X_2. The joint distribution of a_1 and a_2 is the functional $\mu : \mathbb{C}\langle X_1, X_2\rangle \to \mathbb{C}$ defined by

$$\mu(X_{s_1}X_{s_2}\cdots X_{s_m}) := \varphi(a_{s_1}a_{s_2}\cdots a_{s_m})\,,$$

where $s_i \in \{1, 2\}$. We note that the algebra $\mathbb{C}\langle X_1, X_2\rangle$ may be regarded as a *-algebra, where X_1 and X_2 are considered to be selfadjoint. The concept of joint distribution generalizes to several random variables in an obvious way.

Example 1.2.4 Let U be the unitary in Example 1.1.2. The joint distribution of U and U^* at the vector δ is as follows: $\mu(X_{s_1}X_{s_2}\cdots X_{s_m}) = 1$ when the number of 1's is the same as the number of 2's in the 1-2-sequence $(s_1, s_2, \ldots, s_m)$; otherwise $\mu(X_{s_1}X_{s_2}\cdots X_{s_m}) = 0$.

□

From the last example the following definition emerges. Let $(\mathcal{A}, \varphi)$ be a noncommutative probability space with a unital *-algebra $\mathcal{A}$. A noncommutative random variable $u \in \mathcal{A}$ will be called a *Haar unitary* if the joint distribution of u and u^* is the same as in the previous example. A Haar unitary appeared earlier in Example 1.1.2.

Example 1.2.5 For $0 \neq h \in \mathcal{H}$ let $\ell(h)$ be the creation operator defined on the full Fock space $\mathcal{F}(\mathcal{H})$ (defined in (1.1.4)). Compute the joint distribution of $\ell(h)$ and $\ell(h)^*$ at the vacuum vector.

We need to compute

$$\mu(X_{s(n)}X_{s(n-1)}\cdots X_{s(1)}) = \langle a_{s(n)}a_{s(n-1)}\cdots a_{s(1)}\Phi, \Phi\rangle\,,$$

where $s(1), s(2), \dots, s(n) \in \{1, 2\}$ and

$$a_{s(i)} = \begin{cases} \ell(h) & \text{if } s(i) = 1 \\ \ell(h)^* & \text{if } s(i) = 2 \end{cases} \qquad (1 \le i \le n).$$

This computaion has been essentially done in the proof of Theorem 1.1.5, and it is convenient to use the geometric language introduced there. It was found that $\langle a_{s(n)}a_{s(n-1)}\dots a_{s(1)}\Phi, \Phi\rangle$ is 0 or $\|h\|^n$ depending on the polygonal line associated with the long product. When the line ends on the x-axis and does not touch the horizontal line $y = -1$, we get a nonzero expectation value. In particular, $\mu(X_{s(n)}X_{s(n-1)}\cdots X_{s(1)}) = 0$ when n is odd.

For example, $\mu(X_2X_1X_2X_1) = \mu(X_2^2X_1^2) = \|h\|^4$, and in all other cases we have $\mu(X_{s(4)}X_{s(3)}X_{s(2)}X_{s(1)}) = 0$.

□

Example 1.2.6 For $g_1, g_2 \in \mathcal{H}$ let $B(g_1)$ and $B(g_2)$ be the unbounded selfadjoint operators defined by (1.1.9) on the symmetric Fock space $\mathcal{F}_s(\mathcal{H})$. Compute the joint distribution of $B(g_1)$ and $B(g_2)$ at the vacuum vector.

We have to know the expectation value of long products of field operators $B(f_i)$. We benefit from the formula

$$\begin{aligned}&\langle B(f_n)B(f_{n-1})\cdots B(f_1)\Phi, \Phi\rangle \\ &\quad = (-\mathrm{i}\,)^n \frac{\partial^n}{\partial t_n \cdots \partial t_1}\langle W(t_nf_n)\cdots W(t_1f_1)\Phi, \Phi\rangle\Big|_{t_1=\ldots=t_n=0}\,.\end{aligned}$$

Since

$$\begin{aligned}&W(t_nf_n)W(t_{n-1}f_{n-1})\cdots W(t_1f_1) \\ &\quad = W(t_nf_n + t_{n-1}f_{n-1} + \cdots + t_1f_1)\cdot \exp\Big(\mathrm{i}\sum_{j>k} t_jt_k \mathrm{Im}\,\langle f_j, f_k\rangle\Big)\end{aligned}$$

and $\langle W(f)\Phi, \Phi\rangle = \exp(-\|f\|^2/2)$ according to (1.1.8) (in which $h = 0$ should be taken), we have

$$\begin{aligned}&\langle W(t_n f_n)\cdots W(t_1 f_1)\Phi, \Phi\rangle \\ &\quad = \exp\Big(-\frac{1}{2}\sum_{m=1}^{n} t_m^2\|f_m\|^2\Big)\exp\Big(-\sum_{j>k} t_j t_k\langle f_k, f_j\rangle\Big). \qquad (1.2.2)\end{aligned}$$

What we need is the coefficient of $t_1 t_2\cdots t_n$ in the power series expansion. Such a term comes only from the second factor of (1.2.2) and only in the case of an even n. One obtains $t_1 t_2\cdots t_n$ as a product of factors $t_j t_k$ ($j > k$) in many different ways; each of the possibilities is associated with a pair partition of the set $\{1, 2, \ldots, n\}$. Hence

$$\langle B(f_n)B(f_{n-1})\cdots B(f_1)\Phi, \Phi\rangle = \sum\prod_{m=1}^{n/2}\langle f_{k_m}, f_{j_m}\rangle\,, \qquad (1.2.3)$$

where the summation is over all pair partitions $\{\mathcal{V}_1, \mathcal{V}_2, \ldots, \mathcal{V}_{n/2}\}$ of $\{1, 2, \ldots, n\}$ such that $\mathcal{V}_m = \{j_m, k_m\}$ with $j_m > k_m$ $(m = 1, 2, \ldots, n/2)$. Observe that $\langle f_k, f_j\rangle = \langle B(f_j)B(f_k)\Phi, \Phi\rangle$.

In this way we conclude that

$$\mu(X_{s(n)}X_{s(n-1)}\cdots X_{s(1)}) = \begin{cases} 0 & \text{if } n \text{ is odd,} \\ \sum\prod_{m=1}^{n/2}\langle g_{s(k_m)}, g_{s(j_m)}\rangle & \text{if } n \text{ is even,}\end{cases}$$

where the summation is over the pair partitions exactly as in (1.2.3). For example, we have

$$\begin{aligned}\mu(X_1X_2X_1X_2) &= \|g_2\|^2\|g_1\|^2 + \langle g_2, g_1\rangle^2 + |\langle g_2, g_1\rangle|^2\,, \\ \mu(X_1^3X_2) &= 3\|g_1\|^2\langle g_2, g_1\rangle\,, \\ \mu(X_1X_2X_1^2) &= \|g_1\|^2\langle g_2, g_1\rangle + 2\|g_1\|^2\langle g_1, g_2\rangle\,.\end{aligned}$$

□

This is a good place to make a remark about the previous example. Since for every $g \in \mathcal{H}$ we have a noncommutative random variable $B(g)$, the correspondence $g \mapsto B(g)$ is a "noncommutative stochastic process" indexed by $\mathcal{H}$. According to Theorem 1.1.6, the $B(g)$'s are normally distributed. On the other hand, it follows from the computation above that for $g_1 \perp g_2$ we have the factorization

$$\begin{aligned}&\mu(B(f_1)B(f_2)\cdots B(f_n)) \\ &\quad = \mu(\tilde{B}(f_1)\tilde{B}(f_2)\cdots\tilde{B}(f_n))\mu(\hat{B}(f_1)\hat{B}(f_2)\cdots\hat{B}(f_n))\,, \qquad (1.2.4)\end{aligned}$$

where each f_i is g_1 or g_2 and

$$\tilde{B}(f_i) := \begin{cases} B(g_1) & \text{if } f_i = g_1, \\ I & \text{otherwise,} \end{cases} \qquad \hat{B}(f_i) := \begin{cases} B(g_2) & \text{if } f_i = g_2, \\ I & \text{otherwise.} \end{cases}$$

For example, $\mu(B(g_1)^2 B(g_2) B(g_1)^2 B(g_2)) = \mu(B(g_1)^4)\mu(B(g_2)^2)$. It seems that $B(g_1)$ and $B(g_2)$ are independent in a sense, and what we have seems to be a noncommutative analogue of a Gaussian process:

Example 1.2.7 Let $\mathcal{H} := L^2(\mathbb{R}^+)$ and $B_t := B(\chi_{[0,t)})$, where $\chi_{[0,t)}$ stands for the characteristic function of the interval $[0,t)$, $t \geq 0$. Then B_t is a noncommutative analogue of a Gaussian process in the symmetric Fock space $\mathcal{F}_s(L^2(\mathbb{R}^+))$, and it is indexed by $t \geq 0$.

We have

$$\langle B_t B_s \Phi, \Phi \rangle = t \wedge s,$$

and the increments $B_{t(2)} - B_{t(1)}$ and $B_{s(2)} - B_{s(1)}$ are normally distributed and independent in the sense of (1.2.4) whenever $0 \leq t(1) < t(2) \leq s(1) < s(2)$. (1.2.4) enables us to compute all joint moments of the noncommutative random variables $B_{t(2)} - B_{t(1)}$ and $B_{s(2)} - B_{s(1)}$.

□

Next we set the previous example in the full Fock space.

Example 1.2.8 Let $\mathcal{H} := L^2(\mathbb{R}^+)$ and consider the full Fock space $\mathcal{F}(L^2(\mathbb{R}^+))$. If $X_t := \ell(\chi_{[0,t)})^* + \ell(\chi_{[0,t)})$, then we have

$$\langle X_t X_s \Phi, \Phi \rangle = t \wedge s$$

by a simple computation. For $0 \leq t(1) < t(2)$ the increment $X_{t(2)} - X_{t(1)} = \ell(\chi_{[t(1),t(2))})^* + \ell(\chi_{[t(1),t(2))})$ has the semicircular distribution with mean 0 and variance $t(2) - t(1)$ at the vacuum vector. However, it is difficult at the moment to say anything about the joint distribution of two increments $X_{t(2)} - X_{t(1)}$ and $X_{s(2)} - X_{s(1)}$ for $0 \leq t(1) < t(2) \leq s(1) < s(2)$. What we need is a new concept, the subject of the next chapter. It will turn out that the increments are in free relation, which is a particular rule for the computation of the joint distribution; cf. Example 2.2.10.

□

Notes and Remarks. The concept of the distribution of a selfadjoint operator in a vector goes back to the 1920's, when quantum mechanics was created. Vectors are called wave functions in quantum mechanics, and the abstract definition appeared in the famous book "Mathematical Foundation of Quantum Mechanics" published in 1932 in German by John von Neumann. Our Example 1.1.3 was the first and simplest case in which the statistical feature of quantum mechanics was recognized; von Neumann gave Max Born credit for it. Certainly quantum mechanics was one of the main motivations for developing a noncommutative probability theory based on linear functionals of a noncommutative algebra (see [54], for example).

See standard texts such as [188] and [108] for general theory on C^*-algebras (in particular, the two Gelfand-Naimark theorems) and also on von Neumann algebras.

The Fock space appeared in a physical context in the work of V.A. Fock in 1932. We suggest the book of K.R. Parthasarathy [143] for a mathematical treatment of Weyl unitaries, exponential vectors, and several topics related to the symmetric Fock space. Details on quantum statistical mechanics based on creation and annihilation operators on the symmetric Fock space as well as the antisymmetric Fock space are also found in the book of O. Bratteli and D.W. Robinson [44].

The full Fock space over an n-dimensional space is nothing but the ℓ^2 space over the free semigroup generated by n elements. Theorem 1.1.5 has many extensions. One can change the inner product (for which the adjoint of the left creation operator is defined), or the distribution is not taken at the vacuum vector. The q-Gaussian distribution can be obtained in this way. Concerning the relevant Fock space, we refer to [39] and [37]. The density of Example 1.1.8 is from [7] and [41]. Moments and other details of the q-Gaussian distribution are found in [131] and [106]. The formula for the moments is given in the Notes and Remarks to Chap. 2, below. The q-Gaussian distribution is only a small slice in the world of "q" (q-derivation, q-Hermite polynomials, etc.). Another extension of the Fock space formalism is the *interacting Fock space*, see [118]. When $\mathcal{H}$ is the L^2-space over a measure space, on the tensor powers $\mathcal{H}^{\otimes n}$ a modified scalar product is introduced with the help of weight functions $\lambda_n(x_1, \cdots, x_n)$ enjoying certain growth properties, and this leads to the modified Fock space. Then the creation and annihilation operators are introduced in a standard way. The distribution of the selfadjoint operator $\ell^* + \ell$ will be a sort of averaged-out semicircle law in the vacuum state.

Chapter 2

The Free Relation

If two noncommutative random variables a, b belonging to an algebra $\mathcal{A}$ are in a generic relation from the algebraic point of view, then $ab = ba$, or $a = b^2$, or any algebraic equation for a and b should *not* hold. The generic relation is similar to the relation of generators of a free group. When a and b are in such a generic relation there is no joint distribution measure for a and b, but $a+b$ might have a distribution measure. The free relation of a and b defined and studied in this chapter resembles the generic relation of generators of a free structure; however, the formal definition of freeness will go through expectations. The free relation is a very central concept of the analysis of noncommutative random variables, proposed by Voiculescu.

It is tempting to compare the free relation of noncommutative random variables with independence of classical random variables. If a and b are in free relation or stochastically independent, then $\varphi(ab) = \varphi(a)\varphi(b)$. In the case of stochastic independence, $\varphi(abab) = \varphi(a^2)\varphi(b^2)$. However, if a and b are in free relation, then a^2b^2 and $abab$ are different random variables with possibly different expectations. The free relation of a and b provides some rules for computing the expectation of a long product with factors a and b; that is, the joint distribution of a and b is expressed by the moments of a and b. This procedure is different from the simple rule $\varphi(a^k b^m) = \varphi(a^k)\varphi(b^m)$ for independent classical variables.

When a and b are free noncommutative random variables, the distribution of $a + b$ defines the free convolution of the distributions of a and b. It turns out that the proper way of book-keeping for free convolution is not the method of moments but the use of free cumulants (or R-series). The sequence of free cumulants linearizes the free convolution, and in this sense it replaces the logarithm of the Fourier transform in its classical role. The representation of a noncommutative random variable by a formal infinite series of creation operators gives the free cumulants. The transformation between moments and free cumulants is a sort of Möbius inversion procedure on the lattice of non-crossing partitions.

The central limit theorem of classical probability theory leads to the Gaussian law as the limit distribution. Parallel to the classical central limit theorem, the distribution measure of the standardized sum of noncommutative random variables in free relation converges, and the limit distribution is the semicircle law.

The notions of moments and free cumulants as well as the central limit theorem extend to k-tuples of noncommutative random variables. The analogue of the multivariate Gaussian law is the semicircular multivariable. Some noncommutative multivariables are without a direct analogue in the classical theory. The R-diagonal pairs could be of this type.

2.1 The free product of noncommutative probability spaces

Given two groups G_1 and G_2, their free product is a quotient of the free group generated by the set $G_1 \cup G_2$. Namely, we take the free group generated by the disjoint union $G_1 \cup G_2$ (identifying both group units) and we divide by all multiplicative relations holding in G_1 or in G_2. Even if we start with finite groups, their free product will be infinite. The free product of two unital algebras is slightly more complicated, but roughly speaking the free product of algebras is the algebra generated by them, with no relations connecting the two algebras except for identification of the unit elements. To define the free product, a state should be specified on each of the algebras. The free product of noncommutative probability spaces will replace the usual product of measure spaces in Voiculescu's theory. The relation of free product to freeness to be defined in the next section is the same as the relation of product measure space to statistical independence.

Example 2.1.1 Let $G_1 := \{e, f\}$ and $G_2 := \{e, g\}$ be two copies of the group $\mathbb{Z}_2$. (e stands for the unit in both groups.) The free product $G_1 \star G_2$ consists of all alternating finite sequences of f and g. The product is juxtaposition followed by the simplification rule that ff and gg should be canceled out.

□

The *free product of groups* is an associative operation, and we can take the free product of several groups. For axample, the n-fold free product $\mathbb{Z} \star \mathbb{Z} \star \ldots \star \mathbb{Z}$ is the *free group* $\mathbf{F}_n$ with n generators. Next we give the formal definition of the free product of noncommutative probability spaces.

Let $(\mathcal{A}_1, \varphi_1)$ and $(\mathcal{A}_2, \varphi_2)$ be two noncommutative probability spaces and let $\mathbf{1}$ denote the unit elements. (The same notation for both units—a sort of identification.) We decompose $\mathcal{A}_i$ as $\mathbb{C}\mathbf{1} \oplus \mathcal{A}_i^0$, where $\mathcal{A}_i^0 := \{a \in \mathcal{A}_i : \varphi_i(a) = 0\}$. The *free product algebra* is a huge direct sum of tensor products:

$$
\begin{aligned}
\mathcal{B} \; &:= \; \mathbb{C}\mathbf{1} \oplus \mathcal{A}_1^0 \oplus \mathcal{A}_2^0 \oplus (\mathcal{A}_1^0 \otimes \mathcal{A}_2^0) \oplus (\mathcal{A}_2^0 \otimes \mathcal{A}_1^0) \\
&\quad \oplus (\mathcal{A}_1^0 \otimes \mathcal{A}_2^0 \otimes \mathcal{A}_1^0) \oplus (\mathcal{A}_2^0 \otimes \mathcal{A}_1^0 \otimes \mathcal{A}_2^0) \oplus (\mathcal{A}_1^0 \otimes \mathcal{A}_2^0 \otimes \mathcal{A}_1^0 \otimes \mathcal{A}_2^0) \oplus \cdots \\
&= \; \mathbb{C}\mathbf{1} \oplus \bigoplus \Big\{ \mathcal{A}_{i(1)}^0 \otimes \mathcal{A}_{i(2)}^0 \otimes \cdots \otimes \mathcal{A}_{i(n)}^0 : i(k) \in \{1,2\},\, 1 \le k \le n, \\
&\qquad\qquad i(k) \neq i(k+1),\, 1 \le k \le n-1,\, n \in \mathbb{N} \Big\}.
\end{aligned}
$$

Since $\mathcal{A}_i^0$ is a vector space, this definition makes clear the linear structure of $\mathcal{B}$. If $a_i \in \mathcal{A}_i^0$, then their product $a_1 \cdot a_2$ is $a_1 \otimes a_2 \in \mathcal{A}_1^0 \otimes \mathcal{A}_2^0 \subset \mathcal{B}$. More generally, if

(i) $x = a_1 \otimes a_2 \otimes \cdots \otimes a_n \in \mathcal{B}$, $a_k \in \mathcal{A}_{i(k)}^0$, $i(k) \in \{1,2\}$ $(1 \leq k \leq n)$,

(ii) $y = b_1 \otimes b_2 \otimes \cdots \otimes b_m \in \mathcal{B}$, $b_k \in \mathcal{A}_{j(k)}^0$, $j(k) \in \{1,2\}$ $(1 \leq k \leq m)$,

(iii) $i(n) \neq j(1)$,

then

$$x \cdot y = a_1 \otimes a_2 \otimes \cdots \otimes a_n \otimes b_1 \otimes b_2 \otimes \cdots \otimes b_m \,. \tag{2.1.1}$$

If instead of (iii) we have

(iii)′ $i(n) = j(1)$,

then the product cannot be given explicitly, because it is not true in general that $a_n b_1 \in \mathcal{A}_{i(n)}^0$. So we write $a_n b_1 = \lambda \mathbf{1} + a$ with $a \in \mathcal{A}_{i(n)}^0$, and we set

$$\begin{aligned} x \cdot y \quad = \quad & a_1 \otimes a_2 \otimes \cdots \otimes a_{n-1} \otimes a \otimes b_2 \otimes \cdots \otimes b_m \\ & + \lambda\, a_1 \otimes a_2 \otimes \cdots \otimes a_{n-1} \otimes b_2 \otimes b_3 \otimes \cdots \otimes b_m \,. \end{aligned} \tag{2.1.2}$$

This procedure is applied again, since $i(n-1) = j(2)$ (because we have just two indices $1, 2$ here). By means of (2.1.1) and (2.1.2) the multiplication is determined on $\mathcal{B}$, and $\mathcal{B}$ becomes an algebra. The involution is also defined on $\mathcal{B}$ whenever both of the $\mathcal{A}_i$ are *-algebras. The definition is straightforward because $\mathcal{A}_i^0$ is selfadjoint in $\mathcal{A}_i$; that is,

$$x^* = a_n^* \otimes a_{n-1}^* \otimes \cdots \otimes a_1^* \,.$$

Beyond the algebra (or *-algebra) structure, we have a natural linear functional ω on $\mathcal{B}$ which takes the coefficient of $\mathbf{1}$ in the expansion of the element. Namely,

$$\omega : \lambda \mathbf{1} + \sum a_1 \otimes a_2 \otimes \cdots \otimes a_n \mapsto \lambda \,.$$

We call the noncommutative probability space $(\mathcal{B}, \omega)$ the *free product* of the noncommutative probability spaces $(\mathcal{A}_1, \varphi_1)$ and $(\mathcal{A}_2, \varphi_2)$, and use the notation

$$(\mathcal{B}, \omega) = (\mathcal{A}_1, \varphi_1) \star (\mathcal{A}_2, \varphi_2) \,.$$

Furthermore, the canonical embedding of $\mathcal{A}_i$ into $\mathcal{B}$ is $\iota_i : \mathcal{A}_i \to \mathcal{B}$, defined as $a_i \mapsto \varphi_i(a_i)\mathbf{1} \oplus \big(a_i - \varphi_i(a_i)\mathbf{1}\big)$, which pulls back ω to φ_i. The free product of arbitrarily many noncommutative probability spaces may be defined similarly.

Example 2.1.2 Let G_1 and G_2 be two (discrete) groups. The *group algebras* $R(G_i)$ consist of complex functions on G_i with finite support, i.e. finite linear combinations $\sum_h \lambda_i(h)h$ ($\lambda_i(h) \in \mathbb{C}$, $h \in G_i$), $i = 1, 2$. Writing e for both group units, we set

$$\varphi_i\Big(\sum_h \lambda_i(h)h\Big) := \lambda_i(e)\,,$$

and we have two noncommutative probability spaces $(R(G_i), \varphi_i)$, $i = 1, 2$. Then $R(G_i)^0$ is linearly spanned by the group elements different from e. We can see that $R(G_i)^0$ is not closed under multiplication. The free product $(R(G_1), \varphi_1) \star (R(G_2), \varphi_2)$ is $R(G_1 \star G_2)$ endowed with a similarly defined state φ.

□

Example 2.1.3 Let $\mathcal{A}_i := \mathbb{C}\langle X_i \rangle$ be *polynomial algebras.* Define the state φ_i to be the constant term of a polynomial, $i = 1, 2$. Then the construction of the free product is particularly simple, because $\mathcal{A}_i^0$ is not only a linear space but closed under multiplication. $(\mathcal{A}_1, \varphi_1) \star (\mathcal{A}_2, \varphi_2)$ will be the polynomial algebra $\mathbb{C}\langle X_1, X_2 \rangle$ endowed with the similarly defined state.

□

Here we have constructed the free product in the category of algebras or *-algebras. When one starts with C^*-algebras or von Neumann algebras, the above described free product should be endowed with a norm and a completion procedure should be carried out. There is another problem concerning positivity. When φ_1 and φ_2 are states, we want the free product functional ω to be positive. We do not go into the details here, but free product representations and free product von Neumann algebras will be explained in Sec. 7.1.

2.2 The free relation

Let us recall that in the free product $(\mathcal{B}, \omega) = (\mathcal{A}_1, \varphi_1) \star (\mathcal{A}_2, \varphi_2)$ of noncommutative probability spaces the state ω is determined by the condition that it vanishes on all products $a_1 a_2 \cdots a_n$ such that $a_k \in \mathcal{A}_{i(k)}$, $i(k) \neq i(k+1)$, and all factors have 0 expectation. The concept of *free relation* emerges from this observation. The free relation, or *freeness*, is a fundamental concept throughout this monograph.

Let $(\mathcal{A}, \varphi)$ be a noncommutative probability space and let $\mathcal{A}_i$ be subalgebras of $\mathcal{A}$ ($i \in I$). We say that the family $(\mathcal{A}_i)_{i \in I}$ is *in free relation* (or *free*) with respect to φ if, for every $n \in \mathbb{N}$ and $i(1), \dots, i(n) \in I$ such that $i(k) \neq i(k+1)$ $(1 \leq k \leq n-1)$,

$$\varphi(a_1 a_2 \cdots a_n) = 0 \quad \text{whenever} \quad a_k \in \mathcal{A}_{i(k)},\ \varphi(a_k) = 0,\ 1 \leq k \leq n. \tag{2.2.1}$$

(We stress that in this definition non-neighboring $i(k)$'s are allowed to be equal.) The free relation can be formulated equivalently in terms of the free product. Let

$(\mathcal{B}, \omega) = \star_{i \in I}(\mathcal{A}_i, \varphi|_{\mathcal{A}_i})$ and let $\iota_i : \mathcal{A}_i \to \mathcal{B}$ be the canonical embedding. Then the above definition is equivalent to

$$\varphi(a_1 a_2 \cdots a_n) = \omega\big(\iota_{i(1)}(a_1)\iota_{i(2)}(a_2)\cdots\iota_{i(n)}(a_n)\big) \quad \text{for every}$$
$$a_k \in \mathcal{A}_{i(k)},\ i(k) \in I,\ i(k) \neq i(k+1),\ 1 \le k \le n-1,\ n \in \mathbb{N}.$$

By saying that two families of noncommutative random variables $\{a_i : i \in I\}$ and $\{b_j : j \in J\}$ in $\mathcal{A}$ are in free relation, we mean that the subalgebra generated by $\{a_i : i \in I\}$ and that generated by $\{b_j : j \in J\}$ are free (with respect to φ). When $\mathcal{A}$ is a *-algebra, the freeness of $\{a_i\}$ and $\{b_j\}$ is usually defined as above with the generated *-subalgebras, and the term **-freeness* is sometimes referred to. The freeness of several families of random variables is defined similarly.

Example 2.2.1 Let $h_1, h_2 \in \mathcal{H}$, and let $\ell(h_i)$ be the corresponding left creation operators on $\mathcal{F}(\mathcal{H})$. Then $\{\ell(h_1), \ell(h_1)^*\}$ and $\{\ell(h_2), \ell(h_2)^*\}$ are free with respect to the vacuum state if and only if $\langle h_1, h_2\rangle = 0$.

First, if $\ell_1 := \ell(h_1)$ and $\ell_2 := \ell(h_2)$ are free, then $\langle h_1, h_2\rangle = \langle \ell_2^*\ell_1\Phi, \Phi\rangle = 0$ since $\langle \ell_1\Phi, \Phi\rangle = \langle \ell_2\Phi, \Phi\rangle = 0$.

To prove the converse we assume $\langle h_1, h_2\rangle = 0$ and hence $\ell_2^*\ell_1 = 0$, and we want to verify directly that (2.2.1) holds. Since any monomial $p(\ell_i, \ell_i^*)$ of ℓ_i and ℓ_i^* has 0 expectation unless it reduces to $c\mathbf{1}$, we may restrict ourselves to the case when the a_k's are monomials. Due to the relation $\ell_i^*\ell_i = \mathbf{1}$, we may assume that a_k is of the form $\ell_i^n \ell_i^{*m}$, $n + m > 0$. In this way we have reduced the statement to the verification of the relation

$$\big\langle \cdots \big(\ell_1^{n(3)}\ell_1^{*m(3)}\big)\big(\ell_2^{n(2)}\ell_2^{*m(2)}\big)\big(\ell_1^{n(1)}\ell_1^{*m(1)}\big)\Phi, \Phi\big\rangle = 0$$

when $n(k) + m(k) > 0$. It is elementary to analyze that this holds.

More generally, if $\{h_i : i \in I\}$ is an orthogonal family in the Hilbert space $\mathcal{H}$, then the families

$$\{\ell(h_i), \ell(h_i)^*\} \qquad (i \in I)$$

are in free relation with respect to the vacuum expectation. In particular, for any polynomials p_i the operators $\ell(h_i)^* + p_i(\ell(h_i))$ $(i \in I)$ are *-free.

□

Let $(\mathcal{A}_i)_{i \in I}$ be a family of *-subalgebras of $\mathcal{A}$ and assume that they are free with respect to a state φ on $\mathcal{A}$. We need methods for the evaluation of $\varphi(a_1 a_2 \cdots a_n)$ when $a_k \in \mathcal{A}_{i(k)}$ $(i(1), \ldots, i(n) \in I)$ and $i(k) \neq i(k+1)$. We write $a_k = \varphi(a_k)\mathbf{1} + a_k^0$, and

$$\varphi(a_1 a_2 \cdots a_n) = \sum \varphi(a_{\pi(1)})\varphi(a_{\pi(2)})\cdots\varphi(a_{\pi(m)})\varphi(a^0_{\sigma(1)}a^0_{\sigma(2)}\cdots a^0_{\sigma(n-m)}) \qquad (2.2.2)$$

is an identity if the summation is over all ordered partitions $(\{\pi(1), \pi(2), \ldots, \pi(m)\}, \{\sigma(1), \sigma(2), \ldots, \sigma(n-m)\})$ of the set $\{1, 2, \ldots, n\}$ and $0 \le m \le n$. The term corresponding to $m = 0$ vanishes, due to the assumption of freeness. So (2.2.2) gives a recursion. On its right-hand side φ is on products of length less than n. One can obtain a slightly different formula by expanding

$$\varphi\big((a_1 - \varphi(a_1)\mathbf{1})(a_2 - \varphi(a_2)\mathbf{1}) \cdots (a_n - \varphi(a_n)\mathbf{1})\big),$$

which vanishes under the assumption of the free relation. Hence

$$\varphi(a_1 a_2 \cdots a_n) = \sum_{r=1}^{n} \sum_{1 \le k_1 < \ldots < k_r \le n} (-1)^{r+1} \varphi(a_{k_1}) \cdots \varphi(a_{k_r}) \times \varphi(a_1 \cdots \hat{a}_{k_1} \cdots \hat{a}_{k_r} \cdots a_n)\,, \tag{2.2.3}$$

where ˆ indicates terms that are omitted.

Let a and b be in free relation with respect to φ. One can understand the freeness as a rule for computing the expectation of products of a and b. From $\varphi\big((a - \varphi(a)\mathbf{1})(b - \varphi(b)\mathbf{1})\big) = 0$ we obtain $\varphi(ab) = \varphi(a)\varphi(b)$. The factorization of the expectation of free variables may remind us of independence in probability theory. However, the freeness behaves very differently for longer products. By means of the above formulas one computes

$$\varphi(abab) = \varphi(a^2)\varphi(b)^2 + \varphi(a)^2\varphi(b^2) - \varphi(a)^2\varphi(b)^2\,, \tag{2.2.4}$$

and

$$\varphi(ab^2a) = \varphi(a^2)\varphi(b^2)\,.$$

The latter identity shows that the free relation precludes commutativity. Indeed, when $ab = ba$, it follows that

$$\varphi\big((a - \varphi(a)\mathbf{1})^2\big)\varphi\big((b - \varphi(b)\mathbf{1})^2\big) = 0,$$

and the variance of a or that of b must vanish. Formulating this fact in the language of probability theory, we can say that a or b is constant "with probability one".

Let $\mathcal{A}_1$ and $\mathcal{A}_2$ be two *-subalgebras of $\mathcal{A}$. In the algebraic language used here we would say that $\mathcal{A}_1$ and $\mathcal{A}_2$ are *independent* with respect to a state φ on $\mathcal{A}$, if $a_1 a_2 = a_2 a_1$ and $\varphi(a_1 a_2) = \varphi(a_1)\varphi(a_2)$ whenever $a_1 \in \mathcal{A}_1$ and $a_2 \in \mathcal{A}_2$. The independence allows a simple computation of the joint distribution of a_1 and a_2. The same holds true when $\mathcal{A}_1$ and $\mathcal{A}_2$ are in free relation. However, for instance, the computation rule (2.2.4) is more complicated. On the other hand, independence and the free relation may hold at the same time only in trivial cases, as we discussed above.

In the next example (2.2.4) is extended.

Example 2.2.2 Assume that $\{a_1, a_2, a_3\}$ and $\{b_1, b_2\}$ are in free relation with respect to a state φ. Then

$$\begin{aligned}\varphi(a_1 b_1 a_2 b_2 a_3) &= \varphi(a_1 a_3)\varphi(a_2)\varphi(b_1 b_2) + \varphi(a_1 a_2 a_3)\varphi(b_1)\varphi(b_2) \\ &\quad - \varphi(a_1 a_3)\varphi(a_2)\varphi(b_1)\varphi(b_2)\,.\end{aligned}$$

□

Lemma 2.2.3 *Let $(\mathcal{A}_i)_{i\in I}$ be a family of *-subalgebras of $\mathcal{A}$ and assume that they are free with respect to a state φ on $\mathcal{A}$. If $a_k \in \mathcal{A}_{i(k)}$, $i(1), \dots, i(n) \in I$, $i(k) \neq i(k+1)$, and there is a j such that $i(k) \neq i(j)$ for $k \neq j$, then*

$$\varphi(a_1 a_2 \cdots a_n) = \varphi(a_j)\varphi(a_1 a_2 \cdots a_{j-1} a_{j+1} \cdots a_n)\,.$$

Proof: First we assume that $\varphi(a_j) = 0$ and show that $\varphi(a_1 a_2 \cdots a_n) = 0$. We use induction on n. The case $n = 1$ is trivial. In the induction step we benefit from (2.2.2) and show that all summands on the right-hand side vanish. Indeed, $j \in \{\pi(1), \pi(2), \dots, \pi(m)\}$, or $j \in \{\sigma(1), \sigma(2), \dots, \sigma(n-m)\}$. In the first case we refer to $\varphi(a_j) = 0$; in the second case the induction hypothesis is referred to.

When $\varphi(a_j) \neq 0$ we have

$$\begin{aligned}\varphi(a_1 a_2 \cdots a_n) &= \varphi(a_j)\varphi(a_1 a_2 \cdots a_{j-1} a_{j+1} \cdots a_n) \\ &\quad + \varphi(a_1 a_2 \cdots a_{j-1}(a_j - \varphi(a_j)\mathbf{1}) a_{j+1} \cdots a_n),\end{aligned}$$

and the second term on the right-hand side vanishes due to the previous case.

□

For example, a repeated application of Lemma 2.2.3 shows the following:

Example 2.2.4 If (a, b, c, d) is a free family of noncommutative random variables in $(\mathcal{A}, \varphi)$, then

$$\varphi(cbaabcdc) = \varphi(a^2)\varphi(b^2)\varphi(d)\varphi(c^3)\,.$$

□

Lemma 2.2.5 *Let $(\mathcal{A}_i)_{i\in I}$ be a family of *-subalgebras of $\mathcal{A}$, and assume that they are free with respect to a state φ on $\mathcal{A}$. If $x = a_1 a_2 \cdots a_n$ is a product such that $a_k \in \mathcal{A}_{i(k)}$ $(i(k) \in I)$, $i(k) \neq i(k+1)$, and $\varphi(a_k) = 0$, and if $y = b_1 b_2 \cdots b_m$ is such that $b_k \in \mathcal{A}_{j(k)}$ $(j(k) \in I)$, $j(k) \neq j(k+1)$, and $\varphi(b_k) = 0$, then*

$$\varphi(y^* x) = \delta_{nm}\delta_{i(1)j(1)}\delta_{i(2)j(2)} \cdots \delta_{i(n)j(n)}\varphi(b_1^* a_1)\varphi(b_2^* a_2) \cdots \varphi(b_n^* a_n)\,.$$

Proof: We apply induction on $n + m$. Write

$$\varphi(y^* x) = \varphi(b_m^* b_{m-1}^* \cdots b_1^* a_1 a_2 \cdots a_n)\,.$$

When b_1 and a_1 are in different subalgebras, that is, $i(1) \neq j(1)$, the very definition of the free relation gives $\varphi(y^*x) = 0$, and the statement holds true. When they are in the same subalgebra, we write

$$b_1^* a_1 = \varphi(b_1^* a_1)\mathbf{1} + \big(b_1^* a_1 - \varphi(b_1^* a_1)\mathbf{1}\big)$$

and conclude that

$$\varphi(y^* x) = \varphi(b_1^* a_1)\varphi(b_m^* b_{m-1}^* \cdots b_2^* a_2 \cdots a_n)\,.$$

Now the induction hypothesis works.

□

Proposition 2.2.6 *Let $(\mathcal{A}, \varphi) = \star_{i\in I}(\mathcal{A}_i, \varphi_i)$. If φ_i is a tracial state, that is, $\varphi_i(ab) = \varphi_i(ba)$ for every $i \in I$, then φ is tracial.*

Proof: Let x and y be as in Lemma 2.2.5. It is sufficient to prove that $\varphi(y^*x) = \varphi(xy^*)$. This follows from Lemma 2.2.5, because $\varphi(b_i^* a_i) = \varphi(a_i b_i^*)$ holds by assumption.

□

This proposition has a consequence. Dealing with a family of selfadjoint operators which are in free relation with respect to a state φ of a C^*-algebra, we may always assume that φ is tracial. Indeed, the joint distribution of our noncommutative random variables does not change if we pass to the free product of the commutative C^*-algebras generated by each of the selfadjoint operators. According to the proposition, we arrive at a tracial state in this way.

Example 2.2.7 On the full Fock space $\mathcal{F}(\mathcal{H})$, consider the von Neumann algebra $\mathcal{M}$ generated by the operators $\ell(h)^* + \ell(h)$, $h \in \mathcal{H}$. Then the vacuum expectation φ is a tracial state on $\mathcal{M}$.

We may assume that h runs over an orthogonal basis in $\mathcal{H}$. Then we have the freeness of the family $\ell(h) + \ell(h)^*$ of selfadjoint operators, and the previous proposition yields our statement.

□

Proposition 2.2.8 *Let $(\mathcal{A}_i)_{i\in I}$ be a free family of *-subalgebras in $(\mathcal{A}, \varphi)$. If $I = \bigcup_{j\in J} I_j$ is a partition and $\mathcal{B}_j$ is the *-subalgebra generated by $\bigcup_{i\in I_j} \mathcal{A}_i$, then $(\mathcal{B}_j)_{j\in J}$ is free.*

Proof: Let $\mathcal{A}_i^0 := \{a \in \mathcal{A}_i : \varphi(a) = 0\}$ and $\mathcal{B}_j^0 := \{b \in \mathcal{B}_j : \varphi(b) = 0\}$. It is easy to see that each element in $\mathcal{B}_j^0$ is written as a finite sum of products $a_1 a_2 \cdots a_n$ of $a_k \in \mathcal{A}_{i_k}^0$, $i_k \in I_j$, with $i_1 \neq i_2 \neq \ldots \neq i_n$. This fact immediately yields the conclusion.

□

Example 2.2.9 Let I_1 and I_2 be two disjoint intervals in $\mathbb{R}^+$, and let $\mathcal{H}$ be $L^2(\mathbb{R}^+)$. The *-algebra generated by $\{\ell(f) : f \in L^2(\mathbb{R}),\ \mathrm{supp}\, f \subset I_j\}$ is denoted by $\mathcal{A}_j$, $j = 1, 2$. Then the algebras $\mathcal{A}_1$ and $\mathcal{A}_2$ are in free relation in $B(\mathcal{F}(L^2(\mathbb{R})))$ with respect to the vacuum state.

Although a direct proof is possible, we can choose an orthogonal basis $h_i^{(j)}$ $(i = 1, 2, \ldots)$ in $L^2(I_j)$. For the orthogonal family $\{h_i^{(1)} : i \geq 1\} \cup \{h_i^{(2)} : i \geq 1\}$ we apply Example 2.2.1. Then we conclude from Proposition 2.2.8 that the families $\{\ell(h_i^{(1)}) : i \geq 1\}$ and $\{\ell(h_i^{(2)}) : i \geq 1\}$ are *-free. (Stricly speaking, an additional continuity argument is still to be applied to complete our statement.)

□

Example 2.2.10 Let $\mathcal{H} := L^2(\mathbb{R}^+)$ and consider the operators $X_t := \ell(\chi_{[0,t)})^* + \ell(\chi_{[0,t)})$ on the full Fock space $\mathcal{F}(L^2(\mathbb{R}^+))$ as in Example 1.2.8. It follows from Example 2.2.9 that for $0 \leq t(1) < t(2) \leq s(1) < s(2)$ the increments $X_{t(2)} - X_{t(1)}$ and $X_{s(2)} - X_{s(1)}$ are in free relation with respect to the vacuum expectation φ. X_t is an example of the *free Brownian motion.* Remember that φ is tracial on the algebra generated by the X_t's.

□

In the next example matrices built from creation operators are considered.

Example 2.2.11 Let $C^*(\ell(\mathcal{H}))$ denote the C^*-algebra generated by left creation operators on the full Fock space $\mathcal{F}(\mathcal{H})$. Let $\{h_{ij}^k : 1 \leq i, j \leq n,\ k = 1, 2, \ldots\}$ be an orthonormal family of vectors in the Hilbert space $\mathcal{H}$. We consider the algebra $C^*(\ell(\mathcal{H})) \otimes M_n(\mathbb{C})$, the tensor product of $C^*(\ell(\mathcal{H}))$ and the $n \times n$ matrices. In order to view this algebra as a noncommutative probability space, we consider a product state $\psi = \phi \otimes \rho$, where ϕ is the vacuum expectation in $\mathcal{F}(\mathcal{H})$ and ρ is a state on $M_n(\mathbb{C})$ having a diagonal density $\mathbf{Diag}(\lambda_1, \lambda_2, \ldots, \lambda_n)$. Put

$$L_k := \sum_{i,j=1}^{n} \sqrt{\lambda_i}\,\ell(h_{ij}^k) \otimes e_{ij}\,,$$

where (e_{ij}) is the usual system of matrix units in $M_n(\mathbb{C})$. We are going to show that $\mathbb{C}\mathbf{1}_{\mathcal{H}} \otimes M_n(\mathbb{C})$ and the *-algebra generated by $L_1, L_2, \ldots$ are free in the algebra $C^*(\ell(\mathcal{H})) \otimes M_n(\mathbb{C})$ with respect to the state ψ.

The operators L_k are regarded as $n \times n$ matrices with operator entries. For example, when $n = 2$ we have

$$L_k = \begin{bmatrix} \sqrt{\lambda_1}\ell(h_{11}^k) & \sqrt{\lambda_1}\ell(h_{12}^k) \\ \sqrt{\lambda_2}\ell(h_{21}^k) & \sqrt{\lambda_2}\ell(h_{22}^k) \end{bmatrix}.$$

It is easy to see that $L_s^* L_r = \delta_{sr}\mathbf{1}$ by using the assumed orthogonality relations. If $w = L_{i(1)}L_{i(2)} \cdots L_{i(r)}L_{j(1)}^* L_{j(2)}^* \cdots L_{j(s)}^*$ is viewed as a matrix, then all of its entries have 0 expectation under ϕ; hence $\psi(w) = 0$. This says that the joint distribution of the operators $L_1, L_2, \ldots$ with respect to ψ is the same as that of $\ell(f_1), \ell(f_2), \ldots$

in a full Fock space with respect to the vacuum expectation, where the f_k's are orthonormal vectors.

In order to show the above stated free relation, we have to prove that $\psi(W) = 0$ when

$$W = a_0 w_1 a_1 \cdots w_n a_n ,$$

where $a_k = \mathbf{1}_{\mathcal{H}} \otimes e_{pq} - \delta_{pq}\lambda_p \mathbf{1}$ (a_0, a_n may be $\mathbf{1}$) and w_k is of the form of the above w. If W does not contain a substring $L_s^* a_k L_r$, then all of its entries have zero expectation under ϕ, and $\psi(W) = 0$ holds. However, a simple computation yields that $L_s^* a_k L_r = 0$, so $\psi(W) = 0$ in any case.

□

2.3 The free central limit theorem

The classical central limit theorem says that if a_1, a_2, ... are independent and identically distributed random variables with $E(a_i) = 0$ and $E(a_i^2) = 1$, then the distribution of

$$\frac{a_1 + a_2 + \cdots + a_n}{\sqrt{n}} \tag{2.3.1}$$

tends to the standard Gaussian law. The free analogue of this classical result is due to Voiculescu: When the assumption of independence is replaced by the free relation of the noncommutative random variables a_1, a_2, ..., the limit distribution of (2.3.1) is the semicircle law (under certain conditions).

The free central limit theorem is interesting in its own right, however, aside from the fact that the combinatorial proof gives an occasion to introduce the concept of non-crossing partition which is fundamental in the combinatorial approach to freeness. In order to show this result, we first obtain the moments of the semicircle law in a combinatorial form.

A partition $\mathcal{V} = \{V_1, V_2, \dots, V_s\}$ of the set $[k] := \{1, 2, \dots, k\}$ consists of nonempty, pairwise disjoint blocks $V_1, V_2, \dots, V_s$ satisfying $\bigcup_{i=1}^s V_i = [k]$. For the sake of definiteness, we assume that the blocks are indexed in increasing order of their minimum elements, and elements of each block are increasingly ordered as well. For example, $1,5,7/2,6/3,8,10/4,9$ indicates a partition of $[10]$. This is not the only way to represent partitions. We may plot the numbers $1, 2, \dots, k$ horizontally, and successive elements of the same block are joined by arcs.

$\mathcal{V}$ is called a *non-crossing partition* if for $V_i = \{v_1, v_2, \dots, v_p\}$ and $V_j = \{w_1, w_2, \dots, w_q\}$ we have $w_m < v_1 < w_{m+1}$ if and only if $w_m < v_p < w_{m+1}$ $(m = 1, 2, \dots, q-1)$. Non-crossing partitions are represented graphically, as shown in the figure.

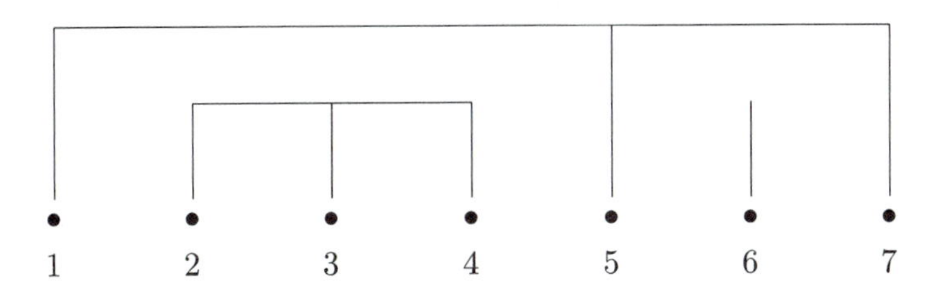

Fig.2.1. Representation of the non-crossing partition $1,5,7/2,3,4/6$.

Another way to represent partitions is the use of the *restricted growth function.* Let $\mathcal{V}$ be a partition of $[k]$. The restricted growth function $\omega : [k] \to [k]$ associates to i the index of the block of $\mathcal{V}$ which contains i. For example, the restricted growth function of the partition $1,5,7/2,3,4/6$ is $1,2,2,2,1,3,1$. The number of crossings in a partition is the number of subwords in the restricted growth function which are of the form a,b,a,b with $a < b$. In the previous example, there is no subword in that form and hence the partition is non-crossing. On the other hand, the partition given by the restricted growth function $1,2,1,2,3,3,2$ has two crossings due to the subword $1,2,1,2$.

Lemma 2.3.1 *The number of non-crossing pair partitions of* $[k]$ *equals the* k*th moment* m_k *of the standard semicircle law* w_2 *given in* (1.1.6).

Proof: The case of odd k is trivial, so we take $2k$ instead of k. Let s_k be the number of non-crossing pair partitions of the set $[2k]$. The element 1 should be paired with an even element, say $2m$. The number of partitions containing $\{1, 2m\}$ is $s_{m-1}s_{k-m}$. Hence

$$s_k = s_{k-1}s_0 + s_{k-2}s_1 + \cdots + s_0 s_{k-1} \tag{2.3.2}$$

($s_0 := 1$). What we have to prove is that s_k equals the Catalan number c_k in (1.1.7). Due to the Taylor expansion we have

$$g(x) := \frac{1}{2}\left(1 - \sqrt{1-4x}\,\right) = \sum_{k=0}^{\infty} \frac{x^{k+1}}{k+1}\binom{2k}{k}, \tag{2.3.3}$$

which is a kind of generator function of the even moments of the semicircle law. The functional equation $g(x)^2 = g(x) - x$ shows that the coefficients of the Taylor expansion must satisfy the recursion (2.3.2), and we conclude that $s_k = c_k$.

□

Theorem 2.3.2 *Let* $a_1, a_2, \dots$ *be a free sequence in the noncommutative probability space* $(\mathcal{A}, \varphi)$. *Assume that* $\varphi(a_i) = 0$ *and* $\varphi(a_i^2) = 1$. *Furthermore, assume that* $\sup_i |\varphi(a_i^k)| < +\infty$ *for all* $k \in \mathbb{N}$. *Then the sequence* $(a_1 + a_2 + \cdots + a_n)/\sqrt{n}$ *converges in distribution to the standard semicircle law* w_2. *That is, for every*

$k \in \mathbb{N}$ the following limit relation holds:

$$\varphi\left(\frac{(a_1+a_2+\cdots+a_n)^k}{n^{k/2}}\right) \to \begin{cases} 0 & \text{if } k \text{ is odd,} \\ \dfrac{1}{1+k/2}\dbinom{k}{k/2} & \text{if } k \text{ is even.} \end{cases}$$

Proof: We have

$$\varphi((a_1+a_2+\cdots+a_n)^k) = \sum \varphi(a_{i(1)}a_{i(2)}\cdots a_{i(k)}),$$

and we are going to group the summands using partitions of $[k] = \{1, 2, \ldots, k\}$. Fix a summand $\varphi(a_{i(1)}a_{i(2)}\cdots a_{i(k)})$, and say that $u \sim v$ if $i(u) = i(v)$. This equivalence relation determines a partition $\mathcal{V} = \{V_1, V_2, \ldots, V_s\}$, where s depends on the summand we chose. (s is the number of different factors in the product $a_{i(1)}a_{i(2)}\cdots a_{i(k)}$.) Many different terms give rise to the same partition. More precisely, a partition $\mathcal{V} = \{V_1, V_2, \ldots, V_s\}$ comes from $n(n-1)\cdots(n-s+1)$ terms. We write

$$\varphi\left(\frac{(a_1+a_2+\cdots+a_n)^k}{n^{k/2}}\right) = \sum_{s=1}^{k}\sum_{\mathcal{V}}\sum \frac{\varphi(a_{i(1)}a_{i(2)}\cdots a_{i(k)})}{n^{k/2}}, \tag{2.3.4}$$

where the second summation is over all partitions $\mathcal{V} = \{V_1, V_2, \ldots, V_s\}$ and the third is over all terms determining this partition in the above scheme.

By repeated use of (2.2.3) it is easy to see that $\varphi(a_{i(1)}a_{i(2)}\cdots a_{i(k)})$ is written as a polynomial in the variables $\varphi(a_i^m)$ ($i \in \{i(1), \ldots, i(k)\}$, $1 \le m \le k$), and the polynomial depends only on the partition $\mathcal{V}$ corresponding to $i(1), \ldots, i(k)$. So the assumption of bounded moments implies that for each $k \in \mathbb{N}$ there is a bound C_k such that $|\varphi(a_{i(1)}a_{i(2)}\cdots a_{i(k)})| \le C_k$ for all choices of $i(1), \ldots, i(k) \in \mathbb{N}$.

We split the summation over s in (2.3.4) into three parts: $s < k/2$, $s > k/2$, and $s = k/2$. For $s < k/2$,

$$\left|\sum_{s<k/2}\sum_{\mathcal{V}}\sum \frac{\varphi(a_{i(1)}a_{i(2)}\cdots a_{i(k)})}{n^{k/2}}\right| \le \sum_{s<k/2}\sum_{\mathcal{V}} \frac{n(n-1)\cdots(n-s+1)}{n^{k/2}} C_k,$$

and this tends to 0 as $n \to \infty$. For $s > k/2$,

$$\sum_{s>k/2}\sum_{\mathcal{V}}\sum \frac{\varphi(a_{i(1)}a_{i(2)}\cdots a_{i(k)})}{n^{k/2}} = 0,$$

because in this case at least one of the V_i's is a singleton and due to Lemma 2.2.3 we have $\varphi(a_{i(1)}a_{i(2)}\cdots a_{i(k)}) = 0$. If k is odd, then $s < k/2$ and $s > k/2$ cover all possibilities and the limit of (2.3.4) is 0. For an even k the term

$$\sum_{\mathcal{V}:s=k/2}\sum \frac{\varphi(a_{i(1)}a_{i(2)}\cdots a_{i(k)})}{n^{k/2}}$$

contributes to the limit. The summation $\sum_{\mathcal{V}:s=k/2}$ is over all pair partitions of $[k]$. Now let $i(1), i(2), \dots, i(k)$ be a string of indices defining a pair partition $\mathcal{V}$ of $[k]$. By mathematical induction on k we show that if $\mathcal{V}$ is crossing then $\varphi(a_{i(1)}a_{i(2)}\cdots a_{i(k)}) = 0$, and if $\mathcal{V}$ is non-crossing then

$$\varphi(a_{i(1)}a_{i(2)}\cdots a_{i(k)}) = \varphi(a_{j(1)}^2)\varphi(a_{j(2)}^2)\cdots\varphi(a_{j(k/2)}^2),$$

where $j(1), j(2), \dots, j(k/2)$ are different indices appearing in $i(1), i(2), \dots, i(k)$. If there is a j such that $i(j) = i(j+1)$, then Lemma 2.2.3 implies that

$$\varphi(a_{i(1)}a_{i(2)}\cdots a_{i(k)}) = \varphi(a_{i(j)}^2)\varphi(a_{i(1)}\cdots a_{i(j-1)}a_{i(j+2)}\cdots a_{i(k)}),$$

and the induction hypothesis works because the pair partition of $\{1, \dots, j-1, j+2, \dots, k\}$ defined by $i(1), \dots, i(j-1), i(j+2), \dots, i(k)$ is crossing if and only if $\mathcal{V}$ is crossing. If there is no j such that $i(j) = i(j+1)$, then the very definition of the freeness gives $\varphi(a_{i(1)}a_{i(2)}\cdots a_{i(k)}) = 0$, and $\mathcal{V}$ is crossing in this case.

What we proved above is that the only contribution to the limit of (2.3.4) is given as

$$\sum \sum_{(j(1),\dots,j(k/2))} \frac{\varphi(a_{j(1)}^2)\varphi(a_{j(2)}^2)\cdots\varphi(a_{j(k/2)}^2)}{n^{k/2}}, \tag{2.3.5}$$

where the first summation is over non-crossing partitions of $[k]$ and the second is over all ordered $(j(1), j(2), \dots, j(k/2))$ whose elements are different numbers from $\{1, 2, \dots, n\}$. Finally, the assumption $\varphi(a_i^2) = 1$ is used to conclude that the limit of (2.3.5) is the number of non-crossing pair partitions of $[k]$, which is exactly the kth moment of the standard semicircle law due to Lemma 2.3.1.

□

Since

$$\frac{1}{n^{k/2}}\left|\left(\sum_{i=1}^{n}\varphi(a_i^2)\right)^{k/2} - \sum_{(j(1),\dots,j(k/2))}\varphi(a_{j(1)}^2)\varphi(a_{j(2)}^2)\cdots\varphi(a_{j(k/2)}^2)\right| \to 0$$

as $n \to \infty$, we observe that the conclusion of Theorem 2.3.2 holds true even if the assumption $\varphi(a_i^2) = 1$ is weakened to $\lim_{n\to\infty}\frac{1}{n}\sum_{i=1}^{n}\varphi(a_i^2) = 1$.

It is noteworthy that the kth moment of the standard normal distribution is the number of all pair partitions of a k-element-set. The normal distribution is the limit distribution in the classical central limit theorem, say for independent identically distributed random variables. Theorem 2.3.2 is analogous, and we call it the *free central limit theorem.* It is natural to say that in free probability theory the semicircle law plays the role of the Gaussian law of classical theory. A different proof using R-series will be given at the end of Sec. 2.4.

2.4 Free convolution of measures

The classical convolution of measures on the real line is strongly related to independence of random variables; namely, the convolution of distributions is the distribution of the sum of independent variables. The free convolution to be discussed in this section is determined by a noncommutative detour. The given distributions are represented by noncommutative random variables in free relation. Those are really noncommuting, because the free relation prevents commutativity. The distribution of their sum gives the additive free convolution. Since the relation of the classical convolution to the free convolution is the same as that of independence to freeness, it is not surprising that semicircle laws form a semigroup with respect to the free convolution. This is again a manifestation of the parallelism between the semicircle law and the Gaussian distribution.

The additive free convolution of measures can be computed by means of certain formal power series, namely R-series. The concept of R-series is regarded as the analogue of the logarithm of the Fourier transform from the classical theory, since it linearizes the additive free convolution. The R-series is defined by a canonical representation of a measure (or more generally, of a noncommutative random variable) by means of a formal series of a creation operator on the Fack space. The convergence of the R-series will be established in the next chapter, where an analytic machinery for the free convolution will appear.

Let μ_1 and μ_2 be two compactly supported probability measures on $\mathbb{R}$. One can find a noncommutative probability space $(\mathcal{A}, \varphi)$ with two elements a_1 and a_2 in free relation such that the noncommutative random variables a_i are distributed according to μ_i, that is,

$$\int x^k \, d\mu_i(x) = \varphi(a_i^k) \qquad (k \in \mathbb{N},\, i = 1, 2).$$

Consider the distribution μ of a_1+a_2. This measure is characterized by the property

$$\int x^k \, d\mu(x) = \varphi\big((a_1 + a_2)^k\big) \qquad (k \in \mathbb{N}), \tag{2.4.1}$$

and it will be called the *additive free convolution* of μ_1 and μ_2, in notation $\mu = \mu_1 \boxplus \mu_2$. In other words, $\mu_1 \boxplus \mu_2$ is the distribution measure of the noncommutative random variable $a_1 + a_2$. Since multiplicative free convolution, another concept of free convolution, is practically not treated in this book, instead of additive free convolution we mostly use the shorter term "free convolution". Recall that the usual (or classical) convolution of μ_1 and μ_2 is achieved by choosing independent usual random variables ξ_1 and ξ_2 distributed according to μ_1 and μ_2, respectively, and then considering the distribution of $\xi_1 + \xi_2$.

The objects $(\mathcal{A}, \varphi), a_1, a_2$ required in the definition of the free convolution always exist. Example 1.2.2 shows that any compactly supported measure can be represented as a noncommutative random variable, so we have $(\mathcal{A}_i, \varphi_i)$ and

$a_i \in \mathcal{A}_i$ such that μ_i and a_i have the same moments. Then the free product $(\mathcal{A}, \varphi) = (\mathcal{A}_1, \varphi_1) \star (\mathcal{A}_2, \varphi_2)$ contains a_1 and a_2. If we wish we can choose $\mathcal{A}_i$ to be a C^*-algebra, and then (after completion) $\mathcal{A}$ is a C^*-algebra as well. When the a_i are selfadjoint, so is $a_1 + a_2$, and the spectral theorem ensures the existence of the measure μ satisfying the condition (2.4.1). This argument shows also that $\operatorname{supp} \mu_i \subset [\alpha_i, \beta_i]$ implies $\operatorname{supp}(\mu_1 \boxplus \mu_2) \subset [\alpha_1 + \alpha_2, \beta_1 + \beta_2]$. To see that μ is independent of the choice of $(\mathcal{A}, \varphi), a_1, a_2$, one has to recognize that (2.4.1) provides a rule for the computation of moments $m_k(\mu) = \int x^k \, d\mu(x)$ of μ from those of μ_1 and μ_2. In fact, there are universal polynomials $P_n(x_1, x_2, \dots, x_n, y_1, y_2, \dots, y_n)$ such that, plugging the first n moments of μ_1 in place of the x_i's and those of μ_2 in place of the y_i's, we obtain the nth moment of $\mu_1 \boxplus \mu_2$. The polynomials are universal in the sense that they are independent of μ_i and $(\mathcal{A}, \varphi), a_1, a_2$. P_n can be computed, at least in principle, by expanding $(a_1 + a_2)^n$ and then using the formula (2.2.3) of the free relation repeatedly. For example,

$$
\begin{aligned}
&m_1(\mu_1 \boxplus \mu_2) = m_1(\mu_1) + m_1(\mu_2) \\
&\qquad \bigl(P_1(x_1, y_1) = x_1 + y_1\bigr), \\
&m_2(\mu_1 \boxplus \mu_2) = m_2(\mu_1) + m_2(\mu_2) + 2m_1(\mu_1)m_1(\mu_2) \\
&\qquad \bigl(P_2(x_1, x_2, y_1, y_2) = x_2 + y_2 + 2x_1y_1\bigr), \\
&m_3(\mu_1 \boxplus \mu_2) = m_3(\mu_1) + m_3(\mu_2) + 3m_1(\mu_1)m_2(\mu_2) + 3m_2(\mu_1)m_1(\mu_2) \\
&\qquad \bigl(P_3(x_1, x_2, x_3, y_1, y_2, y_3) = x_3 + y_3 + 3x_1y_2 + 3x_2y_1\bigr).
\end{aligned}
$$

Up through the third moment of $\mu_1 \boxplus \mu_2$, the computation rule is the same as that for the moments of the usual convolution of μ_1 and μ_2. The difference appears only from the fourth moment:

$$
\begin{aligned}
m_4(\mu_1 \boxplus \mu_2) &= m_4(\mu_1) + m_4(\mu_2) + 4m_3(\mu_1)m_1(\mu_2) + 4m_1(\mu_1)m_3(\mu_2) \\
&\quad + 4m_2(\mu_1)m_2(\mu_2) + 2m_2(\mu_1)m_1(\mu_2)^2 \\
&\quad + 2m_1(\mu_1)^2 m_2(\mu_2) - 2m_1(\mu_1)^2 m_1(\mu_2)^2 \\
\bigl(P_4(x_1, \dots, x_4, y_1, \dots, y_4) &= x_4 + y_4 + 4x_3y_1 + 4x_1y_3 + 4x_2y_2 \\
&\quad + 2x_2y_1^2 + 2x_1^2y_2 - 2x_1^2y_1^2\bigr).
\end{aligned}
$$

It is sometimes convenient to consider the notion of distributions in the abstract sense. A *distribution* μ is abstractly given as a linear functional $\mu : \mathbb{C}\langle X \rangle \to \mathbb{C}$ with $\mu(\mathbf{1}) = 1$, and it may be identified with the *moment* sequence $m_k(\mu) := \mu(X^k)$ $(k \in \mathbb{N})$. Given two distributions μ_i $(i = 1, 2)$ in this abstract sense, we have $(\mathbb{C}\langle X_1 \rangle, \mu_1) \star (\mathbb{C}\langle X_2 \rangle, \mu_2) = (\mathbb{C}\langle X_1, X_2 \rangle, \mu_1 \star \mu_2)$. Since X_1 and X_2 are of course free in this noncommutative probability space, the free convolution $\mu_1 \boxplus \mu_2$ can be defined by

$$
m_k(\mu_1 \boxplus \mu_2) = (\mu_1 \star \mu_2)\bigl((X_1 + X_2)^k\bigr) \qquad (k \in \mathbb{N}).
$$

More information about moments will be given in the next section.

Example 2.4.1 For point measures we have $\delta(x) \boxplus \delta(y) = \delta(x + y)$ $(x, y \in \mathbb{R})$.

Choose an arbitrary noncommutative probability space $(\mathcal{A}, \varphi)$ and let $a_1 = x\mathbf{1}$, $a_2 = y\mathbf{1}$. Then a_1 and a_2 are trivially free, and the distribution of $a_1 + a_2$ is $\delta(x+y)$.

Since a and $\mathbf{1}$ are in free relation in any noncommutative probability space, we can obtain that $\delta(x) \boxplus \mu$ is the translation of the measure μ by $x \in \mathbb{R}$, similarly to the above argument.

□

Example 2.4.2 For semicircle laws we have

$$w_{m_1,r_1} \boxplus w_{m_2,r_2} = w_{m,r}, \quad \text{where} \quad m = m_1 + m_2,\ r^2 = r_1^2 + r_2^2.$$

Theorem 1.1.5 and Example 2.2.1 will be used. We choose $h_i \in \mathcal{H}$ such that h_1 and h_2 are orthogonal. So $\ell(h_1)$ and $\ell(h_2)$ are free with respect to the vacuum expectation in $\mathcal{F}(\mathcal{H})$. If $2\|h_i\| = r_i$, then the distribution of $\ell(h_i)^* + \ell(h_i)$ is w_{r_i} and the distribution of $m_i\mathbf{1} + l(h_i)^* + \ell(h_i)$ is w_{m_i,r_i}. These operators are free with respect to the vacuum, and

$$\begin{aligned}&\big(m_1\mathbf{1} + \ell(h_1)^* + \ell(h_1)\big) + \big(m_2\mathbf{1} + \ell(h_2)^* + \ell(h_2)\big)\\ &\quad = (m_1 + m_2)\mathbf{1} + \ell(h_1 + h_2)^* + \ell(h_1 + h_2)\,. \end{aligned} \tag{2.4.2}$$

Since $\|h_1 + h_2\|^2 = \|h_1\|^2 + \|h_2\|^2$, (2.4.2) is distributed according to the stated semicircle law.

□

Let $\ell = \ell(h)$ be a creation operator on $\mathcal{F}(\mathcal{H})$ with $\|h\| = 1$. If $\sum_{k=0}^{\infty} |\alpha_{k+1}| < \infty$, then the series

$$\ell^* + \sum_{k=0}^{\infty} \alpha_{k+1} \ell^k$$

converges in norm and gives a bounded operator on the full Fock space. It is easy to see that

$$\left\langle \Big(\ell^* + \sum_{k=0}^{\infty} \alpha_{k+1}\ell^k\Big)^n \Phi, \Phi \right\rangle = \alpha_n + p_n(\alpha_1, \alpha_2, \dots, \alpha_{n-1})\,, \tag{2.4.3}$$

where the p_n are certain polynomials. Indeed, in expanding the nth power, a contribution to the expectation value arises only in the case when the total sum of creations is equal to the number of annihilations, that is, the number of factors ℓ^*. If ℓ^k is involved in an n-factor product and $k \ge n$, then we cannot take enough ℓ^* factors to get nonzero. All these terms may be simply neglected. Furthermore, among the permutations of ℓ^{n-1} with $n-1$ copies of ℓ^* factors, only the term $\langle (\ell^*)^{n-1}\ell^{n-1}\Phi, \Phi\rangle$ contributes, and it has the coefficient α_n. This argument justifies (2.4.3), and moreover it allows us to speak about

$$\left\langle \Big(\ell^* + \sum_{k=0}^{\infty} \alpha_{k+1}\ell^k\Big)^n \Phi, \Phi \right\rangle$$

for any sequence (α_{k+1}). In the general case $\ell^* + \sum_{k=0}^{\infty} \alpha_{k+1}\ell^k$ does not correspond to an operator on a Hilbert space, but it is rather regarded as a formal expression. To create a solid foundation of the formal manipulations, we introduce an algebra $\tilde{\mathcal{E}}$ as follows.

$\tilde{\mathcal{E}}$ is the unital algebra of formal infinite sums

$$T = \sum_{n,m=0}^{\infty} c_{m,n}\ell^m \ell^{*n},$$

where the coefficients $c_{m,n}$ are complex numbers and we assume that there exists N such that $c_{m,n} = 0$ whenever $n > N$. The sum of two series is obviously defined. The multiplication is governed by the rule $\ell^*\ell = \mathbf{1}$. For example, $\ell^3\ell^{*2} \times \ell^4\ell^{*2} = \ell^5\ell^{*2}$. The product of two series T and T' is obtained by multiplying them term by term, then simplifying by means of $\ell^*\ell = \mathbf{1}$, and by collecting terms, which means finite addition due to the condition imposed on the coefficients. We consider the linear functional $\tilde{\omega}$ on $\tilde{\mathcal{E}}$ which sends a formal sum T into its coefficient $c_{0,0}$. In this way, $(\tilde{\mathcal{E}}, \tilde{\omega})$ is a noncommutative probability space while the involution is not defined on $\tilde{\mathcal{E}}$.

Let μ be a measure on $\mathbb{R}$ such that all moments of μ exist. The formal sum

$$\ell^* + \sum_{k=0}^{\infty} \alpha_{k+1}\ell^k = \ell^*\Big(\mathbf{1} + \sum_{k=1}^{\infty} \alpha_k \ell^k\Big) \in \tilde{\mathcal{E}}$$

is called the *canonical noncommutative random variable* associated with μ, if its moments under $\tilde{\omega}$ are the same as those of μ, that is,

$$m_n(\mu) = \tilde{\omega}\bigg(\Big(\ell^* + \sum_{k=0}^{\infty} \alpha_{k+1}\ell^k\Big)^n\bigg) \qquad (n \in \mathbb{N}). \tag{2.4.4}$$

In fact, according to (2.4.3), the coefficients of the canonical variable can be computed from the moment sequence of μ recursively. For example,

$$\begin{aligned} &\alpha_1 = m_1(\mu), \quad \alpha_2 = m_2(\mu) - m_1(\mu)^2, \\ &\alpha_3 = m_3(\mu) - 3m_2(\mu)m_1(\mu) + 2m_1(\mu)^3. \end{aligned} \tag{2.4.5}$$

(The general formula for an $m_n(\mu)$ in terms of the α_k's and its inverse formula will be given in the next section.) In a similar manner, we can speak of the canonical random variable associated with any noncommutative random variable a in $(\mathcal{A}, \varphi)$ satisfying (2.4.4) with $\varphi(a^n)$ in place of $m_n(\mu)$.

When the associated canonical random variable is given by the sequence α_k, the formal power series

$$R_\mu(z) := \sum_{k=1}^{\infty} \alpha_k z^k$$

is called the *R-series* of the measure μ. More generally, we can speak of the R-series $R_a(z)$ of any noncommutative random variable a. Here we warn that this differs by a factor of z from Voiculescu's original $\mathcal{R}$-*transform* $\mathcal{R}_\mu(z) := \sum_{k=0}^\infty \alpha_{k+1} z^k$. This is why we use a slightly different term. The use of R-series is notationally a bit more convenient (particularly in the multivariable case, see Sec. 2.5).

Example 2.4.3 If $R_a(z)$ is the R-series of a noncommutative random variable a, then the R-series of λa ($\lambda \in \mathbb{C}$) is $R_a(\lambda z)$.

Let $\sum_{k=1}^\infty \alpha_k z^k$ be the R-series of a. One has to see that λa and the formal sum

$$\ell^* + \sum_{k=0}^\infty \lambda^{k+1} \alpha_{k+1} \ell^k \in \tilde{\mathcal{E}}$$

possess the same nth moments for every n. The argument used to justify (2.4.3) can be continued. In the expansion of the left-hand side of (2.4.3) we have terms like

$$\alpha_{k_1+1}\alpha_{k_2+1}\cdots\alpha_{k_r+1} \times \big\langle (\text{product of } \ell^{k_1}, \ell^{k_2}, \dots, \ell^{k_r} \text{ and } s \text{ factors of } \ell^*)\, \Phi, \Phi \big\rangle . \tag{2.4.6}$$

Here the total number $r+s$ of factors is n, and $k_1 + k_2 + \cdots + k_r = s$ is a necessary condition to have nonzero expectation (total number of creations equals the number of annihilations). Hence $(k_1+1)+(k_2+1)+\cdots+(k_r+1) = n$, and replacing α_{k_i+1} by $\lambda^{k_i+1}\alpha_{k_i+1}$ we observe that (2.4.6) is multiplied by a factor λ^n. In this way we have proved that

$$p_n(\lambda\alpha_1, \lambda^2\alpha_2, \dots, \lambda^{n-1}\alpha_{n-1}) = \lambda^n p_n(\alpha_1, \alpha_2, \dots, \alpha_{n-1}),$$

which is sufficient to conclude the statement. (It is noteworthy that the argument shows that p_n has nonnegative integer coefficients.)

□

The proof in Example 2.4.2 benefits from the fact that the canonical variable associated with the semicircle measure is actually a finite sum, and therefore an operator representation on the Fock space is conveniently used. Moreover, in order to represent free random variables, we may consider the creation operators $\ell(h_1)$ and $\ell(h_2)$ with orthogonal unit vectors. To allow room for similar arguments in the case of arbitrary canonical variables, we need the two-variable extension of the algebra $\tilde{\mathcal{E}}$.

$\tilde{\mathcal{E}}_2$ consists of the infinite sums

$$T = \sum_{m,n=0}^\infty \sum_{i(1),\dots,i(m);j(1),\dots,j(n)} c_{i(1),i(2),\dots,i(m);j(1),j(2),\dots,j(n)} \times \ell_{i(1)}\ell_{i(2)}\cdots\ell_{i(m)}\ell^*_{j(1)}\ell^*_{j(2)}\cdots\ell^*_{j(n)},$$

where $c_{i(1),\ldots,i(m);j(1),\ldots,j(n)} \in \mathbb{C}$ for $i(1),\ldots,i(m),j(1),\ldots,j(n) \in \{1,2\}$, and there exists an $N \in \mathbb{N}$ such that $c_{i(1),\ldots,i(m);j(1),\ldots,j(n)} = 0$ whenever $n > N$. The multiplication is governed by the computational rules $\ell_1^*\ell_1 = \ell_2^*\ell_2 = \mathbf{1}$ and $\ell_1^*\ell_2 = \ell_2^*\ell_1 = 0$. The condition on the coefficients ensures that in the multiplication of two infinite sums we have to add only finitely many numbers. On $\tilde{\mathcal{E}}_2$ we consider the normalized linear functional $\tilde{\omega}_2$ sending each sum to its constant term.

Let $\mathcal{F}(\mathcal{H})$ be the full Fock space over the Hilbert space $\mathcal{H}$, and let h_1 and h_2 be two orthogonal unit vectors in $\mathcal{H}$. If $T \in \tilde{\mathcal{E}}_2$ is a finite sum, then we can obtain an operator $\mathcal{F}(T)$ acting on the Fock space by replacing each ℓ_i by $\ell(h_i)$ and each ℓ_i^* by $\ell(h_i)^*$. It is rather clear that

$$\tilde{\omega}_2(T) = \langle \mathcal{F}(T)\Phi, \Phi\rangle,$$

and on this basis we sometimes say that $\tilde{\omega}_2(\,\cdot\,)$ is the vacuum expectation. As far as finite sums in $\tilde{\mathcal{E}}_2$ are concerned, it is a matter of convenience whether the formal setting of $\tilde{\mathcal{E}}_2$ or the operator setting on $\mathcal{F}(\mathcal{H})$ is considered.

Clearly $(\tilde{\mathcal{E}},\tilde{\omega})$ is embedded into $(\tilde{\mathcal{E}}_2,\tilde{\omega}_2)$ in two different ways. We can replace each ℓ in an element of $\tilde{\mathcal{E}}$ by ℓ_1 or by ℓ_2. The former embedding yields $\tilde{\mathcal{E}}_{(1)} \subset \tilde{\mathcal{E}}_2$ and the latter one gives $\tilde{\mathcal{E}}_{(2)} \subset \tilde{\mathcal{E}}_2$. The argument of Example 2.2.1 goes through, and we have

Example 2.4.4 The subalgebras $\tilde{\mathcal{E}}_{(1)}$ and $\tilde{\mathcal{E}}_{(2)}$ are free in $\tilde{\mathcal{E}}_2$ with respect to the functional $\tilde{\omega}_2$.

□

Theorem 2.4.5 *Let μ_i be probability measures on $\mathbb{R}$ with compact support, $i = 1,2$. If $\mu := \mu_1 \boxplus \mu_2$, then*

$$R_\mu(z) = R_{\mu_1}(z) + R_{\mu_2}(z)$$

in the sense of formal power series.

Proof: Let $R_{\mu_1}(z) = \sum_{k=1}^\infty \alpha_k z^k$ and $R_{\mu_2}(z) = \sum_{k=1}^\infty \beta_k z^k$. We canonically represent μ_i in the algebra $\tilde{\mathcal{E}}_{(i)}$, and use the freeness of those algebras to compute moments of the free convolution $\mu = \mu_1 \boxplus \mu_2$ as

$$m_n(\mu) = \tilde{\omega}_2\bigg(\Big((\ell_1^* + \ell_2^*) + \sum_{k=0}^\infty \alpha_{k+1}\ell_1^k + \sum_{k=0}^\infty \beta_{k+1}\ell_2^k\Big)^n\bigg). \tag{2.4.7}$$

What we need to prove is that this expectation coincides with

$$\tilde{\omega}\bigg(\Big(\ell^* + \sum_{k=0}^\infty \alpha_{k+1}\ell^k + \sum_{k=0}^\infty \beta_{k+1}l^k\Big)^n\bigg). \tag{2.4.8}$$

The expansions of (2.4.7) and (2.4.8) have a similar structure when $(\ell_1^* + \ell_2^*)$ in (2.4.7) is treated as a whole. To each term of the expansion of (2.4.7) there corresponds a term of (2.4.8); they have the same scalar coefficients. The terms of (2.4.7) are monomials of factors $(\ell_1^* + \ell_2^*)$, ℓ_1 and ℓ_2. If those factors are replaced by ℓ^*, ℓ and ℓ, respectively, then the corresponding term of (2.4.8) is obtained. It is sufficient to show that the expectation is invariant under this transformation of monomials. Any monomial of ℓ_1 and ℓ_2 has the vacuum expectation 0, and so does the correponding monomial of (2.4.8), which is a power of ℓ in this case. The same can be said about a monomial of only $(\ell_1^* + \ell_2^*)$. Hence we need to check the invariance of the vacuum expectation for monomials containing $(\ell_1^* + \ell_2^*)$ and ℓ_i $(i = 1, 2)$. This can be done by induction on the number of factors. The point is the transformation of the identities $(\ell_1^* + \ell_2^*)\ell_1 = \mathbf{1}$ and $(\ell_1^* + \ell_2^*)\ell_2 = \mathbf{1}$ into $\ell^*\ell = \mathbf{1}$. By the use of these the length of monomials can be reduced.

□

In fact, the above proof shows that if noncommutative random variables a_1 and a_2 in $(\mathcal{A}, \varphi)$ are in free relation, then

$$R_{a_1+a_2}(z) = R_{a_1}(z) + R_{a_2}(z).$$

In particular,

$$R_{a+\lambda\mathbf{1}}(z) = R_a(z) + \lambda z \tag{2.4.9}$$

for every noncommutative random variable a and every $\lambda \in \mathbb{C}$.

Example 2.4.6 The R-series of the semicircle law $w_{m,r}$ is

$$R_{w_{m,r}}(z) = mz + \frac{r^2}{4}z^2.$$

Indeed, since w_2 is the distribution of $\ell^* + \ell$, we have $R_{w_2}(z) = z^2$. Apply Example 2.4.3 and (2.4.9) to get the R-series of $w_{m,r}$.

□

Normally it is difficult to find explicitly the canonical noncommutative random variable associated to a given probability distribution. Besides the semicircle law, in the next chapter we shall have the canonical representation for some other probability distributions.

Example 2.4.7 Let $p \in \mathcal{A}$ be a projection which is free from the noncommutative random variable $a \in \mathcal{A}$ with respect to the expectation φ. Assume that $\varphi(p) \neq 0$, and let $R_a(z) = \sum_n \alpha_n z^n$ be the R-series of a. Then the R-series of pap in the reduced algebra $p\mathcal{A}p$ with respect to the reduced state $\varphi'(pbp) := \varphi(pbp)/\varphi(p)$ is

$$\sum_n \alpha_n \varphi(p)^{n-1} z^n.$$

Hence the R-series of $\varphi(p)^{-1}pap$ in $(p\mathcal{A}p, \varphi')$ is $\varphi(p)^{-1}R_a(z)$.

We are going to benefit from the matricial model discussed in Example 2.2.11. Let

$$L := \sum_{i,j=1}^{2} \sqrt{\lambda_i}\, \ell(h_{ij}) \otimes e_{ij}\,,$$

where $\lambda_1 := \varphi(p), \lambda_2 := 1 - \varphi(p)$, and the h_{ij} are orthonormal vectors. The noncommutative random variable a is represented by the formal operator sum

$$A := L^* + \sum_{n=0}^{\infty} \alpha_{n+1} L^n\,.$$

(Precisely speaking, A is defined as an element of the noncommutative probability space $(\tilde{\mathcal{E}}_4 \otimes M_2(\mathbb{C}), \tilde{\omega}_4 \otimes \rho)$; see Sec. 2.5 for $(\tilde{\mathcal{E}}_k, \tilde{\omega}_k)$.) According to the quoted example, $\mathbf{1} \otimes e_{11}$ is free from A, and $\psi(\mathbf{1} \otimes e_{11}) = \lambda_1 = \varphi(p)$. Hence the distribution of pap with respect to φ' is the same as that of the $(1,1)$ entry of the matrix A with respect to the vacuum expectation. The $(1,1)$ entry of L^n is a linear combination of terms like $\ell(h_{1i(1)})\ell(h_{i(1)i(2)}) \cdots \ell(h_{i(n-1)1})$, which have nonzero expectation only in the case $i(1) = i(2) = \ldots = i(n-1) = 1$. Therefore, it follows that the $(1,1)$ entry of A has the same distribution as

$$\sqrt{\lambda_1}\ell(h_{11})^* + \sum_{n=0}^{\infty} \alpha_{n+1} \sqrt{\lambda_1^n}\ell(h_{11})^n\,.$$

The distribution of this does not change if we multiply ℓ by t and ℓ^* by t^{-1}. In particular, for $t = \sqrt{\lambda_1} = \sqrt{\varphi(p)}$ we have

$$\ell(h_{11})^* + \sum_{n=0}^{\infty} \alpha_{n+1} \varphi(p)^n \ell(h_{11})^n\,.$$

From this representation the R-series is read out. The second assertion follows from Example 2.4.3.

□

We end the section with another proof of Theorem 2.3.2 by the method of R-series. In the following proof we shall use the fact that when $R_a(z) = \sum_{k=1}^{\infty} \alpha_k z^k$ is the R-series of a noncommutative random variable a in $(\mathcal{A}, \varphi)$, each α_k is a universal polynomial in the moments $\varphi(a), \varphi(a^2), \ldots, \varphi(a^k)$, and conversely $\varphi(a^k)$ is a universal polynomial in $\alpha_1, \alpha_2, \ldots, \alpha_k$. This is an immediate consequence of (2.4.3).

Second proof of Theorem 2.3.2: Let $a_1, a_2, \ldots$ be a free sequence in $(\mathcal{A}, \varphi)$ such that $\varphi(a_i) = 0$, $\sup_i |\varphi(a_i^k)| < +\infty$ for all $k \in \mathbb{N}$, and $\lim_{n\to\infty} \frac{1}{n} \sum_{i=1}^{n} \varphi(a_i^2) = 1$.

The R-series of a_i is written as

$$R_{a_i}(z) = \alpha_{i,2}z^2 + \alpha_{i,3}z^3 + \cdots,$$

where $\alpha_{i,2} = \varphi(a_i^2)$ (see (2.4.5)). In view of Theorem 2.4.5 and Example 2.4.3, the R-series of $(a_1 + a_2 + \cdots + a_n)/\sqrt{n}$ is

$$\sum_{i=1}^{n} R_{a_i}(z/\sqrt{n}) = \sum_{k=2}^{\infty}\left(n^{-k/2}\sum_{i=1}^{n}\alpha_{i,k}\right)z^k.$$

Set $\alpha_k^{(n)} := n^{-k/2}\sum_{i=1}^{k}\alpha_{i,k}$ for $k \geq 2$. By the fact remarked above and the boundedness assumption for the moments of the a_i's, it is clear that $\sup_i |\alpha_{i,k}| < +\infty$ for all $k \geq 2$, so $\alpha_k^{(n)}$ converges to 0 as $n \to \infty$ for any $k \geq 3$. On the other hand, we have $\alpha_2^{(n)} = \frac{1}{n}\sum_{i=1}^{n}\varphi(a_i^2) \to 1$ as $n \to \infty$. Since z^2 is the R-series of the standard semicircle law w_2, the moments of $(a_1 + a_2 + \cdots + a_n)/\sqrt{n}$ converge to those of w_2.

□

2.5 Moments and cumulants

Compactly supported probability measures on the real line can be described both by their moments and by their cumulants. The transformation between moments and cumulants is related to the lattice structure of partitions of finite sets. In the free probabilistic setting the coefficients of the R-series behave like cumulants; moreover the moment-cumulant and cumulant-moment formulas are strongly related to the lattice of non-crossing partitions. Non-crossing partitions appeared already in the combinatorial proof of the free central limit theorem. Their role cannot be overvalued. Besides the Fock space approach, the free cumulants give an equally useful formalism for freeness.

Let ξ_i be a classical random variable and μ_i its distribution measure on $\mathbb{R}$, $i = 1, 2$. The *classical convolution* $\mu_1 * \mu_2$ is the distribution measure of $\xi_1 + \xi_2$ when ξ_1 and ξ_2 are chosen to be independent. The sequence of *moments*

$$m_n(\mu) = \int x^n \, d\mu(x) = E(\xi^n)$$

is conveniently used to describe the convolution (whenever moments exist). The binomial formula allows one to express moments of $\mu_1 * \mu_2$ in terms of those of μ_1 and μ_2. Another approach consists in use of the *cumulant* sequence $s_n(\mu)$, given by

$$\log E(e^{\mathrm{i}\,t\xi}) = \sum_{n=1}^{\infty}\frac{(\mathrm{i}\,t)^n}{n!}s_n(\mu).$$

The convolution is linearized by the cumulant sequence:

$$s_n(\mu_1 * \mu_2) = s_n(\mu_1) + s_n(\mu_2) .$$

(Cumulants are also called *semi-invariants.*) Probability theory knows the relation between the moment and cumulant sequences ([166], Sec. II.12):

$$m_n(\mu) = \sum_{\mathcal{V}} \prod_{i=1}^{k} s_{|V_i|}(\mu) , \tag{2.5.1}$$

where the summation is over all partitions $\mathcal{V} = \{V_1, V_2, \ldots, V_k\}$ of the set $[n]$ and $|V_i|$ denotes the number of elements in V_i. For example,

$$m_3(\mu) = s_3(\mu) + 3s_2(\mu)s_1(\mu) + s_1(\mu)^3 .$$

The inverse relation is similar:

$$s_n(\mu) = \sum_{\mathcal{V}} (-1)^{k-1}(k-1)! \prod_{i=1}^{k} m_{|V_i|}(\mu) , \tag{2.5.2}$$

for example

$$s_4(\mu) = m_4(\mu) - 4m_3(\mu)m_1(\mu) - 3m_2(\mu)^2 + 12m_2(\mu)m_1(\mu)^2 - 6m_1(\mu)^4 .$$

The partitions of the set $[n]$ form a *lattice* L_n. Recall that the partial order on L_n is defined by saying $\mathcal{V}_1 \leq \mathcal{V}_2$ whenever each block of $\mathcal{V}_1$ is contained in a block of $\mathcal{V}_2$. (In other words, one can get $\mathcal{V}_1$ by partitioning the blocks of $\mathcal{V}_2$.) The smallest element of L_n is $\{\{1\}, \{2\}, \ldots, \{n\}\}$, and it will be denoted by $\mathbf{0}_n$. The largest element is $\mathbf{1}_n := \{[n]\}$. The *incidence algebra* is the algebra of real-valued functions on $\{(\mathcal{V}_1, \mathcal{V}_2) \in L_n \times L_n : \mathcal{V}_1 \leq \mathcal{V}_2\}$, where addition is the standard pointwise addition of functions and multiplication is defined by the convolution

$$(F \cdot G)(\mathcal{V}_1, \mathcal{V}_2) := \sum_{\mathcal{V}_1 \leq \mathcal{V} \leq \mathcal{V}_2} F(\mathcal{V}_1, \mathcal{V})G(\mathcal{V}, \mathcal{V}_2) .$$

The unit of this algebra is the *Kronecker δ-function*

$$\delta(\mathcal{V}_1, \mathcal{V}_2) := \begin{cases} 1 & \text{if } \mathcal{V}_1 = \mathcal{V}_2, \\ 0 & \text{otherwise,} \end{cases}$$

and another fundamental function is the *zeta function*

$$\zeta(\mathcal{V}_1, \mathcal{V}_2) := 1 \quad \text{for all} \quad \mathcal{V}_1 \leq \mathcal{V}_2.$$

The zeta function has an inverse, which is called the *Möbius function* and denoted by $\mu(\mathcal{V}_1, \mathcal{V}_2)$. The *Möbius inversion theorem* says that if f and g are functions on L_n, then the following two conditions are equivalent:

$$g(\mathcal{V}_0) = \sum_{\mathcal{V} \le \mathcal{V}_0} f(\mathcal{V}) \quad \text{for every} \quad \mathcal{V}_0 \in L_n, \tag{2.5.3}$$

$$f(\mathcal{V}_0) = \sum_{\mathcal{V} \le \mathcal{V}_0} g(\mathcal{V})\mu(\mathcal{V}, \mathcal{V}_0) \quad \text{for every} \quad \mathcal{V}_0 \in L_n. \tag{2.5.4}$$

In fact, put $F(\mathbf{0}_n, \mathcal{V}) = f(\mathcal{V})$, $G(\mathbf{0}_n, \mathcal{V}) = g(\mathcal{V})$, and $F(\mathcal{V}_1, \mathcal{V}_2) = G(\mathcal{V}_1, \mathcal{V}_2) = 0$ for others. Then (2.5.3) and (2.5.4) are nothing but $G = F \cdot \zeta$ and $F = G \cdot \mu$, respectively, so the above Möbius inversion is a direct consequence of the relation $\zeta \cdot \mu = \mu \cdot \zeta = \delta$.

Given a probability measure μ, set

$$g(\mathcal{V}) := \prod_{V \in \mathcal{V}} m_{|V|}(\mu) \quad \text{and} \quad f(\mathcal{V}) := \prod_{V \in \mathcal{V}} s_{|V|}(\mu) \,.$$

Then (2.5.1) tells us that (2.5.3) holds; that is,

$$g(\mathcal{V}_0) = \sum_{\mathcal{V} \le \mathcal{V}_0} \prod_{V \in \mathcal{V}} s_{|V|}(\mu) \,,$$

because if $\mathcal{V}_0 \in L_n$ then one gets all $\mathcal{V} \le \mathcal{V}_0$ by partitioning the blocks of $\mathcal{V}_0$. Since $s_n(\mu) = f(\mathbf{1}_n)$, in order to deduce (2.5.2) from (2.5.1) we need to know the values $\mu(\mathcal{V}, \mathbf{1}_n)$ of the Möbius function. It is known that

$$\mu(\mathcal{V}, \mathbf{1}_n) = (-1)^{|\mathcal{V}|-1}(|\mathcal{V}| - 1)!,$$

and the formula for $f(\mathbf{1}_n)$ coming from the Möbius inversion agrees with (2.5.2).

The set $NC(n)$ of all non-crossing partitions forms a *lattice* again, and we can consider the *incidence algebra* over $NC(n)$. The *Möbius inversion process* is very general, and it is still available in the setting of non-crossing partitions. Nevertheless, the Möbius function, which is defined again as the inverse of the zeta function, is different. It was computed by G. Kreweras.

Before we explain the Möbius inversion in the setting of the lattice $NC(n)$, it is convenient to introduce the Kreweras complementation on $NC(n)$. Let $\mathcal{V}_1$ and $\mathcal{V}_2$ be two partitions of $[n]$. By the transformation $i \mapsto 2i - 1$ we copy $\mathcal{V}_1$ to the set $\{1, 3, \ldots, 2n - 1\}$, and similarly we copy $\mathcal{V}_2$ to $\{2, 4, \ldots, 2n\}$ by the transformation $i \mapsto 2i$. Combining the two partitions copied on the odd numbers and on the even numbers, we have a partition of $[2n]$, which will be denoted by $\mathcal{V}_1 \sqcup \mathcal{V}_2$. A partition of $[2n]$ is said to be *parity preserving* if every block contains only odd or only even numbers. The partiton $\mathcal{V}_1 \sqcup \mathcal{V}_2$ constructed above is parity preserving.

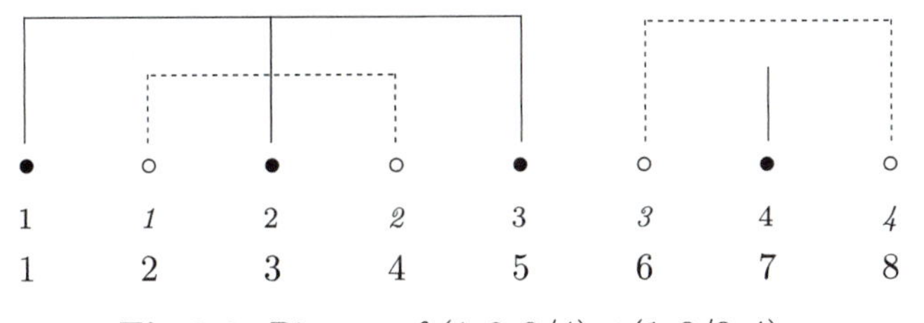

Fig.2.2. Picture of $(1, 2, 3/4) \sqcup (1, 2/3, 4)$.

Given $\mathcal{V} \in NC(n)$, there is the largest among all partitions $\mathcal{V}'$ of $[n]$ such that $\mathcal{V} \sqcup \mathcal{V}'$ is non-crossing. We denote this by $K(\mathcal{V})$ and call it the *Kreweras complement* of $\mathcal{V}$. Hence $\mathcal{V} \sqcup \mathcal{V}' \in NC(2n)$ if and only if $\mathcal{V}' \le K(\mathcal{V})$. The Kreweras complementation is an order anti-automorphism of $NC(n)$; that is, the mapping $\mathcal{V} \mapsto K(\mathcal{V})$ is a bijection on $NC(n)$ and $\mathcal{V}_1 \le \mathcal{V}_2$ is equivalent to $K(\mathcal{V}_2) \le K(\mathcal{V}_1)$.

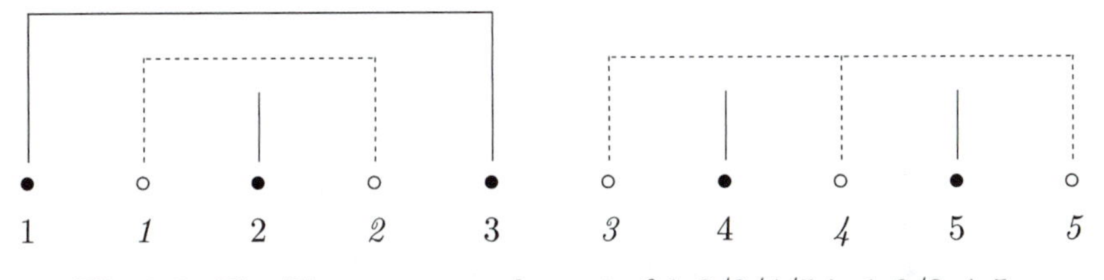

Fig.2.3. The Kreweras complement of $1, 3/2/4/5$ is $1, 2/3, 4, 5$.

For a function F in the incidence algebra over $NC(n)$, the involution $\check{F}$ is defined as $\check{F}(\mathcal{V}_1, \mathcal{V}_2) := F(K(\mathcal{V}_2), K(\mathcal{V}_1))$. Then one has $(F \cdot G)\check{} = \check{G} \cdot \check{F}$. Since $\check{\delta} = \delta$ and $\check{\zeta} = \zeta$, the relation $\zeta \cdot \mu = \mu \cdot \zeta = \delta$ yields $\check{\mu} \cdot \zeta = \zeta \cdot \check{\mu} = \delta$, so $\check{\mu} = \mu$ follows. Here we do not give the formula for $\mu(\mathcal{V}_1, \mathcal{V}_2)$ in full generality, but are content with some particular cases:

$$\mu(\mathbf{0}_n, \mathcal{V}) = \prod_{V \in \mathcal{V}} S_{|V|}, \quad \mu(\mathcal{V}, \mathbf{1}_n) = \prod_{V \in K(\mathcal{V})} S_{|V|}, \tag{2.5.5}$$

where

$$S_k := (-1)^{k-1} \frac{1}{k} \binom{2k-2}{k-1} = (-1)^{k-1} c_{k-1} \qquad (k \in \mathbb{N}).$$

(The c_k's are the Catalan numbers.) The second formula in (2.5.5) is a consequence of the first, thanks to $\check{\mu} = \mu$. On the other hand, the first formula will be determined at the end of this section. The relation $\mu \cdot \zeta = \delta$ implies that

$$\sum_{\mathcal{V} \le \mathcal{V}_0} \mu(\mathbf{0}_n, \mathcal{V}) = \sum_{\mathcal{V} \le \mathcal{V}_0} \prod_{V \in \mathcal{V}} S_{|V|} = 0$$

if $\mathcal{V}_0$ is different from $\mathbf{0}_n$. Equivalently, this says that

$$\sum_{\mathcal{V} \in NC(n)} \prod_{V \in \mathcal{V}} S_{|V|} = 0 \quad \text{for every} \quad n > 1. \tag{2.5.6}$$

After the above detour about the Möbius inversion, we return to the R-series.

Since the R-series linearizes the additive free convolution, we call the coefficients of the R-series $R_\mu(z)$ the *free cumulants* of the measure μ. We write $\alpha_n(\mu)$ for the nth free cumulant, which is the coefficient of z^n in the R-series $R_\mu(z)$. We note that Example 2.4.6 tells us that the free cumulants of the semicircle law with mean m and variance σ^2 coincide with the semi-invariants of the Gaussian law $N(m, \sigma^2)$.

We start from the formula (2.4.4) to express moments in terms of free cumulants. Let Π_n be the set of all sequences $(\varepsilon_1, \varepsilon_2, \ldots, \varepsilon_n)$ such that $\varepsilon_i \in \{-1\} \cup \mathbb{Z}^+$ and $\sum_{i=1}^k \varepsilon_i \geq 0$ for every $1 \leq k \leq n$, with $\sum_{i=1}^k \varepsilon_i = 0$ for $k = n$. (2.4.4) shows that

$$m_n(\mu) = \sum \alpha_{\varepsilon_1+1}\alpha_{\varepsilon_2+1}\cdots\alpha_{\varepsilon_n+1}\langle \ell^{\varepsilon_n}\ell^{\varepsilon_{n-1}}\cdots\ell^{\varepsilon_1}\Phi, \Phi\rangle, \tag{2.5.7}$$

where $\varepsilon_i \in \{-1\} \cup \mathbb{Z}^+$ and ℓ^{-1} is for ℓ^*. The vacuum expectation $\langle \ell^{\varepsilon_n}\ell^{\varepsilon_{n-1}}\cdots\ell^{\varepsilon_1}\Phi, \Phi\rangle$ is different from 0 if and only if $(\varepsilon_1, \varepsilon_2, \ldots, \varepsilon_n) \in \Pi_n$, and then the expectation is 1. Hence the sum (2.5.7) may run over $(\varepsilon_1, \varepsilon_2, \ldots, \varepsilon_n) \in \Pi_n$.

To each $(\varepsilon_1, \varepsilon_2, \ldots, \varepsilon_n) \in \Pi_n$ we can associate a polygonal line from $(0,0)$ to $(n, 0)$ going through the lattice points $(1, \varepsilon_1), (2, \varepsilon_1 + \varepsilon_2), \ldots, (n-1, \varepsilon_1 + \varepsilon_2 + \cdots + \varepsilon_{n-1})$. (The polygonal lines associated with ± 1 sequences are familiar from the proof of Theorem 1.1.5 in Chap. 1.)

Lemma 2.5.1 *There is a one-to-one correspondence between* Π_n *and* $NC(n)$*. The sequence* $(\varepsilon_1, \varepsilon_2, \ldots, \varepsilon_n) \in \Pi_n$ *associated with a non-crossing partition* $\mathcal{V} = \{V_1, V_2, \ldots, V_k\}$ *is described as follows:*

$$\varepsilon_i = \begin{cases} |V_j| - 1 & \textit{if } i \textit{ is the smallest element of } V_j, \\ -1 & \textit{otherwise.} \end{cases}$$

Proof: The proof is an elementary induction.

□

If we rewrite (2.5.7) by means of the bijection described in the previous lemma, then we arrive at

$$m_n(\mu) = \sum_{\mathcal{V}\in NC(n)} \prod_{i=1}^k \alpha_{|V_i|}(\mu), \tag{2.5.8}$$

which shows remarkable similarity to (2.5.1), but the summation here is over the non-crossing partitions $\mathcal{V} = \{V_1, V_2, \ldots, V_k\}$. We call (2.5.8) the *free moment-cumulant formula.* Consider the functions

$$\begin{aligned} g(\mathcal{V}_0) &:= \prod_{V_0\in\mathcal{V}_0} m_{|V_0|}(\mu) = \prod_{V_0\in\mathcal{V}_0} \sum_{\mathcal{U}\in NC(|V_0|)} \prod_{U\in\mathcal{U}} \alpha_{|U|}(\mu) \\ &= \sum_{\mathcal{V}\leq\mathcal{V}_0} \prod_{V\in\mathcal{V}} \alpha_{|V|}(\mu) \end{aligned}$$

and

$$f(\mathcal{V}) := \prod_{V\in\mathcal{V}} \alpha_{|V|}(\mu),$$

where $\mathcal{V}$, $\mathcal{V}_0$ and $\mathcal{U}$ are all non-crossing partitions. We observe that (2.5.3) holds, so the inverse of the free moment-cumulant formula (2.5.8) is

$$\alpha_n(\mu) = \sum_{\mathcal{V}\in NC(n)} \prod_{V\in\mathcal{V}} m_{|V|}(\mu) \prod_{U\in K(\mathcal{V})} S_{|U|}, \tag{2.5.9}$$

as a consequence of the Möbius inversion and (2.5.5).

Example 2.5.2 All semi-invariants of the classical Poisson distribution are the same. So we may call a probability measure μ on $\mathbb{R}$ the *free Poisson distribution* if $\alpha_n(\mu) = \lambda > 0$ for every n. From (2.5.8) we have

$$m_n(\mu) = \sum_{i=1}^{n} Q_n^{(i)} \lambda^i,$$

where $Q_n^{(i)}$ denotes the number of non-crossing partitions of $[n]$ into i blocks, given by

$$Q_n^{(i)} = \frac{1}{n}\binom{n}{i}\binom{n}{i-1}. \tag{2.5.10}$$

It is known that

$$\sum_{i=1}^{n} Q_n^{(i)} = \frac{1}{n+1}\binom{2n}{n}. \tag{2.5.11}$$

This means by Lemma 2.3.1 that the number of non-crossing partitions of $[n]$ is equal to the number of non-crossing pair partitions of $[2n]$. (Indeed, this fact will be seen in the proof of Proposition 2.6.3.) Hence, if μ_1 denotes the free Poisson distribution in the case $\lambda = 1$, then we have $m_n(\mu_1) = m_{2n}(w_2)$ $(n \in \mathbb{N})$, which shows that

$$\mu_1 = \frac{w_2(\sqrt{x})}{\sqrt{x}} \chi_{(0,4]}(x)\, dx = \frac{\sqrt{4x - x^2}}{2\pi x} \chi_{(0,4]}(x)\, dx\,. \tag{2.5.12}$$

This says that if a random variable ξ is distributed according to the standard semicircle law, then the distribution of ξ^2 is μ_1.

The free Poisson distribution is one of the fundamental distributions of free probability theory. The density of the general free Poisson measure will be given in Example 3.3.5. Similarly to the classical Poisson distribition, the parameter λ is the mean and the variance at the same time. For parameter values $\lambda \geq 1$ the measure μ_λ is absolutely continuous with respect to the Lebesgue measure.

□

We want to consider vacuum expectations in $(\tilde{\mathcal{E}}_2, \tilde{\omega}_2)$ like

$$\tilde{\omega}_2(\ell_1^{\varepsilon_n}\ell_2^{\delta_n}\ell_1^{\varepsilon_{n-1}}\cdots\ell_1^{\varepsilon_1}\ell_2^{\delta_1}) = \langle \ell_1^{\varepsilon_n}\ell_2^{\delta_n}\ell_1^{\varepsilon_{n-1}}\cdots\ell_1^{\varepsilon_1}\ell_2^{\delta_1}\Phi, \Phi\rangle\,, \tag{2.5.13}$$

where $\varepsilon_i, \delta_i \in \{-1\} \cup \mathbb{Z}^+$ $(1 \leq i \leq n)$ and ℓ_j^{-1} is for ℓ_j^* $(j = 1, 2)$. (What we have to know about the operators ℓ_1 and ℓ_2 is the relations $\ell_1^*\ell_1 = \ell_2^*\ell_2 = \mathbf{1}$, $\ell_1^*\ell_2 = \ell_2^*\ell_1 = 0$, and $\ell_1^*\Phi = \ell_2^*\Phi = 0$.) Our aim is to find a necessary and sufficient condition for (2.5.13) to be nonzero. Reading a monomial of $\ell_1, \ell_1^*, \ell_2, \ell_2^*, \mathbf{1}$ from right to left, we think that ℓ_1 creates a black particle, ℓ_2 creates a white particle, ℓ_1^* annihilates a black particle, and ℓ_2^* annihilates a white particle. Moreover, the application of a factor $\mathbf{1}$ does not have any effect. Starting from the vacuum Φ, we have to return to the vacuum after a procedure of creations and annihilations. The procedure stops if the annihilation of a particle of a certain colour was not preceded in a previous step by the creation of a particle of the same colour, or if a particle of the same colour was previously created but after that a particle of the different colour was created and not annihilated. Since particles of different colours do not interfere, a necessary condition for (2.5.13) to be nonzero is that

$$(\varepsilon_1, \varepsilon_2, \ldots, \varepsilon_n),\ (\delta_1, \delta_2, \ldots, \delta_n) \in \Pi_n\,. \tag{2.5.14}$$

If this is fulfilled, then by Lemma 2.5.1 we can associate to $(\varepsilon_1, \varepsilon_2, \ldots, \varepsilon_n)$ a non-crossing partition $\mathcal{V}_\varepsilon$ of $[n]$ and to $(\delta_1, \delta_2, \ldots, \delta_n)$ another non-crossing partition $\mathcal{V}_\delta$ of $[n]$. Then we have the partition $\mathcal{V}_\delta \sqcup \mathcal{V}_\varepsilon$ of $[2n]$ which was introduced in defining the Kreweras complement. Recall that $\mathcal{V}_\delta$ is copied on $\{1, 3, \ldots, 2n-1\}$ while $\mathcal{V}_\varepsilon$ is copied on $\{2, 4, \ldots, 2n\}$.

Lemma 2.5.3 *The expectation value* (2.5.13) *is different from* 0 *if and only if* (2.5.14) *holds and the partition* $\mathcal{V}_\delta \sqcup \mathcal{V}_\varepsilon$ *of* $[2n]$ *is non-crossing.*

Proof: We remarked already that $\ell_1^{\varepsilon_n}\ell_2^{\delta_n}\ell_1^{\varepsilon_{n-1}}\cdots\ell_1^{\varepsilon_1}\ell_2^{\delta_1}\Phi = \Phi$ implies (2.5.14) and we have the partition $\mathcal{V}_\delta \sqcup \mathcal{V}_\varepsilon$ at our disposal. We have to show that this partition is non-crossing.

Recall that the zeroth powers of ℓ_1 and ℓ_2 in the string $\ell_1^{\varepsilon_n}\ell_2^{\delta_n}\ell_1^{\varepsilon_{n-1}}\cdots\ell_1^{\varepsilon_1}\ell_2^{\delta_1}$ give singleton blocks in the partition $\mathcal{V} := \mathcal{V}_\delta \sqcup \mathcal{V}_\varepsilon$. If we remove the zeroth powers (i.e. the identity $\mathbf{1}$) from the string and the singletons from the partition $\mathcal{V}$, then the crossing number will not change. If the remaining string contains a factor ℓ_1^*, then it contains also a positive power of ℓ_1, and we find a substring $\ell_1^* \cdots \ell_1^k$ such that $k > 0$ and there are no powers of ℓ_1 in between. If we cannot find powers of ℓ_2 in between, then actually ℓ_1^* and ℓ_1^k are neighbors and we have $\ell_1^*\ell_1^k$. In the corresponding partition the position of this ℓ_1^k is the first in a block of $k+1$ elements, and the position of ℓ_1^* is the second in this block. Now we remove ℓ_1^* from the string and its position from the corresponding partition, and replace ℓ_1^k by ℓ_1^{k-1}. In this way the remaining string still corresponds to the new partition, and the change did not effect the crossing number. If there are ℓ_2's between ℓ_1^* and ℓ_1^k, then we choose a substring $\ell_2^*\ell_2^m$ of $\ell_1^* \cdots \ell_1^k$ and act similarly to the previous step.

When originally $\ell_1^{\varepsilon_n}\ell_2^{\delta_n}\ell_1^{\varepsilon_{n-1}} \cdots \ell_1^{\varepsilon_1}\ell_2^{\delta_1}\Phi = \Phi$, we can continue the above procedure until the long product of ℓ_1's and ℓ_2's becomes empty. Since the crossing number of the partition does not change during the procedure, $\mathcal{V}$ was non-crossing.

The other direction of the assertion is obtained similarly by reversing the reasoning; the change of the string is governed by the partition $\mathcal{V}$. At each step we shrink a block of $\mathcal{V}$ possessing neighboring smallest elements.

□

Proposition 2.5.4 *Assume that $(\mathcal{A}, \varphi)$ is a noncommutative probability space and $a, b \in \mathcal{A}$ are in free relation. Then the moments of ab are expressed by the free cumulants of a and b as follows:*

$$m_n(ab) = \sum_{\mathcal{V}_1, \mathcal{V}_2} \prod_{V_1 \in \mathcal{V}_1} \alpha_{|V_1|}(a) \prod_{V_2 \in \mathcal{V}_2} \alpha_{|V_2|}(b) \,, \tag{2.5.15}$$

where the summation is over all partitions $\mathcal{V}_1$ and $\mathcal{V}_2$ of $[n]$ such that $\mathcal{V}_1 \sqcup \mathcal{V}_2$ is non-crossing.

Proof: Thanks to (2.2.3) it is immediate to see that each $m_n(ab)$ is written as a certain polynomial of the moments of a and b. So we may assume that

$$a = \sum_{k=-1}^{\infty} \alpha_{k+1}(a)\ell_1^k \quad \text{and} \quad b = \sum_{k=-1}^{\infty} \alpha_{k+1}(b)\ell_2^k$$

are the canonical noncommutative random variables in $\tilde{\mathcal{E}}_2$. We use the notation of the previous lemma. (In particular, ℓ_j^{-1} stands for ℓ_j^* and $\alpha_0(a) = \alpha_0(b) = 1$.) We have

$$\begin{aligned} m_n(ab) &= \left\langle \Big(\sum_{k,m} \alpha_{k+1}(a)\alpha_{m+1}(b)\ell_1^k\ell_2^m \Big)^n \Phi, \Phi \right\rangle \\ &= \sum \alpha_{\varepsilon_1+1}(a)\alpha_{\varepsilon_2+1}(a) \cdots \alpha_{\varepsilon_n+1}(a)\alpha_{\delta_1+1}(b)\alpha_{\delta_2+1}(b) \cdots \alpha_{\delta_n+1}(b) \\ &\qquad \times \langle \ell_1^{\varepsilon_n}\ell_2^{\delta_n}\ell_1^{\varepsilon_{n-1}} \cdots \ell_1^{\varepsilon_1}\ell_2^{\delta_1}\Phi, \Phi \rangle, \end{aligned}$$

where the sum is over all sequences $(\varepsilon_1, \varepsilon_2, \ldots, \varepsilon_n)$, $(\delta_1, \delta_2, \ldots, \delta_n)$ described in the previous lemma. Invoking the lemma yields (2.5.15) when the summation is for $\mathcal{V}_1, \mathcal{V}_2$ such that $\mathcal{V}_2 \sqcup \mathcal{V}_1$ is non-crossing. But the non-crossingness of $\mathcal{V}_2 \sqcup \mathcal{V}_1$ is equivalent to that of $\mathcal{V}_1' \sqcup \mathcal{V}_2$ when $\mathcal{V}_1'$ is the transform of $\mathcal{V}_1$ by the cyclic permutation of $[n]$, and clearly $\prod_{V_1 \in \mathcal{V}_1} \alpha_{|V_1|}(a)$ does not change under this transformation. Hence the conclusion follows.

□

Let a and b be as in the above proposition. Let μ_1 be the distribution of a and μ_2 be that of b. Similarly to the idea of additive free convolution, Voiculescu introduced the *multiplicative free convolution* of μ_1 and μ_2 as the distribution μ (in the abstract sense in general) of ab in $(\mathcal{A}, \varphi)$. When μ_1 and μ_2 are compactly supported measures on $\mathbb{R}$ and furthermore μ_1 is supported in $\mathbb{R}^+$, the multiplicative free convolution μ is a compactly supported measure on $\mathbb{R}$. Indeed, one can find a C^*-probability space $(\mathcal{A}, \tau)$ with a tracial state τ and elements $a \geq 0$ and $b = b^*$ in $\mathcal{A}$ having the distributions μ_1 and μ_2, respectively. Then μ can be obtained from the spectral measure of $a^{1/2}ba^{1/2}$ becuase $\tau((ab)^n) = \tau((a^{1/2}ba^{1/2})^n)$; here the uniqueness of μ is guaranteed by Proposition 2.5.4. In particular, if both μ_1 and μ_2 are supported in $\mathbb{R}^+$, then so is μ.

By means of the Kreweras complementation we can reformulate Proposition 2.5.4 as follows:

$$m_n(ab) = \sum_{\mathcal{V} \in NC(n)} \prod_{V \in \mathcal{V}} \alpha_{|V|}(a) \sum_{\mathcal{V}' \leq K(\mathcal{V})} \prod_{V' \in \mathcal{V}'} \alpha_{|V'|}(b).$$

From the moment-cumulant formula (2.5.8) it is immediate that the above sum over $\mathcal{V}' \leq K(\mathcal{V})$ is nothing else but the product $\prod_{U \in K(\mathcal{V})} m_{|U|}(b)$ of moments of b. So

$$m_n(ab) = \sum_{\mathcal{V} \in NC(n)} \prod_{V \in \mathcal{V}} \alpha_{|V|}(a) \prod_{U \in K(\mathcal{V})} m_{|U|}(b). \tag{2.5.16}$$

It seems worthwhile to translate this formula into a context of formal power series. Let $F(X) = \sum_{n=1}^{\infty} a_n X^n$ and $G(X) = \sum_{n=1}^{\infty} b_n X^n$ be two formal power series, and set

$$t_n := \sum_{\mathcal{V} \in NC(n)} \prod_{V \in \mathcal{V}} a_{|V|} \prod_{U \in K(\mathcal{V})} b_{|U|} \qquad (n \in \mathbb{N}).$$

Then we call the series $(F \star G)(X) := \sum_{n=1}^{\infty} t_n X^n$ the *combinatorial convolution* of F and G. We denote by $\mathbb{C}_\infty\langle X \rangle$ the algebra of all formal power series without constant term. The combinatorial convolution is commutative and associative. The series $Sum(X) := X$ is the unit; that is, $Sum \star F = F \star Sum = F$ for every $F \in \mathbb{C}_\infty\langle X \rangle$. (Our notation for $Sum(X) := X$ might seem strange, but in the multivariable case $Sum(X_1, X_2, \ldots, X_n) := X_1 + X_2 + \cdots + X_n$ is a rather natural notation, the case $n = 1$ being a bit degenerate.)

In the new language the formula (2.5.16) reads simply

$$M_{ab} = R_a \star M_b, \tag{2.5.17}$$

where the *moment generating series* $M_b(X) \in \mathbb{C}_\infty\langle X \rangle$ is defined as

$$M_b(X) := \sum_{n=1}^{\infty} m_n(b) X^n.$$

Further evidence of the relevance of the combinatorial convolution is supplied by the fact that the moment-cumulant formulas (2.5.8) and (2.5.9) admit a form of convolution:

$$M_\mu = R_\mu \star Zeta\,, \quad R_\mu = M_\mu \star Moeb\,, \tag{2.5.18}$$

where

$$Zeta(X) := \sum_{n=1}^{\infty} X^n\,, \quad Moeb(X) := \sum_{n=1}^{\infty} (-1)^{n-1} c_{n-1} X^n\,.$$

The condition (2.5.6) is actually equivalent to the relation

$$Moeb \star Zeta = Sum\,, \tag{2.5.19}$$

so $Moeb$ and $Zeta$ are the inverse of each other with respect to the combinatorial convolution.

From (2.5.17) and (2.5.18) we obtain the following:

Theorem 2.5.5 *If a and b are free noncommutative random variables, then*

$$R_{ab} = R_a \star R_b\,.$$

The next theorem gives us a remarkable relation between M_μ and R_μ, which will be useful in Sec. 3.2.

Theorem 2.5.6 *If $F, G \in \mathbb{C}_\infty \langle X \rangle$ satisfy $G = F \star Zeta$, then*

$$F(X(1 + G(X))) = G(X) \quad \text{and} \quad G\Big(\frac{X}{1 + F(X)}\Big) = F(X)$$

as formal power series.

Proof: For coefficients a_n of F and b_n of G the assumption means that

$$b_n = \sum_{\mathcal{V} \in NC(n)} \prod_{V \in \mathcal{V}} a_{|V|} \qquad (n \in \mathbb{N}). \tag{2.5.20}$$

Let $V_1 = \{v(1) = 1, v(2), \ldots, v(k)\}$ be the first block of $\mathcal{V} \in NC(n)$. Then it is obvious that any $\mathcal{V} \in NC(n)$ containing V_1 is the combination of $\{V_1\}$ and partitions taken from $NC([v(j) + 1, v(j + 1) - 1])$ for each $1 \le j \le k$ such that $v(j+1) - v(j) > 1$, where $v(k+1) := n+1$. Since the data relevant to our question are only the block-lengths $|V|$ ($V \in \mathcal{V}$), we may take $NC(v(j+1)-v(j)-1)$ in place of $NC([v(j)+1, v(j+1)-1])$. Hence by (2.5.20) we have

$$
\begin{aligned}
b_n &= \sum_{k=1}^{n} \sum_{v(1)=1<v(2)<\ldots<v(k)} a_k \prod_{\substack{1\le j\le k \\ v(j+1)-v(j)>1}} \left(\sum_{\mathcal{V}\in NC(v(j+1)-v(j)-1)} \prod_{V\in\mathcal{V}} a_{|V|} \right) \\
&= \sum_{k=1}^{n} a_k \sum_{v(1)=1<v(2)<\ldots<v(k)} \prod_{\substack{1\le j\le k \\ v(j+1)-v(j)>1}} b_{v(j+1)-v(j)-1} \\
&= \sum_{k=1}^{n} a_k \sum_{\substack{i(1),\ldots,i(k)\ge 0 \\ i(1)+\cdots+i(k)=n-k}} b_{i(1)} b_{i(2)} \cdots b_{i(k)},
\end{aligned}
$$

with the convention $b_0 := 1$. The above expression implies that

$$
\begin{aligned}
G(X) &= \sum_{n=1}^{\infty} \left(\sum_{k=1}^{n} a_k \sum_{\substack{i(1),\ldots,i(k)\ge 0 \\ i(1)+\cdots+i(k)=n-k}} b_{i(1)} b_{i(2)} \cdots b_{i(k)} \right) X^n \\
&= \sum_{k=1}^{\infty} a_k X^k \left(\sum_{i(1),\ldots,i(k)=0}^{\infty} b_{i(1)} b_{i(2)} \cdots b_{i(k)} \right) X^{i(1)+i(2)+\cdots+i(k)} \\
&= \sum_{k=1}^{\infty} a_k X^k \left(1 + \sum_{i=1}^{\infty} b_i X^i \right)^k \\
&= F(X(1+G(X))) .
\end{aligned}
$$

Next, put $Y := X(1+G(X))$. Since $G(X) = F(Y)$ and $X = Y/(1+F(Y))$, we have $F(Y) = G(Y/(1+F(Y)))$, implying the second equation.

□

It is worth noting that one can use the previous theorem to determine the coefficients of $Moeb$. Indeed, due to (2.5.19) the second equation of the theorem yields

$$
Moeb(X)^2 + Moeb(X) - X = 0,
$$

and hence

$$
Moeb(X) = -\frac{1}{2}(1 - \sqrt{1+4X}) = \sum_{n=1}^{\infty} \frac{(-1)^{n-1}}{n} \binom{2n-2}{n-1} X^n
$$

by (2.3.3). On the other hand, this expression says that (2.5.6) holds true if $S_n = (-1)^{n-1} c_{n-1}$, and it determines the values of $\mu(\mathbf{0}_n, \mathcal{V})$ as in (2.5.5).

2.6 Multivariables

Random vectors are common in classical probability theory. The joint distribution of two classical random variables is the same as the distribution of the random vector formed by the two variables. For two noncommutative random variables a_1 and a_2 in the same noncommutative probability space $(\mathcal{A}, \varphi)$, their joint distribution was introduced in Sec. 1.2. The aim of this section is to extend the concept of free cumulants and the moment-cumulant formula to a k-tuple $(a_1, a_2, \dots, a_k)$ of noncommutative random variables.

Let $\tilde{\mathcal{E}}_k$ consist of the infinite sums

$$T = \sum_{m,n=0}^{\infty} \sum_{i(1),\dots,i(m);j(1),\dots,j(n)} c_{i(1),i(2),\dots,i(m);j(1),j(2),\dots,j(n)} \times \ell_{i(1)}\ell_{i(2)}\cdots\ell_{i(m)}\ell^*_{j(1)}\ell^*_{j(2)}\cdots\ell^*_{j(n)},$$

where $c_{i(1),\dots,i(m);j(1),\dots,j(n)} \in \mathbb{C}$ for $i(1),\dots,i(m),j(1),\dots,j(n) \in [k]$ and there exists an $N \in \mathbb{N}$ such that $c_{i(1),\dots,i(m);j(1),\dots,j(n)} = 0$ whenever $n > N$. The multiplication is governed by the computational rules $\ell_i^*\ell_i = \mathbf{1}$ and $\ell_i^*\ell_j = 0$ for $i \neq j$. The condition on the coefficients ensures that in the multiplication of two infinite sums we have to add only finitely many numbers. On $\tilde{\mathcal{E}}_k$ we consider a normalized linear functional $\tilde{\omega}_k$ sending each infinite sum to its constant term.

Let $(a_1, a_2, \dots, a_k)$ be a k-tuple of noncommutative random variables in a noncommutative probability space $(\mathcal{A}, \varphi)$. The infinite sum

$$T = \mathbf{1} + \sum \alpha(i(1), i(2), \dots, i(m))\ell_{i(m)}\ell_{i(m-1)}\cdots\ell_{i(1)} \in \tilde{\mathcal{E}}_k \tag{2.6.1}$$

is said to be the *canonical representation* of the k-tuple if $(a_1, a_2, \dots, a_k)$ and $(\ell_1^*T, \ell_2^*T, \dots, \ell_k^*T)$ have the same joint distribution, that is, the equality

$$\varphi(a_{i(1)}a_{i(2)}\cdots a_{i(n)}) = \tilde{\omega}_k(\ell^*_{i(1)}T\ell^*_{i(2)}T\cdots\ell^*_{i(n)}T)$$

holds for every $i(1), i(2), \dots, i(n) \in [k]$ and $n \in \mathbb{N}$. (Later it will be shown that the canonical representation exists and is unique.)

Example 2.6.1 A noncommutative random variable in a C^*-probability space is said to be (standard) *semicircular* if it is selfadjoint and has the standard semicircle law. Let a and b be standard semicircular elements in free relation. Then the canonical representation of the pair (a, b) is $T = \mathbf{1} + \ell_1^2 + \ell_2^2$.

Indeed, $\ell_i^*T = \ell_i^* + \ell_i$ $(i = 1, 2)$, and they are free and semicircularly distributed according to Theorem 1.1.5 and Example 2.2.1.

□

The use of infinite power series has proved convenient in the one-variable case, and we are now making the multivariable generalization. We denote by

$\mathbb{C}_\infty\langle X_1, X_2, \ldots, X_k\rangle$ the set of all infinite power series of noncommuting indeterminates $X_1, X_2, \ldots, X_k$ without constant term. When F is such a series and $i(1), i(2), \ldots, i(n) \in [k]$, we use the notation

$$\operatorname{coef}[i(1), i(2), \ldots, i(n)](F)$$

for the coefficient of $X_{i(1)}X_{i(2)}\cdots X_{i(n)}$. If $V = \{h_1 < h_2 < \ldots < h_r\} \subset [n]$, then we set

$$[i(1), i(2), \ldots, i(n)|V] := [i(h_1), i(h_2), \ldots, i(h_r)] .$$

If $\mathcal{V}$ is a partition of $[n]$, then we write

$$\operatorname{coef}[i(1), i(2), \ldots, i(n); \mathcal{V}](F) := \prod_{V\in\mathcal{V}} \operatorname{coef}[i(1), i(2), \ldots, i(n)|V](F) . \tag{2.6.2}$$

We say that $M \in \mathbb{C}_\infty\langle X_1, X_2, \ldots, X_k\rangle$ is the *moment generating series* of a k-tuple $(a_1, a_2, \ldots, a_k)$ when the following relation holds for every $i(1), \ldots, i(n) \in [k]$ and $n \in \mathbb{N}$:

$$\operatorname{coef}[i(1), i(2), \ldots, i(n)](M) = \varphi(a_{i(1)}a_{i(2)}\cdots a_{i(n)}) .$$

We use the notation $M_{(a_1,a_2,\ldots,a_k)}$ for this series. The *R-series* $R_{(a_1,a_2,\ldots,a_k)}$ of $(a_1, a_2, \ldots, a_k)$ is determined by the relation

$$\operatorname{coef}[i(1), i(2), \ldots, i(n)](R) = \alpha\big(i(1), i(2), \ldots, i(n)\big) ,$$

where α is from the canonical representation (2.6.1).

Example 2.6.2 A noncommutative random variable c is said to be (standard) *circular* if it is written in the form $c = (a + \mathrm{i}\, b)/\sqrt{2}$, where a and b are free standard semicircular elements. For such a, b, c the R-series of the pairs (a, b) and (c, c^*) are $X_1^2 + X_2^2$ and $X_1X_2 + X_2X_1$, respectively. Moreover, we have $R_{c^*c} = Zeta$, so the distribution of c^*c is the free Poisson measure μ_1 in (2.5.12), and that of $|c| = (c^*c)^{1/2}$ is

$$\frac{1}{\pi}\sqrt{4 - x^2}\chi_{[0,2]}(x)\, dx . \tag{2.6.3}$$

The pair (a, b) was already treated in the previous example, and its R-series was known.

A representation of a is $\ell(f)^* + \ell(f)$ on the full Fock space with a unit vector f. Similarly, b is represented as $\ell(g)^* + \ell(g)$ with $\|g\| = 1$. The condition $\langle f, g\rangle = 0$ assures the free relation. So

$$\frac{\ell(f)^* + \ell(f)}{\sqrt{2}} + \mathrm{i}\,\frac{\ell(g)^* + \ell(g)}{\sqrt{2}} = \ell\Big(\frac{f - \mathrm{i}\, g}{\sqrt{2}}\Big)^* + \ell\Big(\frac{f + \mathrm{i}\, g}{\sqrt{2}}\Big)$$

represents c. Since $(f + \mathrm{i}\,g)/\sqrt{2}$ and $(f - \mathrm{i}\,g)/\sqrt{2}$ are orthogonal unit vectors, this variable has the same distribution as $\ell_1^* + \ell_2$ in $\tilde{\mathcal{E}}_2$. So the pair (c, c^*) has the same distribution as $(\ell_1^* + \ell_2, \ell_2^* + \ell_1)$. Now it is clear that the canonical representation of the pair is $\mathbf{1} + \ell_1\ell_2 + \ell_2\ell_1$, and consequently $X_1X_2 + X_2X_1$ is the R-series.

For the second assertion, since Example 2.5.2 says that $Zeta$ is the R-series of $(\ell_1^* + \ell_1)^2$, what we need to prove is

$$\tilde{\omega}_2(((\ell_1 + \ell_2^*)(\ell_1^* + \ell_2))^n) = \tilde{\omega}_2((\ell_1^* + \ell_1)^{2n}) \qquad (n \in \mathbb{N}).$$

Now let $a_1 = \ell_1\ell_1^*$, $a_2 = \ell_1\ell_2$, $a_3 = \ell_2^*\ell_1^*$ and $b_1 = \ell_1\ell_1^*$, $b_2 = \ell_1^2$, $b_3 = \ell_1^{*2}$. Then it suffices to prove that for every $(i_1, \dots, i_n) \in \{1, 2, 3\}^n$ $(n \in \mathbb{N})$

$$\tilde{\omega}_2(a_{i_1}a_{i_2}\cdots a_{i_n}) = \tilde{\omega}_2(b_{i_1}b_{i_2}\cdots b_{i_n})\,. \tag{2.6.4}$$

The case $n = 1$ is trivial. Let $n \geq 2$. If $(i_k, i_{k+1}) = (1, 2)$, then a_{i_k} and b_{i_k} can be removed in the respective left-hand and right-hand sides of (2.6.4), because $a_1a_2 = a_2$ and $b_1b_2 = b_2$. If $(i_k, i_{k+1}) = (3, 2)$, then $a_{i_k}a_{i_{k+1}}$ and $b_{i_k}b_{i_{k+1}}$ can be removed in (2.6.4), because $a_3a_2 = b_3b_2 = 1$. Hence the induction hypothesis works in these cases. Otherwise, $(i_k, i_{k+1}) \notin \{(1, 2), (3, 2)\}$ for any $1 \leq k \leq n - 1$, which implies that both sides of (2.6.4) are zero. Thus we have shown that c^*c has the distribution (2.5.12), and it immediately determines the distribution of $|c|$ as (2.6.3).

□

The measure (2.6.3) is called the *quarter-circular distribution.* Moreover, a noncommutative random variable in a C^*-probability space is called a *quarter-circular* element if it is positive and has the quarter-circular distribution.

Proposition 2.6.3 *Let R and M be the R-series and the moment generating series, respectively, of a k-tuple $(a_1, a_2, \dots, a_k)$ of noncommutative random variables. Then the following relation holds for every $1 \leq i(1), i(2), \dots, i(n) \leq k$ and $n \in \mathbb{N}$:*

$$\mathrm{coef}[i(1), i(2), \dots, i(n)](M) = \sum_{\mathcal{V} \in NC(n)} \mathrm{coef}[i(1), i(2), \dots, i(n); \mathcal{V}](R)\,, \tag{2.6.5}$$

where the term in the right-hand side was given in (2.6.2).

Proof: We have to show that

$$\tilde{\omega}_k(\ell_{i(1)}^*T\ell_{i(2)}^*T\cdots\ell_{i(n)}^*T) \tag{2.6.6}$$

equals the right-hand side of (2.6.5). In order to show this, we plug in (2.6.1) in place of T and arrive at the following sum:

$$\sum \alpha_1\alpha_2\cdots\alpha_n\,\tilde{\omega}_k(\ell_{i(1)}^*L_1\ell_{i(2)}^*L_2\cdots\ell_{i(n)}^*L_n)\,, \tag{2.6.7}$$

where $L_1, L_2, \dots, L_n$ are monomials in $\ell_1, \ell_2, \dots, \ell_k, \mathbf{1}$ and α_i is the coefficient of the term L_i in (2.6.1). The sum is over all possibilities, but many of the terms vanish. The rest of the proof consists in analyzing the structure of the non-vanishing terms.

We are dealing with vacuum expectations

$$\langle \ell^*_{i(1)} L_1 \ell^*_{i(2)} L_2 \cdots \ell^*_{i(n)} L_n \Phi, \Phi \rangle \,,$$

which contain n annihilation operators. To be nonzero, a necessary condition is that the total number of creations must be n as well. Moreover, $\ell^*_{i(1)}$ has to be coupled with a factor $\ell_{i(1)}$ included in L_1, or in L_2, ..., or in L_n, $\ell^*_{i(2)}$ has to be coupled with a factor $\ell_{i(2)}$ from L_2, or from L_3, ..., or from L_n, etc. Those couplings establish a pair partition of $[2n]$, $2n$ being the number of factors of the long product and the position of $\ell^*_{i(j)}$ being paired with the position of $\ell_{i(j)}$ annihilated by $\ell^*_{i(j)}$. The vacuum expectation is non-vanishing if and only if this pair partition is non-crossing. Hence (2.6.6) is a sum over the non-crossing pair partitions of $[2n]$. Given such a partition, we can recover the corresponding term as follows: $\ell^*_{i(1)}$ is the first factor under the expectation; the pair of 1 is replaced by $\ell_{i(1)}$; then we choose the smallest among the smaller elements of the pairs, which will be replaced by $\ell^*_{i(2)}$, and the other element of this pair will be replaced by $\ell_{i(2)}$, etc. For example, the expectation

$$\langle \ell^*_{i(1)} \ell^*_{i(2)} \ell^*_{i(3)} \ell_{i(3)} \ell^*_{i(4)} \ell_{i(4)} \ell_{i(2)} \ell^*_{i(5)} \ell_{i(5)} \ell_{i(1)} \Phi, \Phi \rangle$$

corresponds to the non-crossing partition $1, 10/2, 7/3, 4/5, 6/8, 9$. The value of a non-vanishing vacuum expectation is simply 1; hence (2.6.7) reduces to $\sum \alpha_1 \alpha_2 \cdots \alpha_n$, where the summation is over all terms which define non-crossing pair partitions as above. Here $\alpha_1, \alpha_2, \dots, \alpha_n$ are coefficients of the R-series; however, several of them are 1, corresponding to $L_i = \mathbf{1}$.

To complete the proof we need to rewrite the sum $\sum \alpha_1 \alpha_2 \cdots \alpha_n$ over non-crossing pair partitions of $[2n]$ in terms of non-crossing partitions of $[n]$, and we need to identify the term $\alpha_1 \alpha_2 \cdots \alpha_n$ with a term like $\operatorname{coef}[i(1), i(2), \dots, i(n); \mathcal{V}](R)$. So the point is a one-to-one correspondence between the non-crossing pair partitions of $[2n]$ and $NC(n)$.

Let $\{x_1, y_1\}, \{x_2, y_2\}, \dots, \{x_n, y_n\}$ be a non-crossing pair partition of $[2n]$, and assume that $x_i < y_i$ and $x_1 < x_2 < \dots < x_n$. One can define an equivalence relation on $[n]$ by saying that $u \sim v$ if there is no x_i between y_u and y_v. One checks that, starting from a non-crossing pair partition, this equivalence relation generates a non-crossing partition of $[n]$. For example, $1, 10/2, 7/3, 4/5, 6/8, 9$ generates $1, 5/2, 4/3$. Mathematical induction on n is convenient. When $\{x_1, y_1\}, \{x_2, y_2\}, \dots, \{x_n, y_n\}$ is non-crossing, we can find i with $x_i + 1 = y_i$. If we remove the pair $\{x_i, y_i\}$, then the crossing number will not change and after renumbering we can get a non-crossing pair partition of $[2n-2]$. On the other hand, i must be the largest element in a block of the corresponding partition of $[n]$, and, removing i and renumbering, also here we obtain a non-crossing partition of $[n-1]$.

The above procedure can be reversed. Let $\mathcal{V}$ be a non-crossing partition of $[n]$.

For each block $V = \{h_1 < h_2 < \ldots < h_r\}$ set $L_{h_r} := \ell_{i(h_r)}\ell_{i(h_{r-1})}\cdots\ell_{i(h_1)}$, and let $L_i := \mathbf{1}$ when i is not the largest number of any block of $\mathcal{V}$. Then one can easily see that the product $\ell^*_{i(1)}L_1\ell^*_{i(2)}L_2\cdots\ell^*_{i(n)}L_n$ is a non-vanishing term from (2.6.7), and the corresponding non-crossing pair partition of $[2n]$ determines the given $\mathcal{V}$. Moreover, the corresponding term $\alpha_1\alpha_2\cdots\alpha_n$ is equal to

$$\prod_{V\in\mathcal{V}} \alpha\big(i(h_1), i(h_2), \ldots, i(h_r)\big) = \mathrm{coef}[i(1), i(2), \ldots, i(n); \mathcal{V}](R),$$

and (2.6.5) is obtained.

□

The coefficients $\alpha\big(i(1), i(2), \ldots, i(n)\big)$ are called the *free cumulants* of $(a_1, a_2, \ldots, a_k)$, and (2.6.5) may be called the *free moment-cumulant formula* for multivariables. Although we have proved neither the existence nor the uniqueness of the canonical representation yet, they immediately follow from the relation (2.6.5). Indeed, this relation can uniquely determine all coefficients of R from those of M.

To paraphrase the moment-cumulant formula in terms of power series, we introduce the *combinatorial convolution*, which extends the single-variable concept treated in the previous section. Let $F_1, F_2 \in \mathbb{C}_\infty\langle X_1, X_2, \ldots, X_k\rangle$. Then the convolution $F_1 \star F_2$ is determined as

$$\begin{aligned}&\mathrm{coef}[i(1), i(2), \ldots, i(n)](F_1 \star F_2)\\ &= \sum_{\mathcal{V}\in NC(n)} \mathrm{coef}[i(1), i(2), \ldots, i(n); \mathcal{V}](F_1)\mathrm{coef}[i(1), i(2), \ldots, i(n); K(\mathcal{V})](F_2),\end{aligned}$$

where K is the Kreweras complement.

Let $Zeta \in \mathbb{C}_\infty\langle X_1, X_2, \ldots, X_k\rangle$ be such that it is without constant term and all other coefficients are 1. Then the moment-cumulant formula takes the simple form

$$M_{(a_1,a_2,\ldots,a_k)} = R_{(a_1,a_2,\ldots,a_k)} \star Zeta.$$

An important role is played also by the series

$$Moeb(X_1, X_2, \ldots, X_k) := \sum_{n=1}^{\infty} \sum_{i(1),\ldots,i(n)=1}^{k} (-1)^{n-1}c_{n-1}\, X_{i(1)}X_{i(2)}\cdots X_{i(n)}$$

and

$$Sum(X_1, X_2, \ldots, X_k) := X_1 + X_2 + \cdots + X_k.$$

Sum is a unit for the convolution. *Zeta* and *Moeb* are inverse of each other. (In the k-variable case the convolution is not commutative, but *Zeta* and *Moeb* are central elements.) Hence

$$R_{(a_1,a_2,\ldots,a_k)} = M_{(a_1,a_2,\ldots,a_k)} \star Moeb,$$

and one can see again that the R-series and the canonical representation of a k-tuple always exist and are uniquely determined.

All the concepts and notations of convolution, R-series, $Zeta$, Sum, $Moeb$ are extensions of the one-variable case treated in the previous section.

The following is a main property of the multivariate R-series.

Proposition 2.6.4 *Let $(a_1, a_2, \ldots, a_k)$ be a k-tuple and $(b_1, b_2, \ldots, b_m)$ an m-tuple in the same noncommutative probability space. Then $\{a_1, a_2, \ldots, a_k\}$ and $\{b_1, b_2, \ldots, b_m\}$ are free if and only if the R-series of the multivariable $(a_1, a_2, \ldots, a_k, b_1, b_2, \ldots, b_m)$ is*

$$R_{(a_1,a_2,\ldots,a_k)}(X_1, X_2, \ldots, X_k) + R_{(b_1,b_2,\ldots,b_m)}(X_{k+1}, X_{k+2}, \ldots, X_{k+m}).$$

Proof: Let

$$T_a = \mathbf{1} + \sum \alpha\big(i(1), i(2), \ldots, i(n)\big)\ell_{i(n)}\ell_{i(n-1)}\cdots\ell_{i(1)}$$

and

$$T_b = \mathbf{1} + \sum \beta\big(i(1), i(2), \ldots, i(n)\big)\ell_{k+i(n)}\ell_{k+i(n-1)}\cdots\ell_{k+i(1)}$$

be the canonical representations of $(a_1, a_2, \ldots, a_k)$ and $(b_1, b_2, \ldots, b_m)$, respectively. Set

$$\begin{aligned} T := \mathbf{1} &+ \sum \alpha\big(i(1), \ldots, i(n)\big)\ell_{i(n)}\cdots\ell_{i(1)} \\ &+ \sum \beta\big(i(1), \ldots, i(n)\big)\ell_{k+i(n)}\cdots\ell_{k+i(1)}. \end{aligned}$$

Since

$$\ell_i^* T = \begin{cases} \ell_i^* T_a & \text{if } 1 \le i \le k, \\ \ell_i^* T_b & \text{if } k+1 \le i \le k+m, \end{cases}$$

$\{\ell_1^*T, \ldots, \ell_k^*T\}$ and $\{\ell_{k+1}^*T, \ldots, \ell_{k+m}^*T\}$ are free in $\tilde{\mathcal{E}}_{k+m}$. Hence, if $\{a_1, \ldots, a_k\}$ and $\{b_1, \ldots, b_m\}$ are free, then the joint distribution of $(\ell_1^*T, \ldots, \ell_k^*T, \ell_{k+1}^*T, \ldots, \ell_{k+m}^*T)$ is the same as that of $(a_1, \ldots, a_k, b_1, \ldots, b_m)$, so T is the canonical representation of $(a_1, \ldots, a_k, b_1, \ldots, b_m)$.

The proof of the converse is a kind of trick. Choose $(\tilde{a}_1, \ldots, \tilde{a}_k)$ and $(\tilde{b}_1, \ldots, \tilde{b}_m)$ in another noncommutative probability space $(\tilde{\mathcal{A}}, \tilde{\varphi})$ so that they are free and their joint distributions are the same as those of $(a_1, \ldots, a_k)$ and $(b_1, \ldots, b_m)$, respectively. (For instance, take $(\tilde{\mathcal{A}}, \tilde{\varphi}) := (\mathcal{A}_1, \varphi|\mathcal{A}_1) \star (\mathcal{A}_2, \varphi|\mathcal{A}_2)$, where $\mathcal{A}_1$ and $\mathcal{A}_2$ are the subalgebras of $\mathcal{A}$ generated by $\{a_1, \ldots, a_k\}$ and $\{b_1, \ldots, b_m\}$, respectively.) We have

$$\begin{aligned}
&R_{(a_1,\dots,a_k,b_1,\dots,b_m)}(X_1,\dots,X_k,X_{k+1},\dots,X_{k+m}) \\
&\quad = R_{(a_1,\dots,a_k)}(X_1,\dots,X_k) + R_{(b_1,\dots,b_m)}(X_{k+1},\dots,X_{k+m}) \\
&\quad = R_{(\tilde a_1,\dots,\tilde a_k)}(X_1,\dots,X_k) + R_{(\tilde b_1,\dots,\tilde b_m)}(X_{k+1},\dots,X_{k+m}) \\
&\quad = R_{(\tilde a_1,\dots,\tilde a_k,\tilde b_1,\dots,\tilde b_m)}(X_1,\dots,X_k,X_{k+1},\dots,X_{k+m})\,.
\end{aligned}$$

Here the first equality is by assumption and the third is by the first half of the proof. Hence the freeness follows because the R-series determines the joint distribution.

□

Another important result is the multivariable extension of Theorem 2.4.5.

Proposition 2.6.5 *Let $(a_1,a_2,\dots,a_k)$ and $(b_1,b_2,\dots,b_k)$ be two k-tuples in the same noncommutative probability space. If $\{a_1,a_2,\dots,a_k\}$ and $\{b_1,b_2,\dots,b_k\}$ are free, then the R-series of $(a_1+b_1,a_2+b_2,\dots,a_k+b_k)$ is*

$$R_{(a_1,a_2,\dots,a_k)}(X_1,X_2,\dots,X_k) + R_{(b_1,b_2,\dots,b_k)}(X_1,X_2,\dots,X_k)\,.$$

Proof: The idea in the proof of Theorem 2.4.5 can work in the mutivariable case too. Let

$$T_a = \mathbf{1} + \sum \alpha\big(i(1),i(2),\dots,i(n)\big)\ell_{2i(n)-1}\ell_{2i(n-1)-1}\cdots\ell_{2i(1)-1}$$

and

$$T_b = \mathbf{1} + \sum \beta\big(i(1),i(2),\dots,i(n)\big)\ell_{2i(n)}\ell_{2i(n-1)}\cdots\ell_{2i(1)}$$

be the canonical representations of $(a_1,a_2,\dots,a_k)$ and $(b_1,b_2,\dots,b_k)$, respectively. Set

$$T := \mathbf{1} + \sum \big(\alpha(i(1),\dots,i(n)) + \beta(i(1),\dots,i(n))\big)\ell_{i(n)}\ell_{i(n-1)}\cdots\ell_{i(1)}\,.$$

By the freeness assumption, the joint distribution of $(a_1,\dots,a_k,b_1,\dots,b_k)$ is the same as that of $(\ell_1^*T_a,\dots,\ell_{2k-1}^*T_a,\ell_2^*T_b,\dots,\ell_{2k}^*T_b)$, so the joint distributions of $(a_1+b_1,\dots,a_k+b_k)$ and $(\ell_1^*T_a+\ell_2^*T_b,\dots,\ell_{2k-1}^*T_a+\ell_{2k}^*T_b)$ coincide. Furthermore, we can express

$$\begin{aligned}
\ell_i^*T &= \ell_i^* \\
&\quad + \sum \big(\alpha(i(1),\dots,i(n-1),i) + \beta(i(1),\dots,i(n-1),i)\big)\ell_{i(n-1)}\cdots\ell_{i(1)}\,,
\end{aligned}$$

$$\begin{aligned}
\ell_{2i-1}^*T_a + \ell_{2i}^*T_b \quad &= \quad (\ell_{2i-1}^* + \ell_{2i}^*) \\
&\quad + \sum \alpha(i(1),\dots,i(n-1),i)\ell_{2i(n-1)-1}\cdots\ell_{2i(1)-1} \\
&\quad + \sum \beta(i(1),\dots,i(n-1),i)\ell_{2i(n-1)}\cdots\ell_{2i(1)}\,.
\end{aligned}$$

What we have to show is that

$$\tilde{\omega}_k(\ell_{i(1)}^* T \cdots \ell_{i(n)}^* T) = \tilde{\omega}_{2k}((\ell_{2i(1)-1}^* T_a + \ell_{2i(1)}^* T_b) \cdots (\ell_{2i(n)-1}^* T_a + \ell_{2i(n)}^* T_b)) .$$

In view of the above expressions for $\ell_i^* T$ and $\ell_{2i-1}^* T_a + \ell_{2i}^* T_b$, it suffices to show that

$$\tilde{\omega}_k(\ell_{i(1)}^* L_1 \cdots \ell_{i(n)}^* L_n) = \tilde{\omega}_{2k}((\ell_{2i(1)-1}^* + \ell_{2i(1)}^*) \tilde{L}_1 \cdots (\ell_{2i(n)-1}^* + \ell_{2i(n)}^*) \tilde{L}_n) ,$$

where L_j is a monomial of $\ell_1, \ldots, \ell_k, \mathbf{1}$ and $\tilde{L}_j$ is the corresponding monomial of either $\ell_1, \ldots, \ell_{2k-1}, \mathbf{1}$ or $\ell_2, \ldots, \ell_{2k}, \mathbf{1}$. But one can easily check this because $\ell_{2i-1}^* + \ell_{2i}^*$ is canceled as a whole with ℓ_{2i-1} or ℓ_{2i} (cf. the proof of Theorem 2.4.5). □

The linear transformation rule for the multivariate R-series is simply stated in the following proposition:

Proposition 2.6.6 *Let $(a_1, a_2, \ldots, a_k)$ be a k-tuple of noncommutative random variables. If $A = [A_{ij}]_{1 \le i \le m,\, 1 \le j \le k}$ is an $m \times k$ complex matrix and $b_i = \sum_{j=1}^k A_{ij} a_j$ $(1 \le i \le m)$, then*

$$\begin{aligned} &R_{(b_1,b_2,\ldots,b_m)}(X_1, X_2, \ldots, X_m) \\ &\quad = R_{(a_1,a_2,\ldots,a_k)}\Big(\sum_{i=1}^m A_{i1} X_i, \sum_{i=1}^m A_{i2} X_i, \ldots, \sum_{i=1}^m A_{ik} X_i\Big) , \end{aligned}$$

or, for short, $R_{Aa}(X) = R_a(A^t X)$, where ^{t} stands for the transpose.

Proof: Let R and M be as in Proposition 2.6.3 for $(a_1, a_2, \ldots, a_k)$. Let $\hat{M}$ be the moment generating series for $(b_1, b_2, \ldots, b_m)$, and let

$$\hat{R}(X_1, X_2, \ldots, X_m) := R\Big(\sum_{i=1}^m A_{i1} X_i, \sum_{i=1}^m A_{i2} X_i, \ldots, \sum_{i=1}^m A_{ik} X_i\Big) .$$

Let $1 \le i(1), \ldots, i(n) \le m$. Since

$$\begin{aligned} &\mathrm{coef}[i(1), \ldots, i(n)](\hat{R}) \\ &\quad = \sum_{j(1),\ldots,j(n)=1}^{k} A_{i(1)j(1)} \cdots A_{i(n)j(n)}\, \mathrm{coef}[j(1), \ldots, j(n)](R) , \end{aligned}$$

the following holds for every partition $\mathcal{V}$ of $[n]$:

$$\begin{aligned} &\mathrm{coef}[i(1), \ldots, i(n); \mathcal{V}](\hat{R}) \\ &\quad = \sum_{j(1),\ldots,j(n)=1}^{k} A_{i(1)j(1)} \cdots A_{i(n)j(n)}\, \mathrm{coef}[j(1), \ldots, j(n); \mathcal{V}](R) . \end{aligned}$$

Therefore, by (2.6.5) we have

$$\begin{aligned}
&\mathrm{coef}[i(1),\dots,i(n)](\hat{M}) \\
&= \sum_{j(1),\dots,j(n)=1}^{k} A_{i(1)j(1)}\cdots A_{i(n)j(n)}\,\mathrm{coef}[j(1),\dots,j(n)](M) \\
&= \sum_{\mathcal{V}\in NC(n)} \sum_{j(1),\dots,j(n)=1}^{k} A_{i(1)j(1)}\cdots A_{i(n)j(n)}\,\mathrm{coef}[j(1),\dots,j(n);\mathcal{V}](R) \\
&= \sum_{\mathcal{V}\in NC(n)} \mathrm{coef}[i(1),\dots,i(n);\mathcal{V}](\hat{R}),
\end{aligned}$$

which shows that $\hat{R}$ is the R-series of $(b_1, b_2, \dots, b_m)$.

□

Example 2.6.7 Let a and b be free standard semicircular elements. For $-1 < \tau < 1$ set

$$y := \sqrt{\frac{1+\tau}{2}}\,a + \mathrm{i}\sqrt{\frac{1-\tau}{2}}\,b\,.$$

Then the R-series of (y, y^*) is $X_1X_2 + X_2X_1 + \tau X_1^2 + \tau X_2^2$, and the distribution of y^*y is the free Poisson measure μ_1, as in the circular case.

An application of the previous transformation rule to Example 2.6.2 yields

$$R_{(y,y^*)}(X_1, X_2) = \Big(\sqrt{\frac{1+\tau}{2}}(X_1+X_2)\Big)^2 + \Big(\mathrm{i}\sqrt{\frac{1-\tau}{2}}(X_1-X_2)\Big)^2,$$

which equals the required expression.

We can write $y = uc + vc^*$ with a circular element c and

$$u := \frac{\sqrt{1+\tau}+\sqrt{1-\tau}}{2}, \qquad v := \frac{\sqrt{1+\tau}-\sqrt{1-\tau}}{2}.$$

So, for the second assertion we may show that if $y = u(\ell_1^* + \ell_2) + v(\ell_1 + \ell_2^*)$ in $\tilde{\mathcal{E}}_2$, then $\tilde{\omega}_2((y^*y)^n) = \tilde{\omega}_2((\ell_1^* + \ell_1)^{2n})$ ($n \in \mathbb{N}$). The left-hand side is expanded in the following sum:

$$\sum u^p v^{2n-p} \tilde{\omega}_2(a_1 a_2 \cdots a_{2n-1} a_{2n})$$

where $a_i \in \{\ell_1, \ell_1^*, \ell_2, \ell_2^*\}$ and p is the sum of the number of i with $a_{2i-1} \in \{\ell_1, \ell_2^*\}$ and the number of i with $a_{2i} \in \{\ell_1^*, \ell_2\}$. When $\tilde{\omega}_2(a_1a_2\cdots a_{2n}) = 1$ (otherwise this is 0), the product $b_1b_2\cdots b_{2n}$ of ℓ_1, ℓ_1^* is obtained by replacing all ℓ_2, ℓ_2^* in $a_1a_2\cdots a_{2n}$ by ℓ_1, ℓ_1^*, respectively, and then $\tilde{\omega}_2(b_1b_2\cdots b_{2n}) = 1$. Any such term

$a_1a_2\cdots a_{2n}$ is conversely obtained from $b_1b_2\cdots b_{2n}$ by replacing ℓ_1 by ℓ_2 and ℓ_1^* by ℓ_2^* in some positions. Here one can repeatedly replace ℓ_1 in any position by ℓ_2, but the corresponding position where ℓ_1^* must be replaced by ℓ_2^* is uniquely determined, and the coefficients of the replaced ℓ_2, ℓ_2^* are both u or both v. In this way, we observe that the sum of $u^p v^{2n-p}$ for all $a_1a_2\cdots a_{2n}$ corresponding to a fixed $b_1b_2\cdots b_{2n}$ is equal to

$$\sum_{k=0}^{n}\binom{n}{k}u^{2k}v^{2n-2k} = (u^2+v^2)^n = 1\,.$$

For example, the terms $u^p v^{6-p} a_1a_2\cdots a_6$ corresponding to $\ell_1^*\ell_1^*\ell_1^*\ell_1\ell_1\ell_1$ $(n=6)$ are

$$\begin{array}{lll} u^2v^4\ell_1^*\ell_1^*\ell_1^*\ell_1\ell_1\ell_1\,, & u^4v^2\ell_2^*\ell_1^*\ell_1^*\ell_1\ell_1\ell_2\,, & v^6\ell_1^*\ell_2^*\ell_1^*\ell_1\ell_2\ell_1\,, \\ u^4v^2\ell_1^*\ell_1^*\ell_2^*\ell_2\ell_1\ell_1\,, & u^2v^4\ell_2^*\ell_2^*\ell_1^*\ell_1\ell_2\ell_2\,, & u^6\ell_2^*\ell_1^*\ell_2^*\ell_2\ell_1\ell_2\,, \\ u^2v^4\ell_1^*\ell_2^*\ell_2^*\ell_2\ell_2\ell_1\,, & u^4v^2\ell_2^*\ell_2^*\ell_2^*\ell_2\ell_2\ell_2\,, & \end{array}$$

and the sum of these $u^p v^{6-p}$ is $(u^2+v^2)^3 = 1$. Hence the equality

$$\tilde{\omega}_2((y^*y)^n) = \sum_{b_i\in\{\ell_1,\ell_1^*\}} \tilde{\omega}_2(b_1b_2\cdots b_{2n}) = \tilde{\omega}_2((\ell_1^*+\ell_1)^{2n})$$

is obtained.

□

The noncommutative random variable y in this example may be called an *elliptic* element, which contains a circular element (when $\tau = 0$) in Example 2.6.2 and also converges to a standard semicircular law (when $\tau \to 1$).

Let $(\mathcal{A}, \varphi)$ be a C^*-probability space such that φ is a tracial state. A multi-variable $(a_1, a_2, \ldots, a_k)$ of selfadjoint elements in $\mathcal{A}$ is said to be *centered general semicircular* if its R-series is of the form

$$\sum_{i,j=1}^{k} A_{ij}X_iX_j\,.$$

Since $\varphi(a_i) = \alpha(i)$ and $\varphi(a_ia_j) = \alpha(i,j) + \alpha(i)\alpha(j)$ from the moment-cumulant formula, a_i has 0 expectation and the $k\times k$ matrix A is exactly the covariance matrix. In particular, A is symmetric and positive semidefinite. Thanks to Proposition 2.6.6 one can choose an orthogonal matrix S and a free family $(b_1, b_2, \ldots, b_k)$ such that

$$a_i = \sum_{j=1}^{k} S_{ij}b_j \qquad (1\le i\le k)$$

and all b_j are centered semicircular.

One can consider the centered general semicircular multivariables from the multivariate free central limit theorem. Assume that for each $n \in \mathbb{N}$ a multivariable $(b_1^{(n)}, b_2^{(n)}, \dots, b_k^{(n)})$ of selfadjoint elements in $\mathcal{A}$ is given such that its joint distribution is independent of n, the expectation of $b_i^{(n)}$ is 0 and the familiy of $\{b_1^{(n)}, b_2^{(n)}, \dots, b_k^{(n)}\}$ $(n \in \mathbb{N})$ is free. Then the joint distribution of

$$\Big(\frac{1}{\sqrt{n}}\sum_{i=1}^{n} b_1^{(i)}, \frac{1}{\sqrt{n}}\sum_{i=1}^{n} b_2^{(i)}, \dots, \frac{1}{\sqrt{n}}\sum_{i=1}^{n} b_k^{(i)}\Big) \tag{2.6.8}$$

converges as $n \to \infty$ to the joint distribution of the centered general semicircular multivariable determined by the covariance matrix of $(b_1^{(1)}, b_2^{(1)}, \dots, b_k^{(1)})$. An easy proof of this statement comes from the calculation of the R-series of (2.6.8). Propositions 2.6.5 and 2.6.6 imply that this R-series is equal to

$$nR_{(b_1^{(1)}, b_2^{(1)}, \dots, b_k^{(1)})}\Big(\frac{1}{\sqrt{n}}X_1, \frac{1}{\sqrt{n}}X_2, \dots, \frac{1}{\sqrt{n}}X_k\Big),$$

and all coefficients go to 0, except for the quadratic terms, which remain independent of n (cf. the second proof of Theorem 2.3.2 at the end of Sec. 2.4). This means the desired convergence of the joint distribution of (2.6.8) due to the moment-cumulant formula (2.6.5).

Proposition 2.6.6 provides the transformation of the multivariate R-series under linear transformation of the noncommutative random variables. Next we treat compressions of noncommutative random variables and more general expressions in terms of matrix units. Contrary to the case of linear transformation, a hypothesis on a certain free relation will be imposed.

Proposition 2.6.8 *Let $(\mathcal{A}, \varphi)$ be a C^*-probability space and let e_{ij} $(1 \le i, j \le n)$ be a system of matrix units in $\mathcal{A}$. Given $a \in \mathcal{A}$, set $a_{ij} := e_{1i} a e_{j1} \in \mathcal{A}_1 := e_{11}\mathcal{A}e_{11}$. Assume that $\varphi(e_{ij}) = n^{-1}\delta_{ij}$, and consider the family a_{ij} $(1 \le i, j \le n)$ in the noncommutative probability space $(\mathcal{A}_1, \varphi_1)$, $\varphi_1 := \varphi(e_{11})^{-1}\varphi|_{\mathcal{A}_1}$. If a is free from $\{e_{ij} : 1 \le i, j \le n\}$, then the multivariate R-series of the noncommutative random variables $(a_{ij} : 1 \le i, j \le n)$ is of the form*

$$R_{(a_{ij}:1\le i,j\le n)}\big((X_{ij})_{1\le i,j\le n}\big) = n\mathrm{Tr}\, R_a(n^{-1}X),$$

where X_{ij} is the indeterminate corresponding to a_{ij}, the matrix X is formed from the indeterminates, $n^{-1}X$ is formally put in the R-series, and Tr *denotes the formal trace.*

In full detail, the above R-series reads as

$$\sum_{m} \sum_{i(1), i(2), \dots, i(m)=1}^{n} \alpha_m n^{1-m} X_{i(1)i(2)} X_{i(2)i(3)} \cdots X_{i(m)i(1)},$$

if $R_a(z) = \sum_m \alpha_m z^m$. This proposition is not proven here, but a few important particular cases will be discussed.

Example 2.6.9 Let p be a projection in a noncommutative probability space $(\mathcal{A}, \varphi)$, let $a \in \mathcal{A}$, and assume that a and p are free. When $\varphi(p) = 1/n$ for an integer n, we may assume that we are in the setting of the previous proposition and $p = e_{11}$. (If necessary, the given probability space is enlarged by taking the free product with a matrix algebra, or the construction of Example 2.2.11 can be applied.) To compute the R-series of pap with respect to the reduced state φ_1, we have to put 0 in place of the indeterminates X_{ij} whenever i or j is different from 1. In this way,

$$R_{pap}(X_{11}) = \sum_m \alpha_m n^{1-m} X_{11}^m .$$

In particular, if a is free Poisson distributed with parameter λ, then $\varphi(p)^{-1}pap$ is free Poisson as well with parameter $n\lambda$. This can be extended to the case when n is not an integer; in fact, Example 2.4.7 contains our statement. The advantage of the method used there is that it extends to several random variables. Indeed, from Example 2.2.11 one can deduce that $\{pa_1p, pa_2p, \ldots, pa_np\}$ is a free family if $a_1, a_2, \ldots, a_n, p$ are free and p is a projection.

□

Example 2.6.10 In the setting of Proposition 2.6.8, the R-series of the elements $a_1 := e_{11}ae_{21}$ and $a_2 := e_{12}ae_{11}$ with respect to φ_1 is

$$R_{(a_1,a_2)}(X_1, X_2) = \sum_k \alpha_{2k} n^{1-2k} \Big((X_1X_2)^k + (X_2X_1)^k \Big) ,$$

where the α_k's are the free cumulants of a.

Indeed, if a product $X_{i(1)i(2)}X_{i(2)i(3)} \cdots X_{i(m)i(1)}$ is formed only from X_{12} and X_{21}, then the only possibility is an alternating sequence of X_{12} and X_{21} with $m = 2k$. We replace X_{12} and X_{21} by X_1 and X_2, respectively.

□

It is sometimes useful to consider cumulants as multilinear functionals. Let $(\mathcal{A}, \varphi)$ be a noncommutative probability space and $a_1, a_2, \ldots, a_n \in \mathcal{A}$. The value $\phi_n(a_1, a_2, \ldots, a_n)$ of the *cumulant functional* ϕ_n is defined as the coefficient of $X_1X_2 \cdots X_n$ of the R-series of the n-tuple $(a_1, a_2, \ldots, a_n)$; that is,

$$\phi_n(a_1, a_2, \ldots, a_n) := \mathrm{coef}[1, 2, \ldots, n](R_{(a_1,a_2,\ldots,a_n)}) .$$

Actually, by Proposition 2.6.6 one has

$$\phi_k(a_{i(1)}, a_{i(2)}, \ldots, a_{i(k)}) = \mathrm{coef}[i(1), i(2), \ldots, i(k)](R_{(a_1,a_2,\ldots,a_n)}) \tag{2.6.9}$$

for all $k \in \mathbb{N}$ and $1 \leq i(1), i(2), \ldots, i(k) \leq n$. Furthermore, the cumulant functional

$$(a_1, a_2, \ldots, a_n) \mapsto \phi_n(a_1, a_2, \ldots, a_n)$$

is multilinear on $\mathcal{A}^n$. Indeed, if $b_i = \sum_{j=1}^k A_{ij} a_j$ $(1 \leq i \leq n)$, then Proposition 2.6.6 shows that

$$\begin{aligned}
&\phi_n(b_1, b_2, \ldots, b_n) \\
&= \mathrm{coef}[1, 2, \ldots, n]\Big(R_{(a_1, a_2, \ldots, a_k)}\Big(\sum_{i=1}^n A_{i1} X_i, \sum_{i=1}^n A_{i2} X_i, \ldots, \sum_{i=1}^n A_{ik} X_i\Big)\Big) \\
&= \sum_{j(1), \ldots, j(n)=1}^k A_{1j(1)} A_{2j(2)} \cdots A_{nj(n)} \mathrm{coef}[j(1), j(2), \ldots, j(n)](R_{(a_1, a_2, \ldots, a_k)}) \\
&= \sum_{j(1), \ldots, j(n)=1}^k A_{1j(1)} A_{2j(2)} \cdots A_{nj(n)} \phi_n(a_{j(1)}, a_{j(2)}, \ldots, a_{j(n)})
\end{aligned}$$

by (2.6.9). According to the moment-cumulant formula (2.6.5), the cumulant functionals are determined by the recursion

$$\varphi(a_1 a_2 \cdots a_n) = \sum_{\mathcal{V} \in NC(n)} \prod_{V \in \mathcal{V}} \phi_{|V|}(a_1, a_2, \ldots, a_n | V), \tag{2.6.10}$$

where $(a_1, a_2, \ldots, a_n | V) := (a_{i(1)}, a_{i(2)}, \ldots, a_{i(r)})$ for $V = \{i(1) < i(2) < \ldots < i(r)\} \subset [n]$. It follows from Proposition 2.6.4 that if $\{a_1, \ldots, a_k\}$ and $\{a_{k+1}, \ldots, a_n\}$ are free, then

$$\phi_n(a_1, a_2, \ldots, a_n) = 0.$$

Lemma 2.6.11 *The cumulant functionals have the following property for a product:*

$$\begin{aligned}
&\phi_{n-1}(a_1, \ldots, a_{k-1}, a_k a_{k+1}, a_{k+2}, \ldots a_n) \\
&\quad = \phi_n(a_1, \ldots, a_{k-1}, a_k, a_{k+1}, \ldots a_n) \\
&\qquad + \sum_{\mathcal{V}} \prod_{V \in \mathcal{V}} \phi_{|V|}(a_1, \ldots, a_{k-1}, a_k, a_{k+1}, \ldots a_n | V),
\end{aligned}$$

where the summation is over all $\mathcal{V} \in NC(n)$ such that $|\mathcal{V}| = 2$ and $k, k+1$ belong to different blocks.

Proof: We proceed by induction on n. For $n = 2$ we get by (2.6.10)

$$\phi_1(a_1 a_2) = \varphi(a_1 a_2) = \phi_2(a_1, a_2) + \phi_1(a_1)\phi_1(a_2).$$

Now let $n \geq 3$ and assume the property for all $n' < n$. Then it is straightforward to see that for every $\mathcal{U} \in NC(n)$ with $\mathcal{U} \neq \mathbf{1}_{n-1}$

$$\prod_{U \in \mathcal{U}} \phi_{|U|}(a_1, \dots, a_{k-1}, a_k a_{k+1}, a_{k+2}, \dots, a_n|U)$$
$$= \sum_{\substack{\mathcal{V} \in NC(n) \\ \mathcal{V}|_{k=k+1} = \mathcal{U}}} \prod_{V \in \mathcal{V}} \phi_{|V|}(a_1, \dots, a_{k-1}, a_k, a_{k+1}, \dots, a_n|V),$$

where $\mathcal{V}|_{k=k+1} \in NC(n-1)$ is the partition obtained by identifying $k, k+1$ and replacing $k+2, \dots, n$ by $k+1, \dots, n-1$ (here two blocks containing k and $k+1$ must merge if $k, k+1$ belong to different blocks of $\mathcal{V}$). Hence by (2.6.10) we have

$$\begin{aligned}
&\phi_{n-1}(a_1, \dots, a_{k-1}, a_k a_{k+1}, a_{k+2}, \dots, a_n) \\
&\quad = \varphi(a_1 \cdots a_{k-1} a_k a_{k+1} \cdots a_n) \\
&\qquad - \sum_{\substack{\mathcal{U} \in NC(n-1) \\ \mathcal{U} \neq \mathbf{1}_{n-1}}} \prod_{U \in \mathcal{U}} \phi_{|U|}(a_1, \dots, a_{k-1}, a_k a_{k+1}, a_{k+2}, \dots, a_n|U) \\
&\quad = \varphi(a_1 \cdots a_{k-1} a_k a_{k+1} \cdots a_n) \\
&\qquad - \sum_{\substack{\mathcal{U} \in NC(n-1) \\ \mathcal{U} \neq \mathbf{1}_{n-1}}} \sum_{\substack{\mathcal{V} \in NC(n) \\ \mathcal{V}|_{k=k+1} = \mathcal{U}}} \prod_{V \in \mathcal{V}} \phi_{|V|}(a_1, \dots, a_{k-1}, a_k, a_{k+1}, \dots, a_n|V) \\
&\quad = \varphi(a_1 \cdots a_{k-1} a_k a_{k+1} \cdots a_n) \\
&\qquad - \sum_{\substack{\mathcal{V} \in NC(n) \\ \mathcal{V}|_{k=k+1} \neq \mathbf{1}_{n-1}}} \prod_{V \in \mathcal{V}} \phi_{|V|}(a_1, \dots, a_{k-1}, a_k, a_{k+1}, \dots, a_n|V) \\
&\quad = \sum_{\substack{\mathcal{V} \in NC(n) \\ \mathcal{V}|_{k=k+1} = \mathbf{1}_{n-1}}} \prod_{V \in \mathcal{V}} \phi_{|V|}(a_1, \dots, a_{k-1}, a_k, a_{k+1}, \dots, a_n|V),
\end{aligned}$$

which is the desired property for n.

□

Example 2.6.12 Let u be a Haar unitary. Then the R-series of the pair (u, u^*) is

$$R_{(u,u^*)}(X_1, X_2) = \sum_{n=1}^{\infty} (-1)^{n-1} c_{n-1} \Big((X_1 X_2)^n + (X_2 X_1)^n \Big).$$

We compute by recursion that

$$\phi_{2n-1}(u, u^*, \dots, u, u^*, u) = \phi_{2n-1}(u^*, u, \dots, u^*, u, u^*) = 0, \tag{2.6.11}$$

$$\phi_{2n}(u, u^*, \dots, u, u^*) = \phi_{2n}(u^*, u, \dots, u^*, u) = (-1)^{n-1} c_{n-1}. \tag{2.6.12}$$

For $n = 1$ we get $\phi_1(u) = \varphi(u) = 0 = \varphi(u^*) = \phi_1(u^*)$ and $\phi_2(u, u^*) = \varphi(uu^*) - \varphi(u)\varphi(u^*) = 1 = c_0 = \phi_2(u^*, u)$ obviously.

Now assume that the above formulas hold for all $\phi_{2n'-1}$ and $\phi_{2n'}$ when $n' \le n$. We consider $\phi_{2n+1}(u, u^*, \dots, u, u^*, u)$. Applying Lemma 2.6.11, we have

$$\begin{aligned}
&\phi_{2n+1}(u, u^*, \dots, u, u^*, u) \\
&\quad = \phi_{2n}(\mathbf{1}, u, \dots, u, u^*, u) - \sum_{\mathcal{V}} \prod_{V \in \mathcal{V}} \phi_{|V|}(u, u^*, \dots, u, u^*, u|V) \\
&\quad = -\sum_{\mathcal{V}} \prod_{V \in \mathcal{V}} \phi_{|V|}(u, u^*, \dots, u, u^*, u|V),
\end{aligned}$$

where $\phi_{2n}(\mathbf{1}, u, \dots, u, u^*, u) = 0$ because $\mathbf{1}$ is in free relation to anything, and the summation is over all $\mathcal{V} \in NC(2n+1)$ such that $|\mathcal{V}| = 2$ and $1, 2$ belong to different blocks. Such a partition $\mathcal{V}$ must be of the form

$$\mathcal{V} = \big\{\{1, t, t+1, \dots, 2n+1\}, \{2, 3, \dots, t-1\}\big\}.$$

One of the two blocks gives an alternating sequence of u, u^* of odd length, and the corresponding cumulant functional vanishes according to the induction hypothesis. Hence $\phi_{2n+1}(u, u^*, \dots, u, u^*, u) = 0$.

Next we deal with $\phi_{2n+2}(u, u^*, \dots, u, u^*)$. Similarly to the above argument, we form the product of u and u^* in the first two places and obtain

$$\begin{aligned}
&\phi_{2n+2}(u, u^*, \dots, u, u^*) \\
&\quad = -\sum_{\mathcal{V} = \{V_1, V_2\}} \phi_{|V_1|}(u, u^*, \dots, u, u^*|V_1)\phi_{|V_2|}(u, u^*, \dots, u, u^*|V_2),
\end{aligned}$$

where the summation is over all partitions $\mathcal{V} = \big\{\{1, t, t+1, \dots, 2n+2\}, \{2, 3, \dots, t-1\}\big\}$ for $t = 3, 4, \dots, 2n+3$. Due to the induction hypothesis only the even $t = 2m$ survives in the sum, because the others correspond to vanishing cumulants. Hence

$$\begin{aligned}
&\phi_{2n+2}(u, u^*, \dots, u, u^*) \\
&\quad = -\sum_{m=2}^{n+1} \phi_{2(n-m+2)}(u, u^*, \dots, u, u^*)\phi_{2(m-1)}(u, u^*, \dots, u, u^*) \\
&\quad = -\sum_{m=2}^{n+1} (-1)^{n-m+1} c_{n-m+1} (-1)^{m-2} c_{m-2} = (-1)^n c_n .
\end{aligned}$$

In the last step the recursion formula (2.3.2) for the Catalan numbers was used.

We have shown that (2.6.11) and (2.6.12) hold, and we now turn to the coefficients of $R_{(u,u^*)}$. In the light of (2.6.9) the coefficients of the alternating products of X_1 and X_2 are already settled, and it is remarkable that up to this point only the condition $\varphi(u) = 0$ was used from the assumption of u being a Haar unitary. Since the expectations of the product of different numbers of u's and u^*'s are zero, it is a consequence of the moment-cumulant formula that the coefficient of a monomial of different numbers of X_1 and X_2 factors is 0. It remains to prove that $\phi_{2n}(u_1, \ldots, u_{2n}) = 0$ if $(u_1, u_2, \ldots, u_{2n})$ is not an alternating sequence of u, u^* and the number of u terms is n. This can be done by induction on n. For such $(u_1, \ldots, u_{2n})$ choose $1 \leq k < 2n$ such that $u_k = u$ and $u_{k+1} = u^*$ (or $u_k = u^*$ and $u_{k+1} = u$). Lemma 2.6.11 tells us that

$$
\begin{aligned}
0 = & \ \phi_{2n}(u_1, u_2, \ldots, u_{2n}) \\
& + \sum_{\mathcal{V}=\{V_1,V_2\}} \phi_{|V_1|}(u_1, u_2, \ldots, u_{2n}|V_1)\phi_{|V_2|}(u_1, u_2, \ldots, u_{2n}|V_2),
\end{aligned}
$$

where the summation is over all non-crossing partitions $\mathcal{V} = \{V_1, V_2\}$ such that $k \in V_1$ and $k+1 \in V_2$. One can see that under the constraints it is impossible that both $(u_1, u_2, \ldots, u_{2n}|V_1)$ and $(u_1, u_2, \ldots, u_{2n}|V_2)$ are alternating even sequences. So at least one of the corresponding cumulants is 0, according to the induction hypothesis. This gives $\phi_{2n}(u_1, u_2, \ldots, u_{2n}) = 0$.

□

A pair (a, b) of noncommutative random variables is called an *R-diagonal pair* if its R-series is of the form

$$
R_{(a,b)}(X_1, X_2) = \sum_{n=1}^{\infty} \alpha_n \Big((X_1X_2)^n + (X_2X_1)^n \Big).
$$

In this case the one-variable series $\sum_{n=1}^{\infty} \alpha_n X^n$ is called the *determining series* of the R-diagonal pair (a, b). An element a in a C^*-probability space is called an *R-diagonal element* if (a, a^*) is an R-diagonal pair. We know from Examples 2.6.12 and 2.6.2 that R-diagonal elements are common generalizations of Haar unitaries and circular elements. It follows from Proposition 2.6.5 that the sum of free R-diagonal pairs is R-diagonal again. In particular, the sum $a + b$ is R-diagonal when a, b are so and moreover $\{a, a^*\}$ and $\{b, b^*\}$ are in free relation. Example 2.6.10 shows an R-diagonal pair in the setting of compressions.

In the following results, due to Nica and Speicher, $(\mathcal{A}, \varphi)$ is assumed to be a noncommutative probability space such that φ is tracial.

Theorem 2.6.13 *Let a, b, p_1, p_2 be in $(\mathcal{A}, \varphi)$. If (a, b) is an R-diagonal pair and $\{p_1, p_2\}$ is free from $\{a, b\}$, then (ap_1, p_2b) is an R-diagonal pair.*

Proposition 2.6.14 *If (a, b) is an R-diagonal pair in $(\mathcal{A}, \varphi)$, then the determining series of (a, b) is $R_{ab} \star Moeb$. Hence the distribution of (a, b) is determined by that of ab.*

The above theorem and proposition are not proven here, but they have interesting consequences. If a is an R-diagonal element and moreover $\{a, a^*\}$ and $\{b, b^*\}$ are in free relation, then ab is R-diagonal. In particular, if u is a Haar unitary and $h = h^*$ is free from $\{u, u^*\}$ in a C^*-probability space, then uh is an R-diagonal element. In the case where h is quarter-circular, we get a circular element. Furthermore, *R-diagonal elements* are characterized as follows.

Corollary 2.6.15 *An element a in $(\mathcal{A}, \varphi)$ is R-diagonal if and only if there exist a Haar unitary u and a positive element h (in another C^*-probability space) such that h is free from $\{u, u^*\}$ and the distributions of (a, a^*) and (uh, hu^*) are the same.*

Proof: For any R-diagonal element a, choose a Haar unitary u and a positive h such that h is free from $\{u, u^*\}$ and the distribution of h^2 is equal to that of aa^*. Then uh is R-diagonal by Theorem 2.6.13, as noted above. By Proposition 2.6.14 both determining series of (a, a^*) and (uh, hu^*) are $R_\mu \star Moeb$, where μ is the same distribution of aa^* and h^2. Hence the distribution of (a, a^*) is equal to that of (uh, hu^*). The converse is clear.

□

It will be shown in Sec. 4.4 that R-diagonal elements admit a convenient random matrix model.

Notes and Remarks. The definition of free relation was introduced in [196]. Earlier Wai-Mee Ching constructed the free product of von Neumann algebras with cyclic and separating trace vectors in [50]. The free product of C^*-algebras with designated states was first given in [8] and [196] independently. In our language they constructed the free product of noncommutative C^*-probability spaces. In [8] the notion of free relation is implicit, but a sufficient condition for the free product C^*-algebra being simple is important. In the algebraic construction given in Sec. 2.1 one has to check that the functional ω is positive. This comes automatically when one works with representations. Results of [13] show that the free product of von Neumann algebras is often of type III. This is so even for the free product of matrix algebras with respect to non-tracial states; see [65].

The free central limit theorem (Theorem 2.3.2) is due to Voiculescu [196]. Two proofs are supplied; the first, in Sec. 2.2, is based on a combinatorial method. and the second, given at the end of Sec. 2.4, is Voiculescu's original proof using R-series. Previously, Bożejko [36] got a free central limit in a particular case arising from free generators in a discrete group, where the Catalan numbers were found as the limit moments. (A discussion similar to Bożejko's is in the second section of the Overview.)

The number of non-crossing partitions of $[n]$ is equal to the number of non-crossing pair partitions of $[2n]$. This fact is stated in Example 2.5.2 and shown in the proof of Proposition 2.6.3. Indeed, the number (2.5.10) of non-crossing partitions of $[n]$ as well as the identity (2.5.11) were given in [112]. The paper [171] is about non-crossing partitions and contains further references on the subject. The Möbius inversion process grew out of the classical inclusion-exclusion in the lattice of all subsets of a finite set; see [5] for the Möbius inversion theorem and the Möbius function in the lattice L_n, and see the monograph [180] for a comprehensive discussion on incidence algebras. Multiplicative functions on the lattice $NC(n)$ were studied by Speicher in [176], and the form of the Möbius function was computed in [112] (also [176]). The value of the Möbius function $\mu(\mathcal{V}_1, \mathcal{V}_2)$ at a general pair $\mathcal{V}_1 \leq \mathcal{V}_2$ is expressed by decomposing the sublattice $\{\mathcal{V} \in NC(n) : \mathcal{V}_1 \leq \mathcal{V} \leq \mathcal{V}_2\}$ into products of lattices $NC(k)$. Theorem 2.5.6 is from [176].

The additive free convolution, along with the R-series (or $\mathcal{R}$-transform) and the canonical random variables, was introduced in [197], and the multiplicative one in [198]. (For the sake of simplicity, our discussion was restricted here to compactly supported probability measures, where the method of moments is conveniently used.) The notion of *$\mathcal{S}$-transform* plays a role for the multiplicative free convolution similar to the one the R-series plays for the additive free convolution. In [96] Haagerup made an essential simplification in the description of the multiplicative free convolution. In [138] Nica and Speicher gave an alternative approach for the $\mathcal{S}$-transform via the "Fourier transform" of multiplicative functions on $NC(n)$.

It is worth mentioning that the notion of R-series was also discovered in a particular situation from study of random walks on free product groups (see [48], [215]).

The multivariate R-series and Propositions 2.6.3–2.6.6 are from [132], while [176] contains some similar results for multivariables. The combinatorial convolution of formal power series was studied in [136]. Examples 2.2.11 and 2.4.7 are from [168]. The unpublished formula in Proposition 2.6.8 was communicated to the authors by A. Nica and R. Speicher. It can be proven by the methods of the papers [133] and [168].

The concept of R-diagonal was introduced in [137]; Theorem 2.6.13 is a main result there. Proposition 2.6.14 and Corollary 2.6.15 were also shown there. Computation of the R-series of (u, u^*) for a Haar unitary u is from [178], Sec. 3.4. A power of an R-diagonal element is R-diagonal again; see [97], [116] and [111]. In particular, if c is a circular element, then c^k is R-diagonal and

$$\alpha_n = \frac{1}{(k-2)n+1}\binom{(k-1)n}{n-1},$$

the *generalized Catalan numbers* (see [141] for the details).

The nth moment m_n of the standard q-Gaussian distribution that appeared in Example 1.1.8 can be expressed as follows:

$$m_n = \sum_{\mathcal{V}} q^{\mathrm{cr}(\mathcal{V})} = (1-q)^{-n} \sum_{k=-n}^{n} \binom{2n}{n+k} (-1)^k q^{k(k-1)/2},$$

where the first summation is over all pair partitions $\mathcal{V}$ of $[n]$ and $\mathrm{cr}(\mathcal{V})$ stands for the reduced left crossing number of $\mathcal{V}$ defined in [131]. (This paper contains the q-version of Lemma 2.5.1.) The second expression is from [106].

The notion of free products extends to those with amalgamation. The amalgamated free product for C^*-algebras was introduced by Voiculescu [196], where C^*-algebras $\mathcal{A}_i$ have a common C^*-subalgebra $\mathcal{B}$ with conditional expectations $\Phi_i : \mathcal{A}_i \to \mathcal{B}$. The free product of C^*-probability spaces is the case when $\mathcal{B}$ is the scalars $\mathbb{C}\mathbf{1}$ and so the conditional expectations Φ_i are states. As free probability theory corresponds to free products, the amalgamated free probability theory is developed for *amalgamated free products.* Assume that the algebras $\mathcal{A}_i$ are contained in the algebra $\mathcal{A}$, $\mathcal{B}$ is a common subalgebra of the $\mathcal{A}_i$'s, and there is a conditional expectation $E_{\mathcal{B}} : \mathcal{A} \to \mathcal{B}$. The subalgebras $\mathcal{A}_i$ are called *free with amalgamation* over $\mathcal{B}$ if, for every $n \in \mathbb{N}$ and $i(1) \neq i(2) \neq \ldots \neq i(n)$,

$$E_{\mathcal{B}}(a_1 a_2 \cdots a_n) = 0 \quad \text{whenever} \quad a_k \in \mathcal{A}_{i(k)},\ E_{\mathcal{B}}(a_k) = 0,\ 1 \leq k \leq n.$$

In this generalization moments and cumulants of noncommutative random variables are operator-valued, $\mathcal{B}$-valued in the above situation. See [204] and [178] for the details.

Chapter 3

Analytic Function Theory and Infinitely Divisible Laws

This chapter is mostly devoted to the analytic machinery used to deal with free convolutions of measures. When a probability measure on $\mathbb{R}$ is considered as the distribution of a noncommutative random variable, one can associate to it some analytic functions which behave similarly to the logarithm of the Fourier transform in classical probability theory. In the previous chapter the R-series (or $\mathcal{R}$-transform) of a distribution measure was introduced as a formal power series which linearizes the additive free convolution. It will turn out that this formal series is related to the Cauchy transform of the distribution measure, and moreover the series is actually convergent in a complex domain whenever the measure is compactly supported. This relation between the R-series and the Cauchy transform was discovered by Voiculescu, and his original proof used analytic tools with the Toeplitz operator calculus. Two proofs are supplied here; the first, due to Speicher, comes from the combinatorics developed in the previous chapter, and the second is a simpler proof due to Haagerup. Since a probability measure on $\mathbb{R}$ can be recovered from its Cauchy transform by means of the well-known Stieltjes inversion formula, the above relation provides an effective device to determine free convolutions of measures.

In most of the chapter compactly supported measures are treated. Although many of the results hold true without this restriction, the proofs become essentially more transparent for compactly supported measures. By the method of analytic functions, examples will show the apparently strange behavior of the free convolution when one has a comparison with the classical convolution in mind: For example, the free convolution of atomic measures can be continuous.

The aim of the latter half of this chapter is to characterize infinitely divisible laws (with respect to the free convolution). The free Poisson distribution and the semicircle law are typical infinitely divisible measures. The theory of Nevanlinna-Pick functions plays an important role in the description of infinitely divisible laws, as it does in the classical Lévy-Hinčin formula. The compound free Poisson distribution is a natural generalization of the free Poisson distribution, and it is also infinitely divisible.

3.1 Cauchy transform, Poisson integral, and Hilbert transform

This section is a brief survey, for the reader's convenience, on the subjects of the title, which are basic ingredients in potential theory and some aspects of harmonic analysis.

Let μ be a probability measure on $\mathbb{R}$. Its *Cauchy transform*

$$G_\mu(z) := \int_{-\infty}^{\infty} \frac{d\mu(t)}{z-t} \tag{3.1.1}$$

is defined when z lies in the upper half-plane $\mathbb{C}^+ := \{z \in \mathbb{C} : \operatorname{Im} z > 0\}$ (and also in the lower half-plane $\mathbb{C}^- := \{z \in \mathbb{C} : \operatorname{Im} z < 0\}$). The transform $G_\mu(z)$ is an analytic function in $\mathbb{C}^+$ possessing the following properties:

$$G_\mu(\mathbb{C}^+) \subset \mathbb{C}^- \quad \text{and} \quad |G_\mu(z)| \le \frac{1}{\operatorname{Im} z}. \tag{3.1.2}$$

Let the support of μ be bounded. Then $G_\mu(z)$ is analytic in a neighborhood of ∞. Since $(z-t)^{-1} = \sum_{k=0}^{\infty} t^k z^{-k-1}$, it is obvious that $G_\mu(z)$ has the following expansion at $z = \infty$:

$$G_\mu(z) = z^{-1} + \sum_{k=1}^{\infty} m_k(\mu) z^{-k-1}, \tag{3.1.3}$$

where $m_k(\mu) = \int t^k \, d\mu(t)$ $(k \in \mathbb{Z}^+)$.

The *Poisson kernel* P_y is given as

$$P_y(x) := \frac{y}{\pi(x^2+y^2)} \qquad (x \in \mathbb{R},\ y > 0),$$

and it has the following properties:

(1) $P_y(x) > 0$ and $\int_{-\infty}^{\infty} P_y(x)\, dx = 1$.

(2) $P_y \in \bigcap_{1 \le p \le \infty} L^p(\mathbb{R})$.

(3) $P_{y_1} * P_{y_2} = P_{y_1+y_2}$ for $y_1, y_2 > 0$, where $*$ means the usual convolution.

Let ν be a finite positive (or, more generally, signed) measure on $\mathbb{R}$. The *Poisson integral* $P_y * \nu$ is the convolution product of P_y and ν; that is,

$$P_y * \nu(x) = \frac{1}{\pi} \int_{-\infty}^{\infty} \frac{y}{(x-t)^2+y^2}\, d\nu(t) \qquad (x \in \mathbb{R},\ y > 0). \tag{3.1.4}$$

Then $P_y * \nu$ is an integrable function and $\|P_y * \nu\|_1 \le \|\nu\|$, where $\|\nu\|$ is the total variation of ν. For every $f \in L^p(\mathbb{R})$, the Poisson integral $P_y * f$ is defined as (3.1.4) with $f(t)\,dt$ instead of $d\nu(t)$, and it satisfies $\|P_y * f\|_p \le \|f\|_p$, $1 \le p \le \infty$.

The following are some basic facts concerning the Poisson integral.

(4) $F(x + \mathrm{i}\,y) := (P_y * \nu)(x)$ (also $F(x + \mathrm{i}\,y) := (P_y * f)(x)$ for $f \in L^p(\mathbb{R})$) is a harmonic function in $\mathbb{C}^+$.

(5) If $f \in L^p(\mathbb{R})$ with $1 \le p < \infty$, then $\|P_y * f - f\|_p \to 0$ and $(P_y * f)(x) \to f(x)$ a.e. as $y \to +0$.

(6) If f is a bounded continuous function on $\mathbb{R}$, then $P_y * f \to f$ as $y \to +0$ uniformly on every compact set.

(7) If ν is a finite signed measure, then $P_y * \nu \to \nu$ as $y \to +0$ in the w*-topology on the dual space of $C_0(\mathbb{R})$.

The *Hilbert transform* Hf of a function f is given by the principal value integral

$$Hf(x) := \lim_{\varepsilon \to +0} \frac{1}{\pi} \int_{|x-t| \ge \varepsilon} \frac{f(t)}{x - t}\, dt\,,$$

whenever the limit exists for a.e. $x \in \mathbb{R}$. Also let us define $\tilde{P}_y f$ by

$$\tilde{P}_y f(x) := \frac{1}{\pi} \int_{-\infty}^{\infty} \frac{(x-t) f(t)}{(x-t)^2 + y^2}\, dt \qquad (x \in \mathbb{R},\, y > 0).$$

The Hilbert transform has, among others, the following properties.

(8) If $f \in L^p(\mathbb{R})$ with $1 < p < \infty$, then Hf is defined and $\|Hf\|_p \le A_p \|f\|_p$, where A_p is a constant depending on p.

(9) If f is as in (8), then $\tilde{P}_y f = P_y * Hf$, and hence $\|\tilde{P}_y f - Hf\|_p \to 0$ and $\tilde{P}_y f(x) \to Hf(x)$ a.e. as $y \to +0$.

Let μ be a compactly supported probability measure on $\mathbb{R}$. Writing $z = x + \mathrm{i}\,y$, one can split the Cauchy transform (3.1.1) into its real and imaginary parts:

$$G_\mu(x + \mathrm{i}\,y) = \int_{-\infty}^{\infty} \frac{x - t}{(x-t)^2 + y^2}\, d\mu(t) - \mathrm{i} \int_{-\infty}^{\infty} \frac{y}{(x-t)^2 + y^2}\, d\mu(t)\,.$$

Comparing this with (3.1.4) and with the property (7) of the Poisson integral, one can observe that the limit of the imaginary part of $G_\mu(x + \mathrm{i}\,y)$ recovers μ up to a factor $-\pi$. More precisely,

$$\mu = \text{w*-} \lim_{y \to +0} \left[-\frac{1}{\pi} \operatorname{Im} G_\mu(x + \mathrm{i}\,y)\, dx \right].$$

(This relation sometimes bears the name *Stieltjes inversion formula.*) Also, note that $t_0 \in \mathbb{R}$ is an isolated point of the support of μ if and only if $z = t_0$ is a simple pole of $G_\mu(z)$. Moreover, $\mu(\{t_0\})$ is the residue of $G_\mu(z)$ at t_0. When μ has a continuous derivative $f = d\mu/dx$, we obtain

$$f(x) = -\frac{1}{\pi} \lim_{y \to +0} \operatorname{Im} G_\mu(x + \mathrm{i}\, y) \tag{3.1.5}$$

and

$$Hf(x) = \frac{1}{\pi} \lim_{y \to +0} \operatorname{Re} G_\mu(x + \mathrm{i}\, y),$$

due to property (9) of the Hilbert transform.

Example 3.1.1 The Cauchy transform of the semicircle law w_r is

$$G_{w_r}(z) = \frac{2}{r^2}(z - \sqrt{z^2 - r^2}). \tag{3.1.6}$$

(Here the branch of $\sqrt{z^2 - r^2}$ is taken in accordance with (3.1.2).)

The proof of Lemma 2.3.1 contains a formula for the generator function of the even moments of the standard semicircle law. Benefitting from that, we have

$$\frac{1}{2}(1 - \sqrt{1 - 4x}) = \sum_{k=0}^{\infty} \frac{m_{2k}(w_r)}{(r/2)^{2k}} x^{k+1},$$

which holds if $|x|$ is small. Replacing x by $(r/2)^2 z^{-2}$, we arrive at

$$\frac{2}{r^2}(z - \sqrt{z^2 - r^2}) = \sum_{k=0}^{\infty} m_{2k}(w_r) z^{-2k-1}$$

for large $|z|$. Since the left-hand side is analytic in $\mathbb{C}^+$, we get the above expression (3.1.6) in the light of (3.1.3).

□

We can easily take the limits as $y \to +0$ of the real and imaginary parts of (3.1.6):

$$\lim_{y \to +0} \operatorname{Re} G_r(x + \mathrm{i}\, y) = \begin{cases} \frac{2}{r^2} x & \text{if } |x| \le r, \\ \frac{2}{r^2}(x - \sqrt{x^2 - r^2}) & \text{if } |x| > r, \end{cases}$$

$$\lim_{y \to +0} \operatorname{Im} G_r(x + \mathrm{i}\, y) = \begin{cases} -\frac{2}{r^2}\sqrt{r^2 - x^2} & \text{if } |x| \le r, \\ 0 & \text{if } |x| > r. \end{cases}$$

The latter limit is in accordance with (3.1.5).

The R-series of w_r is $R_{w_r}(z) = r^2z^2/4$ by Example 2.4.6. One can easily check that $G_{w_r}(z)$ and $z^{-1}(1 + R_{w_r}(z))$ are the inverse of each other. Indeed, this is a general fact that will be proved in the next section.

3.2 Relation between Cauchy transform and R-series

The R-series introduced in Sec. 2.4 contains all information of the distribution of a noncommutative random variable, and it is an analogue of the logarithm of the Fourier transform in classical probability theory. The essence of this section is the convergence of the formal series in a certain domain and its relation to the Cauchy transform of a measure.

As we mentioned in Sec. 2.4 (see the paragraph before Example 2.4.1), it is convenient to consider abstract distributions which are given as linear functionals $\mu : \mathbb{C}\langle X\rangle \to \mathbb{C}$ with $\mu(\mathbf{1}) = 1$. The set of such distributions is denoted by Σ. If μ is a probability measure on $\mathbb{R}$ all of whose moments exist, then μ is considered as an element of Σ via $\mu(X^k) = m_k(\mu)$ $(k \in \mathbb{Z}^+)$. Note that (3.1.3) is meaningful for any $\mu \in \Sigma$; that is, one can define

$$G_\mu(z) := z^{-1}(1 + M_\mu(z^{-1})) = z^{-1} + \sum_{k=1}^{\infty} \mu(X^k)z^{-k-1} \tag{3.2.1}$$

as a formal power series, where $M_\mu(z) = \sum_{k=1}^{\infty} \mu(X^k)z^k$ is the moment generating series in (2.5.17). Also, the definition of the R-series $R_\mu(z)$ in Sec. 2.4 can be applied to any $\mu \in \Sigma$, and a formal power series $K_\mu(z)$ is defined by

$$K_\mu(z) := z^{-1}(1 + R_\mu(z)) = z^{-1} + \sum_{k=0}^{\infty} \alpha_{k+1}z^k . \tag{3.2.2}$$

Then we have the following important result.

Theorem 3.2.1 *For every $\mu \in \Sigma$, $G_\mu(z)$ and $K_\mu(z)$ given in* (3.2.1) *and* (3.2.2) *are the inverse of each other as formal power series. Thus, in particular, if μ is a compactly supported probability measure on $\mathbb{R}$, then $K_\mu(z)$ is a univalent analytic function from a neighborhood of* 0 *to a neighborhood of ∞ whose inverse is the Cauchy transform $G_\mu(z)$.*

Proof: Theorem 2.5.6 implies that

$$R_\mu(G_\mu(z)) = M_\mu(z^{-1}) \quad \text{and} \quad M_\mu\Bigl(\frac{1}{K_\mu(z)}\Bigr) = R_\mu(z) .$$

Hence we have

$$K_\mu(G_\mu(z)) = \frac{1 + M_\mu(z^{-1})}{G_\mu(z)} = z$$

and

$$G_\mu(K_\mu(z)) = \frac{1 + R_\mu(z)}{K_\mu(z)} = z\,,$$

so the first assertion is shown. Now the second is immediate, because $G_\mu(z)$ is analytic and univalent in a neighborhood of ∞ for any compactly supported measure μ.

□

The above proof is a direct reformulation of Theorem 2.5.6, which is essentially combinatorial, based on the free moment-cumulant formula (2.5.6). It is worth emphasizing that free probability theory possesses both analytic and combinatorial aspects. Voiculescu's original proof of the above theorem is analytic, using Toeplitz operators. The proof presented below is due to Haagerup.

Second Proof of Theorem 3.2.1: Let $\mu \in \Sigma$ and $R_\mu(z) = \sum_{k=1}^\infty \alpha_k z^k$. Then α_k (resp. $\mu(X^k)$) is a polynomial in $\mu(X), \dots, \mu(X^k)$ (resp. $\alpha_1, \dots, \alpha_k$). The formal inverse of the power series $K_\mu(z) = z^{-1} + \sum_{k=0}^\infty \alpha_{k+1} z^k$ is of the form $z^{-1} + \sum_{k=1}^\infty \mu_k z^{-k-1}$, where α_k (resp. μ_k) is a polynomial in $\mu_1, \dots, \mu_k$ (resp. $\alpha_1, \dots, \alpha_k$). Based on these facts, it is enough to assume that $R_\mu(z)$ is a polynomial.

Let $\ell = \ell(h)$ be a creation operator on the full Fock space $\mathcal{F}(\mathcal{H})$, where h is a unit vector, and let Φ be the vacuum vector and φ the vacuum expectation, as usual. Set

$$\Phi(z) := \Phi + \sum_{k=1}^\infty z^k h^{\otimes k}$$

for a family of vectors parametrized by complex numbers z with $|z| < 1$. Assume that $z^{-1}R_\mu(z)$ is a polynomial $P(z) = \sum_{k=0}^N \alpha_{k+1} z^k$, and set

$$S := \ell + P(\ell^*)\,.$$

Note that $S^* = \ell^* + \sum_{k=1}^N \overline{\alpha}_{k+1} \ell^k$ is the canonical noncommutative random variable associated with μ, except that the coeffcients are changed to complex-conjugate. This shows that $\mu(X^k) = \overline{\varphi(S^{*k})} = \varphi(S^k)$, and hence $G_\mu(z) = \sum_{k=0}^\infty \varphi(S^k) z^{-k-1}$. Now compute

$$S\Phi(z) = z^{-1}(\Phi(z) - \Phi) + P(z)\Phi(z)$$

and

$$\big[(z^{-1} + P(z))\mathbf{1} - S\big]\Phi(z) = z^{-1}\Phi\,.$$

Therefore, we have

$$\left[(z^{-1} + P(z))\mathbf{1} - S\right]^{-1}\Phi = z\Phi(z)$$

if $|z| > 0$ is so small that $|z^{-1} + P(z)| > \|S\|$. Taking the inner product with Φ, we obtain $\sum_{k=0}^{\infty}(z^{-1} + P(z))^{-k-1}\varphi(S^k) = z$; that is, $G_\mu(z^{-1} + P(z)) = z$. This is what we wanted, since $z^{-1} + P(z) = K_\mu(z)$.

Note that $\Phi(z)$ is a sort of exponential vector in the context of the full Fock space, cf. (1.1.3).

□

Example 3.2.2 Let $\nu := 2^{-1}(\delta(-1/2) + \delta(1/2))$. Then the free convolution $\mu := \nu \boxplus \nu$ has probability density

$$f_\mu(x) := \begin{cases} \dfrac{1}{\pi\sqrt{1-x^2}} & \text{if } |x| < 1, \\ 0 & \text{otherwise.} \end{cases}$$

(This is called the *arcsine law.*) This example shows that the free convolution of atomic measures can be absolutely continuous. In particular, the free convolution is not distributive with respect to the addition of measures.

First we compute the R-series of ν. The Cauchy transform of ν is $G_\nu(z) = 4z/(4z^2 - 1)$, and to get R_ν we have to solve the equation

$$G_\nu\Big(\frac{1 + R_\nu(u)}{u}\Big) = u\,. \tag{3.2.3}$$

This is easy, and we obtain

$$R_\nu(u) = \frac{-1 + \sqrt{u^2 + 1}}{2}\,.$$

The R-series of μ is $R_\mu = 2R_\nu$, and the Cauchy transform G_μ is the solution of (3.2.3) when ν is replaced by μ. We arrive at

$$G_\mu(z) = \frac{\sqrt{z^2 - 1}}{z^2 - 1}\,,$$

and

$$\lim_{y \to +0} \operatorname{Im} G_\mu(x + \mathrm{i}\,y) = -\pi f_\mu(x)\,.$$

□

The last example is somewhat special; the convolution of atomic measures generally possesses both absolutely continuous and atomic parts.

Example 3.2.3 Let $\nu := (1-\alpha)\delta(-1/2) + \alpha\delta(1/2)$. Then the free convolution $\mu := \nu \boxplus \nu$ is

$$\frac{\sqrt{4\alpha(1-\alpha)-x^2}}{\pi(1-x^2)}\chi(x)\,dx \qquad\qquad (3.2.4)$$
$$+\max\{1-2\alpha,0\}\delta(-1)+\max\{2\alpha-1,0\}\delta(1)\,,$$

where χ is the characteristic function of the interval $\left[-2\sqrt{\alpha(1-\alpha)},\, 2\sqrt{\alpha(1-\alpha)}\right]$.

The computation goes along the lines of the previous example. The Cauchy transform

$$G_\mu(z) = \frac{2\alpha - 1 + \sqrt{z^2 - 4\alpha(1-\alpha)}}{z^2 - 1}$$

has a simple pole at $z = -1$ for $0 \le \alpha < 1/2$ and at $z = 1$ for $1/2 < \alpha \le 1$. The corresponding residues are $1 - 2\alpha$ and $2\alpha - 1$, respectively. In this way, (3.2.4) is obtained.

□

Improving the above computation, it is not difficult to show the following fact: Let μ, ν be compactly supported probability measures and a, b, c the maximum of the supports of $\mu, \nu, \mu \boxplus \nu$, respectively. Then

$$\begin{cases} c = a + b & \text{if } \mu(\{a\}) + \nu(\{b\}) \ge 1, \\ c < a + b & \text{otherwise.} \end{cases}$$

This is another feature of the free convolution which is very different from the classical one.

3.3 Infinitely divisible laws

Infinitely divisible laws with respect to the additive free convolution will be discussed in this section. In the classical theory (related to the ordinary convolution) the theory of Pick-Nevanlinna functions plays an important role, and the Lévy-Hinčin formula is the highlight. The free analogue seems rather similar. Parallel to the Gaussian and Poisson laws that are typical in the classical case, typical infinitely divisible laws in the free case are the Wigner and free Poisson ones.

We first give a short survey on Pick functions. An analytic function f in $\mathbb{C}^+$ such that $\operatorname{Im} f(z) \ge 0$ for all $z \in \mathbb{C}^+$ is called a *Pick function*. Note that a Pick function f satisfies $f(\mathbb{C}^+) \subset \mathbb{C}^+$ unless f is a constant. For $a, b \in \mathbb{R}$, $a < b$, we denote by $\mathcal{P}(a,b)$ the set of Pick functions f which have analytic continuation to $(\mathbb{C} \setminus \mathbb{R}) \cup (a,b)$ so that $f(\overline{z}) = \overline{f(z)}$. The following is a kind of Pick-Nevanlinna theorem, which is tailor-made for our purpose.

Theorem 3.3.1 *Let D be a domain in $\mathbb{C}^+$. For a function $f : D \to \mathbb{C}$ the following conditions are equivalent:*

(i) *f extends to a Pick function.*

(ii) *For every $z_1, \dots, z_n \in D$, the matrix*

$$\left[\frac{f(z_i) - \overline{f(z_j)}}{z_i - \overline{z}_j} \right]_{i,j=1}^{n}$$

is positive semidefinite.

(iii) *There exist $\alpha \in \mathbb{R}$, $\beta \geq 0$, and a positive finite measure ν on $\mathbb{R}$ such that*

$$f(z) = \alpha + \beta z + \int_{-\infty}^{\infty} \frac{1 + xz}{x - z} \, d\nu(x) \qquad (z \in D). \tag{3.3.1}$$

Moreover, if the above conditions hold, then α, β, and ν in (iii) *are unique, and f extends to a Pick function in $\mathcal{P}(a, b)$ if and only if ν is supported on $\mathbb{R} \setminus (a, b)$.*

Corollary 3.3.2 *If f is an analytic function in $\mathbb{C}^+$, then the following conditions are equivalent:*

(i) *f extends to a Pick function in $\mathcal{P}(-\varepsilon, \varepsilon)$ for some $\varepsilon > 0$; that is, f (extended) is analytic in a neighborhood of $(\mathbb{C} \setminus \mathbb{R}) \cup \{0\}$, $f(\overline{z}) = \overline{f(z)}$, and $\operatorname{Im} f(z) \geq 0$ for $z \in \mathbb{C}^+$.*

(ii) *There exist $\alpha \in \mathbb{R}$ and a positive finite measure ν on $\mathbb{R}$ with compact support such that*

$$f(z) = \alpha + \int_{-\infty}^{\infty} \frac{z}{1 - xz} \, d\nu(x) \,.$$

In this case, α and ν in (ii) *are unique and ν is supported on $[-1/\varepsilon, 1/\varepsilon]$.*

Proof: (ii) $\Rightarrow$ (i) is obvious. For (i) $\Rightarrow$ (ii), by Theorem 3.3.1 there exist $\alpha' \in \mathbb{R}$, $\beta \geq 0$ and a positive finite measure ν' on $\mathbb{R} \setminus (-\varepsilon, \varepsilon)$ for which (3.3.1) holds. Let ν'' be the transform of ν' by $x \mapsto x^{-1}$, and define a finite measure ν on $[-1/\varepsilon, 1/\varepsilon]$ by

$$\nu := \beta \delta(0) + (1 + x^2) \nu'' \,.$$

Then we have

$$\begin{aligned} f(z) &= \alpha' + \beta z + \int \frac{1 + x^{-1} z}{x^{-1} - z} \, d\nu''(x) \\ &= \alpha' + \beta z + \int \left(x + \frac{(1 + x^2) z}{1 - xz} \right) d\nu''(x) \\ &= \alpha + \int \frac{z}{1 - xz} \, d\nu(x) \end{aligned}$$

with $\alpha := \alpha' + \int x\,d\nu''(x)$. The last assertion is easily seen from Theorem 3.3.1.

□

From now on we shall apply the Pick-Nevanlinna theory to characterize infinitely divisible laws with respect to the free convolution. Let $\mathcal{M}_0(\mathbb{R})$ denote the set of compactly supported probability measures on $\mathbb{R}$. A measure $\mu \in \mathcal{M}_0(\mathbb{R})$ is said to be $\boxplus$-*infinitely divisible* if for each $n \in \mathbb{N}$ there exists $\mu_n \in \mathcal{M}_0(\mathbb{R})$ such that

$$\mu = \mu_n \boxplus \cdots \boxplus \mu_n \quad (n \text{ times}).$$

In this case, μ_n is unique because $R_{\mu_n} = n^{-1}R_\mu$ by Theorem 2.4.5.

Example 3.3.3 The semicircle law $w_{m,r}$ is $\boxplus$-infinitely divisible because

$$w_{m,r} = (w_{m/n,r/\sqrt{n}})^{\boxplus n},$$

see Example 2.4.2.

□

Lemma 3.3.4 *Let (μ_n) be a sequence in $\mathcal{M}_0(\mathbb{R})$. Then the following conditions are equivalent:*

(i) *The supports of the μ_n are uniformly bounded, and (μ_n) converges in the w^*-topology to a measure $\mu \in \mathcal{M}_0(\mathbb{R})$.*

(ii) *(R_{μ_n}) converges uniformly in some neighborhood of 0 to a function R.*

In this case, $R = R_\mu$.

Proof: (i)$\Rightarrow$(ii). Assume that the supports of μ_n (and hence μ) are included in $[-\alpha, \alpha]$. Let

$$f_n(z) := G_{\mu_n}(z^{-1}) = \int_{-\alpha}^{\alpha} \frac{z}{1 - xz}\,d\mu_n(x).$$

Since $f_n'(z) = \int_{-\alpha}^{\alpha} 1/(1-xz)^2\,d\mu_n(x)$, there exists $\varepsilon > 0$ such that $\operatorname{Re} f_n'(z) > 0$ and $|f_n(z)| > |z|/2$ for $|z| < \varepsilon$ and all n. Since

$$\operatorname{Re}\frac{f_n(z_2) - f_n(z_1)}{z_2 - z_1} = \int_0^1 \operatorname{Re} f_n'(z_1 + t(z_2 - z_1))\,dt > 0 \quad \text{for} \quad |z_1|, |z_2| < \varepsilon,$$

all $f_n(z)$ are univalent in $|z| < \varepsilon$. Furthermore, we have $\{f_n(z) : |z| < \varepsilon\} \supset \{|z| < \varepsilon/2\}$. Thus, if $U := \{|z| > \varepsilon^{-1}\} \cup \{\infty\}$ and $V := \{|z| < \varepsilon/2\}$, then all G_{μ_n} are univalent in U and $G_{\mu_n}(U) \supset V$, that is, $K_{\mu_n}(V) \subset U$. This shows that (K_{μ_n}) in V is a normal family. By (i) we have $m_k(\mu_n) \to m_k(\mu)$ as $n \to \infty$ for all $k \in \mathbb{N}$. Hence the Taylor coefficients of R_{μ_n} converge to the corresponding ones of R_μ, because the coefficient of R_{μ_n} for z^k is a universal polynomial in $m_j(\mu_n)$, $1 \le j \le k$. This implies that (K_{μ_n}) converges uniformly in V to K_μ, as does (R_{μ_n}) to R_μ.

(ii) $\Rightarrow$ (i). Let

$$g_n(z) := \frac{1}{K_{\mu_n}(z)} = \frac{z}{1 + R_{\mu_n}(z)}\,.$$

Since $g_n'(z) = (1 + R_{\mu_n}(z) - zR'_{\mu_n}(z))/(1 + R_{\mu_n}(z))^2$, there exists $\varepsilon > 0$ such that $\operatorname{Re} g_n'(z) > 0$ and $|g_n(z)| > |z|/2$ for $|z| < \varepsilon$ and all n. As in the above proof, all K_{μ_n} are univalent in U, $K_{\mu_n}(U) \supset V$ and $G_{\mu_n}(V) \subset U$, where $U := \{|z| < \varepsilon\}$ and $V := \{|z| > 2/\varepsilon\}$. So it is easy to see that $G_{\mu_n}(z^{-1}) \in \mathcal{P}(-\varepsilon/2, \varepsilon/2)$. This implies by Corollary 3.3.2 that all μ_n are supported on $[-2/\varepsilon, 2/\varepsilon]$. Now let μ be any w*-limit point of (μ_n). Then the above (i) $\Rightarrow$ (ii) shows that $R_\mu = R$. Thus μ is a unique w*-limit point of (μ_n), and so $\mu_n \to \mu$ in the w*-topology.

□

As a consequence of Lemma 3.3.4, we observe that if (μ_n) is a sequence of $\boxplus$-infinitely divisible measures in $\mathcal{M}_0(\mathbb{R})$ with uniformly bounded supports and it converges in the w*-topology to $\mu \in \mathcal{M}_0(\mathbb{R})$, then μ is $\boxplus$-infinitely divisible too.

In Example 2.5.2 the form of the free Poisson distribution μ_1 was determined by a combinatorial argument. For the general free Poisson distribution we have

Example 3.3.5 Let $\lambda > 0$, and let μ_λ be the free Poisson distribution introduced in Example 2.5.2, that is, the distribution of the canonical noncommutative random variable $\ell^* + \sum_{k=0}^\infty \lambda \ell^k$. In order to imitate the Poisson limit theorem, let

$$\mu^{(n)} := \Big(\Big(1 - \frac{\lambda}{n}\Big)\delta(0) + \frac{\lambda}{n}\delta(1)\Big)^{\boxplus n}.$$

Then $(\mu^{(n)})$ converges to μ_λ in the w*-topology. Recall that according to the classical Poisson limit theorem ([76], X.6) this sequence converges to the Poisson law when the free convolution $\boxplus$ is replaced by the classical one. This free analogue might be called the *free Poisson limit theorem*. Moreover, the exact form of μ_λ is

$$\mu_\lambda = \begin{cases} \dfrac{\sqrt{4\lambda - (x - 1 - \lambda)^2}}{2\pi x}\chi(x)\,dx & \text{if } \lambda \ge 1, \\[2ex] (1 - \lambda)\delta(0) + \dfrac{\sqrt{4\lambda - (x - 1 - \lambda)^2}}{2\pi x}\chi(x)\,dx & \text{if } 0 < \lambda < 1, \end{cases} \tag{3.3.2}$$

where χ stands for the characteristic function of the interval $\big[(1 - \sqrt{\lambda})^2, (1 + \sqrt{\lambda})^2\big]$.

The R-series of μ_λ is

$$R_{\mu_\lambda}(z) = \sum_{k=1}^\infty \lambda z^k = \frac{\lambda z}{1 - z},$$

and the Cauchy transform of $\nu_n := (1-\lambda/n)\delta(0) + (\lambda/n)\delta(1)$ is

$$G_{\nu_n}(z) = \Big(1-\frac{\lambda}{n}\Big)\frac{1}{z} + \frac{\lambda}{n}\frac{1}{z-1}.$$

Hence on the basis of Theorem 3.2.1 we obtain the Cauchy transform of μ_λ and the R-series of ν_n as follows:

$$G_{\mu_\lambda}(z) = \frac{z+(1-\lambda)-\sqrt{(z-1-\lambda)^2-4\lambda}}{2z},$$

$$R_{\nu_n}(z) = \frac{z-1-\sqrt{(z-1)^2+4z\lambda/n}}{2} = \frac{\lambda z}{n(1-z)} + O(n^{-2})z.$$

Since $R_{\mu^{(n)}}(z) = nR_{\nu_n}(z)$ has the uniform limit $R_{\mu_\lambda}(z)$ as $n\to\infty$ in a neighborhood of 0, we observe by Lemma 3.3.4 that $(\mu^{(n)})$ has uniformly bounded supports and converges to μ_λ in the w*-topology; in particular, μ_λ is compactly supported. From the above formula for $G_{\mu_\lambda}(z)$ we recover the measure μ_λ by the Stieltjes inversion formula and arrive at the expression (3.3.2). Here, note that, when $0<\lambda<1$, $G_{\mu_\lambda}(z)$ has a pole at $z=0$ and therefore μ_λ has an atom at 0.

Furthermore, one can get the exact form of $\mu^{(n)}$. The Cauchy transform of $\mu^{(n)}$ is computed by Theorem 3.2.1 as follows:

$$G_{\mu^{(n)}}(z) = \frac{(n-2)z+n(1-\lambda)-\sqrt{n^2z^2-2n(n+(n-2)\lambda)z+n^2(1-\lambda)^2}}{2(nz-z^2)}.$$

When $0<\lambda<1$ this has a pole at $z=0$ with residue $1-\lambda$, but $z=n$ is not a pole if n is large. Hence for large n the Stieltjes inversion formula yields

$$\mu^{(n)} = \frac{\sqrt{4\lambda(1-\frac{1}{n})(1-\frac{\lambda}{n})-\big(x-1-(1-\frac{2}{n})\lambda\big)^2}}{2\pi x(1-\frac{x}{n})}\chi_n(x)\,dx \tag{3.3.3}$$

if $\lambda\ge 1$ and

$$(1-\lambda)\delta(0) + \frac{\sqrt{4\lambda(1-\frac{1}{n})(1-\frac{\lambda}{n})-\big(x-1-(1-\frac{2}{n})\lambda\big)^2}}{2\pi x(1-\frac{x}{n})}\chi_n(x)\,dx \tag{3.3.4}$$

when $0<\lambda<1$, where χ_n is the characteristic function of the interval

$$\Big[\Big(1-\sqrt{\lambda(1-\tfrac{1}{n})(1-\tfrac{\lambda}{n})}\Big)^2-\tfrac{\lambda}{n}+(1-\tfrac{1}{n})\tfrac{\lambda^2}{n},$$
$$\Big(1+\sqrt{\lambda(1-\tfrac{1}{n})(1+\tfrac{\lambda}{n})}\Big)^2-\tfrac{\lambda}{n}+(1-\tfrac{1}{n})\tfrac{\lambda^2}{n}\Big].$$

In this way, one can observe that the interval of χ_n converges to that of χ and the density function in (3.3.3) and (3.3.4) uniformly converges to that in (3.3.2). So the convergence $\mu^{(n)} \to \mu_\lambda$ is much better than stated above.

□

For the free Poisson distribution μ_λ the term *Marchenko-Pastur distribution* is also used in connection with random matrices (see Sec. 4.1). Since μ_λ is a limit of certain free convolutions according to the free Poisson limit theorem in the above example, one may expect μ_λ to be infinitely divisible. In fact, $\mu_\lambda \boxplus \mu_{\lambda'} = \mu_{\lambda+\lambda'}$, because $R_{\mu_\lambda} + R_{\mu_{\lambda'}} = R_{\mu_{\lambda+\lambda'}}$. The free Poisson distribution will be described in Example 3.3.10 in a slightly modified form.

A one-parameter family $(\mu_t)_{t\ge 0}$ of measures in $\mathcal{M}_0(\mathbb{R})$ is called a *$\boxplus$-semigroup* if $\mu_t \boxplus \mu_s = \mu_{t+s}$ for all $t, s \ge 0$ (in particular, $\mu_0 = \delta(0)$). The above μ_λ's form a w*-continuous $\boxplus$-semigroup. Also, the family $(w_{2\sqrt{t}})_{t\ge 0}$ of semicircle laws ($w_0 = \delta(0)$) is a w*-continuous $\boxplus$-semigroup by Example 2.4.2. In Examples 1.2.8 and 2.2.10 we considered a noncommutative process $(X_t)_{t\ge 0}$ on the full Fock space. This is distributed according to $(w_{2\sqrt{t}})_{t\ge 0}$ and can be regarded as a *free analogue of Brownian motion.*.

The next theorem characterizes the $\boxplus$-infinitely divisible laws in $\mathcal{M}_0(\mathbb{R})$. Up to now we have used the R-series $R_\mu(z)$, which differs by a factor z from Voiculescu's *$\mathcal{R}$-transform* $\mathcal{R}_\mu(z)$: $R_\mu(z) = z\mathcal{R}_\mu(z)$. To state the theorem the $\mathcal{R}$-transform $\mathcal{R}_\mu$ seems to be more convenient than R_μ, and in the rest of this section $\mathcal{R}_\mu$ will be preferred to R_μ.

Theorem 3.3.6 *The following conditions for $\mu \in \mathcal{M}_0(\mathbb{R})$ are equivalent:*

(i) *μ is $\boxplus$-infinitely divisible.*

(ii) *There exists a w*-continuous $\boxplus$-semigroup $(\mu_t)_{t\ge 0}$ such that $\mu_1 = \mu$.*

(iii) *$\mathcal{R}_\mu$ extends to a Pick function in $\mathcal{P}(-\varepsilon, \varepsilon)$ for some $\varepsilon > 0$.*

(iv) *There exist $\alpha \in \mathbb{R}$ and a positive finite measure ν on $\mathbb{R}$ with compact support such that*

$$\mathcal{R}_\mu(z) = \alpha + \int_{-\infty}^{\infty} \frac{z}{1 - xz}\, d\nu(x) \tag{3.3.5}$$

for all z in a neighborhood of $(\mathbb{C} \setminus \mathbb{R}) \cup \{0\}$.

Moreover, if the above conditions hold, then α and ν in (iv) *are unique and the following hold:*

(1) *$\mathcal{R}_{\mu_t}(z) = t\mathcal{R}_\mu(z)$, $t \ge 0$;*

(2) *$\alpha = m_1(\mu) = \lim_{\varepsilon\to+0} m_1(\mu_\varepsilon)/\varepsilon$;*

(3) *$\nu = \text{w*-}\lim_{\varepsilon\to+0} \nu_\varepsilon$ where $d\nu_\varepsilon := (x^2/\varepsilon) d\mu_\varepsilon$;*

(4) $\nu(\mathbb{R}) = m_2(\mu) - m_1(\mu)^2$, *the variance of* μ.

In classical probability theory, when μ_1, μ_2 are probability measures on $\mathbb{R}$ and $\mu = \mu_1 * \mu_2$ (the usual convolution), we have $\Phi_\mu(x) = \Phi_{\mu_1}(x)\Phi_{\mu_2}(x)$, where Φ_μ is the characteristic function (or the Fourier transform) of μ, i.e. $\Phi_\mu(x) = \int e^{\mathrm{i}\,tx}\,d\mu(t)$, $x \in \mathbb{R}$. If μ is infinitely divisible in the classical sense, then the following *Lévy-Hinčin formula* is known:

$$\log \Phi_\mu(x) = \mathrm{i}\,\alpha x + \int_{-\infty}^{\infty} \left(e^{\mathrm{i}\,tx} - 1 - \frac{\mathrm{i}\,tx}{1+t^2}\right)\frac{1+t^2}{t^2}\,d\nu(t)\,,$$

where $\alpha \in \mathbb{R}$ and ν is a positive finite measure on $\mathbb{R}$. If the $\mathcal{R}$-transform $\mathcal{R}_\mu$ is regarded as the free version of the logarithm of the characteristic function Φ_μ, then the expression (3.3.5) is the free analogue of the Lévy-Hinčin formula.

We need some lemmas to prove Theorem 3.3.6.

Lemma 3.3.7 *Let* $\mathcal{R}(z) = \sum_{k=0}^{\infty} \alpha_{k+1} z^k$ *be a formal power series. Then the following conditions are equivalent:*

(i) *There exists* $\mu \in \mathcal{M}_0(\mathbb{R})$ *such that* $\mathcal{R} = \mathcal{R}_\mu$.

(ii) $\mathcal{R}$ *is convergent in a neighborhood of* 0 *with* $\mathcal{R}(\overline{z}) = \overline{\mathcal{R}(z)}$, *and for every* $z_1, \dots, z_n \in \mathbb{C}^+$ *in a neighborhood of* ∞, *the matrix*

$$\left[\frac{z_i - \overline{z}_j}{z_i - \overline{z}_j + \mathcal{R}(z_i^{-1}) - \mathcal{R}(\overline{z}_j^{-1})}\right]_{i,j=1}^{n}$$

is positive semidefinite.

Furthermore, if $\mathcal{R}$ *is a Pick function in* $\mathcal{P}(-\varepsilon, \varepsilon)$ *for some* $\varepsilon > 0$, *then the above conditions hold.*

Proof: (i) $\Rightarrow$ (ii). Let $\mu \in \mathcal{M}_0(\mathbb{R})$ and $\mathcal{R} = \mathcal{R}_\mu$. Then $F(z) = 1/G_\mu(z)$ is analytic in a neighborhood of $(\mathbb{C} \setminus \mathbb{R}) \cup \{\infty\}$. Theorem 3.2.1 implies that $\mathcal{R}$ is analytic in a neighborhood of 0 and F is univalent in a neighborhood U of ∞ with the inverse $K_\mu(z^{-1}) = z + \mathcal{R}(z^{-1})$. Since $G_\mu(\overline{z}) = \overline{G_\mu(z)}$, we get $F(\overline{z}) = \overline{F(z)}$ and $K_\mu(\overline{z}) = \overline{K_\mu(z)}$. Furthermore, since $G_\mu(\mathbb{C}^+) \subset \mathbb{C}^-$, we get $F(\mathbb{C}^+) \subset \mathbb{C}^+$, so that F is a Pick function. Let $z_1, \dots, z_n \in F(U) \cap \mathbb{C}^+$. Then for some $\zeta_j \in U \cap \mathbb{C}^+$ we have $z_j = F(\zeta_j)$, and so $\zeta_j = z_j + \mathcal{R}(z_j^{-1})$. Therefore we have

$$\left[\frac{z_i - \overline{z}_j}{z_i - \overline{z}_j + \mathcal{R}(z_i^{-1}) - \mathcal{R}(\overline{z}_j^{-1})}\right]_{i,j} = \left[\frac{F(\zeta_i) - F(\overline{\zeta}_j)}{\zeta_i - \overline{\zeta}_j}\right]_{i,j}, \tag{3.3.6}$$

which is positive semidefinite by Theorem 3.3.1.

(ii) $\Rightarrow$ (i). Since $\mathcal{R}$ is analytic in a neighborhood of 0, it is obvious that $\hat{K}(z) := z + \mathcal{R}(z^{-1})$ is a univalent analytic function in a neighborhood V of ∞. Let F be the inverse of $\hat{K}|_V$. A domain $D \subset V \cap \mathbb{C}^+$ can be chosen so that $\hat{K}(D) \subset \mathbb{C}^+$ and the matrix in the condition (ii) is positive semidefinite for every $z_1, \ldots, z_n \in D$. If $\zeta_1, \ldots, \zeta_n \in \hat{K}(D)$ and so $\zeta_j = z_j + \mathcal{R}(z_i^{-1})$ for some $z_j \in D$, then we have (3.3.6) again, and therefore the matrix

$$\left[\frac{F(\zeta_i) - F(\overline{\zeta}_j)}{\zeta_i - \overline{\zeta}_j}\right]_{i,j}$$

is positive semidefinite. Thus F extends to a Pick function by Theorem 3.3.1. Now sst $G(z) := 1/F(z)$. Then, since $\mathcal{R}(\overline{z}) = \overline{\mathcal{R}(z)}$ in a neighborhood of 0, it follows that $G(\overline{z}) = \overline{G(z)}$ in a neighborhood of ∞, so $G(z^{-1}) \in \mathcal{P}(-\varepsilon, \varepsilon)$ for some $\varepsilon > 0$. Hence by Corollary 3.3.2 there exist $\alpha \in \mathbb{R}$ and a positive finite measure μ on $\mathbb{R}$ with compact support such that

$$G(z^{-1}) = \alpha + \int_{-\infty}^{\infty} \frac{z}{1 - xz}\, d\mu(x),$$

that is,

$$G(z) = \alpha + \int_{-\infty}^{\infty} \frac{d\mu(x)}{z - x}.$$

But, since $\lim_{z\to\infty} zG(z) = \lim_{z\to\infty} \hat{K}(z)/z = 1$, we have $\alpha = 0$ and $\mu \in \mathcal{M}_0(\mathbb{R})$, implying $G = G_\mu$. Since G_μ is the inverse of $\hat{K}(z^{-1}) = z^{-1} + \mathcal{R}(z)$, Theorem 3.2.1 shows that $\mathcal{R} = \mathcal{R}_\mu$.

Next, assume that $\mathcal{R}$ belongs to $\mathcal{P}(-\varepsilon, \varepsilon)$ for some $\varepsilon > 0$. Since $\operatorname{Im} z > 0$ implies $\operatorname{Im} \mathcal{R}(z^{-1}) < 0$, $-\mathcal{R}(z^{-1})$ is a Pick function. A neighborhood U of ∞ can be chosen so that

$$\left|\frac{\mathcal{R}(z_1^{-1}) - \mathcal{R}(\overline{z}_2^{-1})}{z_1 - \overline{z}_2}\right| = \frac{1}{|z_1 z_2|}\left|\frac{\mathcal{R}(z_1^{-1}) - \mathcal{R}(\overline{z}_2^{-1})}{z_1^{-1} - \overline{z}_2^{-1}}\right| < \frac{1}{2}$$

for all $z_1, z_2 \in U$. Then for every $z_1, \ldots, z_n \in U \cap \mathbb{C}^+$, the matrix

$$\left[\frac{z_i - \overline{z}_j}{z_i - \overline{z}_j + \mathcal{R}(z_i^{-1}) - \mathcal{R}(\overline{z}_j^{-1})}\right]_{i,j} = \left[1 + \sum_{k=1}^{\infty}\left(-\frac{\mathcal{R}(z_i^{-1}) - \mathcal{R}(\overline{z}_j^{-1})}{z_i - \overline{z}_j}\right)^k\right]_{i,j}$$

is positive semidefinite by Theorem 3.3.1. Here we used the famous Schur theorem that the Hadamard (or entrywise) product of positive semidefinite matrices is positive semidefinite. Hence $\mathcal{R}$ satisfies (ii).

□

Lemma 3.3.8 *Let $(\mu_t)_{t\geq 0}$ be a w*-continuous $\boxplus$-semigroup in $\mathcal{M}_0(\mathbb{R})$, and let $\phi(z) = \mathcal{R}_{\mu_1}(z)$. Then $\mathcal{R}_{\mu_t}(z) = t\phi(z)$ for all $t \geq 0$, and the supports of μ_t $(0 \leq t \leq T)$ are uniformly bounded for any $T > 0$.*

Proof: If $r = m/n$ is rational, then Theorem 2.4.5 gives $n\mathcal{R}_{\mu_r}(z) = \mathcal{R}_{\mu_m}(z) = m\phi(z)$, and so $\mathcal{R}_{\mu_r}(z) = r\phi(z)$. For any $t \geq 0$ let r_n be rational with $r_n \to t$. Since $\mathcal{R}_{\mu_{r_n}}(z) = r_n\phi(z) \to t\phi(z)$ uniformly in a neighborhood of 0 and $\mu_{r_n} \to \mu_t$ in the w*-topology, Lemma 3.3.4 implies that $\mathcal{R}_{\mu_t}(z) = t\phi(z)$. Hence the first assertion is shown. The second is immediate from Lemma 3.3.4 again.

□

The next lemma is given in a slightly more general setting than we need here.

Lemma 3.3.9 *Let $(\mu_t)_{t\geq 0}$ be a w*-continuous $\boxplus$-semigroup in $\mathcal{M}_0(\mathbb{R})$, $\nu \in \mathcal{M}_0(\mathbb{R})$, $G(t,z) := G_{\nu\boxplus\mu_t}(z)$ and $\phi(z) := \mathcal{R}_{\mu_1}(z)$. Then for each $T > 0$ there exists a neighborhood V of ∞ such that the equality*

$$\frac{\partial G}{\partial t}(t,z) + \phi(G(t,z))\frac{\partial G}{\partial z}(t,z) = 0 \tag{3.3.7}$$

holds for all $0 \leq t \leq T$ and $z \in V$.

Proof: Define $\psi(z) := \mathcal{R}_\nu(z)$ and $K(t,z) := K_{\nu\boxplus\mu_t}(z)$. Then $\mathcal{R}_{\nu\boxplus\mu_t}(z) = \psi(z) + t\phi(z)$ by Lemma 3.3.8, and hence $K(t,z) = z^{-1} + \psi(z) + t\phi(z)$. For any $T > 0$, as in the proof (ii)$\Rightarrow$(i) of Lemma 3.3.4, there exists $\varepsilon_T > 0$ such that $K(t,z)$ $(0 \leq t \leq T)$ are univalent in $U := \{|z| < \varepsilon_T\}$ and

$$\bigcap_{0\leq t\leq T} K(t,U) \supset V := \{|z| > 2/\varepsilon_T\}. \tag{3.3.8}$$

Since $G(t,\cdot\,)$ is the inverse of $K(t,\cdot\,)|_U$ for each $0 \leq t \leq T$, it follows that $G(t,z)$ is differentiable in $t \in [0,T]$ for every $z \in K(t,U)$, and

$$G(t,K(t,\zeta)) = \zeta \qquad (0 \leq t \leq T,\ \zeta \in U).$$

Differentiating the above with respect to t yields

$$\frac{\partial G}{\partial t}(t,K(t,\zeta)) + \frac{\partial G}{\partial z}(t,K(t,\zeta))\phi(\zeta) = 0 \qquad (0 \leq t \leq T,\ \zeta \in U).$$

For every $0 \leq t \leq T$ and $z \in V$, since $z = K(t,\zeta)$ and $\zeta = G(t,z)$ for some $\zeta \in U$ by (3.3.8), we get

$$\frac{\partial G}{\partial t}(t,z) + \frac{\partial G}{\partial z}(t,z)\phi(G(t,z)) = 0 \qquad (0 \leq t \leq T,\ z \in V).$$

□

Proof of Theorem 3.3.6: (ii) $\Rightarrow$ (i) is obvious, and (iii) $\Leftrightarrow$ (iv) was given in Corollary 3.3.2. In the following let $\phi := \mathcal{R}_\mu$.

(iii) $\Rightarrow$ (ii). If (iii) is satisfied, then Lemma 3.3.7 implies that for any $t > 0$ there exists $\mu_t \in \mathcal{M}_0(\mathbb{R})$ such that $t\phi = \mathcal{R}_{\mu_t}$. Since $\mathcal{R}_{\mu_s \boxplus \mu_t} = (s+t)\phi = \mathcal{R}_{\mu_{s+t}}$ for $s, t > 0$, it follows that $(\mu_t)_{t \geq 0}$ with $\mu_0 = \delta(0)$ is a $\boxplus$-semigroup, whose w*-continuity is guaranteed by Lemma 3.3.4.

(i) $\Rightarrow$ (ii). Let $r = m/n$ be rational. By assumption there exists $\mu_{1/n} \in \mathcal{M}_0(\mathbb{R})$ such that $\phi = n\mathcal{R}_{\mu_{1/n}}$. Define

$$\mu_r := \mu_{1/n} \boxplus \cdots \boxplus \mu_{1/n} \quad (m \text{ times}).$$

Then $\mathcal{R}_{\mu_r} = m\mathcal{R}_{\mu_{1/n}} = r\phi$. For any $t > 0$, taking rationals r_n with $r_n \to t$ and using Lemma 3.3.4, one can obtain $\mu_t \in \mathcal{M}_0(\mathbb{R})$ such that $t\phi = \mathcal{R}_{\mu_t}$. Then (ii) is seen as in the proof of (iii) $\Rightarrow$ (ii).

(ii) $\Rightarrow$ (iv). Let $G(t,z) := G_{\mu_t}(z)$. Then Lemma 3.3.9 implies that

$$\frac{\partial G}{\partial t}(0, z^{-1}) + \phi(G(0,z^{-1}))\frac{\partial G}{\partial z}(0, z^{-1}) = 0$$

in a neighborhood of 0. Since $G(0,z) = z^{-1}$ and $(\partial G/\partial z)(0,z) = -z^{-2}$, the above equation says that $(\partial G/\partial t)(0,z^{-1}) - z^2\phi(z) = 0$, and we get

$$\begin{aligned}
\phi(z) &= z^{-2} \lim_{\varepsilon \to +0} \frac{1}{\varepsilon}(G(\varepsilon, z^{-1}) - G(0, z^{-1})) \\
&= z^{-2} \lim_{\varepsilon \to +0} \frac{1}{\varepsilon}\left(\int_{-\infty}^{\infty} \frac{z}{1-xz}\, d\mu_\varepsilon(x) - z\right) \\
&= \lim_{\varepsilon \to +0} \frac{1}{\varepsilon} \int_{-\infty}^{\infty} \frac{x}{1-xz}\, d\mu_\varepsilon(x)\,.
\end{aligned}$$

Hence $\phi(0) = \lim_{\varepsilon \to +0} m_1(\mu_\varepsilon)/\varepsilon$ exists. Since $x/(1-xz) = x + x^2z/(1-xz)$, we have

$$\phi(z) = \alpha + \lim_{\varepsilon \to +0} \int_{-\infty}^{\infty} \frac{z}{1-xz}\, d\nu_\varepsilon(x)\,, \tag{3.3.9}$$

where $\alpha := \phi(0)$ and $d\nu_\varepsilon := (x^2/\varepsilon)d\mu_\varepsilon$. By Lemma 3.3.8 the supports of ν_ε $(0 < \varepsilon \leq 1)$ are uniformly bounded. Since (3.3.9) gives

$$\frac{\phi(z) - \phi(0)}{z} = \lim_{\varepsilon \to +0} \int_{-\infty}^{\infty} \frac{d\nu_\varepsilon(x)}{1-xz}\,,$$

the total variations $\nu_\varepsilon(\mathbb{R})$ are bounded as $\varepsilon \to +0$. So one can choose $\varepsilon_n \to +0$ such that (ν_{ε_n}) converges in the w*-topology to some positive finite measure ν on

$\mathbb{R}$ with compact support, for which

$$\phi(z) = \alpha + \int_{-\infty}^{\infty} \frac{z}{1-xz}\, d\nu(x) = \alpha + \sum_{k=0}^{\infty} \Big(\int_{-\infty}^{\infty} x^k \, d\nu(x) \Big) z^{k+1}$$

in a neighborhood of 0. This shows that ν is determined by ϕ only. Hence ν is a unique w*-limit point of ν_ε as $\varepsilon \to +0$, so we obtain $\nu = \text{w*-}\lim_{\varepsilon\to+0} \nu_\varepsilon$ and (3.3.5).

For the second part of the theorem, since $\alpha = \mathcal{R}_\mu(0)$ and $\nu(\mathbb{R}) = \mathcal{R}'_\mu(0)$ by (3.3.5), the first equality of (2) and (4) are immediately checked. On the other hand, (1), the second equality of (2), and (3) were shown in the course of the above proof (ii) $\Rightarrow$ (iv).

□

In the setting of Lemma 3.3.9, Theorem 3.3.6 just proved says that $\phi(z) = \mathcal{R}_{\mu_1}(z)$ is a Pick function in $\mathcal{P}(-\varepsilon, \varepsilon)$ for some $\varepsilon > 0$, and this implies that all terms in the left-hand side of (3.3.7) are analytic in $\mathbb{C}^+$. In this way we observe that the partial differential equation (3.3.7) actually holds for all $t \ge 0$ and $z \in \mathbb{C}^+$.

Example 3.3.10 In Theorem 3.3.6, if $\nu = \gamma\delta(0)$ and so $\mathcal{R}_\mu(z) = \alpha + \gamma z$, then μ is the semicircle law $w_{\alpha, 2\sqrt{\gamma}}$, which is a free analogue of the Gaussian law. On the other hand, if $\alpha = 0$ and $\nu = \gamma\delta(t)$ ($t \ne 0$), then the corresponding $\mu \in \mathcal{M}_0(\mathbb{R})$ satisfies $\mathcal{R}_\mu(z) = \gamma z/(1 - tz)$, which is regarded as a free analogue of the Poisson law. In fact, since $\mathcal{R}_\mu(z) = t\mathcal{R}_{\mu_\lambda}(tz) - \gamma/t$, where μ_λ is given in (3.3.2) with $\lambda = \gamma/t^2$, one can get the exact form of this μ as follows:

$$\mu = \begin{cases} \dfrac{\sqrt{4\gamma - (x-t)^2}}{2\pi(tx+\gamma)} \chi(x)\, dx & \text{if } \gamma \ge t^2, \\[2ex] \Big(1 - \dfrac{\gamma}{t^2}\Big)\delta(-\gamma/t) + \dfrac{\sqrt{4\gamma - (x-t)^2}}{2\pi(tx+\gamma)} \chi(x)\, dx & \text{if } 0 < \gamma < t^2, \end{cases}$$

where χ is the characteristic function of the interval $[t - 2\sqrt{\gamma}, t + 2\sqrt{\gamma}]$. Analogously to the classical case, Theorem 3.3.6 says that any $\boxplus$-infinitely divisible law is a w*-limit of certain free convolutions of a Wigner law and a finite number of modified free Poisson laws.

□

The notion of free Poisson distributions is generalized in the following way. For any $\rho \in \mathcal{M}_0(\mathbb{R})$ and $\lambda > 0$, let

$$\mathcal{R}(z) := \lambda \int \frac{x}{1-xz}\, d\rho(x) = \lambda \int x\, d\rho(x) + \lambda \int \frac{x^2 z}{1-xz}\, d\rho(x). \tag{3.3.10}$$

From Corollary 3.3.2 and Lemma 3.3.7 it follows that $\mathcal{R} = \mathcal{R}_\mu$ for some $\mu \in \mathcal{M}_0(\mathbb{R})$. We call this measure μ a *compound free Poisson distribution* and denote it by $\pi_{\rho,\lambda}$. In fact, $\pi_{\rho,\lambda}$ is the free Poisson distribution μ_λ when $\rho = \delta(1)$. Note that $\mu = \pi_{\rho,\lambda}$

is $\boxplus$-infinitely divisible by Theorem 3.3.6, and its R-series is $R_\mu = \lambda M_\rho$, so that the $\boxplus$-semigroup $(\mu_t)_{t\geq 0}$ in Theorem 3.3.6 is given as $\mu_t = \pi_{\rho,\lambda t}$ $(t > 0)$.

The *free Poisson limit theorem* in Example 3.3.5 is extended as follows.

Proposition 3.3.11 *For any $\rho \in \mathcal{M}_0(\mathbb{R})$ and $\lambda > 0$, define*

$$\mu^{(n)} := \left(\left(1 - \frac{\lambda}{n}\right)\delta(0) + \frac{\lambda}{n}\rho\right)^{\boxplus n}.$$

Then $(\mu^{(n)})$ has uniformly bounded supports, and converges in the w^*-topology to the compound free Poisson distribution $\pi_{\rho,\lambda}$.

Proof: The Cauchy transform of $\nu_n := (1 - \lambda/n)\delta(0) + (\lambda/n)\rho$ is

$$\begin{aligned} G_{\nu_n}(z) &= \left(1 - \frac{\lambda}{n}\right)\frac{1}{z} + \frac{\lambda}{n}G_\rho(z) \\ &= \left(1 - \frac{\lambda}{n}\right)\frac{1}{z} + \frac{\lambda}{nz}\left(1 + M_\rho\left(\frac{1}{z}\right)\right) \\ &= \frac{1}{z} + \frac{\lambda}{nz}M_\rho\left(\frac{1}{z}\right). \end{aligned}$$

Since $R_{\nu_n}(G_{\nu_n}(z)) = M_{\nu_n}(z^{-1})$, we get

$$R_{\nu_n}\left(z + \frac{\lambda}{n}zM_\rho(z)\right) = M_{\nu_n}(z) = \frac{\lambda}{n}M_\rho(z),$$

and hence

$$R_{\mu^{(n)}}\left(z + \frac{1}{n}zR_\mu(z)\right) = R_\mu(z),$$

where $\mu := \pi_{\rho,\lambda}$. From this we can easily see that the coefficients of $R_{\mu^{(n)}}(z)$ converge to the corresponding ones of $R_\mu(z)$ as $n \to \infty$. Hence for each $k \in \mathbb{N}$ the kth moment of $\mu^{(n)}$ converges to that of μ.

Now let α, β be the minimum and the maximum of $\operatorname{supp}\rho$, and set

$$\mu_\alpha^{(n)} := \left(\left(1 - \frac{\lambda}{n}\right)\delta(0) + \frac{\lambda}{n}\delta(\alpha)\right)^{\boxplus n},$$

and similarly $\mu_\beta^{(n)}$ with β in place of α. Then it is clear that

$$\min\operatorname{supp}\mu_\alpha^{(n)} \leq \min\operatorname{supp}\mu^{(n)} \leq \max\operatorname{supp}\mu^{(n)} \leq \max\operatorname{supp}\mu_\beta^{(n)}.$$

It is seen from Example 3.3.5 that the supports of the $\mu_\alpha^{(n)}$'s and $\mu_\beta^{(n)}$'s are uniformly bounded. Hence so are the supports of the $\mu^{(n)}$'s, and the result follows.

□

According to the form (3.3.10) we observe that a $\boxplus$-infinitely divisible $\mu \in \mathcal{M}_0(\mathbb{R})$ having the $\mathcal{R}$-transform (3.3.5) is a compound free Poisson distribution if and only if

$$\int x^{-2}\,d\nu(x) < +\infty\,, \quad \alpha = \int x^{-1}\,d\nu(x)\,.$$

This says that the set of compound free Poisson distributions is rather small in the $\boxplus$-infinitely divisible laws; for example, a semicircle law is not a compound free Poisson distribution. Nevertheless, we have

Proposition 3.3.12 *The following conditions for $\mu \in \mathcal{M}_0(\mathbb{R})$ are equivalent:*

(i) *μ is $\boxplus$-infinitely divisible.*

(ii) *μ is the w^*-limit of a sequence of compound free Poisson distributions with uniformly bounded supports.*

Proof: (ii) $\Rightarrow$ (i) follows from the fact stated after Lemma 3.3.4. To prove the converse, assume that μ has the $\mathcal{R}$-transform (3.3.5) and $(\mu_t)_{t\ge 0}$ is as in Theorem 3.3.6. Since the supports of the μ_t $(0 \le t \le 1)$ are uniformly bounded by Lemma 3.3.8, by (2) and (3) of Theorem 3.3.6 we have

$$\begin{aligned}\mathcal{R}_\mu(z) &= \lim_{\varepsilon\to+0}\left(\frac{1}{\varepsilon}\int x\,d\mu_\varepsilon(x) + \frac{1}{\varepsilon}\int \frac{x^2}{1-xz}\,d\mu_\varepsilon(x)\right)\\ &= \lim_{\varepsilon\to+0}\frac{1}{\varepsilon}\int \frac{x}{1-xz}\,d\mu_\varepsilon(x)\end{aligned}$$

uniformly in a neighborhood of $z = 0$. This implies by Lemma 3.3.4 that $\pi_{\mu_\varepsilon,1/\varepsilon}$ converges to μ in the w*-topology, so (ii) follows.

□

For instance, let (ρ_j) be a sequence of symmetric measures in $\mathcal{M}_0(\mathbb{R})$ such that $\lambda_j^{-1} := \int x^2\,d\rho_j(x) > 0$ and $\operatorname{supp}\rho_j$ tends to $\{0\}$. Then we notice that $\mathcal{R}_{\pi_{\rho_j,\lambda_j}}(z) \to z$ uniformly in a neighborhood of 0, and hence π_{ρ_j,λ_j} converges in the w*-topology to w_2.

Notes and Remarks. Concerning the boundary values of the imaginary and real parts of the Cauchy tranform, as well as properties of the Hilbert transform, the interested reader may consult [110], Chap. VI, or [183], for example.

The relation between the Cauchy transform and the R-series given in Theorem 3.2.1 was established in [197], and the combinatorial proof is from [176]. Another proof of Haagerup is from [96]. In Voiculescu's papers the concept of R-series is different from ours, and it coincides with the $\mathcal{R}$-transform here. (Nica's work on the multivariate case supports very strongly the revised concept of R-series, used also in our discussions.)

It may be worth mentioning that the Pick-Nevanlinna theory of analytic functions plays a fundamental role also in theory of monotone operator functions (see [57]). For details on the Pick-Nevanlinna theory, see [2] and [57]. The classical theory of infinitely divisible distributions is found in standard texts such as [58] and [92]. Theorem 3.3.6 was shown in [197] and [17]. This was generalized to distributions with unbounded support in [119] for the finite variance case and [18] for the general case. In the unboundedly supported case, the transform $\varphi_\mu(z) = F_\mu^{-1}(z) - z$, with F_μ^{-1} being the right inverse of $F_\mu = 1/G_\mu$ in some domain, is more convenient than $\mathcal{R}_\mu$.

The paper [119] describes the reciprocal Cauchy transform and the Voiculescu transform of probability measures of finite variance, and also contains the exact form of the modified free Poisson distribution in Example 3.3.10.

A fine analysis of the analytic aspects of the free central limit theorem was given in [19]. Let $a_1, a_2, \dots$ be a sequence of identically distributed noncommutative random variables in free relation, and let μ_n be the distribution measure of $(a_1 + a_2 + \cdots + a_n)/\sqrt{n}$. Then it turns out that the support of μ_n is an interval $[a_n, b_n]$, and, for n large, the density of μ_n is analytic and tends uniformly to the semicircle.

We did not treat in detail the multiplicative free convolution introduced in [198] when the definition appeared in Sec. 2.4. The papers [17] and [18] treated also the characterizations of infinitely divisible distributions on the circle $\mathbb{T}$ and on $\mathbb{R}_+$ with respect to the multiplicative free convolution.

Example 3.2.3 gives the distribution of $p + q - \mathbf{1}$ when p, q are a free pair of projections having the expectation $\varphi(p) = \varphi(q) = \alpha$. The distributions of some linear combinations of a free family of projections were computed in [3].

The notion of compound free Poisson distributions was introduced in [178] in the framework of operator-valued (or amalgamated) free probability theory. Propositions 3.3.11 and 3.3.12 are given there. For the classical compound Poisson distribution, see [76]. The free Poisson distribution (or the Marchenko-Pastur distribution) in Example 3.3.5 as well as its compound generalization will be represented in the next chapter as the limiting eigenvalue distribution of certain random matrices.

The papers [172], [40] and [91] are about free white noise.

The differential equation (a complex Burger equation) in Lemma 3.3.9 (also the remark after the proof of Theorem 3.3.6) was found by Voiculescu. In the case of $\mu_t = w_{2\sqrt{t}}$ the equation is $\frac{\partial G}{\partial t} + G\frac{\partial G}{\partial z} = 0$, which can be regarded as a free analogue of the heat equation. In the language of this chapter Pastur studied the free convolution $\mu \boxplus w_r$, and he called it the *deformed Wigner law*. For example, a functional equation $G_1(z) = G_0(z - G_1(z))$ was deduced for the Cauchy transform G_1 of the free convolution, where G_0 denotes the Cauchy transform of μ (cf. (B.31) in the book [146]). This functional equation is easily obtained following the proof of Lemma 3.3.9. Let us start with the equation $K(0,z) = K(t,z) - t\phi(z)$, put $G(t,z)$ in place of z and apply $G(0,z)$ to both sides of the equation. In this way we arrive at the equation $G(t,z) = G(0, z - t\phi(G(t,z)))$. For $t = 1$ and $\phi(z) = z$ this is Pastur's relation.

Chapter 4

Random Matrices and Asymptotically Free Relation

Random matrices have been a part of advanced multivariate statistical analysis since the end of the 1920's. When Eugene Wigner proposed in quantum mechanics to replace the selfadjoint Hamiltonian operator in an infinite dimensional Hilbert space by an ensemble of very large Hermitian matrices, and the statistics of experimentally measured energy levels of nuclei were explained in terms of the eigenvalues of random matrices, random matrix theory entered physics.

The space of $n \times n$ random matrices admits a natural linear functional τ_n, defined as

$$\tau_n(X) = \frac{1}{n}\sum_{i=1}^{n} E(X_{ii})$$

for an arbitrary random matrix X (whenever the expectations exist). The random matrices whose entries have all moments form a noncommutative probability space with the tracial functional τ_n. This is the space under study in this chapter, in particular in the limit as $n \to \infty$.

The classical Wigner theorem tells us that the mean eigenvalue density of random symmetric matrices tends to the semicircle law if the matrix size goes to infinity. This result says, in our language, that the distribution of random symmetric matrices converges to the semicircle law. The concept of convergence in distribution appears here. Not only the semicirle law admits a random matrix model; Wishart matrices converge to the Marchenko-Pastur distribution law. Beyond the above mentioned matrix ensembles, there exist further examples, and they all produce particular distributions as the asymptotic eigenvalue density; in particular, random unitaries play important roles. Furthermore, it is noteworthy that the eigenvalue densities of these random matrix ensembles are known to converge not only in expectation but in the almost sure sense for the empirical density. It will be the subject of the next chapter to show that the above mentioned matrix ensembles provide a rich ground for large deviation theorems.

The Wigner theorem is not the only contact point to the semicircle law; in Chap. 2 this law appeared in the free central limit theorem. Hence random matrices have to do with the free relation. The very aim of this chapter is to show that the really pure algebraic concept of free relation of noncommutative random variables can also be modeled by random matrix ensembles if the matrix size tends to infinity. If $U(n)$ and $V(n)$ are independent $n \times n$ random unitaries distributed according to the Haar measure, then they are free asymptotically; that is, the relations of freeness hold not for a finite n but rather in the limit when n goes to infinity. Another example is supplied by a combination of diagonal and symmetric (or selfadjoint) random matrices. If $D(n)$ is a sequence of $n \times n$ diagonal random matrices with the asymptotic spectral density μ and $T(n)$ is a sequence of independent symmetric Gaussian matrices whose asymptotic eigenvalue distribution is the semicircle law w_2, then the sequences $D(n)$ and $T(n)$ are asymptotically free and the asymptotic eigenvalue distribution of $D(n) + T(n)$ is the free convolution $\mu \boxplus w_2$. This fact, recognized by Voiculescu, explains why the semicircle law shows up in two apparently different settings.

Voiculescu proved his asymptotic freeness results first for standard Gaussian matrices, and then he turned to standard unitary matrices together with constant matrices. His difficult proofs can be simplified by working directly with unitary matrices. This approach enables us to prove the results for rather general unitarily invariant random matrix models (together with constant matrices). Moreover, we discuss almost sure convergence, improving Voiculescu's original results.

It is worth emphasizing that the asymptotic freeness of symmetric Gaussian matrices is a far-reaching extension of the Wigner theorem. The latter tells about the limit of the moments of a single Gaussian matrix, and the former describes the limit of joint moments of several independent Gaussian matrices.

4.1 Random matrices and their eigenvalues

In this section we introduce the most frequently used Gaussian matrix ensembles, compute their eigenvalue densities, and study the asymptotics as the matrix size tends to infinity. The selfadjoint Gaussian matrix with independent entries provides a model for the semicircle law. The free Poisson distribution is the limiting eigenvalue distribution of the Wishart matrix. The method of moments and the use of the Fourier transform will be shown.

Although randon matrices were encountered in multivariate mathematical statistics in the 30's, intensive interest in the subject began with the work of Wigner in nuclear physics. Wigner's program was pursued by Porter and Rosenzweig, by Mehta and by Dyson. It seems that today the interest in random matrices is stronger in physics than in mathematics.

First of all, we need to fix some general notation. Let $M_n(\mathbb{C})$ be the space of $n \times n$ complex matrices. We denote by tr_n the normalized trace on $M_n(\mathbb{C})$, and by Tr the usual (non-normalized) trace on matrices. We write $M_n(\mathbb{R})^{sa}$ for the space of $n \times n$ real symmetric matrices and $M_n(\mathbb{C})^{sa}$ for the space of $n \times n$ complex selfadjoint matrices.

Given a probability space $(\Omega, \mathbf{Prob})$, we shall treat random matrices X whose entries X_{ij} have all moments or belong to $\bigcap_{1 \le p < \infty} L^p(\Omega, \mathbf{Prob})$. The set of those $n \times n$ random matrices forms an algebra, which is actually a *-algebra. It becomes a noncommutative probability space endowed with the tracial state

$$\tau_n(X) := \frac{1}{n} \sum_{i=1}^{n} E(X_{ii}) = E(\mathrm{tr}_n(X)) .$$

We shall mostly speak of random matrices without specifying the underlying probability space.

An $n \times n$ random matrix X has n (random) eigenvalues $\lambda_1(X)$, $\lambda_2(X), \dots,$ $\lambda_n(X)$. The random atomic measure

$$\frac{1}{n}\big(\delta(\lambda_1(X)) + \delta(\lambda_2(X)) + \cdots + \delta(\lambda_n(X))\big)$$

is called the *empirical eigenvalue distribution* of X, and the expectation value

$$\mu_X := \frac{1}{n} E\big(\delta(\lambda_1(X)) + \delta(\lambda_2(X)) + \cdots + \delta(\lambda_n(X))\big)$$

is the *mean eigenvalue distribution* of X. The empirical eigenvalue distribution is a random probability measure on $\mathbb{C}$, and in probability theory such a measure is often regarded as a distribution on the space of probability measures.

In case of a symmetric real (or selfadjoint complex) $n \times n$ random matrix T, the real eigenvalues are usually ordered increasingly, $\lambda_1(T) \le \lambda_2(T) \le \ldots \le \lambda_n(T)$. So λ can be regarded as a mapping

$$\lambda : T \mapsto (\lambda_1(T), \lambda_2(T), \dots, \lambda_n(T)) \in \mathbb{R}^n_{\le} ,$$

where

$$\mathbb{R}^n_{\le} := \{(x_1, x_2, \dots, x_n) \in \mathbb{R}^n : x_1 \le x_2 \le \ldots \le x_n\} .$$

The moments of the mean eigenvalue distribution are identical to those of T with respect to τ_n; that is,

$$\int x^m \, d\mu_T(x) = \tau_n(T^m) \qquad (m \in \mathbb{N}). \tag{4.1.1}$$

Example 4.1.1 By an $n \times n$ *standard symmetric Gaussian matrix* we mean a symmetric matrix $T(n) = [T_{ij}(n)]$ such that $\{T_{ij}(n) : 1 \le i \le j \le n\}$ is a family of independent real Gaussian random variables and

$$E(T_{ij}(n)) = 0\,, \quad E(T_{ij}(n)^2) = \frac{1+\delta_{ij}}{n+1}\,.$$

The variances of the entries $T_{ij}(n)$ are chosen such a way that $\tau_n(T(n)^2) = 1$. With respect to the Lebesgue measure

$$dT := \prod_{i \le j} dT_{ij} \quad \text{on} \quad M_n(\mathbb{R})^{sa} \cong \mathbb{R}^{n(n+1)/2}\,, \tag{4.1.2}$$

the density of a standard symmetric Gaussian is

$$p(T) := C_n \exp\Big(-\frac{n+1}{4}\mathrm{Tr}\,T^2\Big), \tag{4.1.3}$$

where

$$C_n = \Big(\frac{4\pi}{n+1}\Big)^{-n/2}\Big(\frac{2\pi}{n+1}\Big)^{-n(n-1)/4} = 2^{-n/2}\Big(\frac{2\pi}{n+1}\Big)^{-n(n+1)/4}. \tag{4.1.4}$$

The measure $p(T)\,dT$ is called the *standard Gaussian measure* on the space of symmetric matrices. In the literature the name *Gaussian orthogonal ensemble* (GOE) is also used.

□

It is remarkable that this measure is invariant under the transformation $T \mapsto OTO^t$ if O is an orthogonal matrix with transpose O^t. In fact, the compact orthogonal group $\mathcal{O}(n)$ has an action $T \mapsto OTO^t$ on the space $M_n(\mathbb{R})^{sa}$. The orthogonal invariance of the standard Gaussian measure follows from the invariance of the measure (4.1.2) and the similarity invariance of the trace.

As it was shown above, the standard Gaussian measure is invariant under the orthogonal transformation and the matrix elements are independent with respect to it. These important features almost characterize the standard Gaussian measure. Namely, any orthogonally invariant measure $q(T)\,dT$ possessing independence of entries is in the form

$$q(T) = \exp\big(-a\mathrm{Tr}\,T^2 + b\mathrm{Tr}\,T + c\big)$$

with certain constants a, b, c ($a, b, c \in \mathbb{R}$, $a > 0$) (cf. [154]).

Let μ be a measure on the space of $n \times n$ real symmetric matrices such that it has a density. The measure μ is invariant under the the action of the orthogonal group $\mathcal{O}(n)$ if and only if the density may be expressed in terms of the eigenvalues:

$$p(T_{11}, T_{12}, \dots, T_{1n}, T_{22}, \dots, T_{2n}, \dots, T_{nn}) = g(\lambda_1, \lambda_2, \dots, \lambda_n)\,,$$

where $\lambda_1, \lambda_2, \ldots, \lambda_n$ is the sequence of increasingly ordered eigenvalues which belongs to $\mathbb{R}^n_\leq$. There exists a measure $\bar{\mu}$ on $\mathbb{R}^n_\leq$ whose inverse image under the mapping λ is the original measure μ. We want to find out the relation between the densities of μ and $\bar{\mu}$, and we consider first the Lebesgue measure (4.1.2).

Lemma 4.1.2 *The measure on* $\mathbb{R}^n_\leq$ *(the space of eigenvalues) induced from the measure* (4.1.2) *has the density*

$$\frac{\pi^{n(n+1)/4}}{\prod_{j=1}^n \Gamma(j/2)} \prod_{i<j} |\lambda_i - \lambda_j| \, .$$

Proof: We recall the diagonalization $T = ODO^t$, where $D := \mathbf{Diag}(\lambda_1, \lambda_2, \ldots, \lambda_n)$ is a diagonal matrix and O is an orthogonal matrix. When $\lambda_1 < \lambda_2 < \ldots < \lambda_n$ (the other degenerate case is of measure 0 and so it is negligible), O is unique if we assume that its first row is nonnegative. Differentiate $T = ODO^t$ and write

$$\begin{aligned} dT &= dO \cdot DO^t + O \cdot dD \cdot O^t + OD \cdot dO^t \\ &= (dD + D \cdot dO^t \cdot O + O^t \cdot dO \cdot D)O^t \\ &= O(dD + D \cdot dM - dM \cdot D)O^t \, , \end{aligned}$$

where we take an infinitesimal matrix $dM := -O^t \cdot dO$ so that $O^tO = I$ gives $dM = dO^t \cdot O = -dM^t$ (i.e., dM is skew-symmetric). From the orthogonal invariance of the measure (4.1.2) we have

$$\prod_{i\leq j} dT_{ij} = \prod_{i \leq j} (dD + D \cdot dM - dM \cdot D)_{ij} = \prod_{i<j} |\lambda_i - \lambda_j| \prod_{i=1}^n d\lambda_i \prod_{i<j} dM_{ij} \, .$$

We may integrate this with respect to $\prod_{i<j} dM_{ij}$ to conclude that the induced measure on $\mathbb{R}^n_\leq$ is

$$C'_n \prod_{i<j} |\lambda_i - \lambda_j| \prod_{i=1}^n d\lambda_i \, .$$

Consider Example 4.1.1 to determine C'_n. The density of the standard Gaussian measure in (4.1.3) is

$$p(T) = C_n \exp\Bigl(-\frac{n+1}{4} \sum_{i=1}^n \lambda_i^2\Bigr) .$$

Hence the following must hold:

$$\frac{C_n C'_n}{n!} \int_{\mathbb{R}^n} \exp\Bigl(-\frac{n+1}{4} \sum_{i=1}^n \lambda_i^2\Bigr) \prod_{i<j} |\lambda_i - \lambda_j| \, d\lambda = 1 \, ,$$

where a factor $n!$ arises because the integration is over the whole $\mathbb{R}^n$ instead of $\mathbb{R}^n_{\leq}$. Below we shall see how to evaluate such an integral, and the constant $C_n C'_n$ is obtained in this way. Since C_n is known to us from (4.1.4), the proof may be concluded.

□

Here we interrupt the discussion on random matrices for a while and treat integrals of

$$\Delta(x) \equiv \Delta(x_1, x_2, \ldots, x_n) := \prod_{i<j} (x_i - x_j) \,. \tag{4.1.5}$$

Such integrals appear often when we pass from a random matrix to its eigenvalue density as in the previous lemma. In 1967 Mehta conjectured that

$$\int_{\mathbb{R}^n} e^{-\|x\|^2/2} |\Delta(x)|^{2\beta} \, dx = (2\pi)^{n/2} \prod_{j=1}^{n} \frac{\Gamma(1+j\beta)}{\Gamma(1+\beta)} \,,$$

where β is a complex number with $\operatorname{Re}\beta > 0$, $dx \equiv dx_1 dx_2 \cdots dx_n$ and $\|x\|^2 := \sum_{i=1}^n x_i^2$. It turned out that his conjecture follows from the *Selberg integral* (see [121], Sec. 17.1) obtained in 1944:

$$\begin{aligned} &\int_0^1 \cdots \int_0^1 |\Delta(y)|^{2\beta} \prod_{i=1}^{n} y_i^{a-1} (1-y_i)^{b-1} \, dy \\ &\qquad = \prod_{j=0}^{n-1} \frac{\Gamma(1+(j+1)\beta)\Gamma(a+j\beta)\Gamma(b+j\beta)}{\Gamma(1+\beta)\Gamma(a+b+(n+j-1)\beta)} \,, \end{aligned} \tag{4.1.6}$$

where $\operatorname{Re} a > 0$, $\operatorname{Re} b > 0$ and $\operatorname{Re}\beta > -\min\{1/n, \operatorname{Re} a/(n-1), \operatorname{Re} b/(n-1)\}$. Indeed, we replace y_i by $1/2 + x_i/2L$ and choose $a = b = \alpha L^2/2 + 1$ ($\alpha > 0$) in (4.1.6) to get

$$\begin{aligned} &\int_{-L}^{L} \cdots \int_{-L}^{L} |\Delta(x)|^{2\beta} \prod_{i=1}^{n} \Big(1 - \frac{x_i^2}{L^2}\Big)^{\alpha L^2/2} dx \\ &\qquad = 2^{n(\alpha L^2 + (n-1)\beta + 1)} L^{n((n-1)\beta+1)} \prod_{j=0}^{n-1} \frac{\Gamma(1+(j+1)\beta)\Gamma(1+\alpha L^2/2+j\beta)^2}{\Gamma(1+\beta)\Gamma(2+\alpha L^2+(n+j-1)\beta)} \,. \end{aligned}$$

The Stirling formula

$$\frac{\Gamma(1+x)}{\sqrt{2\pi}\, x^{x+1/2} e^{-x}} \to 1 \quad \text{as} \quad x \to \infty$$

implies that as $L \to \infty$

$$\frac{\Gamma(1+\alpha L^2/2+j\beta)^2}{\Gamma(2+\alpha L^2+(n+j-1)\beta)}$$

$$\approx \frac{\sqrt{2\pi}(\alpha L^2/2 + j\beta)^{\alpha L^2+2j\beta+1} e^{-2j\beta}}{(\alpha L^2 + (n+j-1)\beta + 1)^{\alpha L^2+(n+j-1)\beta+3/2} e^{-((n+j-1)\beta+1)}}$$

$$\approx \frac{\sqrt{2\pi}(\alpha L^2/2)^{\alpha L^2+2j\beta+1}}{(\alpha L^2)^{\alpha L^2+(n+j-1)\beta+3/2}} = \sqrt{2\pi}\, 2^{-(\alpha L^2+2j\beta+1)} (\alpha L^2)^{-((n-j-1)\beta+1/2)} .$$

Hence the limit as $L \to \infty$ yields

$$\int_{\mathbb{R}^n} e^{-\alpha\|x\|^2/2} |\Delta(x)|^{2\beta}\, dx = (2\pi)^{n/2} \alpha^{-n((n-1)\beta+1)/2} \prod_{j=1}^{n} \frac{\Gamma(1+j\beta)}{\Gamma(1+\beta)}, \tag{4.1.7}$$

which is another form of the Selberg integral. In the case $\beta = 1/2$ (or $\beta = 1$) this formula is related to the real symmetric (or complex selfadjoint) Gaussian matrices. (General $\beta > 0$ will appear in Chap. 5 in some large deviation theorems.)

Similarly, we replace y_i by x_i/L and choose $b = L + 1$ in (4.1.6); then the limit as $L \to \infty$ yields

$$\int_{(\mathbb{R}^+)^n} \prod_{i=1}^{n} x_i^{a-1} \exp\Bigl(-\sum_{i=1}^{n} x_i\Bigr) |\Delta(x)|^{2\beta}\, dx = \prod_{j=0}^{n-1} \frac{\Gamma(1+(j+1)\beta)\Gamma(a+j\beta)}{\Gamma(1+\beta)}, \tag{4.1.8}$$

whenever $\operatorname{Re} a > 0$ and $\operatorname{Re} \beta > 0$.

The next proposition follows directly from Lemma 4.1.2.

Proposition 4.1.3 *If an $n \times n$ random symmetric matrix T has a probability density $g(\lambda_1, \lambda_2, \dots, \lambda_n)$ with respect to $\prod_{i\le j} dT_{ij}$ which is expressed in terms of the increasingly ordered eigenvalues $\lambda_1, \lambda_2, \dots, \lambda_n$, then the joint probability density function of the eigenvalues of T on $\mathbb{R}^n_{\le}$ is*

$$\frac{\pi^{n(n+1)/4}}{\prod_{j=1}^{n} \Gamma(j/2)} g(\lambda_1, \lambda_2, \dots, \lambda_n) \prod_{i<j} |\lambda_i - \lambda_j| .$$

In particular, let $T(n)$ be the $n \times n$ standard symmetric Gaussian matrix. On the convex domain $\mathbb{R}^n_{\le} \subset \mathbb{R}^n$ the *density of the eigenvalues* of $T(n)$ is

$$\frac{2^{-n(n+3)/4}(n+1)^{n(n+1)/4}}{\prod_{j=1}^{n} \Gamma(j/2)} \exp\Bigl(-\frac{n+1}{4} \sum_{i=1}^{n} \lambda_i^2\Bigr) \prod_{i<j} |\lambda_i - \lambda_j| .$$

If the density is taken on the whole $\mathbb{R}^n$, then it is

$$\frac{2^{-n(n+3)/4}(n+1)^{n(n+1)/4}}{n! \prod_{j=1}^{n} \Gamma(j/2)} \exp\Bigl(-\frac{n+1}{4} \sum_{i=1}^{n} \lambda_i^2\Bigr) \prod_{i<j} |\lambda_i - \lambda_j| . \tag{4.1.9}$$

Writing p_n for the density (4.1.9), we have

$$\begin{aligned}\tau_n(T(n)^m) &= \frac{1}{n}\int_{\mathbb{R}^n} (x_1^m + x_2^m + \cdots + x_n^m) p_n(x_1, x_2, \ldots, x_n)\, dx \\ &= \int_{\mathbb{R}^n} x_1^m p_n(x_1, x_2, \ldots, x_n)\, dx\end{aligned}$$

bacause of the symmetry. Now, comparing this with (4.1.1), we see that the probability density function of $\mu_{T(n)}$ is

$$\sigma_n(t) := \int_{-\infty}^{\infty} \cdots \int_{-\infty}^{\infty} p_n(t, x_2, x_3, \ldots, x_n)\, dx_2\, dx_3 \cdots dx_n\,.$$

It is a form of the *Wigner theorem* that $\sigma_n(t)$ tends to the density of the semicircle law w_2 as $n \to \infty$. In other words, the mean eigenvalue distribution of standard symmetric Gaussian matrices is the standard semicircle law asymptotically. We shall use the name Wigner theorem for more general statements as well when the limiting eigenvalue distribution of random matrices of independent entries becomes the semicircle law. The asymptotics of $\sigma_n(t)$ can be treated by very powerful methods from the theory of orthogonal polynomials; however, below we take another route and apply the method of moments.

Theorem 4.1.4 *For $n \in \mathbb{N}$ let $\xi_{ij}(n)$ $(1 \le i \le j \le n)$ be independent real-valued random variables with finite moments. Assume that $E(\xi_{ij}(n)) = 0$ and $E(\xi_{ij}(n)^2) = 1$ for $1 \le i < j \le 1$. If*

$$C_k(n) := \sup_{1 \le i \le j \le n} E(|\xi_{ij}(n)|^k) = O(1) \quad as \quad n \to \infty \qquad (k \in \mathbb{N}),$$

then the mean eigenvalue distribution of the random symmetric matrix $T(n)$ defined by $T_{ij}(n) = \xi_{ij}(n)/\sqrt{n}$ tends to the standard semicircle law w_2 as $n \to \infty$.

Proof: We put ξ_{ij} in place of $\xi_{ij}(n)$ for the sake of simpler writing, and use the convention $\xi_{ji} = \xi_{ij}$. The kth moment of the mean eigenvalue distribution is given as

$$\tau_n(T(n)^k) = \frac{1}{n^{k/2+1}} \sum E(\xi_{m_1 m_2} \xi_{m_2 m_3} \cdots \xi_{m_k m_1})\,, \tag{4.1.10}$$

where the summation is over all $1 \le m_1, m_2, \ldots, m_k \le n$. We shall prove by a combinatorial argument that the limit of the kth moment of the mean eigenvalue distributions is the known kth moment of the semicircle law. We shall rely on the fact that the kth moment of the standard semicircle law equals the number of non-crossing pair partitions of $[k]$ (see Lemma 2.2.1).

We group the terms $E(\xi_{m_1 m_2} \xi_{m_2 m_3} \cdots \xi_{m_k m_1})$ according to the cardinality of the set $\{m_1, m_2, \ldots, m_k\}$. The number of terms such that $\#\{m_1, m_2, \ldots, m_k\} = l$

is less than

$$\binom{n}{l} l^k,$$

and due to the Hölder inequality

$$\begin{aligned}
&|E(\xi_{m_1m_2}(n)\cdots\xi_{m_km_1}(n))| \\
&\qquad \le E(|\xi_{m_1m_2}(n)|^k)^{1/k}\cdots E(|\xi_{m_km_1}(n)|^k)^{1/k} \le C_k(n)\,. \qquad (4.1.11)
\end{aligned}$$

Hence for the sum we have

$$\frac{1}{n^{k/2+1}}\Big|\sum_{(l)} E(\xi_{m_1m_2}\xi_{m_2m_3}\cdots\xi_{m_km_1})\Big| \le \binom{n}{l}\frac{l^kC_k(n)}{n^{k/2+1}}\,,$$

which goes to 0 whenever $l < k/2+1$. On the other hand,

$$E(\xi_{m_1m_2}\xi_{m_2m_3}\cdots\xi_{m_km_1}) = 0$$

if $\#\{m_1,m_2,\dots,m_k\} > k/2+1$, because in this case there must be a factor $\xi_{m_im_{i+1}}$ (with $m_{k+1} := m_1$) such that $m_i \ne m_{i+1}$ which appears without repetition. This fact can be shown by induction on k. Indeed, there is a subscript p such that m_p is without repetition in the sequence $m_1,m_2,\dots,m_k$. When $m_{p-1}\ne m_{p+1}$ either $\xi_{m_{p-1}m_p}$ or $\xi_{m_pm_{p+1}}$ is a desired factor, where the neighbors of m_p is understood in the cyclic order (i.e. mod k) if $p=1$ or $p=k$. When $m_{p-1}=m_{p+1}$ we may apply the induction assumption to a shorter sequence obtained by removing m_{p-1},m_p from $m_1,m_2,\dots,m_k$. In particular, for an odd k the limit of (4.1.10) is 0. In the rest of our argument we consider even k and replace k by $2k$.

We have to show that

$$\frac{1}{n^{k+1}}\sum_{(a),(b)} E(\xi_{m_1m_2}\xi_{m_2m_3}\cdots\xi_{m_{2k}m_1}) \to \frac{1}{k+1}\binom{2k}{k} \quad \text{as} \quad n\to\infty, \qquad (4.1.12)$$

when the summation is over all $1\le m_1,m_2,\dots,m_{2k}\le n$ satisfying the following:

(a) $\#\{m_1,m_2,\dots,m_{2k}\} = k+1$, and

(b) every element of the sequence

$$\{m_1,m_2\},\{m_2,m_3\},\dots,\{m_{2k},m_1\} \qquad (4.1.13)$$

appears at least twice.

In order to prove (4.1.12), we show that if the sequence $m_1,m_2,\dots,m_{2k}$ satisfies (a) and (b), then $m_i\ne m_{i+1}$ for $1\le i\le 2k$ (with $m_{2k+1}:=m_1$), each (unordered) pair appears in (4.1.13) exactly twice, and the pair partition $\mathcal{V}$ of $[2k]$ defined as $\{i,j\} \in \mathcal{V} \Leftrightarrow \{m_i, m_{i+1}\} = \{m_j, m_{j+1}\}$ is non-crossing. This can be done by induction on k. The case $k = 1$ is obvious. Suppose that our statement is true for $k-1$. There is a subscript p such that m_p is without repetition in the sequence $m_1, m_2, \dots, m_{2k}$. Then $m_{p-1} = m_{p+1} \neq m_p$ must hold. Removing m_{p-1}, m_p from $m_1, m_2, \dots, m_{2k}$, we get a shorter sequence $n_1, n_2, \dots, n_{2k-2}$ which contains k different elements. Then the induction assumption can be applied to this shorter sequence, and the pair partition combined by $\{p-1, p\}$ and the non-crossing one corresponding to the shorter sequence is again non-crossing. Hence the statement is true for k as well. Conversely, for any non-crossing pair partition $\mathcal{V}$ of $\{1, 2, \dots, 2k\}$ there are sequences $m_1, m_2, \dots, m_{2k}$ which induce $\mathcal{V}$ as above, and the number of such sequences is $n(n-1)\cdots(n-k)$ for each $\mathcal{V}$. (Prove by induction.)

We thus have

$$\frac{1}{n^{k+1}} \sum_{(a),(b)} E(\xi_{m_1 m_2} \xi_{m_2 m_3} \cdots \xi_{m_{2k} m_1}) = \frac{1}{k+1} \binom{2k}{k} \frac{n(n-1)\cdots(n-k)}{n^{k+1}},$$

which proves (4.1.12).

□

The following interpretation of Theorem 4.1.4 is useful for us. For every $n \in \mathbb{N}$ the random matrix $T(n)$ is a noncommutative random variable with a certain distribution, which is formally a linear functional φ_n on the polynomial algebra $\mathbb{C}\langle X \rangle$ (see Sec. 1.2). If $n \to \infty$, then φ_n goes to the semicircular distribution. In particular, the standard symmetric Gaussian matrices constitute a *random matrix model* for the semicircular distribution.

An $n \times n$ matrix is a noncommutative random variable with respect to the normalized trace functional tr_n. The sequence of standard symmetric Gaussian matrices forms a model for the semicircle law in the almost sure sense: For almost all realizations of the sequence of those matrices we have convergence in distribution to the semicircle law. This is the content of our next result.

Theorem 4.1.5 *Let $T(n) = [\xi_{ij}(n)/\sqrt{n}]$ be as in the previous theorem, with the same assumption. Then the empirical eigenvalue distribution of $T(n)$ converges in distribution almost surely to the semicircle law w_2 as $n \to \infty$; that is,*

$$\lim_{n\to\infty} \mathrm{tr}_n(T(n)^k) = \frac{1}{2\pi} \int_{-2}^{2} x^k \sqrt{4 - x^2}\, dx$$

almost everywhere for every $k \in \mathbb{N}$.

Proof: It is sufficient to show that

$$E\Big(\sum_{n=1}^{\infty} [\mathrm{tr}_n(T(n)^k) - \tau_n(T(n)^k)]^2\Big) < +\infty$$

for $k \in \mathbb{N}$. Indeed, this condition implies that $\sum_{n=1}^{\infty}[\mathrm{tr}_n(T(n)^k) - \tau_n(T(n)^k)]^2$ is finite almost everywhere, and therefore $\mathrm{tr}_n(T(n)^k) - \tau_n(T(n)^k)$ converges to 0 almost surely.

We write

$$\begin{aligned}
&E\big([\mathrm{tr}_n(T(n)^k) - \tau_n(T(n)^k)]^2\big) \\
&\quad = E\big([\mathrm{tr}_n(T(n)^k)]^2\big) - [\tau_n(T(n)^k)]^2 \\
&\quad = \frac{1}{n^{k+2}} \sum Q_n(m_1, \ldots, m_k; m_{k+1}, \ldots, m_{2k}),
\end{aligned}$$

where the summation is over all $1 \le m_1, \ldots, m_k, m_{k+1}, \ldots, m_{2k} \le n$ and

$$\begin{aligned}
&Q_n(m_1, \ldots, m_k; m_{k+1}, \ldots, m_{2k}) \\
&\quad := E\big(\xi_{m_1 m_2}\xi_{m_2 m_3}\cdots\xi_{m_k m_1}\xi_{m_{k+1}m_{k+2}}\xi_{m_{k+2}m_{k+3}}\cdots\xi_{m_{2k}m_{k+1}}\big) \\
&\qquad - E\big(\xi_{m_1 m_2}\xi_{m_2 m_3}\cdots\xi_{m_k m_1}\big)E\big(\xi_{m_{k+1}m_{k+2}}\xi_{m_{k+2}m_{k+3}}\cdots\xi_{m_{2k}m_{k+1}}\big)
\end{aligned}$$

(ξ_{ij} being short for $\xi_{ij}(n)$). As in (4.1.11) we get the estimate

$$|Q_n(m_1, \ldots, m_k; m_{k+1}, \ldots, m_{2k})| \le C_{2k}(n) + C_k(n)^2 = O(1) \tag{4.1.14}$$

as $n \to \infty$.

In order that $Q_n(m_1, \ldots, m_k; m_{k+1}, \ldots, m_{2k})$ be nonzero, it is necessary that every factor ξ_{ij} $(i \ne j)$ of

$$\xi_{m_1 m_2}\xi_{m_2 m_3}\cdots\xi_{m_k m_1}\xi_{m_{k+1}m_{k+2}}\xi_{m_{k+2}m_{k+3}}\cdots\xi_{m_{2k}m_{k+1}}$$

must have an equal pair, and at least one of the first k factors has its pair among the last k factors. Indeed, if there is a ξ_{ij} $(i \ne j)$ which appears only once, then $E(\xi_{ij}) = 0$ is factored out from both terms, and each of the two terms vanishes. If none of the first k factors ξ_{ij} appears among the second k, then

$$\begin{aligned}
&E\big(\xi_{m_1 m_2}\cdots\xi_{m_k m_1}\xi_{m_{k+1}m_{k+2}}\cdots\xi_{m_{2k}m_{k+1}}\big) \\
&\quad = E\big(\xi_{m_1 m_2}\cdots\xi_{m_k m_1}\big)E\big(\xi_{m_{k+1}m_{k+2}}\cdots\xi_{m_{2k}m_{k+1}}\big)
\end{aligned}$$

due to the independence of the ξ_{ij}'s. Therefore, there are $p \in \{1, \ldots, k\}$ and $q \in \{k+1, \ldots, 2k\}$ such that $\{m_p, m_{p+1}\} = \{m_q, m_{q+1}\}$, where m_{p+1} and m_{q+1} are taken in the cyclic order of $(m_1, \ldots, m_k)$ and $(m_{k+1}, \ldots, m_{2k})$, respectively.

Given a nonzero term, we can associate a partition $\mathcal{V}$ of $[2k]$, $p \in \{1, \ldots, k\}$, $q \in \{k+1, \ldots, 2k\}$ and a mapping $\pi : [2k] \to \{\pm 1\}$ such that

(1) $m_i = m_{i+1}$ if $\{i\}$ is a singleton block of $\mathcal{V}$,

(2) $\{m_i, m_{i+1}\} = \{m_j, m_{j+1}\}$ if i and j belong to the same block of $\mathcal{V}$,

(3) $m_i \le m_{i+1}$ if $\pi(i) = 1$ and $m_i \ge m_{i+1}$ if $\pi(i) = -1$,

(4) $\{m_p, m_{p+1}\} = \{m_q, m_{q+1}\}$.

For any quadruple $(\mathcal{V}, \pi; p, q)$ as above we denote by $\Xi_n(\mathcal{V}, \pi; p, q)$ the set of all $(m_1, \dots, m_{2k}) \in [n]^{2k}$ satisfying (1)–(4).

What we want to show is

$$\#\Xi_n(\mathcal{V}, \pi; p, q) \le n^k. \tag{4.1.15}$$

For this we may assume, in view of the cyclicity of $(m_1, \dots, m_k)$ and $(m_{k+1}, \dots, m_{2k})$, that $p = 1$ and $q = k+1$, so that $m_1 = m_{k+1}$ and $m_2 = m_{k+2}$. It is convenient to consider the graph G with $2k-2$ vertices $1\,(=k+1), 2\,(=k+2)$, $3, \dots, k, k+3, \dots, 2k$ and $2k$ edges $[1,2], \dots, [k-1,k], [k,1], [k+1,k+2], \dots, [2k-1, 2k], [2k, k+1]$. It has a double edge joining the vertices $1, 2$.

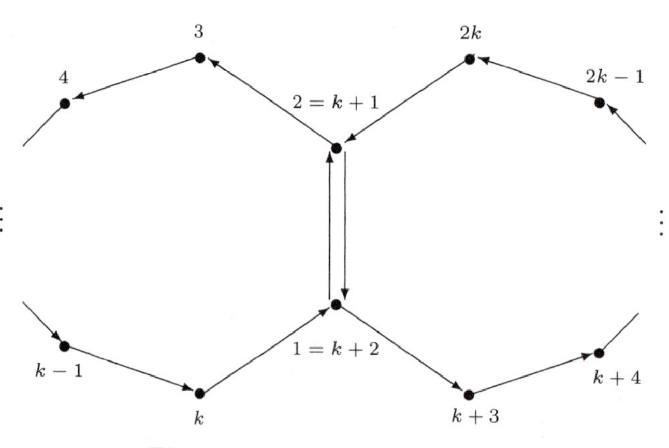

Fig. 4.1. Picture of the graph G.

Let $\tilde{G}$ be the quotient graph obtained by the following procedure: If $\{i\}$ is a singleton block of $\mathcal{V}$, delete the edge $[i, i+1]$ and identify the vertices i and $i+1$; if i and j belong to the same block of $\mathcal{V}$, identify the edges $[i, i+1]$ and $[j, j+1]$ with orientation preserved if $\pi(i)\pi(j) = 1$ and with orientation reversed if $\pi(i)\pi(j) = -1$.

Let l_0 be the number of singleton blocks of $\mathcal{V}$ and l_1 the number of other blocks of $\mathcal{V}$. Then $l_1 \le (2k - l_0)/2 \le k$, and the number of edges of $\tilde{G}$ is equal to l_1. If l_1 is less than k, then $\tilde{G}$ has at most k vertices (because $\tilde{G}$ is connected). Otherwise, $l_1 = k$ and $l_0 = 0$ (so $\mathcal{V}$ is a pair partition), and $\tilde{G}$ has a loop passing through the vertex 1. This implies again that $\tilde{G}$ still has at most k vertices. In this way (4.1.15) is proven, because the freedom for choosing $(m_1, \dots, m_{2k})$ from $\Xi_n(\mathcal{V}, \pi; p, q)$ subject to (1)–(4) is equal to the number of vertices of $\tilde{G}$.

From (4.1.14) and (4.1.15) we have

$$\begin{aligned} &E\big([\mathrm{tr}_n(T(n)^k) - \tau_n(T(n)^k)]^2\big) \\ &\quad = O(n^{-k-2}) \sum_{\mathcal{V}, \pi, p, q} \#\Xi_n(\mathcal{V}, \pi; p, q) = O(n^{-2}) \quad \text{as} \quad n \to \infty, \end{aligned}$$

implying that

$$E\Big(\sum_{n=1}^{\infty}[\mathrm{tr}_n(T(n)^k)-\tau_n(T(n)^k)]^2\Big)<+\infty,$$

which was our goal.

□

Functional analysis prefers complex matrices to real ones. An $n \times n$ complex selfadjoint random matrix $H(n)$ is called a *standard selfadjoint Gaussian matrix* if

(i) $\{\mathrm{Re}\, H_{ij}(n) : 1 \le i \le j \le n\} \cup \{\mathrm{Im}\, H_{ij}(n) : 1 \le i < j \le n\}$ is an independent family of Gaussian random variables, and

(ii) $E(H_{ij}(n)) = 0$ for $1 \le i \le j \le n$, $E(H_{ii}(n)^2) = 1/n$ for $1 \le i \le n$, and $E((\mathrm{Re}\, H_{ij}(n))^2) = E((\mathrm{Im}\, H_{ij}(n))^2) = 1/2n$ for $1 \le i < j \le n$.

We use the word "standard" because $\tau_n(H(n)) = 0$ and $\tau_n(H(n)^2) = 1$. The variance of the entries is chosen in such a way that the distribution of $H(n)$ on $M_n(\mathbb{C})^{sa}$ is invariant under the unitary conjugation $A \mapsto UAU^*$ for $U \in \mathcal{U}(n)$, where $\mathcal{U}(n)$ is the unitary group. This is clear also from the density function

$$C_n \exp\Big(-\frac{n}{2}\mathrm{Tr}\, A^2\Big) \quad \text{with} \quad C_n = 2^{-n/2}\Big(\frac{\pi}{n}\Big)^{-n^2/2} \tag{4.1.16}$$

with respect to the Lebesgue measure

$$dA := \prod_{i=1}^{n} dA_{ii} \prod_{i<j} d(\mathrm{Re}\, A_{ij})\, d(\mathrm{Im}\, A_{ij}) \quad \text{on} \quad M_n(\mathbb{C})^{sa} \cong \mathbb{R}^{n^2}. \tag{4.1.17}$$

The unitary invarince of the distribution is the reason for the nomenclature *Gaussian unitary ensemble* (GUE).

An analogue of Proposition 4.1.3 holds for complex selfadjoint random matrices. The next lemma is enough to show it. The proof is similar to that of Lemma 4.1.2 when we take the diagonalization $A = UDU^*$ ($U \in \mathcal{U}(n)$) and examine the density (4.1.16) making use of the Selberg integral (4.1.7).

Lemma 4.1.6 *The measure on* $\mathbb{R}^n_{\le}$ *(the space of eigenvalues) induced from the measure* (4.1.17) *has the density*

$$\frac{\pi^{n(n-1)/2}}{\prod_{j=1}^{n-1} j!}\prod_{i<j}(\lambda_i-\lambda_j)^2.$$

If the density is taken on the whole $\mathbb{R}^n$*, then* $\prod_{j=1}^{n-1} j!$ *in the above is replaced by* $\prod_{j=1}^{n} j!$.

In particular, when $H(n)$ is the standard selfadjoint Gaussian matrix, the joint probability *density of the eigenvalues* of $H(n)$ on $\mathbb{R}^n$ is

$$\frac{(2\pi)^{-n/2}n^{n^2/2}}{\prod_{j=1}^{n} j!}\exp\Big(-\frac{n}{2}\sum_{i=1}^{n}\lambda_i^2\Big)\prod_{i<j}(\lambda_i-\lambda_j)^2\,. \tag{4.1.18}$$

The standard selfadjoint Gaussian matrices provide another random matrix model for the semicircular distribution; in other words, the mean eigenvalue distribution goes to w_2 as the matrix size goes to infinity. First we present an analytic proof for this fact, and then the combinatorial background will be discussed.

We consider an $n\times n$ random matrix $H_0+H(n)$, where H_0 is a fixed selfadjoint matrix and $H(n)$ is a standard selfadjoint Gaussian matrix. The density of $H_0+H(n)$ with respect to the measure (4.1.17) is

$$C_{H_0}\exp\Big(-\frac{n}{2}\mathrm{Tr}\,A^2+n\mathrm{Tr}\,H_0A\Big)\,.$$

In the following computation, assume that H_0 has the eigenvalues $\delta_1>\delta_2>\ldots>\delta_n$. The Fourier transform of the mean eigenvalue distribution is

$$F(t):=\frac{C_{H_0}}{n}\int\mathrm{Tr}\,e^{\mathrm{i}\,tA}\exp\Big(-\frac{n}{2}\mathrm{Tr}\,A^2+n\mathrm{Tr}\,H_0A\Big)\,dA\,.$$

Since the measure is unitarily invariant, we can first integrate with respect to the Haar probability dU over $\mathcal{U}(n)$:

$$F(t)=\frac{C_{H_0}}{n}\int\mathrm{Tr}\,e^{\mathrm{i}\,tA}\exp\Big(-\frac{n}{2}\mathrm{Tr}\,A^2\Big)\Big(\int\exp(n\mathrm{Tr}\,UAU^*H_0)\,dU\Big)\,dA\,.$$

In this way we are in a position to use an integral formula (see [121], A.5 and also [21], Theorem 7.24 for a more general formula attributed to Harish-Chandra):

$$\int\exp\big(\mathrm{Tr}\,UAU^*B\big)\,dU=\frac{\det[\exp(\lambda_i\rho_j)]}{\Delta(\lambda)\Delta(\rho)}\,,$$

where A,B are $n\times n$ selfadjoint matrices, the λ_i's are the eigenvalues of A, the ρ_j's are those of B, and $\Delta(\lambda)\equiv\Delta(\lambda_1,\lambda_2,\ldots,\lambda_n)$ is in (4.1.5). So by Lemma 4.1.6,

$$\begin{aligned}F(t)&=\frac{C_{H_0}}{n}\int\mathrm{Tr}\,e^{\mathrm{i}\,tA}\exp\Big(-\frac{n}{2}\mathrm{Tr}\,A^2\Big)\frac{\det[\exp(n\lambda_i\delta_j)]}{\Delta(\lambda)\Delta(\delta)}\,dA\\&=\frac{C'_{H_0}}{n}\int\Big(\sum_j e^{\mathrm{i}\,t\lambda_j}\Big)\exp\Big(-\frac{n}{2}\sum_i\lambda_i^2\Big)\Delta(\lambda)^2\frac{\det[\exp(n\lambda_i\delta_j)]}{\Delta(\lambda)\Delta(\delta)}\,d\lambda\\&=\frac{C'_{H_0}}{n\Delta(\delta)}\int\Big(\sum_j e^{\mathrm{i}\,t\lambda_j}\Big)\exp\Big(-\frac{n}{2}\sum_i\lambda_i^2\Big)\Delta(\lambda)\det[\exp(n\lambda_i\delta_j)]\,d\lambda\,,\end{aligned}$$

where the latter integrals are over $\mathbb{R}^n_{\le}$. Integration over $\mathbb{R}^n$ gives a factor $n!$. We expand the determinant by summing over all permutations σ of $[n]$:

$$\frac{C'_{H_0}}{n!\,n\Delta(\delta)}\sum_\sigma \int \Big(\sum_j e^{\mathrm{i}\,t\lambda_j}\Big)\exp\Big(-\frac{n}{2}\sum_i \lambda_i^2\Big)\Delta(\lambda)(-1)^{|\sigma|}\prod_i \exp(n\lambda_i\delta_{\sigma(i)})\,d\lambda\,,$$

and all summands are the same, due to the fact that Δ changes sign when two of its arguments are exchanged. Hence we arrive at

$$F(t) = \frac{C'_{H_0}}{n\Delta(\delta)}\int \Big(\sum_j e^{\mathrm{i}\,t\lambda_j}\Big)\exp\Big(-\frac{n}{2}\sum_i \lambda_i^2 + n\sum_i \lambda_i\delta_i\Big)\Delta(\lambda)\,d\lambda\,.$$

Now we proceed by means of the integral

$$\int \exp\Big(-\frac{n}{2}\sum_i(\lambda_i - a_i)^2\Big)\Delta(\lambda_1,\lambda_2,\ldots,\lambda_n)\,d\lambda = C(n)\Delta(a_1,a_2,\ldots,a_n)$$

and conclude that

$$\begin{aligned} F(t) &= \frac{C''_{H_0}}{n}\exp\Big(\frac{n}{2}\sum_i \delta_i^2\Big)\exp\Big(-\frac{t^2}{2n}\Big) \\ &\quad\times\sum_j e^{\mathrm{i}\,\delta_j t}\frac{\Delta(\delta_1,\ldots,\delta_{j-1},\delta_j+\mathrm{i}\,t/n,\delta_{j+1},\ldots,\delta_n)}{\Delta(\delta_1,\delta_2,\ldots,\delta_n)} \\ &= \frac{1}{n}\exp\Big(-\frac{t^2}{2n}\Big)\sum_j e^{\mathrm{i}\,\delta_j t}\prod_{i\neq j}\frac{\delta_i-\delta_j-\mathrm{i}\,t/n}{\delta_i-\delta_j}, \end{aligned}$$

using the normalization $F(0) = 1$. Since

$$\prod_k \frac{z-\delta_k+\mathrm{i}\,t/n}{z-\delta_k} = 1 + \frac{\mathrm{i}\,t}{n}\sum_j \frac{1}{z-\delta_j}\prod_{i\neq j}\frac{\delta_i-\delta_j-\mathrm{i}\,t/n}{\delta_i-\delta_j},$$

the above sum can be represented by a contour integral on the complex plane:

$$F(t) = \frac{1}{\mathrm{i}\,t}\exp\Big(-\frac{t^2}{2n}\Big)\oint \frac{dz}{2\pi\mathrm{i}}e^{\mathrm{i}\,tz}\prod_k \frac{z-\delta_k+\mathrm{i}\,t/n}{z-\delta_k}$$

if the contour encloses all the eigenvalues δ_i. In this representation we can easily take the limit as $\delta_i \to 0$, and get

$$F_0(t) = \frac{1}{\mathrm{i}\,t}\exp\Big(-\frac{t^2}{2n}\Big)\oint \frac{dz}{2\pi\mathrm{i}}e^{\mathrm{i}\,tz}\Big(1+\frac{\mathrm{i}\,t}{nz}\Big)^n.$$

This integral is evaluated by means of the residue theorem, and the limit as $n \to \infty$ is

$$\lim_{n\to\infty} \exp\Big(-\frac{t^2}{2n}\Big) \sum_{k=0}^{n-1} \binom{n}{k+1} \frac{(\mathrm{i}\,t)^{2k}}{k!\,n^{k+1}} = \sum_{k=0}^{\infty} \frac{1}{k!\,(k+1)!} (\mathrm{i}\,t)^{2k}.$$

Since the moments of the semicircle law are known to us (see Lemma 2.3.1), we recognize that this is the Fourier transform of the semicircle law, which must be the limiting measure. In this way we have shown that the limit distribution of the standard selfadjoint Gaussian matrices is w_2. An explicit combinatorial formula for the moments of the standard selfadjoint Gaussian matrix leads to the same conclusion.

Non-crossing pair partitions of the set $\{1, 2, \ldots, 2k\}$ are in one-to-one correspondence with those permutations in the *symmetric group* $\mathbf{S}_{2k}$ which are of order two and do not have any fixed points. Let Γ_k be the set of all those permutations. For $\gamma \in \Gamma_k$ one defines an equivalence relation on $\{1, 2, \ldots, 2k\}$ by saying that $j \sim_\gamma \gamma(j) + 1$, where addition is mod $2k$. Let $d(\gamma)$ stand for the number of equivalence classes of $\sim_\gamma$. Voiculescu showed that $d(\gamma) \le k+1$, and Shlyakhtenko proved that $d(\gamma) = k + 1$ if and only if the partition corresponding to γ is non-crossing. (Both facts are easily proven by induction.) Now we are ready to state a formula for the *moments of the standard selfadjoint Gaussian matrix* $H(n)$:

$$\tau_n(H(n)^{2k}) = \sum_{\gamma \in \Gamma_k} n^{d(\gamma)-k-1} \tag{4.1.19}$$

If $n \to \infty$, then the limit is the number of non-crossing pair partitions.

The next theorem is a stronger result for more general selfadjoint random matrices.

Theorem 4.1.7 *For $n \in \mathbb{N}$ let $[\xi_{ij}(n)]$ be a selfadjoint random matrix such that the $\xi_{ij}(n)$ are independent for $1 \le i \le j \le n$. Assume that $E(\xi_{ij}(n)) = 0$ and $E(|\xi_{ij}(n)|^2) = 1$ for $1 \le i < j \le n$. Assume further that*

$$C_k(n) := \sup_{1\le i\le j\le n} E(|\xi_{ij}(n)|^k) = O(1) \quad as \quad n \to \infty \qquad (k \in \mathbb{N}).$$

Then the distribution of the random matrix $H(n)$ defined by $H_{ij}(n) := \xi_{ij}(n)/\sqrt{n}$ tends to the standard semicircle law w_2 with respect to the functional τ_n as $n \to \infty$. Moreover, the convergence in distribution holds also with respect to the functional tr_n almost everywhere; that is, the empirical eigenvalue distribution of $H(n)$ converges in distribution almost surely to w_2.

The proof of the first assertion of this theorem may follow the lines of the combinatorial proof of Theorem 4.1.4. When the ξ_{ij} are complex valued, the argument in the case where $\#\{m_1, m_2, \ldots, m_k\} < k/2 + 1$ or $> k/2 + 1$ is the same. But (4.1.12) is to be shown a bit more carefully. The summation is over

(a) and (b), which excludes the possibility of diagonal entries. It turned out that the property (b) holds true in the stronger form that every unordered pair $\{m_1, m_2\}, \{m_2, m_3\}, \ldots, \{m_{2k}, m_1\}$ appears exactly twice. However, the fact that each ordered pair $(m_1, m_2), (m_2, m_3), \ldots, (m_{2k}, m_1)$ shows up exactly once comes also from an induction argument. Hence any factor $\xi_{m_i m_{i+1}}$ may be coupled to another factor $\xi_{m_{i+1} m_i}$. Since the random matrix $H(n)$ is selfadjoint with independent entries, the proof still works. Furthermore, in the same way as in the proof of Theorem 4.1.5 one can show that

$$E\big(|\mathrm{tr}_n(H(n)^k) - \tau_n(H(n)^k)|^2\big) = O(n^{-2}) \quad \text{as} \quad n \to \infty,$$

which yields the second assertion.

An $n \times n$ random matrix W is called a *Wishart matrix* of p degrees of freedom if $W = R^t R$, where R is a real $p \times n$ random matrix with independent standard Gaussian entries. If $p < n$ then W is singular; the statistical literature is mostly confined to the case $p \geq n$. In fact, this is not an essential restriction, because the nonzero eigenvalues of the matrices $R^t R$ and $R R^t$ are the same.

The next lemma contains the *moment generating function* of the Wishart matrix.

Lemma 4.1.8 *Let A be a real $n \times n$ symmetric matrix such that $A \leq I/2$. Then*

$$E(\exp(\mathrm{Tr}\, AW)) = \det(I - 2A)^{-p/2} \tag{4.1.20}$$

for the $n \times n$ regular Wishart matrix W of p degrees of freedom.

Proof: Due to the orthogonal invariance of the Wishart matrix, we may assume that $A = \mathbf{Diag}(a_1, a_2, \ldots, a_n)$. Then $\mathrm{Tr}\, AW = \sum_i a_i W_{ii}$. The random variables W_{ii} are independent and χ^2-distributed with p degrees of freedom. The problem is reduced to computing the moment generating function of χ^2-distributed variables, which is well-known. So

$$\begin{aligned} E\Big(\exp \sum_{i=1}^n a_i W_{ii}\Big) &= \prod_{i=1}^n E(\exp(a_i W_{ii})) \\ &= \prod_{i=1}^n (1 - 2a_i)^{-p/2} = \det(I - 2A)^{-p/2}. \end{aligned}$$

□

It follows from this lemma that the density of the Wishart matrix W with respect to $dT = \prod_{i \leq j} dT_{ij}$ is

$$p(T) := C_{np} \exp\Big(-\frac{1}{2}\mathrm{Tr}\, T\Big)(\det T)^{(p-n-1)/2}, \tag{4.1.21}$$

which is supported on the space $M_n(\mathbb{R})^+$ of $n \times n$ positive semidefinite real symmetric matrices. Indeed, if we compute the moment generating function

$$\int \exp(\mathrm{Tr}\, AT) p(T)\, dT = C_{np} \int \exp\Big(-\frac{1}{2}\mathrm{Tr}\,(I-2A)T\Big)(\det T)^{(p-n-1)/2}\, dT\,,$$

we need to arrive at the right-hand side of (4.1.20). Let us use the substitution $S = BTB$ with $B = (I-2A)^{1/2}$. Due to the orthogonal invariance of dT, to compute the Jacobian of this transformation, it is enough to assume that $B = \mathbf{Diag}(b_1, b_2, \ldots, b_n)$. Then it is easy to check that

$$\det \frac{\partial S}{\partial T} = (b_1 b_2 \cdots b_n)^{n+1} = (\det B)^{n+1}\,.$$

So we have

$$\begin{aligned}
&\int \exp(\mathrm{Tr}\, AT) p(T)\, dT \\
&\quad = C_{np} \int (\det B^{-1} S B^{-1})^{(p-n-1)/2} \exp\Big(-\frac{1}{2}\mathrm{Tr}\, S\Big)\, (\det B)^{-(n+1)}\, dS \\
&\quad = (\det B)^{-p} = \det(I-2A)^{-p/2}\,.
\end{aligned}$$

Now from Proposition 4.1.3 and (4.1.21) we conclude that the joint probability *density of the eigenvalues* of the $n \times n$ regular Wishart matrix of p degrees of freedom is

$$C_{np} \exp\Big(-\frac{1}{2}\sum_{i=1}^{n} \lambda_i\Big) \prod_{i=1}^{n} \lambda_i^{(p-n-1)/2} \prod_{i<j} |\lambda_i - \lambda_j|, \tag{4.1.22}$$

supported on $(\mathbb{R}^+)^n$ $(p \geq n)$. The normalizing constant C_{np} can be determined from the Selberg integral (4.1.8).

The *Marchenko-Pastur distribution* μ_λ appeared historically as the limiting mean eigenvalue distribution of the $n \times n$ random matrices

$$B_p(n) := \sum_{i=1}^{p} P_i\,,$$

where the P_i's are independent rank one projections with unitarily invariant distribution and $p/n \to \lambda$. It is evident that, for $\lambda < 1$, zero is in the spectrum of $B_p(n)$ with multiplicity $n - p$, and this yields the atomic part of the limiting distribution. The measure μ_λ coincides with the free Poisson distribution treated already in Chap. 2 and Chap. 3; see (3.3.2) for the concrete form of the distribution.

Theorem 4.1.9 *Assume that the entries of a real $p(n) \times n$ random matrix $T(n)$ are independent and identically distributed with mean* 0 *and variance* 1 *and with all* *moments bounded. Moreover, assume that* $\lim_{n\to\infty} p(n)/n = \lambda$. *Then the limiting mean eigenvalue distribution of the matrix* $T(n)^t T(n)/n$ *is the Marchenko-Pastur distribution* μ_λ.

The proof of this result can be given by the combinatorial method of moments, similarly to the lines of the proof of Theorem 4.1.4. The key point is the moment formula of the free Poisson distribution given in Example 2.5.2, where the coefficient of λ^i is the number of non-crossing partitions of $[k]$ into i blocks. By a refinement of the above combinatorial method the moments of the limit distribution can be computed as

$$\lim_{n\to\infty} \tau_n\big((n^{-1}T(n)^tT(n))^k\big) = \sum_{i=1}^{k} \frac{1}{k}\binom{k}{i}\binom{k}{i-1}\lambda^i \qquad (k \in \mathbb{N}).$$

Theorem 4.1.9 tells us that the Wishart matrices with appropriate normalization and scaling form a random matrix model for the Marchenko-Pastur distribution.

An $n \times n$ *standard non-selfadjoint Gaussian matrix* $X(n)$ is defined in the following way:

(i) $\{\mathrm{Re}\, X_{ij}(n) : 1 \le i,j \le n\} \cup \{\mathrm{Im}\, X_{ij}(n) : 1 \le i,j \le n\}$ is an independent family of Gaussian random variables, and

(ii) $E(X_{ij}(n)) = 0$ for $1 \le i,j \le n$, and $E((\mathrm{Re}\, X_{ij}(n))^2) = E((\mathrm{Im}\, X_{ij}(n))^2) = 1/2n$ for $1 \le i,j \le n$.

Equivalently, one can define $X(n)$ as

$$X(n) = \frac{H^{(1)}(n) + \mathrm{i}\, H^{(2)}(n)}{\sqrt{2}},$$

where $H^{(1)}(n)$ and $H^{(2)}(n)$ are independent standard selfadjoint Gaussian matrices.

Let $u, v \in \mathbb{R}$ be such that $u^2 + v^2 = 1$, and for $n \in \mathbb{N}$ set

$$Y(n) := uX(n) + vX(n)^*, \tag{4.1.23}$$

which is what we call an *elliptic Gaussian matrix*. This is a kind of interpolation between the standard selfadjoint Gaussian and the standard non-selfadjoint Gaussian cases. Namely, the choices $u = v$ and $v = 0$ recover those examples. The elliptic Gaussian matrix contains some correlations. The correlation coefficient between $\mathrm{Re}\, Y_{ij}(n)$ and $\mathrm{Re}\, Y_{ji}(n)$ is $\tau := 2uv$ $(-1 \le \tau \le 1)$ when $i \ne j$. One can get the elliptic Gaussian matrix as

$$Y(n) = \sqrt{\frac{1+\tau}{2}}\, H^{(1)}(n) + \mathrm{i}\sqrt{\frac{1-\tau}{2}}\, H^{(2)}(n)$$

as well, where $H^{(1)}(n)$ and $H^{(2)}(n)$ are independent standard selfadjoint Gaussian matrices.

We shall need the joint distribution of the eigenvalues of $Y(n)$. The probability measure on $M_n(\mathbb{C})$ induced by $X(n)$ has the density

$$C_n \exp(-n\mathrm{Tr}\, X^*X) \quad \text{with} \quad C_n = \left(\frac{\pi}{n}\right)^{-n^2} \tag{4.1.24}$$

with respect to the Lebesgue measure

$$dX := \prod_{i,j=1}^{n} d(\mathrm{Re}\, X_{ij})\, d(\mathrm{Im}\, X_{ij}) \quad \text{on} \quad M_n(\mathbb{C}) \cong \mathbb{R}^{2n^2}. \tag{4.1.25}$$

The measure induced by $Y(n)$ is the image measure of the above via the transformation $Y = uX + vX^*$. In the case $u \neq \pm v$ this is a non-singular linear transformation on $M_n(\mathbb{C}) \cong \mathbb{R}^{2n^2}$, whose inverse is $X = aY + bY^*$, where

$$a := \frac{u}{u^2 - v^2}, \qquad b := -\frac{v}{u^2 - v^2}.$$

Since the Jacobian $\det(\partial X/\partial Y)$ is clearly a constant, the measure on $M_n(\mathbb{C})$ induced by $Y(n)$ has the density

$$\begin{aligned} &C_n \exp(-n\mathrm{Tr}\, |aX + bX^*|^2) \\ &\quad = C_n \exp\Bigl(-\frac{n}{1-\tau^2}\mathrm{Tr}\,(X^*X - \tau \mathrm{Re}\, X^2)\Bigr), \end{aligned} \tag{4.1.26}$$

where C_n is a new normalizing constant (including the Jacobian).

Now we determine the form of the joint distribution on $\mathbb{C}^n$ (the space of eigenvalues) induced from the distribution (4.1.26).

Lemma 4.1.10 *Assume $u \neq \pm v$ and hence $-1 < \tau < 1$. Then, with respect to $d\zeta \equiv d\zeta_1 d\zeta_2 \cdots d\zeta_n$ ($d\zeta_i$ is the Lebesgue measure on $\mathbb{C}$), the joint probability density of the eigenvalues of $Y(n)$ is*

$$C_n \exp\left[-n\sum_{i=1}^{n}\Bigl(\frac{(\mathrm{Re}\,\zeta_i)^2}{1+\tau} + \frac{(\mathrm{Im}\,\zeta_i)^2}{1-\tau}\Bigr)\right] \prod_{i<j} |\zeta_i - \zeta_j|^2.$$

Proof: The standard case $\tau = 0$ was first discussed by Dyson (see [121], A.35), and the general case follows this pattern. Take a unitary U such that $U^*XU = T$ is upper triangular. Differentiating $X = UTU^*$, we get

$$dX = U(dT + \mathrm{i}\,(dM \cdot T - T \cdot dM))U^*, \tag{4.1.27}$$

where dT is triangular and $dM := -\mathrm{i}\, U^* dU$ is Hermitian. From the n-dimensional freedom of U we may impose n conditions on the variations of U:

$$dM_{ii} = -\mathrm{i}\,(U^* dU)_{ii} = 0 \qquad (1 \le i \le n).$$

Put $dY := U^* \cdot dX \cdot U$, and write (4.1.27) in the matrix elements as follows: for $1 \le i \le j \le n$

$$dY_{ij} = dT_{ij} + \mathrm{i} \sum_{k \le j} T_{kj} dM_{ik} - \mathrm{i} \sum_{l \ge i} T_{il} dM_{lj}\,, \tag{4.1.28}$$

and for $1 \le i < j \le n$

$$dY_{ji} = \mathrm{i}\,(T_{ii} - T_{jj})\overline{dM}_{ij} + \mathrm{i} \sum_{k<i} T_{ki} \overline{dM}_{kj} - \mathrm{i} \sum_{l>j} T_{jl} \overline{dM}_{il}\,. \tag{4.1.29}$$

These give a linear correspondence between $(\mathrm{Re}\, dY_{ij}, \mathrm{Im}\, dY_{ij})_{1 \le i,j \le n}$ and

$$(\mathrm{Re}\, dT_{ij}, \mathrm{Im}\, dT_{ij})_{1 \le i \le j \le n}\,, \quad (\mathrm{Re}\, dM_{ij}, \mathrm{Im}\, dM_{ij})_{1 \le i < j \le n}\,. \tag{4.1.30}$$

When $\prod_{i,j=1}^n (\mathrm{Re}\, dY_{ij})(\mathrm{Im}\, dY_{ij})$ is expanded into products of terms in (4.1.30), one can neglect products other than those where every term in (4.1.30) appears once for each. The element dT_{ij} can arise only from the equation (4.1.28) of dY_{ij}, and so the elements dM_{ij} cannot come from (4.1.28). Then, looking at (4.1.29) successively in the order of $(i,j) = (1,n), (1,n-1), \dots, (1,2), (2,n), \dots, (2,3), \dots, (n-1,n)$, one can see that the element dM_{ij} must arise only from the equation (4.1.29) of dY_{ji}. In this way, to compute $\prod (\mathrm{Re}\, dY_{ij})(\mathrm{Im}\, dY_{ij})$, one may replace (4.1.28) by

$$dY_{ij} = dT_{ij}$$

and (4.1.29) by

$$dY_{ji} = \mathrm{i}\,(T_{ii} - T_{jj})\overline{dM}_{ij}.$$

Hence we have

$$\begin{aligned}
\prod_{i,j=1}^n (\mathrm{Re}\, dX_{ij})(\mathrm{Im}\, dX_{ij}) &= \prod_{i,j=1}^n (\mathrm{Re}\, dY_{ij})(\mathrm{Im}\, dY_{ij}) \\
&= \prod_{i \le j} (\mathrm{Re}\, dT_{ij})(\mathrm{Im}\, dT_{ij}) \prod_{i<j} |T_{ii} - T_{jj}|^2 (\mathrm{Re}\, dM_{ij})(\mathrm{Im}\, dM_{ij}) \\
&= \prod_{i<j} |\zeta_i - \zeta_i|^2 \prod_{i=1}^n d\zeta_i \prod_{i<j} (\mathrm{Re}\, dT_{ij})(\mathrm{Im}\, dT_{ij})(\mathrm{Re}\, dM_{ij})(\mathrm{Im}\, dM_{ij})\,,
\end{aligned}$$

where $\zeta_i := T_{ii}$ are the eigenvalues of X.

Furthermore, we compute

$$\begin{aligned}\operatorname{Tr}|aX+bX^*|^2 &= \frac{1}{1-\tau^2}(\operatorname{Tr}T^*T-\tau\operatorname{Re}\operatorname{Tr}T^2)\\ &= \frac{1}{1-\tau^2}\Big(\sum_{i=1}^n|\zeta_i|^2+\sum_{i<j}|T_{ij}|^2-\tau\operatorname{Re}\sum_{i=1}^n\zeta_i^2\Big)\\ &= \frac{1}{1+\tau}\sum_{i=1}^n(\operatorname{Re}\zeta_i)^2+\frac{1}{1-\tau}\sum_{i=1}^n(\operatorname{Im}\zeta_i)^2+\frac{1}{1-\tau^2}\sum_{i<j}|T_{ij}|^2.\end{aligned}$$

Therefore, the measure (4.1.26) is

$$C_n\exp\Big[-n\sum_{i=1}^n\Big(\frac{(\operatorname{Re}\zeta_i)^2}{1+\tau}+\frac{(\operatorname{Im}\zeta_i)^2}{1-\tau}\Big)\Big]\prod_{i<j}|\zeta_i-\zeta_j|^2\,d\zeta$$
$$\times\exp\Big(-\frac{n}{1-\tau^2}\sum_{i<j}|T_{ij}|^2\Big)\prod_{i<j}(\operatorname{Re}dT_{ij})(\operatorname{Im}dT_{ij})\prod_{i<j}(\operatorname{Re}dM_{ij})(\operatorname{Im}dM_{ij}).$$

Integrating with respect to $\prod(\operatorname{Re}dT_{ij})(\operatorname{Im}dT_{ij})$ and $\prod(\operatorname{Re}dM_{ij})(\operatorname{Im}dM_{ij})$, we conclude that the induced measure on $\mathbb{C}^n$ is written as

$$C_n'\exp\Big[-n\sum_{i=1}^n\Big(\frac{(\operatorname{Re}\zeta_i)^2}{1+\tau}+\frac{(\operatorname{Im}\zeta_i)^2}{1-\tau}\Big)\Big]\prod_{i<j}|\zeta_i-\zeta_j|^2\,d\zeta.$$

□

In particular, the standard non-selfadjoint Gaussian matrix $X(n)$ (in the case $\tau=0$) has the joint eigenvalue density

$$\frac{\pi^{-n}n^{n(n+1)/2}}{\prod_{j=1}^n j!}\exp\Big(-n\sum_{i=1}^n|\zeta_i|^2\Big)\prod_{i<j}|\zeta_i-\zeta_j|^2 \tag{4.1.31}$$

(see [121], Sec. 15.1, for a computation of the above normalizing constant).

Results about the asymptotic freeness in Sec. 4.3 show that the distribution of the elliptic Gaussian matrix $Y(n)$ converges to that of an *elliptic element* discussed in Example 2.6.7. It is worthwhile to form the following definition. Assume that x is a noncommutative random variable and for every $n\in\mathbb{N}$ an $n\times n$ complex matrix $T(n)$ is given. It is said that $T(n)$ is a *random matrix model* of x if the distribution of $(T(n),T(n)^*)$ (with respect to τ_n) converges to the distribition of (x,x^*) as the matrix size goes to infinity. (Earlier we used the concept of random matrix model in the selfadjoint case, and that usage is compatible with this more general definition.) Hence $Y(n)$ is a *random matrix model of the elliptic element*, and in particular the standard non-selfadjoint Gaussian matrices $X(n)$ form a model of the circular element. The elliptic Gaussian matrix will appear also in a large deviation result discussed in Chap. 5.

Unitary random matrices are also important; we shall devote the whole next section to them.

4.2 Random unitary matrices and asymptotic freeness

Let $\mathcal{U}(n)$ be the compact group of $n \times n$ unitary matrices. The Haar probability measure γ_n on $\mathcal{U}(n)$ is two-sided invariant. An $n \times n$ unitary random matrix is said to be *standard* if its distribution is γ_n.

First, we determine the form of the joint distribution of the eigenvalues of the $n \times n$ standard unitary random matrix. Since the eigenvalues of a unitary matrix are on the unit circle $\mathbb{T}$, the joint probability of the eigenvalues is supported on $\mathbb{T}^n$. Most of this section is spent in studying the multiple joint moments of independent standard random unitaries. The concept of asymptotic freeness will emerge from the calculations.

Lemma 4.2.1 *The measure on $\mathbb{T}^n$ induced from the Haar measure γ_n on $\mathcal{U}(n)$ has the density*

$$\frac{1}{n!}\prod_{i<j}|\zeta_i - \zeta_j|^2 = \frac{1}{n!}\prod_{i<j}|e^{\mathrm{i}\,\theta_i} - e^{\mathrm{i}\,\theta_j}|^2$$

with respect to $d\zeta_1 d\zeta_2 \cdots d\zeta_n$, where $d\zeta_i = d\theta_i/2\pi$ for $\zeta_i = e^{\mathrm{i}\,\theta_i}$.

Proof: Consider the differential dU at $U \in \mathcal{U}(n)$; then $dL := -\mathrm{i}\,U^* dU$ is an infinitesimal Hermitian matrix. Applying the invariance of γ_n to $dU = U(\mathrm{i}\,dL) = U(e^{\mathrm{i}\,dL} - I)$, one can see that

$$\gamma_n(dU) = C_n \prod_{i=1}^n dL_{ii} \prod_{i<j} (\mathrm{Re}\, dL_{ij})(\mathrm{Im}\, dL_{ij})\,.$$

Diagonalize U in the form $U = VDV^*$, where V is a unitary and $D = \mathbf{Diag}(e^{\mathrm{i}\,\theta_1}, e^{\mathrm{i}\,\theta_2}, \dots, e^{\mathrm{i}\,\theta_n})$. As in the proof of Lemma 4.1.10 we may require that $dM_{ii} = 0$ $(1 \le i \le n)$ for $dM := -\mathrm{i}\,V^* dV$. Since

$$dL = -\mathrm{i}\,U^* dU = V(-\mathrm{i}\,D^* dD + D^* dMD - dM)V^*\,,$$

we have

$$(V^* dLV)_{ij} = \begin{cases} d\theta_i & \text{if } i = j, \\ (e^{\mathrm{i}\,(\theta_j - \theta_i)} - 1)dM_{ij} & \text{if } i < j, \end{cases}$$

and hence

$$\prod_{i=1}^n dL_{ii} \prod_{i<j} (\mathrm{Re}\, dL_{ij})(\mathrm{Im}\, dL_{ij})$$
$$= \prod_{i<j} |e^{\mathrm{i}\,\theta_i} - e^{\mathrm{i}\,\theta_j}|^2 \prod_{i=1}^n d\theta_i \prod_{i<j} (\mathrm{Re}\, dM_{ij})(\mathrm{Im}\, dM_{ij})\,.$$

This shows that the induced measure on $\mathbb{T}^n$ has the density

$$C'_n \prod_{i<j} |e^{\mathrm{i}\,\theta_i} - e^{\mathrm{i}\,\theta_j}|^2 .$$

To determine C'_n we need to compute the integral

$$\frac{1}{(2\pi)^n} \int_0^{2\pi} \cdots \int_0^{2\pi} \prod_{i<j} |e^{\mathrm{i}\,\theta_i} - e^{\mathrm{i}\,\theta_j}|^2 \, d\theta_1 \cdots d\theta_n . \tag{4.2.1}$$

Since

$$\begin{aligned}
\prod_{i<j} |e^{\mathrm{i}\,\theta_i} - e^{\mathrm{i}\,\theta_j}|^2 &= \prod_{i<j} (e^{\mathrm{i}\,\theta_i} - e^{\mathrm{i}\,\theta_j}) \prod_{i<j} (e^{-\mathrm{i}\,\theta_i} - e^{-\mathrm{i}\,\theta_j}) \\
&= \det\big[e^{\mathrm{i}\,k\theta_i}\big]_{1\le i\le n,\,0\le k\le n-1} \det\big[e^{-\mathrm{i}\,k\theta_j}\big]_{0\le k\le n-1,\,1\le j\le n} \\
&= \det\left[\sum_{k=0}^{n-1} e^{ik(\theta_i-\theta_j)}\right]_{1\le i,j\le n} ,
\end{aligned}$$

one can write

$$\begin{aligned}
\prod_{i<j} |e^{\mathrm{i}\,\theta_i} - e^{\mathrm{i}\,\theta_j}|^2 &= \det\big[S_n(\theta_i - \theta_j)\big]_{1\le i,j\le n} \\
&= \sum_{k=1}^{n} \sum_{\sigma:\sigma(k)=n} \operatorname{sign}(\sigma) \prod_{i=1}^{n} S_n(\theta_i - \theta_{\sigma(i)}) ,
\end{aligned}$$

where $S_n(\theta) := \sum_{k=0}^{n-1} e^{\mathrm{i}\,k\theta}$ and the summation over permutations σ of $[n]$ is decomposed according to $\sigma(k) = n$. Integration with respect to θ_n gives

$$\begin{aligned}
&\frac{1}{2\pi} \int_0^{2\pi} \det\big[S_n(\theta_i - \theta_j)\big]_{1\le i,j\le n} \, d\theta_n \\
&= S_n(0) \sum_{\sigma:\sigma(n)=n} \operatorname{sign}(\sigma) \prod_{i=1}^{n-1} S_n(\theta_i - \theta_{\sigma(i)}) \\
&\quad + \sum_{k=1}^{n-1} \sum_{\sigma:\sigma(k)=n} \operatorname{sign}(\sigma) \prod_{\substack{i=1\\ i\ne k}}^{n-1} S_n(\theta_i - \theta_{\sigma(i)}) \\
&\qquad\qquad \times \frac{1}{2\pi} \int_0^{2\pi} S_n(\theta_k - \theta_n) S_n(\theta_n - \theta_{\sigma(n)}) \, d\theta_n \\
&= n \det\big[S_n(\theta_i - \theta_j)\big]_{1\le i,j\le n-1} \\
&\quad + \sum_{k=1}^{n-1} \sum_{\sigma:\sigma(k)=n} \operatorname{sign}(\sigma) \prod_{\substack{i=1\\ i\ne k}}^{n-1} S_n(\theta_i - \theta_{\sigma(i)}) \cdot S_n(\theta_k - \theta_{\sigma(n)}) ,
\end{aligned}$$

because $S_n(0) = n$ and

$$\frac{1}{2\pi}\int_0^{2\pi} S_n(\theta_1 - \theta)S_n(\theta - \theta_2)\,d\theta = S_n(\theta_1 - \theta_2)\,.$$

For σ such that $\sigma(k) = n$ with $1 \le k \le n-1$, take a permutation σ' of $[n-1]$ as $\sigma'(i) = \sigma(i)$ for $i \in [n-1]$, $i \ne k$, and $\sigma'(k) = \sigma(n)$. Then, since $\mathrm{sign}(\sigma) = -\mathrm{sign}(\sigma')$, we obtain

$$\begin{aligned}\sum_{\sigma:\sigma(k)=n} &\mathrm{sign}(\sigma)\prod_{\substack{i=1\\ i\ne k}}^{n-1} S_n(\theta_i - \theta_{\sigma(i)})\cdot S_n(\theta_k - \theta_{\sigma(n)})\\ &= -\sum_{\sigma'}\mathrm{sign}(\sigma')\prod_{i=1}^{n-1} S_n(\theta_i - \theta_{\sigma'(i)}) = -\det\big[S_n(\theta_i - \theta_j)\big]_{1\le i,j\le n-1},\end{aligned}$$

so that

$$\begin{aligned}\frac{1}{2\pi}\int_0^{2\pi} &\det\big[S_n(\theta_i - \theta_j)\big]_{1\le i,j\le n}\,d\theta_n\\ &= (n-(n-1))\det\big[S_n(\theta_i - \theta_j)\big]_{1\le i,j\le n-1}.\end{aligned}$$

Repeating the above computation yields that the integral (4.2.1) is equal to

$$(n-(n-1))(n-(n-2))\cdots(n-1)S_n(0) = n!\,,$$

as required.

□

For a measurable function $f : \mathcal{U}(n) \to \mathbb{C}$ we have the expectation

$$E(f) := \int f(U)\,d\gamma_n(U)$$

whenever this integral is defined. The standard unitary random matrix U is a noncommutative random variable with respect to the functional τ_n. Due to the invariance of γ_n, $e^{\mathrm{i}\theta}U$ is standard for any $\theta \in \mathbb{R}$, and it follows that the moments $\tau_n(U^k)$ vanish for all $k \in \mathbb{Z}\setminus\{0\}$. Namely, U is a Haar unitary in the sense of Chap. 1 (see Example 1.2.4).

Before we turn to the joint moments of two independent standard random unitaries, we start with the correlation of the matrix entries of a single standard random unitary.

Let $U = [U_{ij}]$ be an $n\times n$ standard unitary matrix. The two-sided invariance of the Haar measure on $\mathcal{U}(n)$ yields that $E(f(U)) = E(f(VUW))$ for every measurable function f and for all unitaries V and W. When $V = \mathbf{Diag}(e^{\mathrm{i}\,\theta_1},\dots,e^{\mathrm{i}\,\theta_n})$ and $W = \mathbf{Diag}(e^{\mathrm{i}\,\psi_1},\dots,e^{\mathrm{i}\,\psi_n})$, we have

$$E(f) = E\big(f([e^{\mathrm{i}\,(\theta_i+\psi_j)}U_{ij}]_{i,j=1}^n)\big) \tag{4.2.2}$$

for every $\theta_i, \psi_j \in \mathbb{R}$. Moreover, choosing permutation matrices V and W, we obtain

$$E(f) = E\big(f([U_{\pi(i)\sigma(j)}]_{i,j=1}^n)\big) \tag{4.2.3}$$

for all permutations π, σ of $[n]$. The next lemma says that most of the multiple moments of the elements U_{ij} vanish.

Lemma 4.2.2 *Let $l \in \mathbb{N}$, $i_1,\dots,i_l,j_1,\dots,j_l \in [n]$ and $k_1,\dots,k_l,m_1,\dots,m_l \in \mathbb{Z}^+$. If either $\sum_{i_r=i}(k_r - m_r) \neq 0$ for some $1 \le i \le n$ or $\sum_{j_r=j}(k_r - m_r) \neq 0$ for some $1 \le j \le n$, then*

$$E\big((U_{i_1j_1}^{k_1}\bar{U}_{i_1j_1}^{m_1})(U_{i_2j_2}^{k_2}\bar{U}_{i_2j_2}^{m_2})\cdots(U_{i_lj_l}^{k_l}\bar{U}_{i_lj_l}^{m_l})\big) = 0\,. \tag{4.2.4}$$

In particular, if $\sum_{r=1}^l(k_r - m_r) \neq 0$ (this is the case when $\sum_{r=1}^l(k_r + m_r)$ is odd), then (4.2.4) holds.

Proof: Suppose that $h := \sum_{i_r=i}(k_r - m_r) \neq 0$. We can apply (4.2.2) above and get

$$E\big((U_{i_1j_1}^{k_1}\bar{U}_{i_1j_1}^{m_1})\cdots(U_{i_lj_l}^{k_l}\bar{U}_{i_lj_l}^{m_l})\big) = e^{\mathrm{i}\,h\theta}E\big((U_{i_1j_1}^{k_1}\bar{U}_{i_1j_1}^{m_1})\cdots(U_{i_lj_l}^{k_l}\bar{U}_{i_lj_l}^{m_l})\big)$$

for every $\theta \in \mathbb{R}$. This gives the statement.

□

The following are examples of multiple moments of the matrix elements U_{ij} of a standard unitary. It is immediate from Lemma 4.2.2 that other multiple moments of U_{ij}'s up to the fourth order are all 0.

Proposition 4.2.3 *For a standard unitary matrix $U = [U_{ij}]$ the following hold:*

$$E(|U_{ij}|^2) = \frac{1}{n} \qquad (1 \le i,j \le n), \tag{4.2.5}$$

$$E(|U_{ij}|^4) = \frac{2}{n(n+1)} \qquad (1 \le i,j \le n), \tag{4.2.6}$$

$$E(|U_{ij}|^2|U_{i'j}|^2) = E(|U_{ij}|^2|U_{ij'}|^2) = \frac{1}{n(n+1)} \qquad (i \neq i',\ j \neq j'), \tag{4.2.7}$$

$$E(|U_{ij}|^2|U_{i'j'}|^2) = \frac{1}{n^2-1} \qquad (i \neq i',\ j \neq j'), \tag{4.2.8}$$

$$E(U_{ij}U_{i'j'}\bar{U}_{ij'}\bar{U}_{i'j}) = -\frac{1}{n(n^2-1)} \qquad (i \neq i',\ j \neq j'). \tag{4.2.9}$$

Proof: The random variables U_{ij} are identically distributed according to (4.2.3). This yields (4.2.5) because $\sum_{j=1}^{n} |U_{ij}|^2 = 1$.

Since the $(1,1)$ entries of U and

$$\left(\begin{bmatrix} \cos\theta & \sin\theta \\ -\sin\theta & \cos\theta \end{bmatrix} \oplus I_{n-2}\right) U$$

are identically distributed, we have

$$\begin{aligned} &E(|U_{11}|^4) \\ &\quad = E(|U_{11}\cos\theta + U_{21}\sin\theta|^4) \\ &\quad = E((|U_{11}|^2\cos^2\theta + |U_{21}|^2\sin^2\theta + (U_{11}\bar{U}_{21} + \bar{U}_{11}U_{21})\cos\theta\sin\theta)^2) \\ &\quad = E(|U_{11}|^4)\cos^4\theta + E(|U_{21}|^4)\sin^4\theta + 4E(|U_{11}|^2|U_{21}|^2)\cos^2\theta\sin^2\theta \\ &\quad = E(|U_{11}|^4)(\cos^4\theta + \sin^4\theta) + 4E(|U_{11}|^2|U_{21}|^2)\cos^2\theta\sin^2\theta\,. \end{aligned}$$

Above we used $E(|U_{11}|^2 U_{11}\bar{U}_{21}) = 0$, etc., according to Lemma 4.2.2. Hence

$$E(|U_{11}|^4) = 2E(|U_{11}|^2|U_{21}|^2) = 2E(|U_{i1}|^2|U_{i'1}|^2) \qquad (i \neq i').$$

Since $\sum_{i,i'=1}^{n} |U_{i1}|^2|U_{i'1}|^2 = 1$, this yields

$$\begin{aligned} 1 &= \sum_{i=1}^{n} E(|U_{i1}|^4) + \sum_{i\neq i'} E(|U_{i1}|^2|U_{i'1}|^2) \\ &= nE(|U_{11}|^4) + \frac{n(n-1)}{2}E(|U_{11}|^4) = \frac{n(n+1)}{2}E(|U_{11}|^4)\,, \end{aligned}$$

so $E(|U_{11}|^4) = 2/n(n+1)$ and $E(|U_{11}|^2|U_{21}|^2) = 1/n(n+1)$. In this way (4.2.6) and (4.2.7) are derived. Apply (4.2.7) to $\sum_{i,i'=1}^{n} |U_{i1}|^2|U_{i'2}|^2 = 1$ to get (4.2.8). The proof of (4.2.9) is similar to that of (4.2.6), and it is left as an exercise.

□

By Proposition 4.2.3 the correlation coefficient between $|U_{ij}|^2$ and $|U_{i'j'}|^2$ is computed as follows:

$$\rho(|U_{ij}|^2, |U_{i'j}|^2) = \rho(|U_{ij}|^2, |U_{ij'}|^2) = -\frac{1}{n-1} \qquad (i \neq i',\ j \neq j'),$$

$$\rho(|U_{ij}|^2, |U_{i'j'}|^2) = \frac{1}{(n-1)^2} \qquad (i \neq i',\ j \neq j').$$

Lemma 4.2.4 *For every $k \in \mathbb{Z}^+$,*

$$E(|U_{ij}|^{2k}) = \binom{n+k-1}{n-1}^{-1} \qquad (1 \leq i, j \leq n).$$

Furthermore, the distribution of U_{ij} is

$$\frac{n-1}{\pi}(1-r^2)^{n-2} r\,dr \cdot d\theta \qquad (\zeta = re^{\mathrm{i}\,\theta},\ 0 \le r \le 1,\ 0 \le \theta \le 2\pi). \tag{4.2.10}$$

Proof: First note that the variables $\langle Ug, h\rangle$ (in particular U_{ij}) have the same distribution for all unit vectors $g, h \in \mathbb{C}^n$. Let $(\zeta_1, \zeta_2, \dots, \zeta_n)^t$ be a column of the Haar distributed random unitary matrix. The unitary transformations act on the sphere

$$\Big\{(z_1, z_2, \dots, z_n) \in \mathbb{C}^n : \sum_i |z_i|^2 = 1\Big\}$$

transitively. The distribution measure induced on this space by the random unit vector $(\zeta_1, \zeta_2, \dots, \zeta_n)$ is invariant under the action of the unitary transformations. The invariant measure is unique, so we may assume that

$$\zeta_j = \frac{\xi_{2j-1} + \mathrm{i}\,\xi_{2j}}{\sqrt{\sum_{i=1}^{2n} \xi_i^2}},$$

where $\xi_1, \xi_2, \dots, \xi_{2n}$ are independent Gaussian variables of $N(0, 1/2)$. We can write ζ_j in a slightly different form as

$$\zeta_j = \frac{u_j \sqrt{\eta_j}}{\sqrt{\sum_{i=1}^{n} \eta_i}}$$

with independent variables $u_1, \dots, u_n, \eta_1, \dots, \eta_n$, where $\eta_j := \xi_{2j-1}^2 + \xi_{2j}^2$ and $u_j\sqrt{\eta_j} = \xi_{2j-1} + \mathrm{i}\,\xi_{2j}$. Note that η_j is exponentially distributed with the density e^{-x} $(x > 0)$ and u_j is uniformly distributed on the unit circle $\mathbb{T}$. So we write $|\zeta_1|^2 = \eta_1/(\eta_1 + \eta)$, where η is independent of η_1 and has the density $x^{n-2}e^{-x}/(n-2)!$ $(x > 0)$. Since

$$\begin{aligned}
\mathbf{Prob}\big(|\zeta_1|^2 \ge x\big) &= \mathbf{Prob}\big(\eta_1 \ge \frac{x}{1-x}\eta\big) \\
&= \int_0^\infty \frac{t^{n-2}}{(n-2)!} e^{-t} \Big(\int_{\frac{x}{1-x}t}^\infty e^{-s}\,ds\Big)\,dt \\
&= (1-x)^{n-1},
\end{aligned}$$

we conclude that the density of $|\zeta_1|^2$ is $(n-1)(1-x)^{n-2}$ and the density of $|\zeta_1|$ is $2(n-1)(1-x^2)^{n-2}x$. Hence the distribution (4.2.10) is obtained.

□

The variable $|\zeta_1|^2$ in the above proof has a beta distribution (4.2.18) with $\alpha = 1$ and $\beta = n-1$; see the end of this section. Furthermore, one can see that the joint distribution of $|\zeta_1|^2, |\zeta_2|^2, \dots, |\zeta_{n-1}|^2$ is uniform on the set

$$\Big\{(x_1, x_2, \dots, x_{n-1}) \in (\mathbb{R}^+)^{n-1} : \sum_i x_i \le 1\Big\}.$$

From this fact, for example, the correlation between $|\zeta_i|^2$ and $|\zeta_j|^2$ is easily computable.

When $U(n)$ is an $n \times n$ standard unitary matrix, it is important for the proof of the next proposition to notice that Lemma 4.2.4 yields

$$E(|U_{ij}(n)|^{2k}) = O(n^{-k}) \quad \text{as} \quad n \to \infty. \tag{4.2.11}$$

This order is the same as the $2k$th moment of the normal distribution $N(0, 1/n)$ with variance $1/n$. Moreover, it is worth noting that

$$\mathbf{Prob}(|\sqrt{n}\, U_{ij}(n)|^2 \geq x) = \Big(1 - \frac{x}{n}\Big)^{n-1} \to e^{-x} \quad \text{as} \quad n \to \infty,$$

and hence the distribution of $\sqrt{n}\, U_{ij}(n)$ tends to the standard complex Gaussian measure (i.e. the distribution of the above $\xi_1 + \mathrm{i}\, \xi_2$). More generally, for each $k \in \mathbb{N}$, if

$$T_{ij}(n) = U_{ij}(n) \quad \text{for} \quad 1 \leq i, j \leq k,$$

then $\sqrt{n}T(n)$ converges in distribution to the $k \times k$ standard Gaussian matrix as $n \to \infty$.

Proposition 4.2.5 *Let $U(n)$ and $V(n)$ be independent $n \times n$ standard unitary random matrices. Then*

$$\lim_{n \to \infty} \tau_n(U(n)^{m_1} V(n)^{m_2} \cdots U(n)^{m_{2l-1}} V(n)^{m_{2l}}) = 0$$

whenever $m_1, m_2, \ldots, m_{2l} \in \mathbb{Z}$ and the matrix under the trace is different from the identity.

Proof: Obviously we may assume that the integers m_r are all nonzero, so we have the product of k unitaries under the trace, $k := |m_1| + |m_2| + \cdots + |m_{2l}|$. We write

$$\begin{aligned} u_{ij}(-1, 1, n) &:= U_{ij}(n)\,, \\ u_{ij}(-1, -1, n) &:= \bar{U}_{ji}(n)\,, \\ u_{ij}(1, 1, n) &:= V_{ij}(n)\,, \\ u_{ij}(1, -1, n) &:= \bar{V}_{ji}(n)\,, \end{aligned}$$

and express the trace of the product of the random unitaries in terms of the matrix elements, keeping the order of the factors:

$$\frac{1}{n} \sum_{i_1, \ldots, i_k = 1}^{n} E\Big(\prod_{h=1}^{k} u_{i_h j_h}(s(h), \varepsilon(h), n)\Big)\,, \tag{4.2.12}$$

where $j_h = i_{h+1}$ $(1 \le h \le k-1)$ and $j_k = i_1$.

When we group together entries of $U(n)$ (indexed by $s(h) = -1$) and those of $V(n)$ (indexed by $s(h) = 1$) in each summand of (4.2.12), the expectation will factor because entries of $U(n)$ are independent of those of $V(n)$. In this way, many of the summands of (4.2.12) vanish according to Lemma 4.2.2; for example, for an odd k, Lemma 4.2.2 implies that the whole expression is zero. We can restrict ourselves to the case of an even k. When k is even, using the Hölder inequality and (4.2.11), we can estimate

$$\begin{aligned} &\Big|E\Big(\prod_{h=1}^{k} u_{i_h j_h}(s(h), \varepsilon(h), n)\Big)\Big| \\ &\qquad \le \prod_{h=1}^{k} E(|u_{i_h j_h}(s(h), \varepsilon(h), n)|^k)^{1/k} = O(n^{-k/2}) \end{aligned} \tag{4.2.13}$$

uniformly for $i_1, j_1, \dots, i_k, j_k$ as $n \to \infty$.

We want to analyze the structure of the nonzero terms of (4.2.12). When the term $E(\prod_{h=1}^{k} u_{i_h j_h}(s(h), \varepsilon(h), n))$ is nonzero, we can apply Lemma 4.2.2 to two factored expectations of products for $s(h) = -1$ and for $s(h) = 1$, and we have, for each $1 \le i, j \le n$ and $\delta \in \{-1, 1\}$,

$$\begin{aligned} \#\{h : i_h = i,\, s(h) = \delta,\, \varepsilon(h) = 1\} &= \#\{h : j_h = i,\, s(h) = \delta,\, \varepsilon(h) = -1\}, \\ \#\{h : j_h = j,\, s(h) = \delta,\, \varepsilon(h) = 1\} &= \#\{h : i_h = j,\, s(h) = \delta,\, \varepsilon(h) = -1\}. \end{aligned}$$

Thus, two pair partitions $\mathcal{U}$ and $\mathcal{V}$ can be chosen so that if $\{h, h'\} \in \mathcal{U}$ then

$$s(h) = s(h'), \quad \varepsilon(h) = 1, \quad \varepsilon(h') = -1, \quad i_h = j_{h'} \,(= i_{h'+1}),$$

and if $\{h, h'\} \in \mathcal{V}$ then

$$s(h) = s(h'), \quad \varepsilon(h) = -1, \quad \varepsilon(h') = 1, \quad i_h = j_{h'} \,(= i_{h'+1}),$$

where either $h < h'$ or $h > h'$ may be chosen. For $\mathcal{U}, \mathcal{V}$ as above let $\mathcal{R}(\mathcal{U}, \mathcal{V})$ denote the equivalence relation on $[k]$ generated by $h \sim h' + 1$ when $\{h, h'\} \in \mathcal{U}$, $\varepsilon(h) = 1$, $\varepsilon(h') = -1$ or when $\{h, h'\} \in \mathcal{V}$, $\varepsilon(h) = -1$, $\varepsilon(h') = 1$. Also, let $\Xi_n(\mathcal{U}, \mathcal{V})$ denote the set of $(i_1, \dots, i_k) \in [n]^k$ subject to the structure determined by $\mathcal{U}$ and $\mathcal{V}$ above. Note that if $(i_1, \dots, i_k) \in \Xi_n(\mathcal{U}, \mathcal{V})$, then i_h must be the same for all h from an equivalence class of $\mathcal{R}(\mathcal{U}, \mathcal{V})$. Therefore $\#\Xi_n(\mathcal{U}, \mathcal{V}) = n^{k_0}$, where k_0 is the number of equivalence classes of $\mathcal{R}(\mathcal{U}, \mathcal{V})$. Suppose that $\{h\}$ is a singleton equivalence class of $\mathcal{R}(\mathcal{U}, \mathcal{V})$. Then we must have both $\{h, h-1\} \in \mathcal{U}$, $\varepsilon(h) = 1$, $\varepsilon(h-1) = -1$ and $\{h, h-1\} \in \mathcal{V}$, $\varepsilon(h) = -1$, $\varepsilon(h-1) = 1$ (where $h-1$ is meant $\bmod k$); however these two conditions cannot happen at the same time. So there is no singleton equivalence class in $\mathcal{R}(\mathcal{U}, \mathcal{V})$. This shows that $k_0 \le k/2$.

From the above argument together with (4.2.13) we obtain

$$\sum_{(i_1,\dots,i_k)\in\Xi_n(\mathcal{U},\mathcal{V})} \Big| E\Big(\prod_{h=1}^{k} u_{i_h j_h}(s(h),\varepsilon(h),n)\Big)\Big| \le n^{k/2} O(n^{-k/2}) = O(1)\,.$$

Summing up for $\mathcal{U},\mathcal{V}$ (the choice of $\mathcal{U},\mathcal{V}$ being independent of n) yields

$$\sum_{i_1,\dots,i_k=1}^{n} \Big| E\Big(\prod_{h=1}^{k} u_{i_h j_h}(s(h),\varepsilon(h),n)\Big)\Big| = O(1)\,,$$

showing that (4.2.12) is $O(n^{-1})$ as $n\to\infty$.

□

Let u and v be free (more precisely, *-free) Haar unitaries. Moreover, for every $n \in \mathbb{N}$ let $U(n)$ and $V(n)$ be independent $n \times n$ standard unitary random matrices. Proposition 4.2.5 tells us that $(U(n), V(n))$ converges in distribution to (u, v). Since u and v are in free relation, we say that $U(n)$ and $V(n)$ are *asymptotically free*. More generally, if $(X(s,n))_{1\le s\le k}$ is a sequence of k-tuples of noncommutative random variables converging in distribution to $(X_s)_{1\le s\le k}$, then $(X(s,n))_{1\le s\le k}$ is said to be asymptotically free if the limiting family $(X_s)_{1\le s\le k}$ is free. The definition of asymptotic freeness extends also to an infinite family; however an infinite family is asymtotically free if and only if all finite subsets are so.

The result and the proof of Proposition 4.2.5 extend to arbitrarily many independent unitaries. Let S be a set, and for $n \in \mathbb{N}$ let $(U(s,n))_{s\in S}$ be an independent family of $n \times n$ standard unitary random matrices. Then, a slight modification of the above proof shows that

$$\lim_{n\to\infty} \tau_n(U(s_1,n)^{m_1} U(s_2,n)^{m_2} \cdots U(s_l,n)^{m_l}) = 0$$

for every $m_1, m_2, \dots, m_l \in \mathbb{Z}\setminus\{0\}$ whenever $s_1, s_2, \dots, s_l \in S$ satisfy $s_1 \ne s_2 \ne \dots \ne s_l$. Thus, the distribution of $(U(s,n))_{s\in S}$ converges to that of free Haar unitaries, so that $(U(s,n))_{s\in S}$ is asymptotically free.

Both the concept of asymptotic freeness and the asymptotic freeness of independent Haar distributed unitaries are due to Voiculescu. This aspect of random matrix theory came into being under the influence of free probability.

Let a and b be free random variables in a C^*-probability space $(\mathcal{A}, \varphi)$. If (d_n) is a sequence converging to 0 in norm, then $a + d_n$ and b are asymptotically free as $n \to \infty$ (obviously). Such a perturbation of the free relation can lead to asymptotic freeness in a trivial way. The critical point in random matrix models is that the freeness comes into being only in the limit, and we do not have the perturbation of some a priori free variables. Random matrices are noncommutative random variables with respect to the functional τ_n, but on the other hand we can study them almost everywhere with respect to the normalized trace tr_n. This happened already in Theorem 4.1.5. In a similar spirit we are going to extend Proposition 4.2.5 as follows.

Theorem 4.2.6 *Let $(U(s,n))_{s\in S}$ be an independent family of $n\times n$ standard unitary random matrices. Let $s_1, s_2, \dots, s_l \in S$ be such that $s_1 \neq s_2 \neq \dots \neq s_l$. Then, for every $m_1, m_2, \dots, m_l \in \mathbb{Z}\setminus\{0\}$ and every $1 \le p < \infty$,*

$$E\big(|\mathrm{tr}_n(U(s_1,n)^{m_1}U(s_2,n)^{m_2}\cdots U(s_l,n)^{m_l})|^p\big)^{1/p} = O(n^{-1}), \tag{4.2.14}$$

and moreover

$$\mathrm{tr}_n(U(s_1,n)^{m_1}U(s_2,n)^{m_2}\cdots U(s_l,n)^{m_l}) \to 0 \tag{4.2.15}$$

almost everywhere as $n\to\infty$.

Proof: First we prove (4.2.14) for the case $p=2$. As in the proof of Proposition 4.2.5, let $k := |m_1|+\cdots+|m_l|$, and when $|m_1|+\cdots+|m_{r-1}|+1 \le h \le |m_1|+\cdots+|m_r|$ $(1\le r\le l)$ set

$$s(h) := s_r\,, \quad \varepsilon(h) := \begin{cases} 1 & \text{if } m_r > 0, \\ -1 & \text{if } m_r < 0. \end{cases}$$

Moreover, set $s(k+h) := s(h)$ and $\varepsilon(k+h) := -\varepsilon(h)$ for $1\le h\le k$. If we write

$$u_{ij}(s,\varepsilon,n) := \begin{cases} U_{ij}(s,n) & \text{if } \varepsilon = 1, \\ \bar U_{ji}(s,n) & \text{if } \varepsilon = -1, \end{cases}$$

then the square of the L^2-norm in (4.2.14) is given as

$$\Big(\frac{1}{n}\Big)^2 \sum_{i_1,\dots,i_k,i_{k+1},\dots,i_{2k}=1}^{n} E\Big(\prod_{h=1}^{2k} u_{i_hj_h}(s(h),\varepsilon(h),n)\Big), \tag{4.2.16}$$

where

$$\begin{cases} j_h = i_{h+1} \ \ (1\le h\le k-1)\,, & j_k = i_1\,, \\ i_h = j_{h+1} \ \ (k+1\le h\le 2k-1)\,, & i_{2k} = j_{k+1}\,. \end{cases}$$

Now we can proceed as in the proof of Proposition 4.2.5. For any nonzero term in (4.2.16) we can choose two pair partitions $\mathcal{U}$ and $\mathcal{V}$ of $[2k]$ such that if $\{h,h'\}\in\mathcal{U}$ then $s(h)=s(h')$, $\varepsilon(h)=1$, $\varepsilon(h')=-1$, $i_h=j_{h'}$, and if $\{h,h'\}\in\mathcal{V}$ then $s(h)=s(h')$, $\varepsilon(h)=-1$, $\varepsilon(h')=1$, $i_h=j_{h'}$. These pair partitions cause many equalities among $i_1,\dots,i_{2k}$ and define the equivalence relation $\mathcal{R}(\mathcal{U},\mathcal{V})$ on $[2k]$ so that $i_h=i_{h'}$ whenever h,h' are in the same equivalence class of $\mathcal{R}(\mathcal{U},\mathcal{V})$. It follows that the number of equivalence classes of $\mathcal{R}(\mathcal{U},\mathcal{V})$ is not more than k, because $\mathcal{R}(\mathcal{U},\mathcal{V})$ contains no singleton equivalence classes. Thus (4.2.16) is $n^{-2}n^kO(n^{-k}) = O(n^{-2})$ as $n\to\infty$, showing (4.2.14) for $p=2$.

What we have just shown implies that

$$E\Big(\sum_{n=1}^{\infty} |\mathrm{tr}_n(U(s_1,n)^{m_1}\cdots U(s_l,n)^{m_l})|^2\Big) < +\infty .$$

This yields the almost sure convergence (4.2.15); see the argument at the beginning of the proof of Theorem 4.1.5.

Finally, to prove (4.2.14) for general p, it is enough to show it when p is an even integer, say $2d$, because $E(|\cdot|^p)^{1/p} \le E(|\cdot|^{2d})^{1/2d}$ for $p \le 2d$. But the above argument works for this case as well, and we can get

$$E\big(|\mathrm{tr}_n(U(s_1,n)^{m_1}\cdots U(s_l,n)^{m_l})|^{2d}\big) = \Big(\frac{1}{n}\Big)^{2d} n^{dk} O(n^{-dk}) = O(n^{-2d})$$

as $n \to \infty$.

□

Independent Haar distributed orthogonal matrices are *asymptotically free* as well. Our discussion on Haar distributed unitary matrices presented in this section can be repeated for Haar distributed orthogonal real matrices. Let $O = [x_{ij}]_{i,j=1}^n$ be an $n \times n$ *standard orthogonal random matrix* distributed according to the Haar probability measure on $\mathcal{O}(n)$. The lemma taking the place of Lemma 4.2.2 says that if $i_1, \dots, i_l, j_1, \dots, j_l \in [n]$ and $E(x_{i_1 j_1} x_{i_2 j_2} \cdots x_{i_k j_k}) \ne 0$, then $\#\{r : i_r = i\}$ and $\#\{r : j_r = j\}$ are even for every $i, j \in [n]$. Lemma 4.2.4 is modified as

$$E(x_{ij}^{2k}) = \prod_{j=0}^{k-1} \frac{1+2j}{n+2j} = O(n^{-k}) \quad \text{as} \quad n \to \infty. \tag{4.2.17}$$

Indeed, to determine the distribution of x_{ij}^2 we may assume that

$$x_{ij}^2 = \frac{\xi_1^2}{\sum_{k=1}^n \xi_k^2}$$

with independent standard Gaussian variables ξ_j. Hence

$$x_{ij}^2 = \frac{\eta_1}{\eta_1 + \eta_2},$$

where η_1 has the χ^2-*distribution* with 1 degree of freedom, η_2 has the same with $n-1$ degrees of freedom, and moreover they are independent. This implies ([161], Chap. 3a) that x_{ij}^2 has a *beta distribution*

$$\frac{\Gamma(\alpha+\beta)}{\Gamma(\alpha)\Gamma(\beta)} x^{\alpha-1}(1-x)^{\beta-1} \tag{4.2.18}$$

with parameters $\alpha = 1/2$ and $\beta = (n-1)/2$. Consequently, (4.2.17) is obtained.

Now the proofs of Proposition 4.2.5 and Theorem 4.2.6 work well for an independent Haar distributed family $(O(s,n))_{s\in S}$ of $n \times n$ random orthogonal matrices. In this way, the distribution of $(O(s,n))_{s\in S}$ converges (in the strong sense for Theorem 4.2.6) to that of free Haar unitaries.

In the next section we shall give more extended results on the asymptotic freeness of standard unitary random matrices, and also some selfadjoint or non-selfadjoint random matrices.

4.3 Asymptotic freeness of some random matrices

The main achievement of the previous section was the asymptotic freeness of sequences of independent Haar distributed unitary (or orthogonal) matrices as the matrix size goes to infinity. In this section similar results are established for some other random matrices: Independent standard unitary matrices and deterministic matrices are asymptotically free when the latter admit the limit distribution; moreover in place of standard unitary matrices we can put selfadjoint or non-selfadjoint random matrices with a unitarily invariant distribution (in particular, standard selfadjoint or standard non-selfadjoint Gaussians are so).

Let S be a set. For $n \in \mathbb{N}$ let $(X(s,n))_{s\in S}$ be a family of $n \times n$ random matrices. $(X(s,n))_{s\in S}$ is said to have the *limit distribution* μ as $n \to \infty$ if μ is a distribution on $\mathbb{C}\langle X_s | s \in S\rangle$ and

$$\mu(X_{s_1} X_{s_2} \cdots X_{s_m}) = \lim_{n\to\infty} \tau_n(X(s_1,n)X(s_2,n)\cdots X(s_m,n))$$

for all $s_1, \ldots, s_m \in S$, where $\mathbb{C}\langle X_s | s \in S\rangle$ denotes the algebra of polynomials in noncommuting indeterminates X_s $(s \in S)$ over $\mathbb{C}$. (Recall that the joint distribition of noncommutative random variables as a functional was introduced in Sec. 1.2). Let $\{S_j : j \in J\}$ be a partition of S. The meaning of the *asymptotic freeness* of $(\{X(s,n) : s \in S_j\})_{j\in J}$ is that on the one hand $(X(s,n))_{s\in S}$ has the limit distribution μ, and on the other hand $(\mathbb{C}\langle X_s | s \in S_j\rangle)_{j\in J}$ is free in $(\mathbb{C}\langle X_s | s \in S\rangle, \mu)$. This notion is concerned with the convergence under the tracial states τ_n. However, it is also natural to consider the almost everywhere convergence under the normalized traces tr_n. Asymptotic freeness almost everywhere is a stronger property of random matrix models; it makes no sense in a general abstract setting of noncommutative random variables.

Given a family $(X(s,n))_{s\in S}$ of $n \times n$ random matrices for every $n \in \mathbb{N}$, we say that $(\{X(s,n) : s \in S_j\})_{j\in J}$ is *asymptotically free almost everywhere* as $n \to \infty$ if $(X(s,n))_{s\in S}$ has the (non-random) limit distribution almost surely, that is, there exists a distribution μ on $\mathbb{C}\langle X_s | s \in S\rangle$ such that for every $s_1, \ldots, s_m \in S$

$$\lim_{n\to\infty} \mathrm{tr}_n(X(s_1,n)X(s_2,n)\cdots X(s_m,n)) = \mu(X_{s_1} X_{s_2} \cdots X_{s_m}) \quad \text{a.s.}$$

and if $(\mathbb{C}\langle X_s | s \in S_j\rangle)_{j\in J}$ is free in $(\mathbb{C}\langle X_s | s \in S\rangle, \mu)$. It is easy to see that this is equivalent to saying that the following two conditions hold:

(i) For each $j \in J$, $(X(s,n))_{s\in S_j}$ has the (non-random) limit distribution almost surely.

(ii) For every collection $j_1, \dots, j_l \in J$ such that $j_1 \neq j_2 \neq \dots \neq j_l$, if $P_r(X_{s_r(1)}, \dots, X_{s_r(m_r)}) \in \mathbb{C}\langle X_s | s \in S_{j_r}\rangle$ satisfies

$$\lim_{n\to\infty} \mathrm{tr}_n\big(P_r(X(s_r(1), n), \dots, X(s_r(m_r), n))\big) = 0 \ \text{ a.s.}$$

for $1 \leq r \leq l$, then

$$\lim_{n\to\infty} \mathrm{tr}_n\Big(\prod_{r=1}^{l} P_r(X(s_r(1), n), \dots, X(s_r(m_r), n))\Big) = 0 \ \text{ a.s.}$$

Indeed, note that (i) and (ii) together imply that the whole $(X(s,n))_{s\in S}$ has the almost sure limit distribution. (Prove by induction.)

For the random matrices $X(s,n)$ treated below we usually have

$$\sup_n \tau_n\big((X(s,n)^* X(s,n))^k\big) < +\infty \qquad (s \in S,\ k \in \mathbb{N}).$$

This is the case if $X(s,n)^* X(s,n)$ has a limit distribution which is given as a compactly supported probability measure on $\mathbb{R}$. The above boundedness property implies that each sequence $\{\mathrm{tr}_n(\cdots)\}$ in (i) and (ii) is uniformly integrable, and the almost everywhere convergence yields the convergence of the expectations. Restricted to uniformly integrable matrix models, the definition of asymptotic freeness almost everywhere is actually a stronger property than plain asymptotic freeness. In the rest of the section we are concerned with asymptotic freeness in the almost everywhere sense.

Theorem 4.3.1 *Let $(U(s,n))_{s\in S}$ be an independent family of $n \times n$ standard unitary random matrices. Let $(D(t,n))_{t\in T}$ be a family of $n \times n$ constant (i.e. non-random) matrices such that $\sup_n \|D(t,n)\| < +\infty$ for each $t \in T$ ($\|\cdot\|$ denotes the operator norm) and $(D(t,n), D(t,n)^*)_{t\in T}$ has the limit distribution. Then the family*

$$\Big(\big(\{U(s,n), U(s,n)^*\}\big)_{s\in S},\ \{D(t,n), D(t,n)^* : t \in T\}\Big)$$

is asymptotically free almost everywhere as $n \to \infty$.

Proof: Theorem 4.2.6 says in particular that $(U(s,n), U(s,n)^*)$ has the almost sure limit distribution. As is readily seen, we may assume without loss of generality that $\{(D(t,n))_{n\in\mathbb{N}} : t \in T\}$ forms a *-subalgebra of $\prod_{n\in\mathbb{N}} M_n(\mathbb{C})$. (In fact, the *-subalgebra of $\prod_{n\in\mathbb{N}} M_n(\mathbb{C})$ generated by $(D(t,n))_{n\in\mathbb{N}}$ $(t \in T)$ and the identity

$(I_n)_{n\in\mathbb{N}}$ may be considered as T itself.) Then it suffices (cf. the proof of Theorem 4.2.6) to prove that if $s_1,\dots,s_l \in S$, $m_1,\dots,m_l \in \mathbb{Z}\setminus\{0\}$ and $t_1,\dots,t_l \in T$ are such that for $1 \le r \le l$ either

$$\mathrm{tr}_n(D(t_r,n)) = 0 \quad (n\in\mathbb{N})$$

or

$$D(t_r,n) = I_n \quad (n\in\mathbb{N}) \quad \text{and} \quad s_r \ne s_{r+1} \quad (\text{with } s_{l+1} = s_1),$$

then

$$E\big(|\mathrm{tr}_n(U(s_1,n)^{m_1}D(t_1,n)U(s_2,n)^{m_2}D(t_2,n)\cdots U(s_l,n)^{m_l}D(t_l,n))|^2\big) = O(n^{-2}) \quad \text{as} \quad n\to\infty. \tag{4.3.1}$$

(Here, note that the assumption $\mathrm{tr}_n(D(t_r,n)) = 0$ $(n\in\mathbb{N})$ can be imposed instead of $\lim_n \mathrm{tr}_n(D(t_r,n)) = 0$, because we may replace $D(t_r,n)$ by $D(t_r,n) - \mathrm{tr}_n(D(t_r,n))I_n$.)

In addition to the notations in the proof of Theorem 4.2.6, set $k(r) := |m_1| + \cdots + |m_r|$, $k(l+r) := k + k(r)$, and $t_{l+r} := t_r$ for $1 \le r \le l$. Then the left-hand side of (4.3.1) is expressed as

$$\left(\frac{1}{n}\right)^2 \sum_{i_1,\dots,i_{2k}=1}^{n} \sum_{j_{k(1)},\dots,j_{k(l)},j_{k(l+1)+1},\dots,j_{k(2l)+1}=1}^{n} \Big(\prod_{r=1}^{l} d_{j_{k(r)}i_{k(r)+1}}(t_r,n)\Big) \times \Big(\prod_{r=l+1}^{2l} \bar{d}_{i_{k(r)}j_{k(r)+1}}(t_r,n)\Big) E\Big(\prod_{h=1}^{2k} u_{i_h j_h}(s(h),\varepsilon(h),n)\Big), \tag{4.3.2}$$

where $i_{k(l)+1} = i_1$, $j_{k(2l)+1} = j_{k+1}$, and

$$\begin{cases} j_h = i_{h+1} & \text{for } h\in\{1,\dots,k\}\setminus\{k(1),\dots,k(l)\}, \\ i_h = j_{h+1} & \text{for } h\in\{k+1,\dots,2k\}\setminus\{k(l+1),\dots,k(2l)\}. \end{cases} \tag{4.3.3}$$

For any nonzero term of (4.3.2), choose two pair partitions $\mathcal{U},\mathcal{V}$ of $[2k]$ as in the proof of Theorem 4.2.6. These $\mathcal{U},\mathcal{V}$ together with (4.3.3) give many equalities among $i_1,\dots,i_{2k}$, so the equivalence relation $\mathcal{R}(\mathcal{U},\mathcal{V})$ on $[2k]$ is defined as in the proof of Theorem 4.2.6. However, singleton equivalence classes may appear in this case. If $\{h\}$ is a singleton equivalence class of $\mathcal{R}(\mathcal{U},\mathcal{V})$, then the following must hold: When $1 \le h \le k$, either $\{h,h-1\}\in\mathcal{U}$, $\varepsilon(h) = 1$, $\varepsilon(h-1) = -1$ or $\{h,h-1\}\in\mathcal{V}$, $\varepsilon(h) = -1$, $\varepsilon(h-1) = 1$, and for some $1 \le r \le l$

$$h = k(r)+1, \; s_r = s_{r+1}, \; j_{k(r)} = i_{k(r)+1}, \; \mathrm{tr}_n(D(t_r,n)) = 0 \, (n\in\mathbb{N}), \tag{4.3.4}$$

where $h \in [k]$ and $r \in [l]$ are understood in the cyclic order. On the other hand, when $k+1 \le h \le 2k$, either $\{h, h+1\} \in \mathcal{U}$, $\varepsilon(h) = 1$, $\varepsilon(h+1) = -1$ or $\{h, h+1\} \in \mathcal{V}$, $\varepsilon(h) = -1$, $\varepsilon(h+1) = 1$, and for some $l+1 \le r \le 2l$

$$h = k(r)\,, \quad s_r = s_{r+1}\,, \quad i_{k(r)} = j_{k(r)+1}\,, \quad \mathrm{tr}_n(D(t_r, n)) = 0 \ (n \in \mathbb{N}), \tag{4.3.5}$$

where $h \in \{k+1, \dots, 2k\}$ and $r \in \{l+1, \dots, 2l\}$ are understood in the cyclic order.

Now, fix $\mathcal{U}, \mathcal{V}$ and let $h(1), \dots, h(k_0)$ be the representatives from the equivalence classes of $\mathcal{R}(\mathcal{U}, \mathcal{V})$ where $h(1), \dots, h(l_0)$ are from the singleton equivalence classes $(0 \le l_0 \le k_0)$. It is obvious that

$$k_0 \le l_0 + \frac{2k - l_0}{2} = k + \frac{l_0}{2}\,.$$

When $(i_1, \dots, i_{2k})$ is subject to $\mathcal{R}(\mathcal{U}, \mathcal{V})$, the terms of (4.3.2) are determined by $(\iota_1, \dots, \iota_{k_0}) := (i_{h(1)}, \dots, i_{h(k_0)})$, so that we can set

$$Q_n(\iota_1, \dots, \iota_{k_0}) := E\Big(\prod_{h=1}^{2k} u_{i_h j_h}(s(h), \varepsilon(h), n)\Big)\,,$$

$$C_n(\iota_1, \dots, \iota_{k_0}) := \Big(\prod_{r=1}^{l} d_{j_{k(r)} i_{k(r)+1}}(t_r, n)\Big)\Big(\prod_{r=l+1}^{2l} \bar{d}_{i_{k(r)} j_{k(r)+1}}(t_r, n)\Big)\,,$$

where $j_1, \dots, j_{2k}$ are determined subject to $\mathcal{U}, \mathcal{V}$, that is, $j_{h'} = i_h$ if $\varepsilon(h') = -1$ and $\{h, h'\} \in \mathcal{U}$, or if $\varepsilon(h') = 1$ and $\{h, h'\} \in \mathcal{V}$. Then it remains to prove that for any partition $\mathcal{W}$ of $\{1, \dots, k_0\}$ we have

$$\sum_{(\iota_1, \dots, \iota_{k_0}) : \mathcal{W}} C_n(\iota_1, \dots, \iota_{k_0}) Q_n(\iota_1, \dots, \iota_{k_0}) = O(1) \quad \text{as} \quad n \to \infty, \tag{4.3.6}$$

where the summation is over $(\iota_1, \dots, \iota_{k_0}) \in [n]^{k_0}$ such that $\iota_p = \iota_q$ if p, q are in the same block of $\mathcal{W}$ and otherwise $\iota_p \ne \iota_q$. Indeed, the sum in (4.3.2) can be divided into finite disjoint portions, each of which is written as the sum in (4.3.6) subject to some possible triple $(\mathcal{U}, \mathcal{V}, \mathcal{W})$.

First, assume $l_0 \le 1$, and so $k_0 \le k$. Since $C_n(\iota_1, \dots, \iota_{k_0})$ are uniformly bounded and by (4.2.13)

$$\sum_{\iota_1, \dots, \iota_{k_0} = 1}^{n} |Q_n(\iota_1, \dots, \iota_{k_0})| = n^{k_0} O(n^{-k}) = O(1)\,,$$

(4.3.6) holds for any $\mathcal{W}$.

Next assume $2 \le l_0 \le 3$, and so $k_0 \le k+1$. If $\#\mathcal{W} \le k_0 - 1$ ($\#\mathcal{W}$ denotes the number of blocks of $\mathcal{W}$), then

$$\sum_{(\iota_1,\dots,\iota_{k_0}):\mathcal{W}} |Q_n(\iota_1,\dots,\iota_{k_0})| = n^{k_0-1}O(n^{-k}) = O(1).$$

Assume that $\#\mathcal{W} = k_0$, so "$(\iota_1,\dots,\iota_{k_0}) : \mathcal{W}$" means that $\iota_1,\dots,\iota_{k_0}$ are all distinct. Choose $\{h\} \in \mathcal{R}(\mathcal{U},\mathcal{V})$ because $l_0 \ge 2$; then according to (4.3.4) and (4.3.5) either

$$d_{j_{k(r)}i_{k(r)+1}}(t_r,n)\ (= d_{i_h i_h}(t_r,n)) \quad \text{for some} \quad 1 \le r \le l$$

or

$$\bar{d}_{i_{k(r)}j_{k(r)+1}}(t_r,n)\ (= \bar{d}_{i_h i_h}(t_r,n)) \quad \text{for some} \quad l+1 \le r \le 2l$$

appears only once in the product $C_n(\iota_1,\dots,\iota_{k_0})$. So we may write

$$C_n(\iota_1,\dots,\iota_{k_0}) = \tilde{d}_{\iota_1\iota_1}(t_{r_1},n)\tilde{C}_n(\iota_2,\dots,\iota_{k_0}) \tag{4.3.7}$$

for some $1 \le r_1 \le 2l$, where $\tilde{d}_{\iota\iota}(\cdot,\cdot)$ means $d_{\iota\iota}(\cdot,\cdot)$ or $\bar{d}_{\iota\iota}(\cdot,\cdot)$. Note that the permutation invariance (4.2.3) implies that $Q_n(\iota_1,\dots,\iota_{k_0})$ has a constant value for all distinct $\iota_1,\dots,\iota_{k_0}$. Hence we have

$$\begin{aligned}
&\left|\sum_{(\iota_1,\dots,\iota_{k_0}):\mathcal{W}} C_n(\iota_1,\dots,\iota_{k_0})Q_n(\iota_1,\dots,\iota_{k_0})\right| \\
&\quad = \left|\sum_{(\iota_2,\dots,\iota_{k_0}):\mathcal{W}} \Big(\sum_{\iota_1\ne\iota_2,\dots,\iota_{k_0}} \tilde{d}_{\iota_1\iota_1}(t_{r_1},n)\Big)\tilde{C}_n(\iota_2,\dots,\iota_{k_0})Q_n(\iota_1,\dots,\iota_{k_0})\right| \\
&\quad = n^{k_0-1}O(n^{-k}) = O(1),
\end{aligned} \tag{4.3.8}$$

because

$$\sum_{\iota_1\ne\iota_2,\dots,\iota_{k_0}} d_{\iota_1\iota_1}(t_{r_1},n) = -\sum_{\iota_1=\iota_2,\dots,\iota_{k_0}} d_{\iota_1\iota_1}(t_{r_1},n) = O(1)$$

due to $\mathrm{tr}_n(D(t_{r_1},n)) = 0$ ($n \in \mathbb{N}$).

Now assume $4 \le l_0 \le 5$, and so $k_0 \le k+2$. As above we get (4.3.6) for any $\mathcal{W}$ with $\#\mathcal{W} \le k_0 - 2$. When $\#\mathcal{W} = k_0 - 1$, we can choose $\alpha \in \{1,\dots,l_0\}$ such that $\{\alpha\}$ is a singleton block of $\mathcal{W}$. Then either $d_{\iota_\alpha\iota_\alpha}(t_r,n)$ or $\bar{d}_{\iota_\alpha\iota_\alpha}(t_r,n)$ appears only once in $C_n(\iota_1,\dots,\iota_{k_0})$. Letting $\alpha = 1$ and writing $C_n(\iota_1,\dots,\iota_{k_0})$ as (4.3.7), we get the same expression as (4.3.8), so that (4.3.6) holds because $\#\mathcal{W} = k_0 - 1$ yields $\#\{(\iota_2,\dots,\iota_{k_0}) : \mathcal{W}\} = O(n^{k_0-2})$. When $\#\mathcal{W} = k_0$, we can write

$$C_n(\iota_1,\dots,\iota_{k_0}) = \tilde{d}_{\iota_1\iota_1}(t_{r_1},n)\tilde{d}_{\iota_2\iota_2}(t_{r_2},n)\tilde{C}_n(\iota_3,\dots,\iota_{k_0})$$

for some $1 \le r_1, r_2 \le 2l$. Since $Q_n(\iota_1, \dots, \iota_{k_0})$ is constant for all distinct $\iota_1, \dots, \iota_{k_0}$, we have

$$\begin{aligned} &\Big| \sum_{(\iota_1,\dots,\iota_{k_0}):\mathcal{W}} C_n(\iota_1, \dots, \iota_{k_0}) Q_n(\iota_1, \dots, \iota_{k_0}) \Big| \\ &\quad = \Big| \sum_{(\iota_3,\dots,\iota_{k_0}):\mathcal{W}} \Big(\sum_{\iota_2 \ne \iota_3,\dots,\iota_{k_0}} \tilde{d}_{\iota_2\iota_2}(t_{r_2}, n) \sum_{\iota_1 \ne \iota_2,\dots,\iota_{k_0}} \tilde{d}_{\iota_1\iota_1}(t_{r_1}, n) \Big) \\ &\qquad\qquad \times \tilde{C}_n(\iota_3, \dots, \iota_{k_0}) Q_n(\iota_1, \dots, \iota_{k_0}) \Big| \\ &\quad = n^{k_0 - 2} O(n^{-k}) = O(1) \,. \end{aligned}$$

We can proceed similarly in the case $6 \le l_0 \le 7$ and so on, and the proof is completed.

□

It is noteworthy that there is no restriction on the type of constant matrices $D(t, n)$ in Theorem 4.3.1; the assumptions on them are only the boundedness (for separate t) and the existence of the limit distribution. Furthermore, the latter assumption was not used in the proof of (4.3.1), and we actually proved that the estimate (4.3.1) is uniform for $D(t, n)$ whenever $\|D(t, n)\| \le R$ for some $R > 0$. We here record this fact as a lemma for later use in Chap. 6.

Lemma 4.3.2 *Let $(U(s, n))_{s \in S}$ be an independent family of $n \times n$ standard unitary random matrices. Let $s_1, \dots, s_l \in S$, $m_1, \dots, m_l \in \mathbb{Z} \setminus \{0\}$ and $R > 0$. Then*

$$\begin{aligned} &E\big(|\mathrm{tr}_n(U(s_1, n)^{m_1} D_1(n) U(s_2, n)^{m_2} D_2(n) \cdots U(s_l, n)^{m_l} D_l(n))|^2\big) \\ &\quad = O(n^{-2}) \quad \text{as} \quad n \to \infty \end{aligned}$$

uniformly for the choice of any $D_r(n) \in M_n(\mathbb{C})$ $(1 \le r \le l)$ such that for $1 \le r \le l$ either

$$\mathrm{tr}_n(D_r(n)) = 0 \quad \text{and} \quad \|D_r(n)\| \le R \quad (n \in \mathbb{N})$$

or

$$D_r(n) = I_n \quad (n \in \mathbb{N}) \quad \text{and} \quad s_r \ne s_{r+1} \quad (\text{with } s_{l+1} = s_1).$$

□

Example 4.3.3 If the unitary $V(n) \in M_n(\mathbb{C})$ permutes the basis vectors cyclically and v is a Haar unitary, then $(V(n), V(n)^*)$ converges to (v, v^*) in distribution. In fact, we have $V(n)^n = I_n$ and $\mathrm{tr}_n(V(n)^k) = 0$ for $k = 1, 2, \dots, n-1$. This implies that $\lim_n \mathrm{tr}_n(V(n)^k) = 0$ for all $k \in \mathbb{Z} \setminus \{0\}$. Further, let $U(n)$ be an $n \times n$ standard random unitary. Theorem 4.3.1 implies that

$$\lim_{n \to \infty} \mathrm{tr}_n(U(n)^{m_1} V(n)^{m_2} \cdots U(n)^{m_{2l-1}} V(n)^{m_{2l}}) = 0$$

almost surely for every $m_1, m_2, \ldots, m_{2l} \in \mathbb{Z} \setminus \{0\}$. Therefore $V(n)$ and $U(n)$ are asymptotically free almost everywhere. This tells us that if $(U(n))_{n\in\mathbb{N}}$ is chosen from $\prod_{n\in\mathbb{N}} \mathcal{U}(n)$ randomly (according to the Haar probability), then $(U(n), V(n))$ converges with probability 1 to the free pair of Haar unitaries.

□

Next we prepare a technical lemma whose proof requires the notions of the weak topology and the Lévy metric on the space of probability measures. Let X be a Polish space (i.e. a complete separable metric space), $C_b(X)$ the space of bounded continuous functions on X, and $\mathcal{M}(X)$ the space of probability Borel meausres on X. The *weak topology* on $\mathcal{M}(X)$ is defined from the dual pair $\mu(f) := \int f\,d\mu$ ($\mu \in \mathcal{M}(X)$, $f \in C_b(X)$); that is, $\mu_n \to \mu$ weakly means $\mu_n(f) \to \mu(f)$ for all $f \in C_b(X)$. The so-called *Lévy metric* on $\mathcal{M}(X)$ is defined as

$$\begin{aligned}\rho(\mu,\nu) := \inf\{\delta > 0 : \mu(F) \le \nu(F^\delta) + \delta \text{ and } \nu(F) \le \mu(F^\delta) + \delta \\ \text{for every closed } F \subset X\}\end{aligned} \tag{4.3.9}$$

($\mu, \nu \in \mathcal{M}(X)$), where $F^\delta := \{x \in X : d(x,F) < \delta\}$. A basic fact ([56], Sec. 3.2) is that the metric ρ in (4.3.9) is compatible with the weak topology on $\mathcal{M}(X)$ and $(\mathcal{M}(X), \rho)$ is a Polish space.

In particular, let $X = \mathbb{R}$. In this case, several equivalent conditions for weak convergence are known. Indeed, the weak convergence $\mu_n \to \mu$ is equivalent to each of the following:

(i) the w*-convergence: $\mu_n(f) \to \mu(f)$ for all $f \in C_0(\mathbb{R})$, the space of continuous functions vanishing at infinity;

(ii) $\int e^{\mathrm{i}\,tx}\,d\mu_n(x) \to \int e^{\mathrm{i}\,tx}\,d\mu(x)$ for all $t \in \mathbb{R}$;

(iii) $\mu_n((-\infty, t]) \to \mu((-\infty, t])$ for all $t \in \mathbb{R}$ at which $\mu((-\infty, t])$ is continuous.

When each μ_n has all moments and μ is compactly supported, we note ([33], Theorem 30.2) that if μ_n converges in distribution (i.e. in moments, $m_k(\mu_n) \to m_k(\mu)$ for all $k \in \mathbb{N}$), then $\mu_n \to \mu$ weakly; however, the converse is not necessarily true unless the supports of the μ_n are uniformly bounded.

Lemma 4.3.4 *For every $p \ge 1$, $R > 0$ and $\varepsilon > 0$, there exist $k_0 \in \mathbb{N}$ and $\delta > 0$ such that, for every $n \in \mathbb{N}$ and every $\lambda, \eta \in \mathbb{R}^n_{\le}$, if $|\eta_i| \le R$ $(1 \le i \le n)$ and $|\sum_{i=1}^n \lambda_i^k - \sum_{i=1}^n \eta_i^k| \le n\delta$ for all $1 \le k \le k_0$, then $\sum_{i=1}^n |\lambda_i - \eta_i|^p \le n\varepsilon$.*

Proof: Write $\kappa_\lambda := \frac{1}{n}\sum_{i=1}^n \delta(\lambda_i)$ $(\in \mathcal{M}(\mathbb{R}))$, $\|\lambda\|_\infty := \max_i |\lambda_i|$ and $\|\lambda\|_p := (\frac{1}{n}\sum_i |\lambda_i|^p)^{1/p}$ for $\lambda \in \mathbb{R}^n$. We may assume that $R = 1$. First we prove the assertion under the assumption that $\|\lambda\|_\infty \le 1$ as well as $\|\eta\|_\infty \le 1$. From the above remarked fact on the weak topology, it suffices to show that $\|\lambda - \eta\|_p$ is arbitrarily small (independently of n) as $\rho(\kappa_\lambda, \kappa_\eta) \to 0$. Let $m \in \mathbb{N}$ and divide $[-1, 1]$ into $J(l) := [(l-1)m^{-3}, lm^{-3})$ $(-m^3 + 1 \le l \le m^3 - 1)$ and $J(m^3) := [1 - m^{-3}, 1]$. For each $j = -m+1, \ldots, m$, a number l_j can be chosen from $\{(j-1)m^2 + 1, \ldots, jm^2\}$

so that $\kappa_\lambda(J(l_j)) \le m^{-2}$. Take closed intervals $F_{-m} := [-1, (l_{-m+1} - 1)m^{-3}]$, $F_j := [l_j m^{-3}, (l_{j+1} - 1)m^{-3}]$ $(-m < j < m)$, and $F_m := [l_m m^{-3}, 1]$. Then the length of F_j is less than $2m^{-1}$. If $\rho(\kappa_\lambda, \kappa_\eta) < \delta := m^{-3}/2$, then for $-m \le j \le m$ we get $\kappa_\lambda(F_j) \le \kappa_\eta(F_j^\delta) + \delta$; that is,

$$\#\{i : \lambda_i \in F_j\} \le \#\{i : \eta_i \in F_j^\delta\} + n\delta\,.$$

Hence we can select λ_i $(i \in I_j)$ from $\{\lambda_1, \dots, \lambda_n\} \cap F_j$ so that

$$\#\{i : \lambda_i \in F_j\} - n\delta \le \#I_j \le \#\{i : \eta_i \in F_j^\delta\}\,.$$

Since the F_j^δ's are disjoint, we have a permutation π on $[n]$ such that $\eta_{\pi(i)} \in F_j^\delta$ for all $i \in I_j$ $(-m \le j \le m)$. When the coordinates of $\lambda, \eta \in \mathbb{R}^n$ are increasingly arranged, the following is well-known:

$$\sum_{i=1}^n |\lambda_i - \eta_i|^p \le \sum_{i=1}^n |\lambda_i - \eta_{\pi(i)}|^p$$

for every permutation π. Since

$$\begin{aligned} \frac{1}{n}\sum_{j=-m}^m \#I_j &\ge \frac{1}{n}\sum_{j=-m}^m \#\{i : \lambda_i \in F_j\} - (2m+1)\delta \\ &\ge 1 - \kappa_\lambda\Bigl(\bigcup_{j=-m+1}^m J(l_j)\Bigr) - 2m^{-1} \ge 1 - 4m^{-1}\,, \end{aligned}$$

we get

$$\begin{aligned} \frac{1}{n}\sum_{i=1}^n |\lambda_i - \eta_i|^p &\le \frac{1}{n}\sum_{j=-m}^m \sum_{i \in I_j} |\lambda_i - \eta_{\pi(i)}|^p + 4m^{-1}2^p \\ &\le (2m^{-1} + \delta)^p + 2^{p+2}m^{-1} \le (3^p + 2^{p+2})m^{-1}\,. \end{aligned}$$

By the assertion proved above, given $\varepsilon > 0$ there exist $k_1 \in \mathbb{N}$ and $\delta > 0$ with $\delta < \varepsilon/2$ such that if $\lambda, \eta \in \mathbb{R}^n_\le$ satisfy $\|\lambda\|_\infty, \|\eta\|_\infty \le 1$ and $|m_k(\kappa_\lambda) - m_k(\kappa_\eta)| \le 2\delta$ for $1 \le k \le k_1$, then $\|\lambda - \eta\|_p \le \varepsilon/2$. For any $\lambda \in \mathbb{R}^n_\le$ let $\tilde\lambda$ be given as

$$\tilde\lambda_i := \begin{cases} -1 & \text{if } \lambda_i < -1, \\ \lambda_i & \text{if } -1 \le \lambda_i \le 1, \\ 1 & \text{if } \lambda_i > 1. \end{cases}$$

Then

$$|m_k(\kappa_\lambda) - m_k(\kappa_{\tilde\lambda})| \le \frac{1}{n}\sum_{|\lambda_i|>1} (|\lambda_i|^k - 1) \le \Bigl(\sup_{t \ge 1} \frac{t^k - 1}{t^r}\Bigr) m_r(\kappa_\lambda)$$

and

$$\frac{1}{n}\sum_{i=1}^{n}|\lambda_i-\tilde{\lambda}_i|^p=\frac{1}{n}\sum_{|\lambda_i|>1}(|\lambda_i|-1)^p\le\left(\sup_{t\ge1}\frac{t-1}{t^{r/p}}\right)^p m_r(\kappa_\lambda)$$

for any $r\in\mathbb{N}$. Since $\sup_{t\ge1}(t^k-1)/t^r\to0$ as $r\to\infty$, we can choose k_0 ($\ge k_1$) such that

$$\sup_{t\ge1}\frac{t^k-1}{t^{k_0}}\le\sup_{t\ge1}\frac{t^k-1}{t^{k_0/p}}\le\frac{\delta}{1+\delta}\qquad(1\le k\le k_1).$$

If $\lambda,\eta\in\mathbb{R}^n_\le$ with $\|\eta\|_\infty\le1$ and $|m_k(\kappa_\lambda)-m_k(\kappa_\eta)|\le\delta$ for $1\le k\le k_0$, then we have

$$|m_k(\kappa_\lambda)-m_k(\kappa_{\tilde{\lambda}})|\le\frac{\delta}{1+\delta}(m_{k_0}(\kappa_\eta)+\delta)\le\delta\qquad(1\le k\le k_1),$$

$$\frac{1}{n}\sum_{i=1}^{n}|\lambda_i-\tilde{\lambda}_i|^p\le\left(\frac{\delta}{1+\delta}\right)^p(m_{k_0}(\kappa_\eta)+\delta)\le\delta^p\,.$$

These estimates imply that $|m_k(\kappa_{\tilde{\lambda}})-m_k(\kappa_\eta)|\le2\delta$ $(1\le k\le k_1)$ and

$$\|\lambda-\eta\|_p\le\|\lambda-\tilde{\lambda}\|_p+\|\tilde{\lambda}-\eta\|_p\le\delta+\varepsilon/2\le\varepsilon\,.$$

□

We say that an $n\times n$ selfadjoint random matrix T is *unitarily invariant* if the distribution on $M_n(\mathbb{C})^{sa}$ of T is equal to that of the unitary transformation VTV^* for any $V\in\mathcal{U}(n)$. A function of a standard selfadjoint random Gaussian matrix is unitarily invariant. We are in a position to show the asymptotic freeness of an independent family of unitarily invariant selfadjoint random matrices and non-random matrices.

In the proof we shall use, besides the operator norm $\|A\|$, the Schatten p-norm (with respect to tr_n) $\|A\|_p:=\mathrm{tr}_n(|A|^p)^{1/p}$ for $A\in M_n(\mathbb{C})$, where $1\le p<\infty$. Note that $|\mathrm{tr}_n(A)|\le\|A\|_1\le\|A\|_p\le\|A\|$. The Hölder inequality $\|AB\|_r\le\|A\|_p\|B\|_q$ when $1/r=1/p+1/q$ is particularly useful.

Theorem 4.3.5 *Let $(H(s,n))_{s\in S}$ be an independent family of $n\times n$ unitarily invariant selfadjoint random matrices, and let $(D(t,n))_{t\in T}$ be as in Theorem 4.3.1. If $H(s,n)$ converges in distribution (with respect to tr_n) almost surely to a compactly supported $\rho_s\in\mathcal{M}(\mathbb{R})$ for each $s\in S$, then the family*

$$\Big((\{H(s,n)\})_{s\in S},\ \{D(t,n),D(t,n)^*:t\in T\}\Big)$$

is asymptotically free almost everywhere as $n\to\infty$.

Proof: Take the diagonalization

$$H(s,n) = U(s,n)\Lambda(s,n)U(s,n)^*,$$

where $U(s,n)$ is a unitary random matrix and

$$\Lambda(s,n) = \mathbf{Diag}(\lambda_1(s,n),\ldots,\lambda_n(s,n))$$

is a diagonal random matrix such that $\lambda_1(s,n) \le \lambda_2(s,n) \le \ldots \le \lambda_n(s,n)$. We can make $(\{U(s,n),\Lambda(s,n)\})_{s\in S}$ an independent family. Choose an independent family $(V(s,n))_{s\in S}$ of standard unitary matrices which are also independent of $(\{U(s,n),\Lambda(s,n)\})_{s\in S}$. Then $(V(s,n)U(s,n))_{s\in S}$ becomes an independent family of standard unitary matrices, and $V(s,n)H(s,n)V(s,n)^*$ has the same distribution as $H(s,n)$ due to the unitary invariance. In this way, we may assume without loss of generality that $(U(s,n))_{s\in S}$ is an independent family of standard unitary matrices.

By assumption, for each $s \in S$ the empirical eigenvalue distribution $\frac{1}{n}\sum_{i=1}^n \delta(\lambda_i(s,n))$ converges in distribution to a compactly supported measure ρ_s almost surely as $n\to\infty$. We can choose (non-random) $\xi_1(s,n) \le \xi_2(s,n) \le \ldots \le \xi_n(s,n)$ in $\operatorname{supp}\rho_s$ such that $\frac{1}{n}\sum_{i=1}^n \delta(\xi_i(s,n)) \to \rho_s$ in distribution as $n\to\infty$. Now set $\Xi(s,n) := \mathbf{Diag}(\xi_1(s,n),\ldots,\xi_n(s,n))$. Then for every $s\in S$ Lemma 4.3.4 implies that

$$\lim_{n\to\infty} \|\Lambda(s,n) - \Xi(s,n)\|_p = 0 \ \text{ a.s.} \qquad (p\ge 1). \tag{4.3.10}$$

For any $m\in\mathbb{N}$ and $p\ge 1$, using the Hölder inequality we get

$$\begin{aligned}&\|\Lambda(s,n)^m - \Xi(s,n)^m\|_p \\ &\qquad \le \sum_{j=1}^m \|\Lambda(s,n)\|_{mp}^{m-j}\|\Lambda(s,n)-\Xi(s,n)\|_{mp}\|\Xi(s,n)\|^{j-1} \to 0 \ \text{ a.s.}\end{aligned}$$

due to (4.3.10). Hence for any polynomial P and $p\ge 1$ we have

$$\lim_{n\to\infty} \|P(\Lambda(s,n)) - P(\Xi(s,n))\|_p = 0 \ \text{ a.s.} \tag{4.3.11}$$

To prove the result, we may assume as in the proof of Theorem 4.3.1 that $\{(D(t,n))_{n\in\mathbb{N}} : t\in T\}$ forms a *-subalgebra of $\prod_{n\in\mathbb{N}} M_n(\mathbb{C})$. We have to prove that if $s_1,\ldots,s_l\in S$, $P_1,\ldots,P_l\in\mathbb{C}\langle X\rangle$ and $t_1,\ldots,t_l\in T$ are such that for $1\le r\le l$

$$\lim_{n\to\infty} \operatorname{tr}_n(P_r(H(s_r,n))) = 0 \ \text{ a.s.} \tag{4.3.12}$$

and either

$$\lim_{n\to\infty} \operatorname{tr}_n(D(t_r,n)) = 0$$

or

$$D(t_r, n) = I_n \quad (n \in \mathbb{N}) \quad \text{and} \quad s_r \neq s_{r+1},$$

then

$$\lim_{n\to\infty} \mathrm{tr}_n\Big(\prod_{r=1}^{l} P_r(H(s_r, n))D(t_r, n)\Big) = 0 \ \text{ a.s.},$$

that is,

$$\lim_{n\to\infty} \mathrm{tr}_n\Big(\prod_{r=1}^{l} U(s_r, n)P_r(\Lambda(s_r, n))U(s_r, n)^* D(t_r, n)\Big) = 0 \ \text{ a.s.} \tag{4.3.13}$$

Using the Hölder inequality again, we get

$$\begin{aligned}
&\Big|\mathrm{tr}_n\Big(\prod_{r=1}^{l} U(s_r, n)P_r(\Lambda(s_r, n))U(s_r, n)^* D(t_r, n)\Big) \\
&\qquad\qquad - \mathrm{tr}_n\Big(\prod_{r=1}^{l} U(s_r, n)P_r(\Xi(s_r, n))U(s_r, n)^* D(t_r, n)\Big)\Big| \\
&\quad \le \sum_{m=1}^{l}\Big(\prod_{r=1}^{m-1} \|P_r(\Xi(s_r, n))\|\Big)\|P_m(\Lambda(s_m, n)) - P_m(\Xi(s_m, n))\|_l \\
&\qquad\qquad \times\Big(\prod_{r=m+1}^{l} \|P_r(\Lambda(s_r, n))\|_l\Big)\prod_{r=1}^{l} \|D(t_r, n)\| \\
&\quad \to 0 \ \text{ a.s.}
\end{aligned}$$

by (4.3.11). On the other hand, since $\mathrm{tr}_n(P_r(\Xi(s_r, n))) \to 0$ by (4.3.11) and (4.3.12), Theorem 4.3.1 implies that

$$\lim_{n\to\infty} \mathrm{tr}_n\Big(\prod_{r=1}^{l} U(s_r, n)P_r(\Xi(s_r, n))U(s_r, n)^* D(t_r, n)\Big) = 0 \ \text{ a.s.}$$

Here it should be remarked that each term $P_r(\Xi(s_r, n))$ is separated from $D(t_{r'}, n)$ by $U(s_r, n)$ or $U(s_r, n)^*$, so we do not need to assume the existence of the limit distribution of $(D(t, n))_{t\in T}$ and $(\Xi(s, n))_{s\in S}$ combined. Thus (4.3.13) follows.

□

By Theorems 4.1.7 and 4.3.5 we have

Corollary 4.3.6 *Let $(H(s, n))_{s\in S}$ be an independent family of $n \times n$ standard selfadjoint Gaussian matrices, and let $(D(t, n))_{t\in T}$ be as in Theorem 4.3.1. Then the family*

$$\Big((\{H(s, n)\})_{s\in S},\ \{D(t, n), D(t, n)^* : t \in T\}\Big)$$

is asymptotically free almost everywhere as $n \to \infty$.

The corollary contains the asymptotic freeness of independent standard selfadjoint Gaussian matrices in the strong sense. As we noted at the begining of the section, this implies asymptotic freeness in the mean:

$$\tau_n(H(s(1),n)H(s(2),n)\cdots H(s(m),n)) \to \tau(a_{s(1)}a_{s(2)}\cdots a_{s(m)}),$$

where $(a_s)_{s\in S}$ is a free family of semicircular elements. Since asymptotic freeness is a description of joint moments of independent selfadjoint Gaussian matrices, it could be regarded as a deep extension of the Wigner theorem. Thorbjørnsen [189] generalized the combinatorial formula (4.1.19) to several independent matrices:

$$\begin{aligned}&\tau_n(H(s(1),n)H(s(2),n)\cdots H(s(m),n)) \\ &\quad = \sum_{\gamma\in\Gamma(s(1),s(2),\dots,s(m))} n^{d(\gamma)-k-1}, \end{aligned} \tag{4.3.14}$$

where $\Gamma(s(1),s(2),\dots,s(m))$ is the set of permutations γ in the *symmetric group* $\mathbf{S}_m$ which satisfy the conditions $\gamma(j)\neq j$, $\gamma(\gamma(j))=j$ and $s(j)=s(\gamma(j))$ for all $1\le j\le m$. Since $\tau(a_{s(1)}a_{s(2)}\cdots a_{s(m)}) = \#\Gamma(s(1),s(2),\dots,s(m))$, the asymptotic freeness follows from (4.3.14).

Continuing the last discussion of the previous section, one can conclude that Theorem 4.3.1 holds analogously when $(U(s,n))_{s\in S}$ is replaced by an independent family $(O(s,n))_{s\in S}$ of standard orthogonal random matrices. In this way, one can similarly prove the previous theorem in the case when $(H(s,n))_{s\in S}$ is an independent family of real symmetric random matrices having an orthogonally invariant distribution and an almost sure limit distribution (in particular, the $H(s,n)$ could be independent standard real symmetric Gaussian matrices). Furthermore, one can apply the method in the proof of Theorem 4.3.5 to extend Theorem 4.3.1 itself to the case when $(U(s,n))_{s\in S}$ is an independent family of unitary random matrices such that the distribuion of $U(s,n)$ is equal to that of $VU(s,n)V^*$ for any $V\in\mathcal{U}(n)$ and such that $U(s,n)$ converges in distribution almost surely to a measure on $\mathbb{T}$. (Such random unitaries are considered in Sec. 5.4.)

Example 4.3.7 Let $T(n)$ be a sequence of real symmetric random matrices such that the $T_{ij}(n)$ $(1\le i\le j\le n)$ are independent Gaussian random variables with mean m/n and variance $(1+\delta_{ij})\sigma^2/n$. Then the almost sure limiting eigenvalue density of $T(n)$ is the semicircle law with mean 0 and variance σ^2.

Indeed, we decompose $T(n)$ as the sum of a non-random matrix $D(n)$ with all entries equal to m/n and a centered Gaussian random matrix $T_0(n)$. The asymptotic eigenvalue distribution of the latter is the semicircle law w, due to Theorem 4.1.5. The first summand is m times a rank one projection, so the limiting eigenvalue distribution is δ_0. Since $D(n)$ and $T_0(n)$ are asymptotically free almost everywhere, the limiting eigenvalue distribution of $T(n)$ is $\delta_0 \boxplus w = w$.

□

A consequence of Corollary 4.3.6 is that a sequence of standard non-selfadjoint Gaussian matrices forms a matrix model of the *circular distribution.* An $n \times n$ standard non-selfadjoint Gaussian matrix $X(n)$ is given in the form $X(n) = (H^{(1)}(n) + \mathrm{i}\, H^{(2)}(n))/\sqrt{2}$, where $H^{(1)}(n)$ and $H^{(2)}(n)$ are independent standard selfadjoint Gaussian matrices. Corollary 4.3.6 shows that $(H^{(1)}(n), H^{(2)}(n))$ has the limit distribution almost surely, and it is the distribution of the free pair (a, b) of standard semicircular elements. So the distribution of $(X(n), X(n)^*)$ converges to that of (c, c^*), where $c = (a + \mathrm{i}\, b)/\sqrt{2}$ is a circular element. A free family of circular elements may be easily represented in terms of creation operators in the full Fock space. Let $\mathcal{H}$ be a Hilbert space with an orthonormal basis $\{f_s\}_{s\in S} \cup \{g_s\}_{s\in S}$. Set $c_s := \ell(f_s)^* + \ell(g_s)$ on the Fock space $\mathcal{F}(\mathcal{H})$. When the vacuum state is considered, $(c_s)_{s\in S}$ constitutes a free family of circular elements, as is seen from the proof of Example 2.6.2. The following is the asymptotic freeness of standard non-selfadjoint Gaussian matrices.

Corollary 4.3.8 *Let $(X(s,n))_{s\in S}$ be an independent family of $n \times n$ standard non-selfadjoint Gaussian matrices, and let $(D(t,n))_{t\in T}$ be as in Theorem 4.3.1. Then the family*

$$\Big((\{X(s,n), X(s,n)^*\})_{s\in S},\ \{D(t,n), D(t,n)^* : t \in T\}\Big)$$

is asymptotically free almost everywhere as $n \to \infty$. Moreover, the limit distribution of $(X(s,n), X(s,n)^*)_{s\in S}$ is the distribution of a free family $(c_s, c_s^*)_{s\in S}$ of circular elements given above.

Proof: The family $(X(s,n))_{s\in S}$ can be written as

$$X(s,n) = \frac{H^{(1)}(s,n) + \mathrm{i}\, H^{(2)}(s,n)}{\sqrt{2}},$$

where $(H^{(1)}(s,n))_{s\in S} \cup (H^{(2)}(s,n))_{s\in S}$ is an independent family of $n \times n$ standard selfadjoint Gaussians. Corollary 4.3.6 says that

$$\Big((\{H^{(1)}(s,n)\})_{s\in S},\ (\{H^{(2)}(s,n)\})_{s\in S},\ \{D(t,n), D(t,n)^* : t \in T\}\Big)$$

is asymptotically free almost everywhere. In view of Proposition 2.2.8 this implies that so is the family

$$\Big((\{H^{(1)}(s,n), H^{(2)}(s,n)\})_{s\in S},\ \{D(t,n), D(t,n)^* : t \in T\}\Big),$$

which means the first assertion. The assertion on the limit distribution is immediate from the discussion just before the corollary.

□

By an obvious alteration in the statement and proof, the corollary easily extends to elliptic Gaussian random matrices.

The free convolution may be used to compute the asymptotic eigenvalue distribution of the sum of independent random matrices. Informally, the next result says that if μ is the eigenvalue distribution of a huge selfadjoint random matrix A and similarly ν is the eigenvalue distribution of B, then $\mu \boxplus \nu$ is nearly the eigenvalue distribution of $A+B$ when A and B are in generic position.

Proposition 4.3.9 *For $n \in \mathbb{N}$ let $A(n)$ and $B(n)$ be $n \times n$ selfadjoint random matrices, and let $U(n)$ be an $n \times n$ standard unitary random matrix independent of $A(n), B(n)$. Assume that the empirical eigenvalue distribution of $A(n)$ (resp. $B(n)$) converges in distribution almost surely to a compactly supported probability measure μ (resp. ν). Then $(A(n), U(n)B(n)U(n)^*)$ is asymptotically free almost everywhere, and the limiting eigenvalue distribution of $A(n)+U(n)B(n)U(n)^*$ is the free convolution $\mu \boxplus \nu$.*

Proof: Write

$$\begin{aligned} A(n) &= V_1(n)\mathbf{Diag}(a_1(n), \ldots, a_n(n))V_1(n)^*\,, \\ B(n) &= V_2(n)\mathbf{Diag}(b_1(n), \ldots, b_n(n))V_2(n)^*\,, \end{aligned}$$

where $a_1(n) \le \ldots \le a_n(n)$, $b_1(n) \le \ldots \le b_n(n)$ and $V_1(n), V_2(n)$ are random unitaries. The assumption guarantees that $U(n)$ is independent of $V_1(n), V_2(n)$. Hence $V_1(n)^*U(n)V_2(n)$ is still a standard unitary matrix. So we may assume that $A(n), B(n)$ are diagonal random matrices whose diagonals are ordered increasingly (though $U(n)$ is no longer independent of $A(n), B(n)$). As in the proof of Theorem 4.3.5, choose (non-random) $\varphi_1(n) \le \varphi_2(n) \le \ldots \le \varphi_n(n)$ in $\operatorname{supp}\mu$ such that $\frac{1}{n}\sum_{i=1}^n \delta(\varphi_i(n)) \to \mu$ in distribution as $n \to \infty$, and set $\Phi(n) := \mathbf{Diag}(\varphi_1(n), \ldots, \varphi_n(n))$. Similarly set $\Psi(n) := \mathbf{Diag}(\psi_1(n), \ldots, \psi_n(n))$ for ν. Then Lemma 4.3.4 implies that $\lim_n \|A(n) - \Phi(n)\|_p = 0$ a.s. and $\lim_n \|B(n) - \Psi(n)\|_p = 0$ a.s. for all $p \ge 1$. Now the proof using Theorem 4.3.1 goes through similarly to the proof of Theorem 4.3.5, so we omit the details. The second assertion is immediate from the first.

□

Let $H(n)$ be a sequence of real symmetric or complex selfadjoint Gaussian random matrices which constitutes a standard matrix model of the semicircle law w_r. Furthermore, let $A(n)$ be a sequence of diagonal random matrices with increasingly ordered diagonals such that the empirical density of $A(n)$ converges to a compactly supported measure μ almost surely. Since $H(n)$ is diagonalized by a standard orthogonal or unitary matrix, it follows as in Proposition 4.3.9 that the almost sure limiting eigenvalue distribution of $A(n)+H(n)$ is the free convolution $\mu \boxplus w_r$. The measure $\mu \boxplus w_r$ can be computed in principle by the R-series. Pastur called this diagonal perturbation of the Gaussian matrix the *deformed Wigner ensemble* and its limiting eigenvalue distribution the *deformed semicircle law*. Historically, this was much earlier than the breakthrough of free probability.

We say that an $n \times n$ random matrix T is *bi-unitarily invariant* if the distribution on $M_n(\mathbb{C})$ of T is equal to that of V_1TV_2 for any $V_1, V_2 \in \mathcal{U}(n)$. Let $M_n(\mathbb{C})^+$ denote the set of positive semidefinite matrices in $M_n(\mathbb{C})$. For a non-singular $X \in M_n(\mathbb{C})$

one has a unique polar decomposition $X = UH$ with $U \in \mathcal{U}(n)$ and $H = |X| = (X^*X)^{1/2} \in M_n(\mathbb{C})^+$.

Lemma 4.3.10 *An $n \times n$ random matrix T is bi-unitarily invariant if and only if its distribution on $M_n(\mathbb{C})$ is equal to that of a random matrix of the form UH such that*

(1) *U is an $n \times n$ standard unitray random matrix,*

(2) *H is an $n \times n$ unitarily invariant positive semidefinite random matrix, and*

(3) *U and H are independent.*

Proof: Let U and H be as stated in (1)–(3). For any $V_1, V_2 \in \mathcal{U}(n)$ it follows that the distributions on $M_n(\mathbb{C})$ of $V_1(UH)V_2 = (V_1UV_2)(V_2^*HV_2)$ and UH are the same. Hence UH is bi-unitarily invariant.

Conversely, assume that T is a bi-unitarily invariant random matrix defined on a probability space $(\Omega, \mathbf{P})$. Here we write the underlying probability space explicitly to make the proof precise. Let $T = U_0H$ be the polar decomposition with a unitary random matrix U_0 and $H = (T^*T)^{1/2}$. Note that H is unique while U_0 is not. The bi-unitary invariance of T implies the unitary invariance of H. Now choose a standard unitary matrix V on another probability space $(\Omega', \mathbf{P}')$ and define a unitary random matrix $U(\omega', \omega) := V(\omega')U_0(\omega)$ on $(\Omega' \times \Omega, \mathbf{P}' \otimes \mathbf{P})$. It is immediate to see that the distributions of T and $VT = UH$ are the same. For any Borel sets $\Gamma \subset \mathcal{U}(n)$ and $\Xi \subset M_n(\mathbb{C})$ we have

$$
\begin{aligned}
&(\mathbf{P}' \otimes \mathbf{P})(U \in \Gamma, H \in \Xi) \\
&\quad = \int \Big(\int \chi_\Gamma(V(\omega')U_0(\omega))\, d\mathbf{P}'(\omega') \Big) \chi_\Xi(H(\omega))\, d\mathbf{P}(\omega) \\
&\quad = \gamma_n(\Gamma)\mathbf{P}(H \in \Xi),
\end{aligned}
$$

where γ_n is the Haar measure on $\mathcal{U}(n)$. This shows that U is Haar distributed and U, H are independent. Hence the required properties of U, H are shown.

□

When a bi-unitarily invariant random matrix T is non-singular almost surely, then the polar decomposition $T = UH$ is unique (almost surely), and it is easily verified that U and H themselves satisfy the properties (1)–(3) above.

Theorem 4.3.11 *Let $(X(s,n))_{s \in S}$ be an independent family of $n \times n$ bi-unitarily invariant random matrices, and let $(D(t,n))_{t \in T}$ be as in Theorem 4.3.1. If $X(s,n)^*X(s,n)$ converges in distribution almost surely to a compactly supported $\rho_s \in \mathcal{M}(\mathbb{R}^+)$ for each $s \in S$, then the family*

$$
\Big((\{X(s,n), X(s,n)^*\})_{s \in S},\ \{D(t,n), D(t,n)^* : t \in T\} \Big)
$$

is asymptotically free almost everywhere as $n \to \infty$. Moreover, the almost sure limit distribution of $(X(s,n), X(s,n)^*)_{s \in S}$ is the distribution of a free family $(x_s, x_s^*)_{s \in S}$ of R-diagonal elements, where $x_s^*x_s$ has the distribution ρ_s.

Proof: According to Lemma 4.3.10 we may write $X(s,n) = U(s,n)H(s,n)$, where $U(s,n)$ and $H(s,n)$ are as stated in (1)–(3) of the lemma. Furthermore, as in the proof of Theorem 4.3.5, we may write $H(s,n) = V(s,n)\Lambda(s,n)V(s,n)^*$, where $V(s,n)$ is a standard unitary matrix and $\Lambda(s,n)$ is a diagonal random matrix with increasingly ordered diagonals. Here we can make $(U(s,n))_{s\in S} \cup (V(s,n))_{s\in S}$ an independent family. Since $H(s,n)^2 = X(s,n)^*X(s,n)$, it follows (use Lemma 4.3.4) that $H(s,n)$ converges in distribution almost surely to the image measure of ρ_s by $t \mapsto t^{1/2}$ for each $s \in S$. Hence the method in proving Theorem 4.3.5 can be applied to show that

$$\Big((\{U(s,n)\})_{s\in S},\ (\{H(s,n)\})_{s\in S},\ \{D(t,n), D(t,n)^* : t \in T\}\Big)$$

is asymptotically free almost everywhere. Furthermore, this implies that $(X(s,n), X(s,n)^*)$ converges in distribution to $(u_s h_s, h_s u_s^*)$ almost surely, where u_s is a Haar unitary and h_s is a positive element such that h_s is free from $\{u_s, u_s^*\}$ and h_s^2 has the distribution ρ_s. Hence we have the conclusion by Corollary 2.6.15. □

Let x be a noncommutative random variable and $T(n)$ an $n \times n$ random matrix for every $n \in \mathbb{N}$. We say that $T(n)$ is an *almost everywhere random matrix model* of x if the distribution of $(T(n), T(n)^*)$ with respect to tr_n converges almost surely to that of (x, x^*) as the matrix size n goes to infinity. The asymptotic freeness results in this section tell us how to construct almost everywhere random matrix models for certain free families of noncommutative random variables. In particular, independent standard selfadjoint or non-selfadjoint Gaussian matrices provide models of free semicircular or circular elements. More examples of random matrix models will be discussed in the next section.

4.4 Random matrix models of noncommutative random variables

In this section we study some particular noncommutative random variables and their random matrix models. The polar decomposition of an R-diagonal element is discussed, and R-diagonal elements are characterized by the existence of an almost sure bi-unitarily invariant random matrix model. In particular, we treat a specified random matrix model of an R-diagonal element x when x^*x has the free Poisson distribution. A random matrix model for the compound free Poisson distribution is also given. The random matrix models we discuss have limits almost everywhere.

First, we show that in the polar decomposition of a circular element the unitary and positive parts are free. A random matrix model is useful in the proof of the next lemma.

Lemma 4.4.1 *Let x be a circular element, and let v be a Haar unitary in a C^*-probability space. If x and $\{v, v^*\}$ are in free relation, then vx is circular.*

Proof: Let $X(n)$ be an $n \times n$ standard non-selfadjoint Gaussian matrix, which is a random matrix model of x. If $D(n) := \mathbf{Diag}(1, e^{2\pi i/n}, \dots, e^{2\pi(n-1)i/n})$, then $(D(n), D(n)^*)$ obviously converges to (v, v^*) in distribution. According to Corollary 4.3.8, $(\{X(n), X(n)^*\}, \{D(n), D(n)^*\})$ is asymptotically free, and hence $D(n)X(n)$ is a random matrix model of vx. Since $D(n)X(n)$ is a standard non-selfadjoint Gaussian again, we obtain the conclusion.

□

Hereafter in this book we shall sometimes work with a noncommutative probability space over a von Neumann algebra. A *W^*-probability space* $(\mathcal{M}, \varphi)$ is a noncommutative probability space consisting of a von Neumann algebra $\mathcal{M}$ and a faithful normal state φ on $\mathcal{M}$. In particular, when τ is a faithful normal tracial state on $\mathcal{M}$, we call $(\mathcal{M}, \tau)$ a *tracial W^*-probability space.* (See Sec. 7.1 for a brief survey on von Neumann algebras.)

Proposition 4.4.2 *Let x be a noncommutative random variable in a tracial W^*-probability space, and let $x = uh$ be the polar decomposition. Then x is a circular element if and only if the following three conditions hold:*

(i) *u is a Haar unitary.*

(ii) *h is a quarter-circular element.*

(iii) *h is free from $\{u, u^*\}$.*

Proof: Assume that x is circular. The ondition (ii) was shown in Example 2.6.2. Hence the kernel of h is $\{0\}$, and so u is a unitary. Let $y_\delta := x(\delta\mathbf{1} + x^*x)^{-1/2}$ for $\delta > 0$. We have, for all $p \geq 1$,

$$\|u - y_\delta\|_p = \|\mathbf{1} - h(\delta\mathbf{1} + h^2)^{-1/2}\|_p \to 0 \quad \text{as} \quad \delta \to +0,$$

where $\|a\|_p := \tau(|a|^p)^{1/p}$ (the Schatten p-norm with respect to τ) for $a \in \mathcal{M}$. Then, appealing to the Hölder inequality, we can get

$$\lim_{\delta \to +0} \|P(h, u, u^*) - P(h, y_\delta, y_\delta^*)\|_1 = 0$$

for any polynomial P of three noncommuting indeterminates. Furthermore, h and y_δ are approximated in norm by polynomials of x, x^*. Therefore, the distribution of (h, u, u^*) is determined from that of (x, x^*), so it is independent of the choice of a circular element x. Since $e^{i\theta}x$ is also circular (consider the representation $x = \ell(f)^* + \ell(g)$, cf. Example 2.6.2), we must have $\tau((e^{i\theta}u)^k) = \tau(u^k)$ for all $k \in \mathbb{Z}$. This yields (i).

To show (iii), let $\mathcal{A}$ and $\mathcal{B}$ be the algebraic *-subalgebras generated by x and by h, respectively. A *-automorphism α on $\mathcal{A}$ is determined by $\alpha(x) = -x$. Since the distribution of (x, x^*) is invariant under α, $\tau = \tau \circ \alpha$ holds on $\mathcal{A}$ and hence α extends to $\mathcal{A}''$ (i.e. the closure of $\mathcal{A}$ in the strong operator topology). Note that $\mathcal{A}'' \supset \mathcal{B} \cup \{u\}$. We have $\alpha(y) = y$ for all $y \in \mathcal{B}$, and $\alpha(u) = -u$. So

$\tau(uy) = \tau(\alpha(uy)) = -\tau(uy)$, and $\tau(uy)$ must vanish for every $y \in \mathcal{B}$. Now choose a Haar unitary $v \in \mathcal{M}$ which is free from $\mathcal{A}''$. (This is always possible if we enlarge $\mathcal{M}$ by taking a free product.) Set $w := vu$. It is not difficult to see that

$$\tau(y_0 w^{k_1} y_1 w^{k_2} \cdots y_{n-1} w^{k_n} y_n) = 0 \tag{4.4.1}$$

whenever $k_j \in \mathbb{Z}\backslash\{0\}$ and $y_j \in \mathcal{B}$ with $\tau(y_j) = 0$ (but y_0, y_n may be $\mathbf{1}$). Indeed, the array inside the tracial state can be rewritten in the form $z_0 v_1 z_1 v_2 \cdots v_m z_m$ with $v_i \in \{v, v^*\}$ and $z_i \in \{y_0, y_n, u, uy_j, y_j u^*, uy_j u^*\}$. Hence (4.4.1) is a consequence of the free relation of v and $\mathcal{A}''$, so that w and $\mathcal{B}$ are free. In this way we conclude that vx has the polar decomposition wh with free w and h. Since vx is circular due to the previous lemma, (h, u, u^*) and (h, w, w^*) have the same distribution by the fact proved in the first paragraph. So (iii) follows.

The converse is obvious, because the distribution of (x, x^*) is fixed by the conditions (i)–(iii).

□

On the other hand, a semicircular element a has the polar decomposition $a = sh$ such that the positive part h is quarter-circular and the selfadjoint unitary part s has the distribution $\frac{1}{2}(\delta(-1) + \delta(1))$. Note that s and h are not free, because they commute.

Example 4.4.3 Let a be a *semicircular element* and v a Haar unitary in a tracial W^*-probability space. If a and v are in free relation, then va is *circular.*

Indeed, let $a = sh$ be the polar decomposition mentioned above. When one can show that vs is a Haar unitary and it is free from h, the conclusion follows from Proposition 4.4.2 applied to $va = (vs)h$. Since s and v are free, it is clear that $\tau((vs)^k) = 0$ for all $k \in \mathbb{Z} \setminus \{0\}$. For the freeness of vs and h, we proceed as in the proof of Proposition 4.4.2 and show that

$$\tau(y_0 (vs)^{k_1} y_1 (vs)^{k_2} \cdots y_{n-1} (vs)^{k_n} y_n) = 0 \tag{4.4.2}$$

whenever $k_j \in \mathbb{Z}\backslash\{0\}$ and y_j are polynomials of h with $\tau(y_j) = 0$ (but y_0, y_n may be $\mathbf{1}$). One can rewrite the array inside the trace in (4.4.2) as $z_0 v_1 z_1 v_2 \cdots v_m z_m$, where $v_i \in \{v, v^*\}$ and $z_i \in \{s, y_j, sy_j\}$ thanks to $sy_j = y_j s$. Here note that $\tau(sh^n) = 0$ for all $n \in \mathbb{N}$; in fact, $\tau(sh^n) = \tau(sa^n) = 0$ for even n and $\tau(sh^n) = \tau(a^n) = 0$ for odd n. This gives $\tau(sy_j) = 0$ for any y_j. Hence (4.4.2) follows from the free relation of a and v.

□

It is tempting to view the above example as the free analogue of the following probabilistic statement: If ξ and η are independent random variables such that ξ is a real Gaussian and η is uniformly distributed on the unit circle $\mathbb{T}$, then $\eta\xi$ has a rotation-invariant normal distribution.

R-diagonal elements (with an additional kernel assumption) in a tracial W^*-probability space can be characterized in terms of their polar decomposition.

Proposition 4.4.4 *Let x be an element in a tracial W^*-probability space such that $\ker x = \{0\}$. Let $x = uh$ be the polar decomposition (u is a unique unitary, because $\ker x = \{0\}$). Then x is an R-diagonal element if and only if u is a Haar unitary and h is free from $\{u, u^*\}$.*

Proof: Assume that x is R-diagonal, and choose a Haar unitary u' and a positive element h' as in Corollary 2.6.15. Here we may assume that x, u', h' are in the same tracial W^*-probability space $(\mathcal{M}, \tau)$. Let $\mathcal{A}$ and $\mathcal{B}$ be the algebraic $*$-subalgebras generated by x and $u'h'$, respectively. Then the $*$-isomorphism $\alpha : \mathcal{A} \to \mathcal{B}$ sending x to $u'h'$ preserves τ, because the distributions of (x, x^*) and $(u'h', h'u'^*)$ are the same. Hence α extends to a τ-preserving isomorphism of $\mathcal{A}''$ onto $\mathcal{B}''$. Since the distribution of h' has no atom at 0, u' and h' are in $\mathcal{A}''$, so we can set $u := \alpha^{-1}(u')$ and $h := \alpha^{-1}(h')$. Now it is clear that u is a Haar unitary, h is free from $\{u, u^*\}$, and $x = uh$ (this is the polar decomposition, from its uniqueness).

□

Note that due to use of Corollary 2.6.15 the previous proof is not fully contained in the book. Of course, Proposition 4.4.2 becomes a particular case.

The next proposition says that R-diagonal elements can also be characterized in terms of random matrix models.

Theorem 4.4.5 *Let x be an element in a C^*-probability space $(\mathcal{A}, \varphi)$ with a tracial state φ. Then x is R-diagonal if and only if x admits an almost everywhere random matrix model $X(n)$ which is bi-unitarily invariant.*

Proof: Assume that x is R-diagonal. By Corollary 2.6.15 we may write $x = uh$ with a Haar unitary u and a positive element h free from $\{u, u^*\}$. Choose a uniformly bounded sequence of constant diagonal positive matrices $D(n)$ ($n \in \mathbb{N}$) whose limit distribution is the distribution of h. Moreover, choose two independent $n \times n$ standard unitary random matrices $U(n), V(n)$, and define $X(n) := U(n)V(n)D(n)V(n)^*$. Then the bi-unitary invariance of $X(n)$ is immediate, and Theorem 4.3.1 says that

$$(\{U(n), U(n)^*\}, \{V(n), V(n)^*\}, \{D(n)\})$$

is asymptotically free almost everywhere. Hence $(U(n), U(n)^*, V(n)D(n)V(n)^*)$ converges in distribution to (u, u^*, h) almost surely, so $X(n)$ is an almost everywhere random matrix model of x.

The converse implication is included in Theorem 4.3.11.

□

In particular, the above discussions give more light on the polar decomposition of a circular element via the random matrix model. Let $X(n)$ be a standard non-selfadjoint Gaussian matrix. Since $X(n)$ is bi-unitarily invariant, it has the polar decomposition $X(n) = U(n)H(n)$ with independent $U(n)$ and $H(n)$. The limit $n \to \infty$ in distribution yields the polar decomposition $x = uh$ of a circular element x as stated in Proposition 4.4.2. ($X(n)$ is an almost everywhere random matrix model of the circular distribution, as Corollary 4.3.8 includes this fact.)

Next, we shall discuss the *complexification of Wishart matrices*, which provides random matrix models of R-diagonal elements related with the Marchenko-Pastur distribution (or the free Poisson distribution). For $p \geq n$ let $Z(n)$ be a complex $p \times n$ random matrix such that $\operatorname{Re} Z_{ij}(n)$ and $\operatorname{Im} Z_{ij}(n)$ $(1 \leq i \leq p,\ 1 \leq j \leq n)$ are independent with the same distribution $N(0, 1/2n)$. The random matrix $Z(n)^*Z(n)$ and its square root are unitarily invariant. Further, let $U(n)$ be an $n \times n$ standard unitary random matrix independent of $Z(n)$. Define

$$H(n) := (Z(n)^*Z(n))^{1/2}, \quad T(n) := U(n)H(n). \tag{4.4.3}$$

The complex analogue of Theorem 4.1.9 is true: The mean eigenvalue density of $H(n) = Z(n)^*Z(n)$ converges to the free Poisson distribution μ_λ as $p/n \to \lambda > 0$. It is known ([211]) that the convergence is not only in expectation but almost everywhere. (Indeed, this is included in Proposition 4.4.11 below.) By construction it follows that $T(n)$ is bi-unitarily invariant. Hence, when $n \to \infty$ and $p/n \to \lambda$, Theorem 4.3.11 tells us that the complexified Wishart matrix $T(n)$ is an almost everywhere random matrix model of an R-diagonal element $x = uh$, where h^2 has the distribution μ_λ. (Our term "complexified Wishart" is certainly not optimal; complexified is understood in the sense that the eigenvalues are complex since the matrix is not selfadjoint.)

We compute the joint distribution of the eigenvalues of the random matrix $T(n)$, and we record some auxiliary results for later use. From now on we refer to the measure

$$d\Lambda_n(A) := 2^{n(n-1)/2} \prod_{i=1}^{n} dA_{ii} \prod_{i<j} d(\operatorname{Re} A_{ij})\, d(\operatorname{Im} A_{ij})$$

as the Lebesgue measure on the space $M_n(\mathbb{C})^{sa}$ of selfadjoint matrices. (Note that this measure differs slightly from the measure (4.1.17) in a normalizing constant; this normalization arises when we take the isometry between $M_n(\mathbb{C})^{sa}$ with the Hilbert-Schmidt norm and $\mathbb{R}^{n^2}$ with the Euclidean norm.) On the other hand, the measure (4.1.25) is induced via the isometry between $M_n(\mathbb{C})$ and $\mathbb{R}^{2n^2}$. We denote this Lebesgue measure on $M_n(\mathbb{C})$ by $\hat{\Lambda}_n$.

For $A \in M_n(\mathbb{C})^{sa}$ let $A = VDV^*$ be the diagonalization, where $V \in \mathcal{U}(n)$ and $D = \mathbf{Diag}(t_1, t_2, \ldots, t_n)$ with $t_1 \leq t_2 \leq \ldots \leq t_n$. Except for a negligible set, V is unique up to a diagonal unitary factor. So one can change the coordinates as follows: $A \in M_n(\mathbb{C})^{sa} \leftrightarrow (V, D) \in \mathcal{U}(n)/T \times \mathbb{R}^n_{\leq}$, where $\mathcal{U}(n)/T$ is the homogeneous space divided by the torus of diagonal unitaries. Let $\dot{\gamma}_n$ denote the probability measure on $\mathcal{U}(n)/T$ induced from the Haar probability measure γ_n on $\mathcal{U}(n)$.

Lemma 4.4.6 *The measure Λ_n is transformed into the product measure*

$$\dot{\gamma}_n \otimes \left(\frac{(2\pi)^{n(n-1)/2}}{\prod_{j=1}^{n-1} j!} \prod_{i<j} (t_i - t_j)^2 \prod_{i=1}^{n} dt_i \right)$$

on $\mathcal{U}(n)/T \times \mathbb{R}^n_{\leq}$ under the above change of coordinates.

Proof: The proof starts on the same lines as that of Lemma 4.1.2, but an additional argument is needed. With an infinitesimal Hermitian $dM := -\mathrm{i}\, V^* dV$ we get

$$dA = V(dD + \mathrm{i}\,(dM \cdot D - D \cdot dM))V^*,$$

so that

$$d\Lambda_n(A) = \mathrm{const} \cdot \prod_{i<j} (t_i - t_j)^2 \prod_{i=1}^n dt_i \prod_{i<j} d(\mathrm{Re}\, M_{ij})\, d(\mathrm{Im}\, M_{ij}) \, .$$

Thanks to the unitary invariance of Λ_n, the measure

$$\prod_{i<j} d(\mathrm{Re}\, M_{ij})\, d(\mathrm{Im}\, M_{ij}) \quad \text{on} \quad \mathcal{U}(n)/T$$

is invariant under left unitary multiplication. Since $\dot{\gamma}_n$ is a unique probability measure on $\mathcal{U}(n)/T$ invariant in this sense, the above measure must be $\mathrm{const} \cdot \dot{\gamma}_n$. Thus the conclusion follows because the normalizing constant is $2^{n(n-1)/2}$ times the constant in Lemma 4.1.6.

□

Let $\Lambda_{+,n}$ be the measure on $M_n(\mathbb{C})^+$ induced from $\hat{\Lambda}_n$ on $M_n(\mathbb{C})$ via the map $X \mapsto X^*X$. (One can also use the measure induced via $X \mapsto |X|$, but the former is more convenient.) Consider the map (or the coordinate change)

$$X \in M_n(\mathbb{C}) \mapsto (U, X^*X) \in \mathcal{U}(n) \times M_n(\mathbb{C})^+,$$

where U is the unitary part of X. (Note that the singular matrices are negligible with respect to $\hat{\Lambda}_n$.) The next lemma shows that $\hat{\Lambda}_n$ on $M_n(\mathbb{C})$ corresponds (up to a constant) to the product of γ_n on $\mathcal{U}(n)$ and the restriction of Λ_n on $M_n(\mathbb{C})^+$ under this map.

Lemma 4.4.7 *The measure $\hat{\Lambda}_n$ is transformed to the product measure $\gamma_n \otimes \Lambda_{+,n}$ under the above coordinate change $X \mapsto (U, X^*X)$. Furthermore, the measure $\Lambda_{+,n}$ is a constant multiple of the restriction of Λ_n on $M_n(\mathbb{C})^+$ as follows:*

$$\Lambda_{+,n} = C_n \Lambda_n|_{M_n(\mathbb{C})^+} \quad \textit{with} \quad C_n = \frac{\pi^{n(n+1)/2}}{2^{n(n-1)/2} \prod_{j=1}^{n-1} j!} \, .$$

Proof: One can further take the coordinate change of X^*X by the diagonalization. So write $X^*X = VDV^*$ and $X = UVD^{1/2}V^*$, where $U \in \mathcal{U}(n)$, $V \in \mathcal{U}(n)/T$ and $D = \mathbf{Diag}(t_1, \dots, t_n)$ with $0 < t_1 \le \dots \le t_n$ (the case $t_1 = 0$ is negligible). We differentiate $X = UVD^{1/2}V^*$ to obtain

$$dX = UV\big[\tfrac{1}{2}D^{-1/2}dD + \mathrm{i}\,\big((V^* \cdot dL \cdot V)D^{1/2} + (dM \cdot D^{1/2} - D^{1/2} \cdot dM)\big)\big]V^*$$

with infinitesimal Hermitians $dL := -\mathrm{i}\,U^*dU$ and $dM := -\mathrm{i}\,V^*dV$. Hence we compute

$$\begin{aligned} d\hat{\Lambda}_n(A) &= \frac{1}{2^n}\prod_{i=1}^n (V^*\cdot dL\cdot V)_{ii}\prod_{i<j}\mathrm{Re}(V^*\cdot dL\cdot V)_{ij}\mathrm{Im}(V^*\cdot dL\cdot V)_{ij} \\ &\qquad \times\prod_{i<j}(\mathrm{Re}\,dM_{ij})(\mathrm{Im}\,dM_{ij})\prod_{i<j}(t_i-t_j)^2\prod_{i=1}^n dt_i \\ &= \frac{1}{2^n}\prod_{i=1}^n dL_{ii}\prod_{i<j} d(\mathrm{Re}\,L_{ij})\,d(\mathrm{Im}\,L_{ij}) \\ &\qquad \times\prod_{i<j} d(\mathrm{Re}\,M_{ij})\,d(\mathrm{Im}\,M_{ij})\prod_{i<j}(t_i-t_j)^2\prod_{i=1}^n dt_i\,, \end{aligned}$$

and the bi-unitary invariance of $\hat{\Lambda}_n$ implies

$$\prod_{i=1}^n dL_{ii}\prod_{i<j} d(\mathrm{Re}\,L_{ij})\,d(\mathrm{Im}\,L_{ij}) = \mathrm{const}\cdot d\gamma_n(U)\,,$$

$$\prod_{i<j} d(\mathrm{Re}\,M_{ij})\,d(\mathrm{Im}\,M_{ij}) = \mathrm{const}\cdot d\dot{\gamma}_n(V)\,.$$

Therefore, we infer that $\hat{\Lambda}_n$ is transformed to the measure

$$\gamma_n\otimes\dot{\gamma}_n\otimes\left(C_n'\prod_{i<j}(t_i-t_j)^2\prod_{i=1}^n dt_i\right)$$

on $\mathcal{U}(n)\times\mathcal{U}(n)/T\times(\mathbb{R}^+)^n_{\le}$ under the map $X\mapsto(U,V,D)$. To determine the constant C_n', use the $n\times n$ standard non-selfadjoint Gaussian matrix having the density (4.1.24), and compute

$$\begin{aligned} 1 &= \left(\frac{\pi}{n}\right)^{-n^2} C_n'\int_0^\infty\cdots\int_0^\infty \exp\left(-n\sum_{i=1}^n t_i\right)\prod_{i<j}(t_i-t_j)^2\,dt_1\cdots dt_n \\ &= \pi^{-n^2}C_n'\int_0^\infty\cdots\int_0^\infty \exp\left(-\sum_{i=1}^n x_i\right)\prod_{i<j}(x_i-x_j)^2\,dx_1\cdots dx_n \\ &= \pi^{-n^2}C_n'\left(\prod_{j=1}^{n-1} j!\right)^2 \end{aligned}$$

by the Selberg integral ([121], p. 354). Hence, under the coordinate change $M_n(\mathbb{C})^+\leftrightarrow\mathcal{U}(n)/T\times(\mathbb{R}^+)^n_{\le}$ the measure $\Lambda_{+,n}$ is written as

$$\dot{\gamma}_n\otimes\left(\frac{\pi^{n^2}}{\left(\prod_{j=1}^{n-1} j!\right)^2}\prod_{i<j}(t_i-t_j)^2\prod_{i=1}^n dt_i\right). \tag{4.4.4}$$

Comparing this with the expression of Λ_n in Lemma 4.4.6 gives the desired conclusion.

□

It is noteworthy that taking the map $X \mapsto |X|$ (instead of $X \mapsto X^*X$) we find the following induced measure on $M_n(\mathbb{C})^+$:

$$\dot{\gamma}_n \otimes \left(\frac{2^n \pi^{n^2}}{\left(\prod_{j=1}^{n-1} j!\right)^2} \prod_{i<j} (t_i^2 - t_j^2)^2 \prod_{i=1}^n t_i \prod_{i=1}^n dt_i \right)$$

under the coordinate change $M_n(\mathbb{C})^+ \leftrightarrow \mathcal{U}(n)/T \times (\mathbb{R}^+)^n_{\leq}$.

Lemma 4.4.8 *Let $p \geq n$, and let $T(n)$ be the random matrix given in (4.4.3). Then the joint probability density of the eigenvalues of $T(n)$ with respect to the Lebesgue measure $d\zeta_1 \cdots d\zeta_n$ is*

$$C_{np} \exp\Big(-n \sum_{i=1}^n |\zeta_i|^2 \Big) \prod_{i=1}^n |\zeta_i|^{2(p-n)} \prod_{i<j} |\zeta_i - \zeta_j|^2 .$$

Proof: Let R be a complex $p \times n$ random matrix with independent standard Gaussian $\mathrm{Re}\, R_{ij}$, $\mathrm{Im}\, R_{ij}$, and set $W := R^* R$, a complex Wishart matrix. Lemma 4.1.8 is slightly modified in the complex case, and we have

$$E(\exp(\mathrm{Tr}\, AW)) = \det(I - 2A)^{-p}$$

for every $A \in M_n(\mathbb{C})^{sa}$ with $A \leq I/2$. The argument in deriving the density (4.1.21) works in the present case as well, and the density on $M_n(\mathbb{C})^+$ of W with respect to $d\Lambda_n$ is computed as

$$p(H) = C_{np} \exp\Big(-\frac{1}{2} \mathrm{Tr}\, H \Big) (\det H)^{p-n} .$$

Hence the density of $H(n)^2 = Z(n)^* Z(n)$ in (4.4.3) is

$$C_{np} \exp(-n \mathrm{Tr}\, H)(\det H)^{p-n}, \tag{4.4.5}$$

with a different constant C_{np}.

Since $U(n)$ and $Z(n)$ are independent, the distribution of $T(n)$ is written as

$$d\gamma_n(U) \otimes \big(C_{np} \exp(-n \mathrm{Tr}\, H)(\det H)^{p-n}\, d\Lambda_n(H) \big)$$

under the change of coordinates $X \in M_n(\mathbb{C}) \mapsto (U, X^*X) \in \mathcal{U}(n) \times M_n(\mathbb{C})^+$, where U is the unitary part of X. We conclude from Lemma 4.4.7 that the distribution of $T(n)$ has the density

$$C'_{np} \exp(-n \mathrm{Tr}\, X^*X)(\det X^*X)^{p-n}$$

with respect to $dX = d\hat{\Lambda}_n(X)$.

Now one can proceed as in the proof of Lemma 4.1.10. After a triangulation the joint eigenvalue density is obtained as stated. The proof is left to the reader. □

When R is a $p \times n$ random matrix with independent Gaussian $\mathrm{Re}\, R_{ij}$, $\mathrm{Im}\, R_{ij}$ of $N(0, 1/2n)$ and B is a $p \times p$ selfadjoint random matrix independent of R, we have an $n \times n$ selfadjoint random matrix R^*BR called a *compound Wishart matrix.*

In the rest of this section we show that a suitably arranged sequence of compound Wishart matrices is a random matrix model of the *compound free Poisson distribution* introduced in Sec. 3.3. First, the asymptotic freeness almost everywhere is shown for an independent family of compound Wishart matrices.

Let S be a set. For $s \in S$ and $n \in \mathbb{N}$ let $p(s,n) \in \mathbb{N}$ be such that

$$p(s,n)/n \to \lambda_s \in (0, \infty) \quad \text{as} \quad n \to \infty.$$

Let $Z(s,n)$ be a $p(s,n) \times n$ random matrix such that $\mathrm{Re}\, Z_{ij}(s,n)$ and $\mathrm{Im}\, Z_{ij}(s,n)$ $(1 \le i \le p(s,n),\ 1 \le j \le n)$ are independent with the same distribution $N(0, 1/2n)$. Let $B(s,n)$ be a $p(s,n) \times p(s,n)$ selfadjoint random matrix. Assume that $(Z(s,n))_{s \in S} \cup (B(s,n))_{s \in S}$ is an independent family for each $n \in \mathbb{N}$. For each $s \in S$ assume further that $B(s,n)$ converges in distribution almost surely to a compactly supported $\rho_s \in \mathcal{M}(\mathbb{R})$.

Proposition 4.4.9 *Let $(Z(s,n))_{s \in S}$ and $(B(s,n))_{s \in S}$ be given as above, and let $(D(t,n))_{t \in T}$ be as in Theorem* 4.3.1. *Then*

$$\Big(\big(\{Z(s,n)^* B(s,n) Z(s,n)\} \big)_{s \in S}, \ \{D(t,n), D(t,n)^* : t \in T\} \Big)$$

is asymptotically free almost everywhere as $n \to \infty$.

Proof: First we treat the case where $p(s,n) = n$ for all $s \in S$, and so the $Z(s,n)$'s are $n \times n$ standard non-selfadjoint Gaussian matrices denoted by $X(s,n)$. Write

$$B(s,n) = V(s,n) \mathbf{Diag}(b_1(s,n), \dots, b_n(s,n)) V(s,n)^*,$$

where $b_1(s,n) \le \dots \le b_n(s,n)$ and $V(s,n)$ is a random unitary. By assumption, $(X(s,n))_{s \in S} \cup (V(s,n))_{s \in S}$ can be an indepedent family, so that $(V(s,n)^* X(s,n))_{s \in S}$ is still an independent family of standard non-selfadjoint Gaussians. Hence we may assume that $B(s,n)$ is diagonal with increasingly ordered diagonals (though it is no longer independent of $X(s,n)$). Since $(X(s,n)^* B(s,n) X(s,n))_{s \in S}$ is an independent family of $n \times n$ unitarily invariant selfadjoint random matrices, Theorem 4.3.5 yields the result once we show that $X(s,n)^* B(s,n) X(s,n)$ converges in distribution almost surely to a compactly supported measure for each $s \in S$. To prove this, choose (non-random) $\xi_1(s,n) \le \dots \le \xi_n(s,n)$ in $\mathrm{supp}\, \rho_s$ such that $\frac{1}{n} \sum_{i=1}^n \delta(\xi_i(s,n)) \to \rho_s$ as

$n \to \infty$, and set $\Xi(s,n) := \mathbf{Diag}(\xi_1(s,n), \dots, \xi_n(s,n))$. Then by Lemma 4.3.4 we get $\lim_{n\to\infty} \|B(s,n) - \Xi(s,n)\|_p = 0$ a.s. for all $p \geq 1$. So, as in the proof of Theorem 4.3.5, it is enough to consider $X(s,n)^*\Xi(s,n)X(s,n)$ instead of $X(s,n)^*B(s,n)X(s,n)$. Corollary 4.3.6 shows that $(\{X(s,n), X(s,n)^*\}, \{\Xi(s,n)\})$ is asymptotically free almost everywhere (for each $s \in S$). This implies the existence of the almost sure limit distribution of $X(s,n)^*B(s,n)X(s,n)$, and we have the conclusion in the case of $p(s,n) = n$.

For the general case, let $X(s,r,n)$ ($s \in S$, $r \in \mathbb{N}$) be independent $n \times n$ standard non-selfadjoint Gaussians, and for each $s \in S$ choose $d_s \in \mathbb{N}$ such that $p(s,n) \leq d_s n$ for all n. As above, we may assume that $B(s,n) = \mathbf{Diag}(b_1(s,n), \dots, b_{p(s,n)}(s,n))$ with $b_1(s,n) \leq \dots \leq b_{p(s,n)}(s,n)$. Set

$$B(s,r,n) := \mathbf{Diag}(b_{(r-1)n+1}(s,n), \dots, b_{rn}(s,n)) \qquad (1 \leq r \leq d_s),$$

where $b_i(s,n) := 0$ for $i > p(s,n)$. It is obvious that the distribution of $B(s,r,n)$ converges almost surely to some part (with renormalization) of ρ_s for any $s \in S$ and $1 \leq r \leq d_s$. Furthermore, we may write

$$Z(s,n)^*B(s,n)Z(s,n) = \sum_{r=1}^{d_s} X(s,r,n)^*B(s,r,n)X(s,r,n)\,,$$

and the above case says that

$$\Big(\big(\{X(s,r,n)^*B(s,r,n)X(s,r,n)\}\big)_{s\in S,\,1\leq r\leq d_s},\ \{D(t,n), D(t,n)^* : t \in T\}\Big)$$

is asymptotically free almost everywhere. This yields the conclusion.

□

Below we treat a single sequence of compound Wishart matrices, and write $Z(n)$, $B(n)$, λ, ρ without index s. Our aim is to show that the almost sure limit distribution of $Z(n)^*B(n)Z(n)$ is the compound free Poisson distribution $\pi_{\rho,\lambda}$. (In particular, the distribution of the complex Wishart matrix $Z(n)^*Z(n)$ converges almost surely to the free Poisson distribution μ_λ.) As it is clear from the above proof that the limit distribution is determined only by ρ and λ, we may assume that the $B(n)$'s are real constant diagonal matrices and their diagonals are ordered increasingly.

Since $p(n)/n \to \lambda \in (0,\infty)$, one can choose $d(n) \in \mathbb{N}$ such that $n, p(n) \leq d(n)$, $n/d(n) \to \theta$ and $p(n)/d(n) \to \lambda\theta$ for some $0 < \theta \leq 1$ (also $\lambda\theta \leq 1$). Let $X(n)$ be a $d(n) \times d(n)$ standard non-selfadjoint Gaussian matrix, $P(n) := \mathbf{Diag}(1, \dots, 1, 0, \dots, 0)$ with n ones and $d(n) - n$ zeros, and $\tilde{B}(n) := B(n) \oplus 0_{d(n)-p(n)}$. Then we can write

$$Z(n)^*B(n)Z(n) \oplus 0_{d(n)-n} = \frac{d(n)}{n} P(n)X(n)^*\tilde{B}(n)X(n)P(n)\,, \tag{4.4.6}$$

where the factor $d(n)/n$ arises because the variance of the real and imaginary parts of $X_{ij}(n)$ is $1/2d(n)$ while that of $Z_{ij}(n)$ is $1/2n$. By Corollary 4.3.8 the limit distribution of $P(n)X(n)^*\tilde{B}(n)X(n)P(n)$ is the distribution of pc^*bcp in a C^*-probability space $(\mathcal{A}, \varphi)$, where $c, p, b \in \mathcal{A}$ are such that

(i) c is a circular element,

(ii) p is a projection with $\varphi(p) = \theta$,

(iii) b is selfadjoint, $bp = pb$, and the distribution of b is $\lambda\theta\rho + (1 - \lambda\theta)\delta(0)$, and

(iv) $\{c, c^*\}$ and $\{p, b\}$ are free.

In this way, we conclude from (4.4.6) that the distribution of the $n \times n$ random matrix $Z(n)^*B(n)Z(n)$ converges almost surely to that of $\varphi(p)^{-1}pc^*bcp$ in $(p\mathcal{A}p, \varphi(p)^{-1}\varphi|_{p\mathcal{A}p})$. Note that the choice of $0 < \theta \le 1$ is not essential in the above construction. In particular, when $p(n) \le n$ for all $n \in \mathbb{N}$, we may take $d(n) = n$ (so $\theta = 1$ and $p = \mathbf{1}$); otherwise we need $\theta < 1$.

Lemma 4.4.10 *In a C^*-probability space $(\mathcal{A}, \varphi)$ let c be a circular element, let p be a projection with $\varphi(p) > 0$, and let $b_1, \ldots, b_N$ be selfadjoint elements such that $b_ip = pb_i$ and $b_ib_j = 0$ for $i \ne j$. If $\{c, c^*\}$ and $\{b_1, \ldots, b_N, p\}$ are free, then $\{pcb_1c^*p, \ldots, pcb_Nc^*p\}$ is a free family in $(p\mathcal{A}p, \varphi(p)^{-1}\varphi|_{p\mathcal{A}p})$. (A related result was shown in Example* 2.6.9.)

Proof: One can easily see that there are real constant diagonal matrices $P(n)$, $B_1(n), \ldots, B_N(n)$ of size n such that $P(n)^2 = P(n)$, $B_i(n)B_j(n) = 0$ for $i \ne j$ and the limit distribution of $(P(n), B_1(n), \ldots, B_N(n))$ is equal to the distribution of $(p, b_1, \ldots, b_N)$ in $(\mathcal{A}, \varphi)$. Let $X(n)$ be an $n \times n$ standard non-selfadjoint Gaussian. Since by Corollary 4.3.8

$$(\{X(n), X(n)^*\}, \{P(n), B_1(n), \ldots, B_N(n)\})$$

is asymptotically free, the distribution of $(pcb_ic^*p)_{1\le i\le N}$ in $(\mathcal{A}, \varphi)$ is equal to the limit distribution of $(P(n)X(n)^*B_i(n)X(n)P(n))_{1\le i\le N}$. Set $p(n) := \operatorname{rank} P(n)$, so that $p(n)/n \to \varphi(p)$. From the disjointness of $B_i(n)$ we can write, for $1 \le i \le N$,

$$P(n)X(n)B_i(n)X(n)^*P(n) = Z_i(n)^*B_i(n)Z_i(n) \oplus 0_{n-p(n)}\,, \tag{4.4.7}$$

where $(Z_i(n))_{1\le i\le N}$ is an independent family of $p(n) \times n$ Gaussian matrices as in Proposition 4.4.9. The proposition says that $(Z_i(n)^*B_i(n)Z_i(n))_{1\le i\le N}$ is asymptotically free. Hence by (4.4.7) we have the freeness of $(pcb_ic^*p)_{1\le i\le N}$ in $(p\mathcal{A}p, \varphi(p)^{-1}\varphi|_{p\mathcal{A}p})$.

□

Proposition 4.4.11 *The almost sure limit distribution of $Z(n)^*B(n)Z(n)$ is the compound free Poisson distribution $\pi_{\rho,\lambda}$.*

Proof: Let c, p, b be as in the above (i)–(iv) (where a W^*-probability space can be taken for $(\mathcal{A}, \varphi)$). According to the above argument it suffices to prove that $\pi_{\rho,\lambda}$

is the distribution of $\varphi(p)^{-1}pc^*bcp$ in $(p\mathcal{A}p, \varphi(p)^{-1}\varphi|_{p\mathcal{A}p})$. First, when $q \in \mathcal{A}$ is a projection, the distribution of $\varphi(p)^{-1}pc^*qcp \in p\mathcal{A}p$ is the free Poisson distribution μ_λ with $\lambda = \varphi(q)/\varphi(p)$ (with $\mu_0 = \delta(0)$). Indeed, this can be readily seen by looking at (4.4.6), where $B(n)$ is a diagonal projection $Q(n)$ such that $\operatorname{rank} Q(n)/d(n) \to \varphi(q)$ or $\operatorname{rank} Q(n)/n \to \varphi(q)/\varphi(p)$. (This follows from Example 2.4.7 as well.) Next assume that the spectrum of b is finite, so $b = \sum_{i=1}^N \xi_i q_i$ is the spectral decomposition and the distribution ν_b of b is given by $\nu_b = \sum_{i=1}^N \varphi(q_i)\delta(\xi_i)$. Since $\{pc^*q_1cp, \dots, pc^*q_Ncp\}$ is a free family by Lemma 4.4.10, we have

$$\begin{aligned}
\mathcal{R}_{\varphi(p)^{-1}pc^*bcp}(z) &= \sum_{i=1}^N \xi_i \mathcal{R}_{\varphi(p)^{-1}pc^*q_icp}(\xi_i z) \\
&= \sum_{i=1}^N \xi_i \frac{\varphi(q_i)/\varphi(p)}{1-\xi_i z} = \frac{1}{\varphi(p)} \int \frac{x}{1-xz}\, d\nu_b(x)\,.
\end{aligned}$$

For a general b, choose a sequence b_j with finite spectra such that $b_j p = p b_j$, $\|b_j - b\| \to 0$ and $\{p, b_j\}$ is free from $\{c, c^*\}$. Since $\|pc^*b_jcp - pc^*bcp\| \to 0$, it follows that $\mathcal{R}_{\varphi(p)^{-1}pc^*b_jcp}(z) \to \mathcal{R}_{\varphi(p)^{-1}pc^*bcp}(z)$ uniformly in a neighborhood of $z = 0$ (cf. Lemma 3.3.4). Hence we have

$$\begin{aligned}
\mathcal{R}_{\varphi(p)^{-1}pc^*bcp}(z) &= \lim_{j\to\infty} \frac{1}{\varphi(p)} \int \frac{x}{1-xz}\, d\nu_{b_j}(x) \\
&= \frac{1}{\varphi(p)} \int \frac{x}{1-xz}\, d\nu_b(x) = \lambda \int \frac{x}{1-xz}\, d\rho(x),
\end{aligned}$$

because $\varphi(p) = \theta$ and $\nu_b = \lambda\theta\rho + (1-\lambda\theta)\delta(0)$. By definition (Sec. 3.3) this means that the distribution of $\varphi(p)^{-1}pc^*bcp$ in $(p\mathcal{A}p, \varphi(p)^{-1}\varphi|_{p\mathcal{A}p})$ is $\pi_{\rho,\lambda}$.

□

In the above argument one can replace a circular element by a semicircular element. Let c, p, b be as above, and let a be a standard semicircular element free from $\{p, b\}$ in $(\mathcal{A}, \varphi)$. Then both distributions of $\varphi(p)^{-1}pc^*bcp$ and $\varphi(p)^{-1}pabap$ in $(p\mathcal{A}p, \varphi(p)^{-1}\varphi|_{p\mathcal{A}p})$ are the same $\pi_{\rho,\lambda}$. Indeed, p is free from c^*bc and also from aba. This can be seen from Proposition 4.4.9 and a similar result where the $Z(s,n)$'s are replaced by independent standard selfadjoint Gaussians. Hence it is enough to check that the distributions of c^*bc and aba are the same. When c, a, b are realized in a tracial W^*-probability space, these distributions are equal to those of $|c^*|\,b\,|c^*|$ and $|a|\,b\,|a|$, respectively. Since $|a|$ as well as $|c^*|$ is a quarter-circular free from b, we have the conclusion.

Notes and Remarks. There are several surveys about the relation of random matrices to physics; the first one was written by Wigner himself [214]. More recent ones are [191] and [95].

The classic papers of Wigner appeared in 1955 and in 1958 on the eigenvalue distribution of random matrices. His result provided the asymptotic mean eigenvalue density, first in [212] for matrices taking the entries ± 1, each with probability one-half, then in [213] for entries distributed symmetrically about zero. Later Arnold [6] proved that the limit holds almost everywhere, and not only in expectation. In fact, Wigner's main interest was not exactly the eigenvalue distribution but the distribution of the difference of consecutive eigenvalues. Let $\begin{bmatrix} a & b \\ b & c \end{bmatrix}$ be a random matrix and let $\lambda_1 \geq \lambda_2$ be its eigenvalues. Then the "spacing" $\lambda_1 - \lambda_2$ is $\sqrt{(a-c)^2 + 4b^2}$. If a, b and c are independent Gaussian random variables, then under certain conditions on the dispersions $(\lambda_1 - \lambda_2)^2$ has the χ^2- or exponential distribution. So the density of $\lambda_1 - \lambda_2$ is

$$p(x) = \begin{cases} \frac{\pi}{2} x \exp(-\frac{\pi}{4} x^2) & \text{for } x > 0, \\ 0 & \text{otherwise.} \end{cases}$$

It was expected by Wigner, the *Wigner Surmise*, that if the matrix size tends to infinity, then the difference of the consecutive eigenvalues of a Gaussian matrix has this limit distribution. Actually, Wigner's surmise is inaccurate, cf. [121], Sec. 1.5.

Mehta's book [121] and Girko's monograph [87] are the most comprehensive sources on random matrices. Theorem 4.1.5 is a special case of [6] with a simpler proof, and an analytic proof in the Gaussian case is found in [146], Appendix B. The computation of the Fourier transform related to the random matrix $H_0 + H(n)$ (above Theorem 4.1.7) is from [45]. The Laplace transform of the mean eigenvalue density of a standard selfadjoint Gaussian matrix is expressed in terms of confluent hypergeometric functions in [98].

A more detailed discussion on the *Wishart matrix* is in the book [4]. In particular, the Jacobian of the transformation $S = BTB$ used after Lemma 4.1.8 is also computed there. A combinatorial proof of Theorem 4.1.9 is in [142], and it seems that the moments of the Marchenko-Pastur distribution were explicitly given there first. A stronger form of the theorem was obtained earlier by Wachter [211]. It is not easy to give a precise historical account on the Wishart and random covariance matrices. [98] contains several interesting old references and a recursion for the moments of the complex Wishart matrix.

It was shown in [79] that all eigenvalues of the $n \times n$ standard symmetric Gaussian matrix are in $[-c, c]$ for $c > 2$ with probability $1 - o(1)$ as $n \to \infty$. There are several other papers concerned with the asymptotics of the largest eigenvalue or the norm (also the spectral radius) of symmetric or non-symmetric random matrices; see [83], [84], [10] for instance. A main result of [98] with applications to theory of exact C^*-algebras is a very far-reaching generalization of the fact that the largest eigenvalue of the standard selfadjoint Gaussian matrix converges to 2 almost surely.

According to [121], complex non-selfadjoint Gaussian matrices were first studied by Ginibre [85] in 1965.

The *asymptotic free* property of random matrices was established by Voiculescu [201] in the case of Gaussian random matrices together with diagonal constant matrices. Dykema [61] proved the same result in the case of general (non-Gaussian) random matrices together with block-diagonal constant matrices (with bounded block-size). The inclusion of constant matrices in these results is useful in applications to von Neumann algebra theory (in particular, to problems on free group factors); see [199], [155], [157], [158], [63]. In [201] Voiculescu obtained the asymptotic freeness of standard unitary random matrices as well by considering the unitary part in the polar decomposition of non-selfadjoint Gaussian matrices. Also, Speicher [174] used a similar method to prove Proposition 4.3.9 in the case of non-random $A(n)$ and $B(n)$. Voiculescu [208] strengthened his asymptotic freeness result so that the restriction on the type of constant matrices was removed. Another proof of the asymptotic freeness is in [167]. A different approach using Feynman diagrams was treated in [216] to obtain asymptotic freeness for unitary random matrices (plus constant matrices). Our approach here is opposite to Voiculescu's. In Sec. 4.3 we first treated the asymptotic freeness of the standard unitary random matrices, and then went to the case of unitarily invariant selfadjoint random matrices (in particular, standard selfadjoint Gaussians) via the diagonalization process. In [189] Thorbjørnsen deduced from (4.3.14) that

$$|\tau_n(H(s(1),n)H(s(2),n)\dots H(s(m),n)) - \tau(a_{s(1)}a_{s(2)}\cdots a_{s(m)})| = O(n^{-2}),$$

and used this relation to prove almost sure convergence.

Lemma 4.3.4 was given in [203], and it will be used in the next chapter too. Proposition 4.4.2 is from [199]; its combinatorial proof is found in [12]. A detailed discussion on the distribution of matrices under the polar decomposition is found in the book [87]; Lemma 4.4.7 is from Sec. 1.1 of that book, with modifications.

Proposition 4.4.4 is from [137]. The relation of bi-unitarily invariant matrix models and R-diagonal elements is new. In Theorem 4.4.5 the assumption of almost sure convergence is necessary. If $U(n)$ is a Haar distributed random unitary and ξ is a random variable independent of $U(n)$ with the property $\mathbf{Prob}(\xi = \alpha) = \mathbf{Prob}(\xi = \beta) = 1/2$ $(\alpha \neq \beta)$, then $\xi U(n)$ is bi-unitarily invariant and has a limit distribution in the sense of joint momemts. However, the limit x is not R-diagonal since $x^*x = xx^*$. This example was communicated to the authors by U. Haagerup. Similarly, let $H(n)$ be a standard selfadjoint Gaussian matrix and ξ a real random variable independent of $H(n)$, taken as above. Then $(H(n), \xi I_n)$ has the limit distribution in the sense of mean convergence, but the free relation does not appear in the limit. The concept of asymptotic freeness almost everywhere should be more appropriate than the plain asymptotic freeness when unitarily invariant random matrices are concerned.

Propositions 4.4.9 and 4.4.11 on compound Wishart matrices are new, while results similar to Lemma 4.4.10 were given in [136]. The authors thank R. Speicher for bringing compound Wishart matrices and compound free Poisson distributions to their attention.

Chapter 5

Large Deviations for Random Matrices

The concept of entropy originated from thermodynamics and became a mathematical notion in the work of Gibbs and Boltzmann. Later it got importance in information theory and in the statistical problem of testing hypothesis. The entropy $-\int f(x)\log f(x)\,dx$ of a probability density f appears mostly in limit theorems.

Even the *central limit theorem* of probability theory is understandable in terms of entropy. Let $\xi_1, \xi_2, \ldots$ be a sequence of independent identically distributed random variables of mean 0. Then the random variables $\eta_n = (\xi_1 + \xi_2 + \cdots + \xi_n)/\sqrt{n}$ have the same variance and their entropy is increasing (when n runs over the powers of 2). The limiting Gaussian variable has maximal entropy among distributions of given variance.

The reason for the observation that entropy shows up in so many limit problems is the fact that this quantity often governs the asymptotics of probabilities. This is very clear in the large deviation theory, which concerns limit theorems with exponential convergence. The rate of the convergence is described by an entropy functional.

Voiculescu recognized that the free relation may be modeled asymptotically by independent random matrices, and studied the asymptotics of the Boltzmann-Gibbs entropy of large random matrices. In this way he arrived at the appropriate entropy concept from the point of view of the free relation. In fact, the same quantity has been used in potential theory under the name logarithmic energy. However, Voiculescu's work opened a completely new perspective for the logarithmic energy, and the new terminology "free entropy" expresses the new aspects.

The free entropy appears as an important component of the rate function in large deviation theorems for random matrices. Wigner's original result was the convergence of the mean eigenvalue distribution of a certain Gaussian random matrix to the semicircle law. His result was improved by showing that the empirical eigenvalue distribution converges almost everywhere. The large deviation result was hinted at in the work of Voiculescu, but first proven by Ben Arous and Guionnet. According to this result the convergence is exponentially fast. Besides the selfadjoint Gaussian case we treat some non-selfadjoint random matrices and some unitary random matrices. In all cases the limiting eigenvalue distribution is determined by minimization of the rate function. The minus of the rate function is the weighted logarithmic energy familiar in potential theory, and the minimization problem can be solved by means of a general theorem of Mhaskar and Saff.

This chapter explains the background of Voiculescu's free entropy for a single (selfadjoint) noncommutative random variable. The logarithmic double integral is in the rate function and therefore it could be expressed as the limit of volumes of approximating matrices. In fact, this will be the way towards the extension of the concept of free entropy to several noncommutative random vairables, which will be the subject of the next chapter.

5.1 Boltzmann entropy and large deviations

In this section the flavour of large deviation theory is given after a concise discussion of the Boltzmann-Gibbs or differential entropy and relative entropy. We approach large deviations from the law of large numbers and observe that the rate of the exponential convergence is strongly related to an entropy functional.

Let μ be a measure on $\mathbb{R}^n$ with density $f(x) \equiv f(x_1, x_2, \dots, x_k)$. The *Boltzmann-Gibbs* (or *differential*) *entropy* of μ is defined as

$$S(\mu) = S(f) := -\int f(x) \log f(x)\, dx$$

(whenever this has a meaning). In particular, if μ has a *multivariate normal distribution* $N(m, \Sigma)$ whose density is

$$\frac{1}{(2\pi)^{n/2}|\Sigma|^{1/2}} \exp\bigl(-\tfrac{1}{2}\langle \Sigma^{-1}(x-m), (x-m)\rangle\bigr) \tag{5.1.1}$$

with the mean vector m and the covariance matrix Σ, then its entropy is equal to

$$\frac{n}{2}\log(2\pi e) + \frac{1}{2}\log|\Sigma|\,,$$

where $|\Sigma|$ denotes the determinant of Σ.

To explain the behavior of the Boltzmann-Gibbs entropy we shall use the apparently more complicated concept of relative entropy. If μ_i $(i = 1, 2)$ are measures on $\mathbb{R}^n$ with densities $f_i(x)$, then the *relative entropy* of μ_1 with respect to μ_2 is defined as

$$S(\mu_1, \mu_2) = S(f_1, f_2) := \int f_1(x)\bigl(\log f_1(x) - \log f_2(x)\bigr)\, dx\,. \tag{5.1.2}$$

This quantity is nonnegative when μ_1 and μ_2 are probability measures. It is worth noting that $S(\mu_1, \mu_2)$ is not symmetric in its two variables; μ_1 and μ_2 play really different roles.

Assume now that μ_1 has the mean vector m and the covariance matrix C, and moreover μ_2 is normal with the density (5.1.1). Then

$$S(\mu_1, \mu_2) = -S(\mu_1) + \frac{n}{2} \log(2\pi) + \frac{1}{2} \log |\Sigma| + \frac{1}{2} \mathrm{Tr}(\Sigma^{-1} C) .$$

So we can see that the Boltzmann-Gibbs entropy is essentially the relative entropy with respect to a Gaussian distribution with the same mean and covariance up to a sign and an additive constant depending on dimension and covariance. Moreover, the nonnegativity of $S(\mu_1, \mu_2)$ yields the inequality

$$S(\mu_1) \le \frac{n}{2} \log(2\pi e) + \frac{1}{2} \log |C| \tag{5.1.3}$$

if $\Sigma = C$ is chosen. Thus we have proved the known fact that the normal density (5.1.1) maximizes the Boltzmann-Gibbs entropy if the mean vector m and the covariance matrix Σ are fixed.

Another simple observation is that $S(\mu_1, \mu_2) = -S(\mu_1)$ if μ_2 is chosen to be the Lebesgue measure on $\mathbb{R}^n$. This fact is less useful, because the Lebesgue measure is infinite on $\mathbb{R}^n$. However, if μ_1 is compactly supported, then the Lebesgue measure on the support can be normalized into a probability measure.

The relative entropy functional often appears as the rate function in large deviation theorems. Given a complete separable metric space $\mathcal{X}$, assume that a sequence (ν_n) of probability measures on $\mathcal{X}$ converges to a point measure $\delta(x_0)$ at $x_0 \in \mathcal{X}$. (It will be shown below that the setting of the law of large numbers provides such a situation.) For an open set G whose closure does not contain x_0, we have $\nu_n(G) \to 0$. In order to speak of large deviation it is necessary that this convergence should be exponential. When

$$\liminf_{n\to\infty} L_n \log \nu_n(G) \ge -\inf\{I(x) : x \in G\} \tag{5.1.4}$$

for some sequence $0 < L_n \to 0$ and for a nonnegative function I on $\mathcal{X}$, there is room for the large deviation theory. Given a nonnegative and lower semicontinuous function I on $\mathcal{X}$, we say that (ν_n) satisfies the *large deviation principle* with the *rate function* I (in the scale L_n) if for every Borel subset Γ of $\mathcal{X}$ we have

$$\begin{aligned} -\inf\{I(x) : x \in \Gamma^\circ\} &\le \liminf_{n\to\infty} L_n \log \nu_n(\Gamma) \\ &\le \limsup_{n\to\infty} L_n \log \nu_n(\Gamma) \le -\inf\{I(x) : x \in \overline{\Gamma}\} , \end{aligned}$$

where Γ° and $\overline{\Gamma}$ denote the interior and closure of Γ. This is equivalent to saying that (5.1.4) holds for every open set $G \subset \mathcal{X}$ and

$$\limsup_{n\to\infty} L_n \log \nu_n(F) \le -\inf\{I(x) : x \in F\} \tag{5.1.5}$$

holds for every closed set $F \subset \mathcal{X}$. If the latter condition (5.1.5) holds only for compact F (and the former condition is satisfied as well), then the *weak large deviation principle* is said to hold. When the level sets $\{x \in \mathcal{X} : I(x) \le c\}$ are compact for all $c \ge 0$, it is often said that I is a *good rate function*.

Let $\eta_1, \eta_2, \dots$ be random variables and ν_n the distribution of η_n. Assume that the limit

$$F(\lambda) := \lim_{n\to\infty} \frac{1}{n} \log E(\exp n\lambda\eta_n) \tag{5.1.6}$$

exists and is finite for every $\lambda \in \mathbb{R}$, and moreover assume that $F(\lambda)$ is a differentiable function of λ. Under these conditions the *Ellis theorem* ([73], Theorem II.6.1) tells us that the large deviation principle holds for (ν_n) with $L_n = n^{-1}$ and with the rate function

$$I(t) := \sup\{t\lambda - F(\lambda) : \lambda \in \mathbb{R}\},$$

which is the convex conjugate of (5.1.6). In particular, when η_n has the normal distribution with mean 0 and variance σ^2/n, we have $F(\lambda) = \sigma^2\lambda^2/2$ and $I(t) = t^2/2\sigma^2$. In this case, the rough meaning of the large deviation result is

$$\mathbf{Prob}(|\eta_n| \ge u) \approx \exp(-nu^2/2\sigma^2) \quad \text{for} \quad u > 0.$$

Next, the so-called *level-2 large deviation theorem* will be explained for a sequence $\xi_1, \xi_2, \dots$ of independent identically distributed random variables. In brief, this theorem states the exponential convergence of the sequence

$$\frac{\delta(\xi_1) + \delta(\xi_2) + \cdots + \delta(\xi_n)}{n} \tag{5.1.7}$$

of atomic random measures to the point measure δ_ν, where ν is the common distribution of the random variables ξ_n. Note that the convergence itself without the additional features is just the law of large numbers. The complete separable metric space in the present case is the space $\mathcal{X} = \mathcal{M}(\mathbb{R})$ of probability Borel measures on $\mathbb{R}$. $\mathcal{M}(\mathbb{R})$ is endowed with the weak topology. ($\mathcal{M}(\mathbb{R})$ is a metrizable space; for instance, the so-called Lévy metric is compatible with the weak topology, see (4.3.9).) The random measure (5.1.7) is regarded as a random variable with values in $\mathcal{M}(\mathbb{R})$. The *large deviation principle* consists of two conditions; one refers to an open set $G \subset \mathcal{M}(\mathbb{R})$ and the other one tells about a closed set $F \subset \mathcal{M}(\mathbb{R})$. Namely,

$$\limsup_{n\to\infty} \frac{1}{n} \log \mathbf{Prob}\left(\frac{\delta(\xi_1) + \delta(\xi_2) + \cdots + \delta(\xi_n)}{n} \in F\right) \le -\inf\{I(\mu) : \mu \in F\},$$

$$\liminf_{n\to\infty} \frac{1}{n} \log \mathbf{Prob}\left(\frac{\delta(\xi_1) + \delta(\xi_2) + \cdots + \delta(\xi_n)}{n} \in G\right) \ge -\inf\{I(\mu) : \mu \in G\},$$

where $I(\mu) := S(\mu, \nu)$ is the *relative entropy* of μ with respect to ν, defined by

$$S(\mu, \nu) := \int \frac{d\mu}{d\nu} \log \frac{d\mu}{d\nu}\, d\nu = \int \log \frac{d\mu}{d\nu}\, d\mu$$

if μ is absolutely continuous with respect to ν and $d\mu/d\nu$ is the Radon-Nikodým derivative; otherwise $S(\mu, \nu) := +\infty$. This is written as (5.1.2) whenever μ, ν have the densities. The relative entropy $S(\cdot\,, \nu)$ is a weakly lower semicontinuous, strictly convex functional, and its level sets are known to be compact. In this way, the large deviation principle in the sense of the above definition holds with the good rate function $S(\cdot\,, \nu)$. This is the fundamental example of a level-2 large deviation theorem, due to Sanov ([56], Sec. 3.2).

Let μ and ν be probability measures on $\mathbb{R}$ and assume that μ is compactly supported. Let ν^n denote the n-fold product measure $\nu \otimes \nu \otimes \cdots \otimes \nu$ and $m_k(\mu)$ the kth moment of μ, i.e. $m_k(\mu) = \int x^k \, d\mu(x)$ for $k \in \mathbb{N}$. For $x = (x_1, \dots, x_n) \in \mathbb{R}^n$ we denote by $\kappa_n(x)$ the discrete measure $\frac{1}{n}\bigl(\delta(x_1) + \delta(x_2) + \cdots + \delta(x_n)\bigr)$. Then we can apply the Sanov theorem to show the following:

Proposition 5.1.1 *Assume that μ and ν are probability measures on $\mathbb{R}$ and μ is supported in a compact subset K of $\mathbb{R}$. Then*

$$\begin{aligned}
&\lim_{\substack{r\to\infty \\ \varepsilon\to+0}} \limsup_{n\to\infty} \frac{1}{n} \log \nu^n\bigl(\{x \in \mathbb{R}^n : |m_k(\kappa_n(x)) - m_k(\mu)| \le \varepsilon,\ k \le r\}\bigr) \\
&\quad = \lim_{\substack{r\to\infty \\ \varepsilon\to+0}} \liminf_{n\to\infty} \frac{1}{n} \log \nu^n\bigl(\{x \in \mathbb{R}^n : |m_k(\kappa_n(x)) - m_k(\mu)| \le \varepsilon,\ k \le r\}\bigr) \\
&\quad = \lim_{\substack{r\to\infty \\ \varepsilon\to+0}} \lim_{n\to\infty} \frac{1}{n} \log \nu^n\bigl(\{x \in K^n : |m_k(\kappa_n(x)) - m_k(\mu)| \le \varepsilon,\ k \le r\}\bigr) \\
&\quad = -S(\mu, \nu)\,.
\end{aligned}$$

Proof: First, we prove the last equality by application of the Sanov theorem to the conditional measure $\nu_K := \nu/\nu(K)$ in the space $\mathcal{M}(K)$. For $r \in \mathbb{N}$ and $\varepsilon > 0$ the set

$$F(r, \varepsilon) := \bigl\{\kappa \in \mathcal{M}(K) : |m_k(\kappa) - m_k(\mu)| \le \varepsilon,\ k \le r\bigr\}$$

is closed in $\mathcal{M}(K)$, and, replacing $\le \varepsilon$ by $< \varepsilon$, we get an open set $G(r, \varepsilon)$. The Sanov theorem says that

$$\begin{aligned}
&\limsup_{n\to\infty} \frac{1}{n} \log \nu_K^n\bigl(\{x \in K^n : \kappa_n(x) \in F(r, \varepsilon)\}\bigr) \\
&\quad \le -\inf\bigl\{S(\kappa, \nu_K) : \kappa \in F(r, \varepsilon)\bigr\}\,,
\end{aligned}$$

and the opposite inequality also holds if lim sup is replaced by lim inf and $F(r, \varepsilon)$ by $G(r, \varepsilon)$. If $\inf\bigl\{S(\kappa, \nu_K) : \kappa \in F(r, \varepsilon)\bigr\} = +\infty$ for some $r = r_1$ and $\varepsilon = \varepsilon_1$, then

$$\begin{aligned}
&\lim_{n\to\infty} \frac{1}{n} \log \nu_K^n\bigl(\{x \in K^n : \kappa_n(x) \in F(r, \varepsilon)\}\bigr) \\
&\quad = -\inf\bigl\{S(\kappa, \nu_K) : \kappa \in F(r, \varepsilon)\bigr\} = -\infty
\end{aligned}$$

for every $r \geq r_1$ and $\varepsilon \leq \varepsilon_1$. So assume that $\inf\{S(\kappa, \nu_K) : \kappa \in F(r, \varepsilon)\} < +\infty$ for all r and ε. Then, by the convexity of relative entropy, it is easy to check that

$$\inf\{S(\kappa, \nu_K) : \kappa \in F(r, \varepsilon)\} = \inf\{S(\kappa, \nu_K) : \kappa \in G(r, \varepsilon)\}.$$

Since $S(\kappa, \nu_K) = S(\kappa, \nu) + \log \nu(K)$ for $\kappa \in \mathcal{M}(K)$, we have

$$\lim_{n\to\infty} \frac{1}{n} \log \nu^n(\{x \in K^n : \kappa_n(x) \in F(r, \varepsilon)\}) = -\inf\{S(\kappa, \nu) : \kappa \in F(r, \varepsilon)\}.$$

Note that the sets $G(r, \varepsilon)$ constitute a neighborhood base of μ in $\mathcal{M}(K)$. Hence the infimum in the above tends to $S(\mu, \nu)$ as $r \to \infty$ and $\varepsilon \to +0$, so the last equality holds true.

Next, one can apply the Sanov theorem to a measurable subset

$$\Gamma(r, \varepsilon) := \{\kappa \in \mathcal{M}(\mathbb{R}) : |m_k(\kappa) - m_k(\mu)| \leq \varepsilon,\ k \leq r\}$$

of $\mathcal{M}(\mathbb{R})$, to get

$$\limsup_{n\to\infty} \frac{1}{n} \log \nu^n(\{x \in \mathbb{R}^n : \kappa_n(x) \in \Gamma(r, \varepsilon)\}) \leq -\inf\{S(\kappa, \nu) : \kappa \in \overline{\Gamma(r, \varepsilon)}\}.$$

So it remains to show that

$$\lim_{\substack{r\to\infty \\ \varepsilon\to+0}} \inf\{S(\kappa, \nu) : \kappa \in \overline{\Gamma(r, \varepsilon)}\} = S(\mu, \nu).$$

But, by the weak lower semicontinuity of $S(\,\cdot\,, \nu)$, it suffices to see that $\kappa_j \to \mu$ weakly for any $\kappa_j \in \Gamma(r_j, \varepsilon_j)$ with $r_j \to \infty$ and $\varepsilon_j \to +0$. This is a consequence of the known fact on convergence of probability measures mentioned just before Lemma 4.3.4.

□

For the above ν we can choose a Gaussian measure with the variance σ^2 of μ. In this case the double limit in the previous proposition is $S(\mu) - \frac{1}{2}\log(2\pi e \sigma^2)$. The measure μ being compactly supported, we may choose ν as the normalized Lebesgue measure of a large compact interval. The normalizing constant c contributes to both sides a term $\log c$, so they are eliminated. Therefore, we have

$$\begin{aligned}\lim_{\substack{r\to\infty \\ \varepsilon\to+0}} \lim_{n\to\infty} \frac{1}{n} \log \lambda^n(\{x \in [-R, R]^n : |m_k(\kappa_n(x)) - m_k(\mu)| \leq \varepsilon,\ k \leq r\}) \\ = S(\mu), \qquad (5.1.8)\end{aligned}$$

where R is large enough and λ^n is the n-dimensional Lebesgue measure. This formula is understood as follows. For a fixed measure μ on $\mathbb{R}$ we consider the set $G_n(r, \varepsilon)$ of all $x \in \mathbb{R}^n$ such that $\kappa_n(x)$ approximates μ in the moments up to r and to the extent ε. Then the entropy of μ governs the asymptotics of $\lambda^n(G_n(r,\varepsilon))$ when ε is very small and r is very large.

Let us emphasize that in Proposition 5.1.1 the reference measure ν^n is of product type. One can regard the content of the next section in such a way that instead of a product measure we use the joint distribution of the eigenvalues of a certain random matrix such as a standard symmetric Gaussian matrix. The eigenvalues of such a Gaussian matrix are not independent; physicists would say they repel each other.

5.2 Entropy and random matrices

In this section we deal with the analogue of Proposition 5.1.1 when the measure ν^n is different. To motivate the choice of the measure we first solve an entropy maximization problem. When the mean and covariance are fixed, the Gaussian random variable has maximal Boltzmann-Gibbs entropy. The analogue of this statement for random matrices will be formulated in terms of the tracial functional τ_n, given as

$$\tau_n(X) := \frac{1}{n}\sum_{i=1}^{n} E(X_{ii})$$

for an $n \times n$ random matrix X (whenever the expectations exist). The entropy of an $n\times n$ symmetric random matrix $[\xi_{ij}]_{i,j=1}^n$ is understood as the Boltzmann-Gibbs entropy of the joint distribution of the random variables $\{\xi_{ij} : 1 \le i \le j \le n\}$, that is, the upper diagonal part of the matrix. Then the symmetric Gaussian matrix maximizes the entropy under a constraint on $\tau_n(X^2)$. This is the content of the first proposition of this section. Later on, the joint distribution of the eigenvalues of the symmetric Gaussian matrix will be chosen to be the reference measure on $\mathbb{R}^n$. In this way we can arrive at Voiculescu's entropy along the lines of Proposition 5.1.1.

Proposition 5.2.1 *The entropy on the set of $n\times n$ symmetric random matrices T satisfying the conditions $\tau_n(T) = m$ and $\tau_n((T-mI)^2) = \sigma^2$ is maximized by the matrix $[\xi_{ij}]$ which has Gaussian independent entries and*

$$E(\xi_{ij}) = \delta_{ij} m\,, \quad E((\xi_{ii}-m)^2) = \frac{2\sigma^2}{n+1} = 2E(\xi_{ij}^2) \quad \textit{for} \quad i \ne j\,.$$

For $m = 0$ and $\sigma = 1$ the maximizer is the standard symmetric Gaussian matrix.

Proof: We have to maximize the Boltzmann-Gibbs entropy of the random variables $\{\xi_{ij} : 1 \le i \le j \le n\}$ under the constraints

$$\sum_{i=1}^{n} E(\xi_{ii}) = nm, \quad \sum_{i=1}^{n} E((\xi_{ii}-m)^2) + 2\sum_{i<j} E(\xi_{ij}^2) = n\sigma^2\,.$$

We may assume that $m = 0$, and we know from the additivity and subadditivity of the entropy that the maximum entropy is attained when the variables ξ_{ij} are independent. Let m_{ij} be the mean and v_{ij} the variance of ξ_{ij}. By (5.1.3) the entropy of the joint distribution is estimated from above by that of a normal distribution. Up to an additive and a multiplicative constant the entropy is $\sum_{i\leq j} \log v_{ij}$, and we trivially have

$$\sum_{i\leq j} \log v_{ij} \leq \sum_{i\leq j} \log E(\xi_{ij}^2)\,.$$

The arithmetic-geometric mean inequality gives

$$\begin{aligned} 2^{n(n-1)/2} \prod_{i\leq j} E(\xi_{ij}^2) &= \prod_{i=1}^{n} E(\xi_{ii}^2) \prod_{i<j} 2E(\xi_{ij}^2) \\ &\leq \left(\frac{\sum_{i=1}^n E(\xi_{ii}^2) + \sum_{i<j} 2E(\xi_{ij}^2)}{n(n+1)/2} \right)^{n(n+1)/2} = \left(\frac{n\sigma^2}{n(n+1)/2} \right)^{n(n+1)/2}. \end{aligned}$$

Analysis of all these inequalities shows that they are saturated when (and only when) ξ_{ij} has the normal distribution with mean zero and $E(\xi_{ii}^2) = 2E(\xi_{ij}^2)$ for $i \neq j$.

□

The distribution of the maximizer in the previous proposition will be denoted by $N^{(n)}(m, \sigma^2)$. In particular, $N^{(n)}(0,1)$ is the distribution of the $n \times n$ standard symmetric Gaussian matrix familiar from Example 4.1.1. We return to the subject of maximal entropy matrix ensembles after the proof of Voiculescu's theorem.

It was Voiculescu's idea to modify the formula of Proposition 5.1.1 and to use it as a definition of another kind of entropy of a measure μ. The modified formula is written as

$$\begin{aligned} \lim_{\substack{r\to\infty \\ \varepsilon\to+0}} \limsup_{n\to\infty} \frac{1}{n^2} \log \nu_n\big(\{A \in M_n(\mathbb{R})^{sa} : \\ |\mathrm{tr}_n(A^k) - m_k(\mu)| \leq \varepsilon,\ k \leq r\}\big), \end{aligned} \tag{5.2.1}$$

where ν_n must be a measure on $M_n(\mathbb{R})^{sa}$. Observe that the original formula of Proposition 5.1.1 is recovered if we require A to be diagonal and replace the normalization $1/n^2$ by $1/n$. The Gaussian measure $N^{(n)}(0,\sigma^2)$ is a natural candidate for ν_n. The advantage of the Gaussian measure is the invariance under orthogonal transformations, which allows the eigenvalues of A to enter. If $(\lambda_1, \lambda_2, \ldots, \lambda_n)$ are the eigenvalues of A, then $\tau(A^k) = \frac{1}{n}\sum_{i=1}^n \lambda_i^k = m_k(\kappa_n(\lambda))$ and

$$\begin{aligned} &\nu_n\big(\{A \in M_n(\mathbb{R})^{sa} : |\mathrm{tr}_n(A^k) - m_k(\mu)| \leq \varepsilon,\ k \leq r\}\big) \\ &\quad = \bar{\nu}_n\big(\{\lambda \in \mathbb{R}^n : |m_k(\kappa_n(\lambda)) - m_k(\mu)| \leq \varepsilon,\ k \leq r\}\big), \end{aligned}$$

where $\bar{\nu}_n$ is the measure on $\mathbb{R}^n$ induced by ν_n.

According to the density (4.1.9) the measure $N^{(n)}(0,\sigma^2)$ on $M_n(\mathbb{R})^{sa}$ (the real symmetric matrices) induces the joint probability density

$$\frac{1}{Z_n}\exp\Big(-\frac{n+1}{4\sigma^2}\sum_{i=1}^n \lambda_i^2\Big)\prod_{i<j}|\lambda_i-\lambda_j|$$

on $\mathbb{R}^n$ (the space of the eigenvalues), where Z_n is the normalization constant. Slightly more generally, we consider the probability measure on $\mathbb{R}^n$ having the probability density

$$\frac{1}{Z_{\beta,\sigma}^{(n)}}\exp\Big(-\frac{n+1}{4\sigma^2}\sum_{i=1}^n \lambda_i^2\Big)\prod_{i<j}|\lambda_i-\lambda_j|^{2\beta}, \tag{5.2.2}$$

where β is a fixed positive number.

Theorem 5.2.2 *Let $\bar\nu_n$ denote the probability measure with density* (5.2.2) *on $\mathbb{R}^n$. If μ is a compactly supported probability measure on $\mathbb{R}$, then*

$$\begin{aligned}&\lim_{\substack{r\to\infty\\ \varepsilon\to+0}}\limsup_{n\to\infty}\frac{1}{n^2}\log\bar\nu_n\big(\{\lambda\in\mathbb{R}^n: |m_k(\kappa_n(\lambda))-m_k(\mu)|\le\varepsilon,\ k\le r\}\big)\\ &\quad=\beta\iint\log|x-y|\,d\mu(x)\,d\mu(y)-\frac{1}{4\sigma^2}\int x^2\,d\mu(x)-\frac{\beta}{2}\log(2\beta\sigma^2)+\frac{3\beta}{4}.\end{aligned}$$

The same holds true when lim sup *is replaced by* lim inf.

Proof: First, we deal with the asymptotics of the normalization constant $Z_{\beta,\sigma}^{(n)}$. Rewriting (4.1.7), we get

$$\begin{aligned}Z_{\beta,\sigma}^{(n)} &= \int_{-\infty}^{\infty}\cdots\int_{-\infty}^{\infty}\exp\Big(-\frac{n+1}{4\sigma^2}\sum_{i=1}^n x_i^2\Big)\prod_{i<j}|x_i-x_j|^{2\beta}\,dx_1\cdots dx_n\\ &= (2\pi)^{n/2}\Big(\frac{n+1}{2\sigma^2}\Big)^{-n(\beta(n-1)+1)/2}\prod_{j=1}^n\frac{\Gamma(1+j\beta)}{\Gamma(1+\beta)}.\end{aligned} \tag{5.2.3}$$

The asymptotics of $\Gamma(1+j\beta)$ is provided by the *Stirling formula*

$$\log\Gamma(1+x)=x\log x-x+\frac12\log x+\frac12\log(2\pi)+o(1).$$

Hence

$$\begin{aligned}&\lim_{n\to\infty}\frac{1}{n^2}\log Z_{\beta,\sigma}^{(n)}\\ &\quad=\lim_{n\to\infty}\Big[\frac{\beta}{2}\log(2\sigma^2)-\frac{\beta}{2}\log(n+1)+\frac{1}{n^2}\sum_{j=1}^n\beta j\log\beta j-\frac{\beta}{2}\Big]\end{aligned}$$

$$
\begin{aligned}
&= \frac{\beta}{2}(\log(2\sigma^2) - 1) + \lim_{n\to\infty} \frac{\beta}{n} \sum_{j=1}^{n} \frac{j}{n} \log \frac{\beta j}{n} \\
&= \frac{\beta}{2}(\log(2\sigma^2) - 1) + \beta \int_0^1 x \log \beta x \, dx \\
&= \frac{\beta}{2} \log(2\beta\sigma^2) - \frac{3\beta}{4},
\end{aligned}
$$

and we write B for this number.

Let $\Gamma_n(\mu; r, \varepsilon)$ denote the set whose measure $\bar{\nu}_n$ is taken in the statement of the theorem. Then it suffices to show the following two estimates:

$$
\begin{aligned}
&\liminf_{n\to\infty} \frac{1}{n^2} \log \bar{\nu}_n\big(\Gamma_n(\mu; r, \varepsilon)\big) \\
&\quad \geq \beta \iint \log|x - y| \, d\mu(x) \, d\mu(y) - \frac{1}{4\sigma^2} \int x^2 \, d\mu(x) - B
\end{aligned} \tag{5.2.4}
$$

for every $r \in \mathbb{N}$ and $\varepsilon > 0$, and

$$
\begin{aligned}
&\lim_{\substack{r\to\infty \\ \varepsilon\to+0}} \limsup_{n\to\infty} \frac{1}{n^2} \log \bar{\nu}_n\big(\Gamma_n(\mu; r, \varepsilon)\big) \\
&\quad \leq \beta \iint \log|x - y| \, d\mu(x) \, d\mu(y) - \frac{1}{4\sigma^2} \int x^2 \, d\mu(x) - B\,.
\end{aligned} \tag{5.2.5}
$$

To show the lower estimate (5.2.4), a suitable regularization argument enables us to assume that the support of μ is an interval $[a, b]$ and μ has a continuous density $f > 0$ on $[a, b]$, so that $\delta \leq f(x) \leq \delta^{-1}$ $(a \leq x \leq b)$ for some $\delta > 0$. The details of this argument are omitted here; we shall give them in the proof of a large deviation theorem in Sec. 5.4 (see the proof of Lemma 5.4.6). For each $n \in \mathbb{N}$ let $a < a_1^{(n)} < b_1^{(n)} < a_2^{(n)} < \ldots < a_n^{(n)} < b_n^{(n)} = b$ be such that

$$
\int_a^{a_i^{(n)}} f(x) \, dx = \frac{i - \frac{1}{2}}{n}, \qquad \int_a^{b_i^{(n)}} f(x) \, dx = \frac{i}{n} \qquad (1 \leq i \leq n).
$$

Then it is immediate that

$$
\frac{\delta}{2n} \leq b_i^{(n)} - a_i^{(n)} \leq \frac{1}{2n\delta} \qquad (1 \leq i \leq n). \tag{5.2.6}
$$

Furthermore, if we define

$$
\Delta_n := \left\{(\lambda_1, \lambda_2, \ldots, \lambda_n) \in \mathbb{R}^n : a_i^{(n)} \leq \lambda_i \leq b_i^{(n)}, \ 1 \leq i \leq n\right\},
$$

then it is obvious that

$$
\Delta_n \subset \Gamma_n(\mu; r, \varepsilon)
$$

for all n large enough. We want to estimate the measure $\bar{\nu}_n$ of Δ_n. Recall that $\bar{\nu}_n(\Delta_n)$ is the integral of the function (5.2.2) over Δ_n. On Δ_n we have

$$\lambda_j - \lambda_i \geq a_j^{(n)} - b_i^{(n)} \qquad (i < j)$$

and

$$\exp\Big(-\frac{n+1}{4\sigma^2}\lambda_i^2\Big) \geq \exp\Big(-\frac{n+1}{4\sigma^2}(b_i^{(n)})^2\Big).$$

Hence

$$\bar{\nu}_n(\Delta_n) \geq \frac{1}{Z_{\beta,\sigma}^{(n)}} \exp\Big(-\frac{n+1}{4\sigma^2}\sum_{i=1}^n (b_i^{(n)})^2\Big) \prod_{i<j}(a_j^{(n)} - b_i^{(n)})^{2\beta} \prod_{i=1}^n (b_i^{(n)} - a_i^{(n)}).$$

Now we can establish the asymptotics of each factor. The normalization constant was treated already. Furthermore, let $g : [0,1] \to [a,b]$ be the inverse function of $t \mapsto \int_a^t f(x)\,dx$. Since $a_j^{(n)} = g((j-\frac{1}{2})/n)$ and $b_j^{(n)} = g(j/n)$, we have

$$\lim_{n\to\infty} \frac{1}{n^2}\Big(-\frac{n+1}{4\sigma^2}\sum_i (b_i^{(n)})^2\Big) = -\frac{1}{4\sigma^2}\int_0^1 g(t)^2\,dt = -\frac{1}{4\sigma^2}\int x^2 f(x)\,dx,$$

$$\begin{aligned}\lim_{n\to\infty} \frac{1}{n^2} \log \prod_{i<j}(a_j^{(n)} - b_i^{(n)})^{2\beta} &= 2\beta \iint_{0\leq s\leq t\leq 1} \log(g(t) - g(s))\,ds\,dt \\ &= \beta \iint f(x)f(y)\log|x-y|\,dx\,dy,\end{aligned}$$

and by (5.2.6)

$$\lim_{n\to\infty} \frac{1}{n^2}\log\prod_{i=1}^n (b_i^{(n)} - a_i^{(n)}) = 0.$$

Therefore (5.2.4) is obtained.

To deal with the upper estimate (5.2.5) we apply Lemma 4.3.4. Assume that $\operatorname{supp}\mu \subset [-R,R]$, and fix $\varepsilon > 0$ and $0 < \alpha < 1/2$. Set $\theta := (\alpha + 2\alpha^2)/(\alpha+2)$ and take k_0, δ according to Lemma 4.3.4 for $p = 2$, $\varepsilon^2\theta$ in place of ε, and the above R. For any large n we can choose $\eta^{(n)} \in \Gamma_n(\mu; r, \delta/2)$ such that $|\eta_i^{(n)}| \leq R$. Put

$$\Omega_n := \Big\{\lambda \in \mathbb{R}^n : \min_\pi \sum_{i=1}^n (\lambda_i - \eta_{\pi(i)}^{(n)})^2 \leq n\varepsilon^2\theta\Big\},$$

where the minimum is taken for all permutations π on $[n]$. Then

$$\Gamma_n(\mu; r, \delta/2) \subset \Omega_n$$

whenever $r \geq k_0$. This inclusion allows us to estimate $\bar{\nu}_n(\Omega_n)$ to find an upper bound for $\bar{\nu}_n\big(\Gamma_n(\mu; r, \delta/2)\big)$. The Lebesgue measure of Ω_n is majorized by $n!$ times the n-dimensional volume of a ball of radius $\varepsilon\sqrt{n\theta}$, that is,

$$\int \cdots \int_{\Omega_n} dx_1 \cdots dx_n \leq n! \frac{\pi^{n/2}}{\Gamma(n/2+1)} (n\varepsilon^2\theta)^{n/2}.$$

The values of the density (5.2.2) are to be majorized on Ω_n. When M_n is a bound for $\prod_{i<j} |\lambda_i - \lambda_j|^{2\beta}$ and M_n' is a bound for $\exp\big(-(n+1)/4\sigma^2 \sum_{i=1}^n \lambda_i^2\big)$ on Ω_n, we obviously have

$$\bar{\nu}_n(\Omega_n) \leq \frac{n!\,(\pi n\varepsilon^2\theta)^{n/2}}{\Gamma(n/2+1)} \cdot \frac{1}{Z_{\beta,\sigma}^{(n)}} M_n M_n'$$

and

$$\lim_{n\to\infty} \frac{1}{n^2} \log \frac{n!\,(\pi n\theta\varepsilon^2)^{n/2}}{\Gamma(n/2+1)} = 0.$$

For $\lambda, \eta \in \mathbb{R}^n$ satisfying $\sum_i (\lambda_i - \eta_i)^2 \leq n\varepsilon^2\theta$, let $v_i := |\lambda_i - \eta_i|$ and $\delta_i := \theta^{-1} v_i^2$. Then

$$\sum_i \delta_i \leq n\varepsilon^2, \qquad (1+2\alpha^{-1})v_i^2 = (1+2\alpha)\delta_i,$$

and we have

$$\begin{aligned}
|\lambda_i - \lambda_j|^2 &\leq (|\eta_i - \eta_j| + v_i + v_j)^2 \\
&= (\eta_i - \eta_j)^2 + v_i^2 + v_j^2 + 2(|\eta_i - \eta_j| v_i + |\eta_i - \eta_j| v_j + v_i v_j) \\
&\leq (\eta_i - \eta_j)^2 + v_i^2 + v_j^2 + \alpha(\eta_i - \eta_j)^2 + \alpha^{-1} v_i^2 \\
&\qquad + \alpha(\eta_i - \eta_j)^2 + \alpha^{-1} v_j^2 + v_i^2 + v_j^2 \\
&\leq (1+2\alpha)(\eta_i - \eta_j)^2 + (1+2\alpha^{-1})(v_i^2 + v_j^2) \\
&= (1+2\alpha)((\eta_i - \eta_j)^2 + \delta_i + \delta_j) \\
&\leq (1+2\alpha)((\eta_i - \eta_j)^2 + \varepsilon)(1 + (\delta_i + \delta_j)/\varepsilon) \\
&\leq (1+2\alpha)((\eta_i - \eta_j)^2 + \varepsilon) e^{(\delta_i + \delta_j)/\varepsilon},
\end{aligned}$$

and taking the product of these inequalities yields

$$\prod_{i<j} |\lambda_i - \lambda_j|^{2\beta} \leq (1+2\alpha)^{\beta n(n+2)/2} e^{2\beta n^2 \varepsilon} \prod_{i<j} \big((\eta_i - \eta_j)^2 + \varepsilon\big)^{\beta}.$$

Therefore,

$$\limsup_{n\to\infty} \frac{1}{n^2} \log M_n$$
$$\leq \frac{\beta}{2} \log(1+2\alpha) + 2\beta\varepsilon + \beta \limsup_{n\to\infty} \frac{1}{n^2} \log \prod_{i<j} \left((\eta_i^{(n)} - \eta_j^{(n)})^2 + \varepsilon \right).$$

Here the first two terms are arbitrarily small as α, ε are small. The third term approximates

$$\frac{\beta}{2} \iint \log \left((x-y)^2 + \varepsilon \right) d\mu(x) d\mu(y)$$

as the discrete measure $\frac{1}{n}\left(\delta(\eta_1^{(n)}) + \cdots + \delta(\eta_n^{(n)})\right)$ approximates the measure μ (when r is large and ε small).

On the other hand, for $\lambda, \eta \in \mathbb{R}^n$ as above, we have

$$\eta_i^2 \leq (|\lambda_i| + v_i)^2 \leq (1+\alpha)\lambda_i^2 + (1+\alpha^{-1})v_i^2 \leq (1+\alpha)\lambda_i^2 + (1+2\alpha)\delta_i,$$

and so

$$-\sum_i \lambda_i^2 \leq \frac{2n\varepsilon^2}{1+\alpha} - \frac{1}{1+\alpha} \sum_i \eta_i^2 .$$

This implies that

$$\limsup_{n\to\infty} \frac{1}{n^2} \log M_n' \leq \frac{2\varepsilon^2}{(1+\alpha)4\sigma^2} - \frac{1}{(1+\alpha)4\sigma^2} \liminf_{n\to\infty} \frac{1}{n} \sum_i (\eta_i^{(n)})^2 .$$

In the limit as $r \to \infty$ and $\varepsilon \to +0$ the term

$$-\frac{1}{4\sigma^2} \int x^2 \, d\mu(x)$$

appears. In this way we obtain (5.2.5), completing the proof of the theorem.

□

In the above discussions one can replace real symmetric matrices by complex selfadjoint ones. The difference between the real and complex cases is not essential; use of the complex field might be closer to a functional analytic viewpoint. Let X be a selfadjoint random matrix. The entropy $S(X)$ is defined as the Boltzmann-Gibbs entropy of the joint probability density of the real random variables $\{X_{ii} : 1 \leq i \leq n\} \cup \{\operatorname{Re} X_{ij} : 1 \leq i < j \leq n\} \cup \{\operatorname{Im} X_{ij} : 1 \leq i < j \leq n\}$. The lines of the proof of Proposition 5.2.1 are easily adapted to the complex case, and the entropy of the $n \times n$ standard selfadjoint Gaussian matrix (introduced in Sec. 4.1) is maximal among $n \times n$ selfadjoint random matrices X satisfying $\tau(X) = 0$ and $\tau(X^2) = 1$. Furthermore, the joint eigenvalue distribution of the $n \times n$ standard selfadjoint Gaussian matrix has the probability density (4.1.18). When we choose this as the reference measure $\bar{\nu}_n$, Theorem 5.2.2 holds with $\beta = 1$ and $\sigma^2 = 1/2$, while $\beta = 1/2$ and $\sigma^2 = 1$ when $\bar{\nu}_n$ arises from the $n \times n$ standard symmetric real Gaussian matrix.

The previous theorem motivated Voiculescu to regard the double integral

$$\Sigma(\mu) := \iint \log|x - y| \, d\mu(x) \, d\mu(y) \tag{5.2.7}$$

as a kind of entropy of a probability measure μ on $\mathbb{R}$ (or $\mathbb{C}$). In fact, he used the term *free entropy*. The use of the adjective "free" is not explained by the theorem. Later on, this entropy will be extended to the case of several noncommutative random variables, and it will turn out that the new entropy is additive for variables in free relation, while the Boltzmann-Gibbs entropy is known to be additive for independent variables. It is already time to warn that this free entropy is not related to the free entropy of thermodynamics. Note that when μ is compactly supported, the integral (5.2.7) always exists, although it can be $-\infty$, for example if μ has an atom.

Now return to *maximum entropy matrix ensembles*. As is seen from the previous section, the Boltzmann-Gibbs entropy $S(X)$ of a selfadjoint random matrix X is written as the relative entropy

$$S(X) = -S(\mu, \Lambda)\,,$$

where μ is the distribution measure on $M_n(\mathbb{C})^{sa}$ of X and $\Lambda = dA$ is the product Lebesgue measure (4.1.17) on $M_n(\mathbb{C})^{sa} \cong \mathbb{R}^{n^2}$. Let $V : \mathbb{R} \to \mathbb{R}$ be a continuous function. We aim to maximize $S(X)$ under the constraint $\tau(V(X)) = c$, where $V(X)$ is understood in the sense of functional calculus for selfadjoint matrices. In terms of eigenvalues the constraint is $\sum_i V(\lambda_i) = nc$.

First we prove that the maximizer is among the unitary conjugation-invariant measures on $M_n(\mathbb{C})^{sa}$. Indeed, if

$$X_0 := \int UXU^* \, dU\,, \quad \mu_0 := \int \mu(U \cdot U^*) \, dU$$

(integrations are with respect to the normalized Haar measure on $\mathcal{U}(n)$), then by the lower semicontinuity and convexity of relative entropy we have

$$S(\mu_0, \Lambda) \le \int S(\mu(U \cdot U^*), \Lambda) \, dU = S(\mu, \Lambda),$$

and hence $S(X_0) \geq S(X)$. Now let X be a unitarily invariant selfadjoint random matrix with a joint eigenvalue density $f(\lambda)\Delta(\lambda)^2$ (see (4.1.5) for $\Delta(\lambda)$). Then

$$\tau(V(X)) = \frac{1}{n}\int \Big(\sum_i V(\lambda_i)\Big) f(\lambda)\Delta(\lambda)^2\, d\lambda\,.$$

So the maximization of $S(X)$ when $\tau(V(X))$ is fixed is equivalent to the minimization of

$$\int f(\lambda)\Delta(\lambda)^2(\log f(\lambda)\Delta(\lambda)^2 - \log \Delta(\lambda)^2)\, d\lambda = \int (f(\lambda)\log f(\lambda))\Delta(\lambda)^2\, d\lambda$$

when $\int(\sum_i V(\lambda_i))f(\lambda)\Delta(\lambda)^2\, d\lambda$ is fixed. It is known that the maximizer has the form

$$f(\lambda) = \frac{1}{Z}\exp\Big(-t\sum_i V(\lambda_i)\Big)\,,$$

where Z is only for normalization and the real parameter t should be chosen so that the constraint is satisfied. Going back from the eigenvalue space to the matrix space (cf. Lemma 4.1.6), we find that the maximizer is of the form

$$Z^{-1}\exp\big(-t\mathrm{Tr}\, V(A)\big)\, dA.$$

The measure

$$\frac{1}{Z_n}\exp\big(-n\mathrm{Tr}\, V(A)\big)\, dA \tag{5.2.8}$$

on the space $M_n(\mathbb{C})^{sa}$ is also called the *orthogonal polynomial ensemble.* The reason for this is the fact that the correlation functions of the eigenvalues are conveniently expressed in terms of the orthogonal polynomials with respect to the weight function $w_V(x) := \exp(-V(x))$ defined on $\mathbb{R}$.

5.3 Logarithmic energy and free entropy

Voiculescu's entropy $\Sigma(\mu) = \iint \log|x-y|\, d\mu(x)\, d\mu(y)$ was introduced only recently, but with a different sign it is a classical quantity in two-dimensional potential theory. This section is devoted to properties of the free entropy, which is nothing else but the negative logarithmic energy. The main subject is the maximization of free entropy under different constraints.

We start with an example from *electrostatics.* Let $[-a,a]$ be a compact interval in $\mathbb{R}$. If μ is a probability measure on $[-a,a]$, then the double integral

$$\iint \log\frac{1}{|x-y|}\, d\mu(x)\, d\mu(y) \tag{5.3.1}$$

is called the *logarithmic energy* of μ. The integral always exists, but it can be $+\infty$ if μ has an atom, for example. In a physical picture the measure μ may be thought of as the distribution of electric charges along the interval

$$[-a, a] = \{(x, y) \in \mathbb{R}^2 : -a \leq x \leq a,\ y = 0\}$$

in a two-dimensional universe. The integral (5.3.1) yields the *Coulomb energy* of the two-dimensional electrostatic field due to electrostatic repulsion if the repulson force between charges is proportional to the inverse of the distance. This follows from the two-dimensional Coulomb law. To have a connection with the "real" 3-dimensional electrostatics, we regard the interval as the cross section of the infinite strip

$$\{(x, y, z) \in \mathbb{R}^3 : -a \leq x \leq a,\ y = 0,\ z \in \mathbb{R}\}$$

on which the distribution density of charges does not depend on the z coordinate. To get the true physical energy of a real charge density, one should replace $-\log|x-y|$ by $1/|x-y|$ in the integral.

When the interval is an ideal conductor, the equilibrium charge distribution is the minimizer of the energy functional. The advantage of the 2-dimensional x-y model is the mathematical convenience of the complex variable $x + \mathrm{i}\, y$. The *complex potential* of a (two-dimensional) electrostatic field is an analytic function

$$F(x + \mathrm{i}\, y) = u(x, y) + \mathrm{i}\, v(x, y)$$

(in the domain free of charges), for which the potential function $v(x, y)$ of the given electrostatic field is the imaginary part. The level lines $v(x, y) = C$ are the equipotential lines of the given field, and $u(x, y) = C$ gives the force lines. (These two curves are orthogonal to each other, thanks to the Cauchy-Riemann relations.)

The equilibrium distribution on $[-a, a]$ is the so-called *arcsine distribution*

$$h(x) := \begin{cases} \dfrac{1}{\pi\sqrt{a^2 - x^2}} & \text{if } -a < x < a, \\ 0 & \text{otherwise.} \end{cases}$$

The corresponding complex potential

$$F(z) := -\frac{\mathrm{i}}{\pi\sqrt{z^2 - a^2}} \qquad (z = x + \mathrm{i}\, y)$$

is analytic in the complement of $\{x + \mathrm{i}\, y : -a \leq x \leq a,\ y = 0\}$. We have

$$\lim_{y \to 0} \operatorname{Im} F(x + \mathrm{i}\, y) = \begin{cases} 0 & \text{if } |x| < a, \\ -\dfrac{1}{\pi\sqrt{x^2 - a^2}} & \text{if } |x| > a, \end{cases}$$

and observe that the interval $[-a, a]$ is equipotential, indeed. Therefore the limit

$$\lim_{y \to 0} \operatorname{Re} F(x + \mathrm{i}\, y) = h(x)$$

gives the equilibrium distribution. Note that $F(z)$ is i/π times the Cauchy transform of the measure $\mu = h(x)\, dx$ (see Sec. 3.1).

It is instructive to approach the above problem by discretization. Let $x_0, x_1, \dots, x_n, x_{n+1} \in [-1, 1]$, and minimize the function

$$T_0(x_0, x_1, \dots, x_n, x_{n+1}) := -\sum_{i<j} \log |x_i - x_j| \, .$$

We think that there are n equal charges at the positions x_i, and T_0 takes its minimum at the positions $x_i = t_i$ of equilibrium. Our physical intuition says that in the equilibrium there are charges at ± 1 due to the repulsion. Hence equivalently we maximize the function

$$T(x_1, x_2, \dots, x_n) := -\sum_{i<j} \log |x_i - x_j| - \sum_i \log(1 - x_i) - \sum_i \log(1 + x_i) \, .$$

It is no more difficult to deal with the more general function

$$\begin{aligned} &T^{\alpha,\beta}(x_1, x_2, \dots, x_n) \\ &= -\sum_{i<j} \log |x_i - x_j| - \sum_i \log(1 - x_i)^{(\alpha+1)/2} - \sum_i \log(1 + x_i)^{(\beta+1)/2} \, . \quad (5.3.2) \end{aligned}$$

From the conditions $\partial T^{\alpha,\beta} / \partial x_i = 0$ one can deduce a differential equation for the function $f_n(x) = (x - t_1)(x - t_2) \cdots (x - t_n)$, where $t_1, t_2, \dots, t_n$ are the minimum positions of the variables. It turns out ([187], Sec. 6.7) that f_n is the *Jacobi polynomial* $P_n^{\alpha,\beta}(x)$ if $\alpha, \beta > -1$. The Jacobi polynomials $P_n^{\alpha,\beta}(x)$ are orthogonal with respect to the measure $(1 - x)^\alpha (1 + x)^\beta \, dx$ on $[-1, 1]$. If $x_{1,n}, x_{2,n}, \dots, x_{n,n}$ stands for the zeros t_i of $P_n^{\alpha,\beta}(x)$, then it is plausible on the basis of the electrostatic model that the sequence of atomic measures $\frac{1}{n}\big(\delta(x_{1,n}) + \delta(x_{2,n}) + \cdots + \delta(x_{n,n})\big)$ tends to the arcsine law weakly, independently of α and β. This is due to the fact that the arcsine law is the equilibrium measure on $[-1, 1]$ and the constants α and β affect only $2n$ terms in (5.3.2) while the total nunber of terms is about $n^2/2$. Actually, the convergence to the semicircle law is known in theory of orthogonal polynomials.

It is convenient to extend the logarithmic energy (5.3.1) to signed measures. The so-called *logarithmic energy* $E(\nu)$ of a signed measure ν on $\mathbb{C}$ is given as

$$E(\nu) := \iint \log \frac{1}{|x - y|} \, d\nu(x) \, d\nu(y)$$

whenever

$$\iint \left| \log \frac{1}{|x - y|} \right| d|\nu|(x) \, d|\nu|(y) < +\infty$$

($|\nu|$ denotes the total variation of ν); otherwise put $E(\nu) := +\infty$. The logarithmic energy plays a role in potential theory. For a probability measure μ the free entropy given in (5.2.7) differs in sign from $E(\mu)$. The following lemma from potential theory (cf. [114], Theorem 1.16 or [165], Lemma I.1.8) will be useful in this section. The proof is provided for convenience.

Lemma 5.3.1 *Let ν is a compactly supported signed measure on $\mathbb{C}$ such that $\nu(1) = 0$. Then $E(\nu) \geq 0$, and $E(\nu) = 0$ if and only if $\nu = 0$.*

Proof: Recall that a real symmetric kernel $L(x, y)$ is called *negative definite* if

$$\sum_i \sum_j c_i c_j L(x_i, x_j) \leq 0 \tag{5.3.3}$$

whenever real numbers $c_1, \ldots, c_n$ satisfy $\sum_{i=1}^n c_i = 0$. It follows by approximation that for a continuous negative definite kernel $L(x, y)$ one gets

$$\iint L(x, y)\, d\nu(x)\, d\nu(y) \leq 0$$

if ν is a compactly supported signed measure such that $\nu(1) = 0$. Indeed, approximating ν by atomic measures, one can have a double integral which reduces to a double sum of the form (5.3.3).

The *logarithmic kernel* $K(x, y) := \log|x - y|$ has a singularity at $x = y$, and to avoid this we set, for positive ε,

$$K_\varepsilon(x, y) := \log(\varepsilon + |x - y|) = \int_0^\infty \Big(\frac{1}{1+t} - \frac{1}{t + \varepsilon + |x - y|} \Big)\, dt\,.$$

This kernel $K_\varepsilon(x, y)$ is the integral of negative definite kernels ([20], Chap. 3), and it is negative definite by itself. So we have

$$\iint K_\varepsilon(x, y)\, d\nu(x)\, d\nu(y) \leq 0\,,$$

and we take $\varepsilon \to +0$ to conclude that $E(\nu) \geq 0$ whenever $K(x, y)$ is integrable with respect to $|\nu| \otimes |\nu|$ (otherwise, $E(\nu) = +\infty$ by definition).

Now assume $E(\nu) = 0$. For $0 < \varepsilon < R < +\infty$ we get

$$\begin{aligned}
&-\int_\varepsilon^R \Big(\iint \frac{1}{t + |x - y|}\, d\nu(x)\, d\nu(y) \Big) dt \\
&= \int_\varepsilon^R \Big(\iint \Big(\frac{1}{1+t} - \frac{1}{t + |x - y|} \Big)\, d\nu(x)\, d\nu(y) \Big) dt \\
&= \iint \Big(\log(\varepsilon + |x - y|) + \log \frac{1 + R}{(1+\varepsilon)(R + |x - y|)} \Big)\, d\nu(x)\, d\nu(y)
\end{aligned}$$

by the Fubini theorem. Here note ([20]) that $(t+|x-y|)^{-1}$ is a *positive definite kernel* for any $t>0$, and hence

$$\iint \frac{1}{t+|x-y|}\,d\nu(x)\,d\nu(y) \ge 0 \qquad (t>0).$$

So we can take the limit of the above as $\varepsilon \to +0$ and $R \to +\infty$ to obtain

$$\int_0^\infty \left(\iint \frac{1}{t+|x-y|}\,d\nu(x)\,d\nu(y) \right) dt = 0\,,$$

which implies that

$$\iint \frac{1}{t+|x-y|}\,d\nu(x)\,d\nu(y) = 0 \qquad (t>0).$$

Taking the expansion

$$\frac{1}{t+|x-y|} = \sum_{n=0}^\infty \frac{(-1)^n}{t^{n+1}} |x-y|^n$$

in a neighborhood of $t=\infty$, we have

$$\iint |x-y|^{2n}\,d\nu(x)\,d\nu(y) = 0 \qquad (n=0,1,2,\ldots).$$

This means that

$$\sum_{i,j=0}^n (-1)^{i+j} \binom{n}{i} \binom{n}{j} \int x^i \bar{x}^j\,d\nu(x) \int x^{n-i}\bar{x}^{n-j}\,d\nu(x) = 0$$

for all n. Now we can easily show by induction that $\int x^i \bar{x}^j\,d\nu(x) = 0$ for all $i,j=0,1,2,\ldots$, which is enough to conclude that $\nu = 0$.

□

Proposition 5.3.2 *The free entropy functional $\Sigma(\mu)$ is weakly upper semicontinuous and concave on the set of probability measures restricted on any compact subset of $\mathbb{C}$. Moreover, it is strictly concave in the sense that $\Sigma(\lambda\mu_1 + (1-\lambda)\mu_2) > \lambda\Sigma(\mu_1) + (1-\lambda)\Sigma(\mu_2)$ if $0<\lambda<1$ and μ_1, μ_2 are compactly supported probability measures such that $\mu_1 \ne \mu_2$, $\Sigma(\mu_1) > -\infty$ and $\Sigma(\mu_2) > -\infty$.*

Proof: Let $K_\varepsilon(x,y)$ be the kernel given in the previous proof. The weak upper semicontinuity follows because $\Sigma(\mu)$ is written as

$$\Sigma(\mu) = \inf_{\varepsilon>0} \iint K_\varepsilon(x,y)\,d\mu(x)\,d\mu(y)$$

and the above double integral is continuous in the weak topology when the support of μ is restricted on a compact subset.

To prove the strictly concavity, let $\mu_1 \neq \mu_2$ be compactly supported measures such that $\Sigma(\mu_1) > -\infty$ and $\Sigma(\mu_2) > -\infty$. First we show that

$$E(\mu_1, \mu_2) := \iint \log \frac{1}{|x-y|} \, d\mu_1(x) \, d\mu_2(y)$$

is finite. Since the kernel $K_\varepsilon(x,y)$ is negative definite, we get

$$\begin{aligned} 0 &\geq \iint K_\varepsilon(x,y) \, d(\mu_1 - \mu_2)(x) \, d(\mu_1 - \mu_2)(y) \\ &\geq \Sigma(\mu_1) + \Sigma(\mu_2) - 2 \iint K_\varepsilon(x,y) \, d\mu_1(x) \, d\mu_2(y) \,. \end{aligned}$$

Letting $\varepsilon \to +0$ yields $\Sigma(\mu_1) + \Sigma(\mu_2) + 2E(\mu_1, \mu_2) \leq 0$, so $E(\mu_1, \mu_2) < +\infty$. Now we are in the situation where $E(\mu_1)$, $E(\mu_2)$ and $E(\mu_1, \mu_2)$ are all finite. Then we have for, $0 < \lambda < 1$,

$$E(\lambda\mu_1 + (1-\lambda)\mu_2) = E(\mu_2) + 2\lambda E(\mu_2, \mu_1 - \mu_2) + \lambda^2 E(\mu_1 - \mu_2),$$

and by Lemma 5.3.1

$$\frac{d^2}{d\lambda^2} E(\lambda\mu_1 + (1-\lambda)\mu_2) = E(\mu_1 - \mu_2) > 0 \,.$$

This implies that $\Sigma(\mu)$ is strictly concave (hence also concave).

□

Let S be a closed subset in $\mathbb{R}$ (or $\mathbb{C}$). Let $\mathcal{M}(S)$ denote the set of all probability measures whose support $\operatorname{supp}\mu$ is included in S. Moreover, let $w : S \to [0, \infty)$ be a *weight function*, which is assumed for simplicity to satisfy the following conditions:

(a) w is continuous on S.

(b) $S_0 := \{x \in S : w(x) > 0\}$ has positive (inner logarithmic) *capacity*, that is, $E(\mu) < +\infty$ for some probability measure μ such that $\operatorname{supp}\mu \subset S_0$.

(c) $|x|w(x) \to 0$ as $x \in S$, $|x| \to \infty$, when S is unbounded.

Let $Q(x) := -\log w(x)$ and define the *weighted energy integral* (or *weighted potential*)

$$E_Q(\mu) := \iint \Big(\log \frac{1}{|x-y|} + Q(x) + Q(y)\Big) \, d\mu(x) \, d\mu(y) \,. \tag{5.3.4}$$

The terminology was explained at the beginning of this section where the relation to electrostatics was discussed. One observes that $E_Q(\mu) > -\infty$ is well-defined, thanks to the above assumptions. Then the next theorem, due to Mhaskar and Saff ([165], Theorem I.1.3), is fundamental in the theory of weighted potentials, and it is proved by the adaptation of the classical Frostman method.

Theorem 5.3.3 *With the above assumptions, there exists a unique $\mu_0 \in \mathcal{M}(S)$ such that*

$$E_Q(\mu_0) = \inf\{E_Q(\mu) : \mu \in \mathcal{M}(S)\}.$$

Then $E_Q(\mu_0)$ is finite, μ_0 has finite logarithmic energy, and $\operatorname{supp}\mu_0$ is compact. Furthermore, the minimizer μ_0 is characterized as $\mu_0 \in \mathcal{M}(S)$ with compact support such that for some real number B the following holds:

$$\int \log|x-y|\,d\mu_0(y) \begin{cases} = Q(x) - B & \text{if } x \in \operatorname{supp}\mu_0, \\ \le Q(x) - B & \text{if } x \in S \setminus \operatorname{supp}\mu_0. \end{cases}$$

In this case, $B = E_Q(\mu_0) - \int Q\,d\mu_0$.

When S is a compact set (having positive capacity), a unique minimizer μ_S for $E(\mu)$ (or maximizer of $\Sigma(\mu)$) on $\mathcal{M}(S)$ is sometimes called the *equilibrium measure* on S. For instance, as we mentioned at the beginning of this section, the arcsine law

$$h(x) := \frac{1}{\pi\sqrt{a-x^2}}\chi_{(-a,a)}(x)$$

is the equilibrium measure on $[-a,a]$, and $\Sigma(h) = \log\frac{a}{2}$. In fact,

$$\int_{-a}^{a} h(y)\log|x-y|\,dy = \log\frac{a}{2} \qquad (-a \le x \le a).$$

Next we solve the maximization problems for free entropy under constraints on the pth moment. Since the maximization of free entropy is equivalent to the minimization of a weighted energy integral, we can benefit from the previous theorem.

For $p, r > 0$ the probability density

$$v_r^{(p)}(x) := \begin{cases} \dfrac{p}{\pi r^p}\displaystyle\int_{|x|}^{r} \frac{t^{p-1}}{\sqrt{t^2-x^2}}\,dt & \text{if } -r \le x \le r, \\ 0 & \text{otherwise,} \end{cases}$$

is called the *Ullman distribution*, and will play a role below. For instance,

$$v_r^{(1)}(x) = \chi_{[-r,r]}(x)\frac{1}{\pi r}\log\frac{r+\sqrt{r^2-x^2}}{|x|},$$

$$\begin{aligned}
v_r^{(2)}(x) &= w_r(x)\,,\\
v_r^{(3)}(x) &= \chi_{[-r,r]}(x)\frac{3}{2\pi r^3}\left(x^2\log\frac{r+\sqrt{r^2-x^2}}{|x|}+r\sqrt{r^2-x^2}\right),\\
v_r^{(4)}(x) &= \chi_{[-r,r]}(x)\frac{4(2x^2+r^2)}{3\pi r^4}\sqrt{r^2-x^2}\,.
\end{aligned}$$

Note that $\operatorname{supp} v_r^{(p)} = [-r,r]$, $v_r^{(p)}(\pm r) = 0$ and

$$\begin{aligned}
\int_{-r}^{r}|x|^p v_r^{(p)}(x)\,dx &= \frac{2p}{\pi r^p}\int_0^r t^{p-1}\left(\int_0^t \frac{x^p}{\sqrt{t^2-x^2}}\,dx\right)dt\\
&= \frac{2p}{\pi r^p}\int_0^r t^{2p-1}dt\int_0^1 \frac{x^p}{\sqrt{1-x^2}}\,dx = \alpha_p r^p\,, \qquad (5.3.5)
\end{aligned}$$

where

$$\alpha_p := \frac{1}{\pi}\int_0^1 \frac{x^p}{\sqrt{1-x^2}}\,dx = \frac{\Gamma(\frac{p+1}{2})}{2\sqrt{\pi}\Gamma(\frac{p}{2}+1)}\,.$$

For instance, $\alpha_1 = 1/\pi$, $\alpha_2 = 1/4$, $\alpha_3 = 2/3\pi$ and $\alpha_4 = 3/16$. Moreover, when $0 < p \le 1$, $v_r^{(p)}(x)$ has a singularity at $x = 0$.

Let ξ and η be independent random variables such that ξ has the standard *arcsine distribution* and η has the *beta* $(p,1)$ *distribution*, i.e. $\mathbf{Prob}(\eta < t) = t^p$ $(t \ge 0)$. Then the distribution of $\xi\eta$ is $v_1^{(p)}$. This might be useful when one computes moments of the Ullman distribution, although above we computed them directly. Note that when $p \to \infty$ the beta $(p,1)$ distribution converges to the point measure $\delta(1)$, so the Ullman distribution $v_1^{(p)}$ coverges to the standard arcsine law.

Now we are in a position to maximize the free entropy of $\mu \in \mathcal{M}(\mathbb{R})$ when the pth moment is fixed.

Proposition 5.3.4 *Let $p, r > 0$ and α_p be as above. Among the probability measures μ on $\mathbb{R}$ with $\int |x|^p d\mu(x) \le \alpha_p r^p$ (or $= \alpha_p r^p$), the Ullman distribution $v_r^{(p)}$ has maximal free entropy and $\Sigma(v_r^{(p)}) = \log(r/2) - 1/2p$. Furthermore, $v_r^{(p)}$ is the unique maximizer of the functional*

$$\Sigma(\mu) - \frac{1}{p\alpha_p r^p}\int |x|^p\,d\mu(x) \quad on \quad \mathcal{M}(\mathbb{R}).$$

Proof: First note that $\Sigma(\mu) < +\infty$ is well-defined whenever the pth moment of μ is finite, because $\log|x-y| \le C(|x|^p + |y|^p)$ for some constant $C > 0$. Also note that the constraint $\int |x|^p d\mu(x) \le \alpha_p r^p$ is equivalent to $\int |x|^p d\mu(x) = \alpha_p r^p$. Indeed, if we take the dilation $\tilde{\mu} := D_\lambda\mu$, i.e. $\tilde{\mu}(B) = \mu(\lambda^{-1}B)$ for $B \subset \mathbb{R}$, where $\lambda > 0$, then $\int |x|^p d\tilde{\mu}(x) = \lambda^p \int |x|^p d\mu(x)$ and $\Sigma(\tilde{\mu}) = \Sigma(\mu) + \log\lambda$.

It is known ([165], Sec. IV.5) that

$$\int_{-r}^{r} v_r^{(p)}(y)\log|x-y|\,dy \begin{cases} = \dfrac{|x|^p}{2p\alpha_p r^p} + \log\dfrac{r}{2} - \dfrac{1}{p} & \text{if } |x| \le r, \\ < \dfrac{|x|^p}{2p\alpha_p r^p} + \log\dfrac{r}{2} - \dfrac{1}{p} & \text{if } |x| > r. \end{cases} \tag{5.3.6}$$

We set $Q(x) := (2p\alpha_p r^p)^{-1}|x|^p$ on $\mathbb{R}$ and apply Theorem 5.3.3. Then (5.3.4) may be written as

$$E_Q(\mu) = -\Sigma(\mu) + \frac{1}{p\alpha_p r^p}\int |x|^p d\mu(x)\,,$$

so in minimizing $E_Q(\mu)$ we may restrict $\mu \in \mathcal{M}(\mathbb{R})$ to those with finite pth moment. For such μ set $\lambda := \left((1/\alpha_p r^p)\int |x|^p d\mu(x)\right)^{1/p}$ and $\tilde{\mu} := D_{1/\lambda}\mu$. Then we have $\int |x|^p d\tilde{\mu}(x) = \alpha_p r^p$ and

$$E_Q(\mu) = -\Sigma(\tilde{\mu}) + \frac{\lambda^p}{p} - \log\lambda\,.$$

Since $\lambda^p/p - \log\lambda$ $(\lambda > 0)$ takes the minimum $1/p$ at $\lambda = 1$, it follows that the minimization problem for $E_Q(\mu)$ on $\mathcal{M}(\mathbb{R})$ is equivalent to the maximization problem for $\Sigma(\mu)$ on $\{\mu \in \mathcal{M}(\mathbb{R}) : \int |x|^p d\mu(x) = \alpha_p r^p\}$. Hence, according to (5.3.6), Theorem 5.3.3 implies that $v_r^{(p)}$ is the unique maximizer for $\Sigma(\mu)$ with $\int |x|^p d\mu(x) = \alpha_p r^p$. Furthermore, by (5.3.5) and (5.3.6) we get

$$\Sigma(v_r^{(p)}) = \log\frac{r}{2} - \frac{1}{2p}\,.$$

□

In particular, when $p = 2$, the *semicircle law* $w_{m,r}$ has maximal free entropy among the probability measures on $\mathbb{R}$ with variance $\le r^2/4$, and

$$\Sigma(w_{m,r}) = \log\frac{r}{2} - \frac{1}{4}\,.$$

This fact is one of the reasons why the semicircle law is regarded as the free analogue of the normal distribution; recall (see Sec. 5.1) that the latter maximizes the differential (or Boltzmann-Gibbs) entropy when the variance is fixed.

The maximizer $v_r^{(p)}$ in Proposition 5.3.4 is symmetric. Let $\mathcal{M}_s(\mathbb{R})$ denote the set of all symmetric probability measures μ on $\mathbb{R}$. Define a bijective correspondence between $\mathcal{M}_s(\mathbb{R})$ and $\mathcal{M}(\mathbb{R}^+)$ by

$$T : \mathcal{M}_s(\mathbb{R}) \to \mathcal{M}(\mathbb{R}^+)\,, \quad (T\mu)(B) := \mu(\sigma^{-1}B)\,,$$

where $\sigma : \mathbb{R} \to \mathbb{R}^+$, $\sigma(x) := x^2$. Note that if $\mu \in \mathcal{M}_s(\mathbb{R})$ has density f satisfying $f(x) = f(-x)$, then $T\mu$ has the density Tf given by

$$(Tf)(x) := \frac{f(\sqrt{x})}{\sqrt{x}}.$$

We have the following properties:

(1) $\int |x|^{2p} d\mu(x) = \int x^p d(T\mu)(x)$ for $p > 0$.

(2) $2\Sigma(\mu) = \Sigma(T\mu)$ whenever $\Sigma(\mu)$ or $\Sigma(T\mu)$ is well-defined.

Indeed, (1) is immediate and (2) comes from an elementary computation.

Now we are going to maximize the free entropy on the set of probability measures on $\mathbb{R}^+$ with a given pth moment. By means of the transformation T this extremum problem is reduced to the one previously treated in Proposition 5.3.4. For $p, r > 0$ define the probability density $u_r^{(p)}$ by $u_r^{(p)}(x) := v_{\sqrt{2r}}^{(2p)}(\sqrt{x})/\sqrt{x}$; more concretely,

$$u_r^{(p)}(x) = \begin{cases} \dfrac{p}{\pi r^p} \displaystyle\int_{x/2}^{r} \frac{t^{p-1}}{\sqrt{2tx - x^2}}\, dt & \text{if } 0 \le x \le 2r, \\ 0 & \text{otherwise.} \end{cases}$$

For instance,

$$\begin{aligned} u_r^{(1)}(x) &= \chi_{[0,2r]}(x) \frac{\sqrt{2rx - x^2}}{\pi r x}, \\ u_r^{(2)}(x) &= \chi_{[0,2r]}(x) \frac{2(x+r)}{3\pi r^2 x} \sqrt{2rx - x^2}. \end{aligned} \tag{5.3.7}$$

Proposition 5.3.5 *Let $p, r > 0$ and $\tilde{\alpha}_p := 2^{p-1}\Gamma(p + \frac{1}{2})/\sqrt{\pi}\Gamma(p+1)$. Among the probability measures μ supported in $\mathbb{R}^+$ with $\int x^p d\mu(x) \le \tilde{\alpha}_p r^p$ (or $= \tilde{\alpha}_p r^p$), the distribution $u_r^{(p)}$ has maximal free entropy and $\Sigma(u_r^{(p)}) = \log(r/2) - 1/2p$. Furthermore, $u_r^{(p)}$ is a unique maximizer of the functional*

$$\Sigma(\mu) - \frac{1}{p\tilde{\alpha}_p r^p} \int x^p \, d\mu(x) \quad \text{on} \quad \mathcal{M}(\mathbb{R}^+).$$

In particular, since $\tilde{\alpha}_1 = 1/2$, the case $p = 1$ says that for each $m > 0$ the distribution $u_{2m}^{(1)}$ has maximal free entropy among the probability measures on $\mathbb{R}^+$ with mean $\le m$. In the case of the classical differential entropy, the exponential distribution is the maximizer when the support is $\mathbb{R}^+$ and the mean is given. So the family of distributions (5.3.7) is considered as the free analogue of the exponential distributions. Note that (5.3.7) for $r = 2$ is the Marchenko-Pastur distribution μ_1 which appeared in connection with random matrices (see Sec. 4.1).

As to probability measures on $\mathbb{C}$, the maximization of free entropy under constraints of the pth moment is solved as follows. For $p, R > 0$ define the distribution $\lambda_R^{(p)}$ supported on the disk $\{\zeta \in \mathbb{C} : |\zeta| \le R\}$ by

$$\lambda_R^{(p)} := \frac{p}{2\pi R^p}\, d\theta \cdot r^{p-1}\chi_{[0,R]}(r)\, dr \qquad (\zeta = re^{\mathrm{i}\,\theta}).$$

In particular, when $p = 2$, $\lambda_R^{(2)}$ is the uniform distribution on $\{\zeta \in \mathbb{C} : |\zeta| \le R\}$ that is called the *circular law.*

Proposition 5.3.6 *Let $p, R > 0$. Among $\mu \in \mathcal{M}(\mathbb{C})$ with $\int |\zeta|^p\, d\mu(\zeta) \le R^p/2$ (or $= R^p/2$), the distribution $\lambda_R^{(p)}$ has maximal free entropy and $\Sigma(\lambda_R^{(p)}) = \log R - 1/2p$. Furthermore, $\lambda_R^{(p)}$ is the unique maximizer of the functional*

$$\Sigma(\mu) - \frac{2}{pR^p}\int |\zeta|^p\, d\mu(\zeta) \quad \textit{on} \quad \mathcal{M}(\mathbb{C}).$$

Proof: Since

$$\frac{1}{2\pi}\int_0^{2\pi} \log|\zeta - re^{\mathrm{i}\,\theta}|\, d\theta = \begin{cases} \log r & \text{if } |\zeta| \le r, \\ \log|\zeta| & \text{if } |\zeta| > r, \end{cases} \tag{5.3.8}$$

it is easy to compute

$$\int \log|\zeta - \eta|\, d\lambda_R^{(p)}(\eta) = \begin{cases} \dfrac{|\zeta|^p}{pR^p} + \log R - \dfrac{1}{p} & \text{if } |\zeta| \le R, \\ \log|\zeta| < \dfrac{|\zeta|^p}{pR^p} + \log R - \dfrac{1}{p} & \text{if } |\zeta| > R. \end{cases}$$

For any $\mu \in \mathcal{M}(\mathbb{C})$ with $\int |\zeta|^p\, d\mu(\zeta) < +\infty$, set $\lambda := \left((2/R^p)\int |\zeta|^p\, d\mu(\zeta)\right)^{1/p}$ and $\tilde{\mu} := D_{1/\lambda}\mu$. Then $\int |\zeta|^p\, d\tilde{\mu}(\zeta) = R^p/2$ and

$$-\Sigma(\mu) + \frac{2}{pR^p}\int |\zeta|^p\, d\mu(\zeta) = -\Sigma(\tilde{\mu}) + \frac{\lambda^p}{p} - \log\lambda\,.$$

Hence, as in the proof of Proposition 5.3.4, we can apply Theorem 5.3.3 to get the result.

□

The Marchenko-Pastur distribution (alias the free Poisson distribution) μ_λ ($\lambda > 0$) was given in Example 3.3.5. Next we prove that μ_λ with $\lambda \ge 1$ is a maximizer of a certain free entropy functional.

Proposition 5.3.7 *When $\lambda \geq 1$, the Marchenko-Pastur distribution μ_λ is a unique maximizer of the functional*

$$\Sigma(\mu) - \int (x - (\lambda - 1)\log x)\, d\mu(x) \quad on \quad \mathcal{M}(\mathbb{R}^+). \tag{5.3.9}$$

Furthermore,

$$\Sigma(\mu_\lambda) = \frac{1}{2}\int (x - (\lambda - 1)\log x)\, d\mu_\lambda(x) + \frac{1}{2}(\lambda \log \lambda - 1 - \lambda)\,. \tag{5.3.10}$$

Proof: From the proof of Example 3.3.5 we know that the Cauchy transform of μ_λ is

$$G_{\mu_\lambda}(z) = \frac{z + (1-\lambda) - \sqrt{(z - 1 - \lambda)^2 - 4\lambda}}{2z}\,.$$

Hence the Hilbert transform of

$$f_\lambda(x) := \frac{\sqrt{4\lambda - (x - 1 - \lambda)^2}}{2\pi x}\chi_{\left[(1-\sqrt{\lambda})^2,\,(1+\sqrt{\lambda})^2\right]}(x)$$

is

$$Hf_\lambda(x) = \frac{1}{\pi}\lim_{y\to+0}\operatorname{Re} G_{\mu_\lambda}(x + \mathrm{i}\,y) = \frac{1}{2\pi}\Big(1 - \frac{\lambda - 1}{x}\Big)$$

if $(1-\sqrt{\lambda})^2 < x < (1+\sqrt{\lambda})^2$. Put

$$F(x) := \int f_\lambda(y)\log|x - y|\, dy\,.$$

Since $F'(x) = \pi H f_\lambda(x)$ in the sense of distributions in $\big((1-\sqrt{\lambda})^2, (1+\sqrt{\lambda})^2\big)$, we have

$$F(x) = \frac{1}{2}(x - (\lambda - 1)\log x) + C$$

for $x \in \big[(1-\sqrt{\lambda})^2, (1+\sqrt{\lambda})^2\big]$, where C is a constant. On the other hand, $F(x)$ is differentiable in the usual sense in $\mathbb{R}^+ \setminus \big[(1-\sqrt{\lambda})^2, (1+\sqrt{\lambda})^2\big]$, and

$$\begin{aligned} F'(x) &= \int \frac{f_\lambda(y)}{x - y}\, dy = G_{\mu_\lambda}(x) \\ &= \frac{x + (1-\lambda) - \sqrt{(x - 1 - \lambda)^2 - 4\lambda}}{2x} < \frac{1}{2}\Big(1 - \frac{\lambda - 1}{x}\Big) \end{aligned}$$

for $x \in \mathbb{R}^+ \setminus \left[(1-\sqrt{\lambda})^2, (1+\sqrt{\lambda})^2\right]$. Since $F(x)$ is continuous at $x = (1 \pm \sqrt{\lambda})^2$, we have

$$\int \log|x-y|\, d\mu_\lambda(y) \begin{cases} = \frac{1}{2}(x-(\lambda-1)\log x) + C & \text{if } x \in \operatorname{supp}\mu_\lambda, \\ < \frac{1}{2}(x-(\lambda-1)\log x) + C & \text{if } x \in \mathbb{R}^+ \setminus \operatorname{supp}\mu_\lambda. \end{cases} \tag{5.3.11}$$

We can apply Theorem 5.3.3 to get the first assertion, because the weight function $w(x) := e^{-x/2}x^{(\lambda-1)/2}$ on $\mathbb{R}^+$ satisfies (i)–(iii) stated above Theorem 5.3.3.

Taking $x = 1+\lambda$ in (5.3.11), we get

$$C = \int_{(1-\sqrt{\lambda})^2}^{(1+\sqrt{\lambda})^2} \frac{\sqrt{4\lambda-(y-1-\lambda)^2}}{2\pi y} \log|y-1-\lambda|\, dy \\ -\frac{1}{2}(1+\lambda-(\lambda-1)\log(1+\lambda)).$$

We compute

$$\begin{aligned} &\int_{(1-\sqrt{\lambda})^2}^{(1+\sqrt{\lambda})^2} \frac{\sqrt{4\lambda-(y-1-\lambda)^2}}{2\pi y} \log|y-1-\lambda|\, dy \\ &\quad = \log 2\sqrt{\lambda} + \frac{\sqrt{\lambda}}{\pi} \int_{-1}^{1} \frac{\sqrt{1-x^2}}{\frac{1+\lambda}{2\sqrt{\lambda}} + x} \log|x|\, dx \\ &\quad = \log 2\sqrt{\lambda} + \frac{1+\lambda}{\pi} \int_0^1 \frac{\sqrt{1-x^2}}{\frac{(1+\lambda)^2}{4\lambda} - x^2} \log x\, dx \\ &\quad = \log 2\sqrt{\lambda} + \frac{1+\lambda}{\pi}\Big(\frac{\pi(b-1)}{2b}\log\frac{2b-1}{2b} - \frac{\pi}{2b}\log 2\Big), \end{aligned}$$

where $b := (1+\lambda)/2$. The last equality in the above is due to the integral formula ([115], Theorem 4.3)

$$\int_0^1 \frac{\sqrt{1-x^2}}{\alpha - x^2} \log x\, dx = \frac{\pi(b-1)}{2b}\log\frac{2b-1}{2b} - \frac{\pi}{2b}\log 2,$$

where $\alpha \in \mathbb{R} \setminus [0,1)$ and $b = \alpha + \operatorname{sign}\alpha\sqrt{\alpha^2-\alpha}$. Therefore,

$$C = \frac{1}{2}(\lambda\log\lambda - 1 - \lambda),$$

so that (5.3.10) is shown.

□

For $0 \le \rho < R$ let $m_{\rho,R}$ denote the uniform distribution on the annulus $\{\zeta \in \mathbb{C} : \rho \le |\zeta| \le R\}$, that is,

$$m_{\rho,R} := \frac{1}{\pi(R^2-\rho^2)} d\theta \cdot r\chi_{[\rho,R]}(r)\, dr \qquad (\zeta = re^{\mathrm{i}\theta}).$$

We may call this the *annular law.* The next proposition says that the measure $m_{\sqrt{\lambda-1},\sqrt{\lambda}}$ with $\lambda \ge 1$ is a maximizer of a functional which can be regarded as the complexification of (5.3.9).

Proposition 5.3.8 *When $\lambda \ge 1$, the annular law $m_{\sqrt{\lambda-1},\sqrt{\lambda}}$ is the unique maximizer of the functional*

$$\Sigma(\mu) - \int (|\zeta|^2 - 2(\lambda-1)\log|\zeta|)\, d\mu(\zeta) \quad on \quad \mathcal{M}(\mathbb{C}).$$

Moreover,

$$\Sigma(m_{\sqrt{\lambda-1},\sqrt{\lambda}}) = -\frac{3}{4} + \frac{1}{2}(\lambda + \log\lambda + (\lambda-1)^2 \log(1-\lambda^{-1}))\,. \tag{5.3.12}$$

Proof: From (5.3.8) it is easy to compute

$$\int \log|\zeta-\eta|\, dm_{\sqrt{\lambda-1},\sqrt{\lambda}}(\eta) = \begin{cases} \frac{1}{2}(|\zeta|^2 - 2(\lambda-1)\log|\zeta| + \lambda\log\lambda - \lambda) & \text{if } \sqrt{\lambda-1} \le |\zeta| \le \sqrt{\lambda}, \\ \frac{1}{2}(\lambda\log\lambda - (\lambda-1)\log(\lambda-1) - 1) & \text{if } |\zeta| < \sqrt{\lambda-1}, \\ \log|\zeta| & \text{if } |\zeta| > \sqrt{\lambda}. \end{cases} \tag{5.3.13}$$

Also, the following are readily checked:

$$\begin{aligned} &\lambda\log\lambda - (\lambda-1)\log(\lambda-1) - 1 \\ &\qquad < |\zeta|^2 - 2(\lambda-1)\log|\zeta| + \lambda\log\lambda - \lambda \quad \text{for} \quad |\zeta| < \sqrt{\lambda-1}, \end{aligned}$$

$$2\log|\zeta| < |\zeta|^2 - 2(\lambda-1)\log|\zeta| + \lambda\log\lambda - \lambda \quad \text{for} \quad |\zeta| > \sqrt{\lambda}.$$

Hence the first assertion is obtained by Theorem 5.3.3. A direct computation from (5.3.13) gives (5.3.12).

□

In the rest of this section let us treat three more results on maximization of the free entropy. The Poisson kernel plays an essential role in analytic function theory in the unit disk. For $\alpha \in \mathbb{C}$, $|\alpha| < 1$, we have the *Poisson kernel measure* p_α supported on the unit circle $\mathbb{T}$, defined by

$$p_\alpha := \frac{1-|\alpha|^2}{|\zeta-\alpha|^2}\, d\zeta \qquad (\zeta = e^{\mathrm{i}\theta},\ d\zeta = d\theta/2\pi). \tag{5.3.14}$$

The next maximization result is related to this.

Proposition 5.3.9 *For any $\alpha \in \mathbb{C}$, $|\alpha| < 1$, the Poisson kernel measure p_α is the unique maximizer of the functional*

$$\Sigma(\mu) - \int \log |\zeta\bar{\alpha} - 1|^2 \, d\mu(\zeta) \quad on \quad \mathcal{M}(\overline{\mathbb{D}}),$$

where $\overline{\mathbb{D}} := \{\zeta \in \mathbb{C} : |\zeta| \le 1\}$. Furthermore, p_α maximizes $\Sigma(\mu)$ among $\mu \in \mathcal{M}(\overline{\mathbb{D}})$ with $\int \log |\zeta\bar{\alpha} - 1| \, d\mu(\zeta) = \log(1 - |\alpha|^2)$, and $\Sigma(p_\alpha) = \log(1 - |\alpha|^2)$.

Proof: By noting that $\log |\zeta|$ is harmonic in $\mathbb{C} \setminus \{0\}$, it is easy to check that

$$\int \log |\zeta - \eta| \, dp_\alpha(\eta) = \log |\zeta\bar{\alpha} - 1| \qquad (\zeta \in \overline{\mathbb{D}}).$$

Hence Theorem 5.3.3 implies the first assertion. Furthermore,

$$\Sigma(p_\alpha) = \int \log |\zeta\bar{\alpha} - 1| \, dp_\alpha(\zeta) = \log(1 - |\alpha|^2) \, .$$

Next, assume that $\mu \in \mathcal{M}(\overline{\mathbb{D}})$ and $\int \log |\zeta\bar{\alpha} - 1| \, d\mu(\zeta) = \log(1 - |\alpha|^2)$. Then, by Lemma 5.3.1,

$$\begin{aligned} 0 &\le E(\mu - p_\alpha) \\ &= -\Sigma(\mu) - \Sigma(p_\alpha) + 2 \iint \log |\zeta - \eta| \, dp_\alpha(\eta) \, d\mu(\zeta) \\ &= -\Sigma(\mu) + \log(1 - |\alpha|^2) \, , \end{aligned}$$

so $\Sigma(\mu) \le \log(1 - |\alpha|^2)$, and equality occurs if and only if $\mu = p_\alpha$.

□

In particular, when $\alpha = 0$, the above proposition gives a known fact: the Lebesgue probability measure on $\mathbb{T}$ is the equilibrium measure on $\overline{\mathbb{D}}$ (or $\mathbb{T}$).

Define a one-parameter family of probability measures ρ_λ $(0 < \lambda \le \infty)$ on $\mathbb{T}$ by

$$\rho_\lambda := \begin{cases} \dfrac{1}{2\pi}\left(1 + \dfrac{2}{\lambda}\cos\theta\right) d\theta & \text{if } 2 \le \lambda \le \infty, \\[2ex] \dfrac{2}{\pi\lambda}\cos\dfrac{\theta}{2}\sqrt{\dfrac{\lambda}{2} - \sin^2\dfrac{\theta}{2}}\,\chi_{[-a,a]}(\theta) \, d\theta & \text{if } 0 < \lambda < 2, \end{cases} \tag{5.3.15}$$

where $a := 2\arcsin\sqrt{\lambda/2}$. In the following we show that these distributions are maximizers of the free entropy of $\mu \in \mathcal{M}(\mathbb{T})$ when the mean $\int \zeta \, d\mu(\zeta)$ is fixed.

Proposition 5.3.10 *For any $0 < \lambda \le \infty$ the distribution ρ_λ is a unique maximizer of the functional*

$$\Sigma(\mu) + \frac{2}{\lambda}\int \operatorname{Re}\zeta \, d\mu(\zeta) \quad on \quad \mathcal{M}(\mathbb{T}).$$

Furthermore, when $2 \le \lambda \le \infty$, ρ_λ *maximizes* $\Sigma(\mu)$ *among* $\mu \in \mathcal{M}(\mathbb{T})$ *with* $\int \mathrm{Re}\,\zeta\,d\mu(\zeta)$ *(or* $\int \zeta\,d\mu(\zeta)$*)* $= 1/\lambda$, *and* $\Sigma(\rho_\lambda) = -1/\lambda^2$. *When* $0 < \lambda < 2$, ρ_λ *maximizes* $\Sigma(\mu)$ *among* $\mu \in \mathcal{M}(\mathbb{T})$ *with* $\int \mathrm{Re}\,\zeta\,d\mu(\zeta)$ *(or* $\int \zeta\,d\mu(\zeta)$*)* $= 1 - \lambda/4$, *and* $\Sigma(\rho_\lambda) = \frac{1}{2}\log(\lambda/2) - 1/4$.

Proof: When $2 \le \lambda \le \infty$ the computation is straightforward. For $\zeta = e^{\mathrm{i}\,t}$,

$$\begin{aligned}
\int \log|\zeta - \eta|\,d\rho_\lambda(\eta) &= \frac{1}{2\pi}\int_0^{2\pi}\Big(1 + \frac{2}{\lambda}\cos(\theta + t)\Big)\log|1 - e^{\mathrm{i}\,\theta}|\,d\theta \\
&= \frac{1}{2\pi\lambda}\int_0^{2\pi}\cos(\theta + t)\log 2(1 - \cos\theta)\,d\theta \\
&= \frac{\cos t}{2\pi\lambda}\int_0^{2\pi}\cos\theta\log(1 - \cos\theta)\,d\theta \\
&= \frac{\cos t}{\pi\lambda}\int_{-1}^{1}\frac{t}{\sqrt{1 - t^2}}\log(1 - t)\,dt \\
&= -\frac{\cos t}{\pi\lambda}\int_{-1}^{1}\sqrt{\frac{1 + t}{1 - t}}\,dt = -\frac{1}{\lambda}\cos t\,.
\end{aligned}$$

This implies the first assertion by Theorem 5.3.3. Moreover,

$$\Sigma(\rho_\lambda) = -\frac{1}{2\pi\lambda}\int_0^{2\pi}\cos t\Big(1 + \frac{2}{\lambda}\cos t\Big)\,dt = -\frac{1}{\lambda^2}\,.$$

If $\mu \in \mathcal{M}(\mathbb{T})$ satisfies $\int \mathrm{Re}\,\zeta\,d\mu(\zeta) = 1/\lambda$, then by Lemma 5.3.1

$$\begin{aligned}
0 \le E(\mu - \rho_\lambda) &= -\Sigma(\mu) - \Sigma(\rho_\lambda) + 2\iint \log|\zeta - \eta|\,d\rho_\lambda(\eta)\,d\mu(\zeta) \\
&= -\Sigma(\mu) + \frac{1}{\lambda^2} - \frac{2}{\lambda}\int \mathrm{Re}\,\zeta\,d\mu(\zeta) = -\Sigma(\mu) - \frac{1}{\lambda^2}\,,
\end{aligned}$$

so $\Sigma(\mu) \le -1/\lambda^2$, and equality occurs if and only if $\mu = \rho_\lambda$.

When $0 < \lambda < 2$ the computation is quite involved. We use the technique of the "Hilbert transform" on the circle. Put $\alpha := \sqrt{\lambda/2}$ and $\beta := 2\arcsin\alpha$. Define, for $\zeta = e^{\mathrm{i}\,t}$,

$$\begin{aligned}
F(t) &:= \int \log|\zeta - \eta|\,d\rho_\lambda(\eta) \\
&= \frac{2}{\pi\lambda}\int_{-\beta}^{\beta}\cos\frac{\theta}{2}\sqrt{\frac{\lambda}{2} - \sin^2\frac{\theta}{2}}\,\log|1 - e^{\mathrm{i}\,(\theta - t)}|\,d\theta \\
&= \frac{2}{\pi\lambda}\int_{-\beta}^{\beta}\cos\frac{\theta}{2}\sqrt{\frac{\lambda}{2} - \sin^2\frac{\theta}{2}}\,\log\Big(2\Big|\sin\frac{\theta - t}{2}\Big|\Big)\,d\theta\,.
\end{aligned}$$

When $|t| < \alpha$, the differential of $F(t)$ in the sense of distributions in $(-\alpha, \alpha)$ is given as

$$F'(t) = \frac{1}{\pi\lambda} \int_{-\beta}^{\beta} \cos\frac{\theta}{2} \sqrt{\frac{\lambda}{2} - \sin^2\frac{\theta}{2}} \cot\frac{\theta - t}{2}\, d\theta\,.$$

(This, as well as the integrals below, is understood as a principal value integral.) We proceed to compute

$$\begin{aligned} F'(t) &= \frac{2}{\pi\lambda} \int_{-\alpha}^{\alpha} \sqrt{\alpha^2 - x^2}\, \frac{\sqrt{1-x^2}\cos\frac{t}{2} + x\sin\frac{t}{2}}{x\cos\frac{t}{2} - \sqrt{1-x^2}\sin\frac{t}{2}}\, dx \\ &= \frac{1}{\pi\lambda} \int_{-\alpha}^{\alpha} \sqrt{\alpha^2 - x^2}\, \frac{\sin t + 4x\sqrt{1-x^2}}{x^2 - \sin^2\frac{t}{2}}\, dx \\ &= \frac{\sin t}{\pi\lambda} \int_{-\alpha}^{\alpha} \frac{\sqrt{\alpha^2 - x^2}}{x^2 - \sin^2\frac{t}{2}}\, dx\,. \end{aligned}$$

Since the above principal value integral is equal to $-\pi$ (cf. [121], p. 74), we have $F'(t) = -\lambda^{-1}\sin t$, and hence

$$F(t) = -\frac{1}{\lambda}\cos t + \frac{1}{2}\log\frac{\lambda}{2} + \frac{1}{\lambda} - \frac{1}{2} \qquad (|t| < \alpha),$$

because

$$\begin{aligned} F(0) &= \frac{2}{\pi\lambda} \int_{-\beta}^{\beta} \cos\frac{\theta}{2} \sqrt{\frac{\lambda}{2} - \sin^2\frac{\theta}{2}} \log\left(2\left|\sin\frac{\theta}{2}\right|\right) d\theta \\ &= \frac{8}{\pi\lambda} \int_0^{\alpha} \sqrt{\alpha^2 - x^2}\log(2x)\, dx \\ &= \frac{4}{\pi} \int_0^1 \sqrt{1-x^2}\log(2\alpha x)\, dx = \frac{1}{2}\log\frac{\lambda}{2} - \frac{1}{2}\,. \end{aligned}$$

On the other hand, when $|t| > \alpha$, $F(t)$ is differentiable in the usual sense and

$$\begin{aligned} F'(t) &= \frac{\sin t}{2\pi\alpha^2} \int_{-\alpha}^{\alpha} \frac{\sqrt{\alpha^2 - x^2}}{x^2 - \sin^2\frac{t}{2}}\, dx \\ &< \frac{\sin t}{2\pi\alpha^2} \int_{-\alpha}^{\alpha} \frac{\sqrt{\alpha^2 - x^2}}{x^2 - \alpha^2}\, dx = -\frac{1}{\lambda}\sin t\,. \end{aligned}$$

Therefore, since $F(t)$ is continuous at $t = \pm\alpha$, we obtain

$$\int \log|\zeta - \eta|\, d\rho_\lambda(\eta) \begin{cases} = -\frac{1}{\lambda}\mathrm{Re}\,\zeta + \frac{1}{2}\log\frac{\lambda}{2} + \frac{1}{\lambda} - \frac{1}{2} & \text{if } \zeta \in \operatorname{supp}\rho_\lambda, \\ < -\frac{1}{\lambda}\mathrm{Re}\,\zeta + \frac{1}{2}\log\frac{\lambda}{2} + \frac{1}{\lambda} - \frac{1}{2} & \text{if } \zeta \in \mathbb{T} \setminus \operatorname{supp}\rho_\lambda. \end{cases}$$

This implies the first assertion by Theorem 5.3.3. Moreover,

$$\begin{aligned}\Sigma(\rho_\lambda) &= -\frac{2}{\pi\lambda^2}\int_{-\beta}^{\beta}\cos t\cos\frac{t}{2}\sqrt{\frac{\lambda}{2}-\sin^2\frac{t}{2}}\,dt+\frac{1}{2}\log\frac{\lambda}{2}+\frac{1}{\lambda}-\frac{1}{2}\\ &= -\frac{8}{\pi\lambda^2}\int_0^{\alpha}(1-2x^2)\sqrt{\alpha^2-x^2}\,dx+\frac{1}{2}\log\frac{\lambda}{2}+\frac{1}{\lambda}-\frac{1}{2}\\ &= \frac{1}{2}\log\frac{\lambda}{2}-\frac{1}{4}.\end{aligned}$$

If $\mu \in \mathcal{M}(\mathbb{T})$ satisfies $\int \operatorname{Re}\zeta\,d\mu(\zeta) = 1-\lambda/4$, then by Lemma 5.3.1

$$\begin{aligned}0 &\le E(\mu-\rho_\lambda)\\ &= -\Sigma(\mu)-\Sigma(\rho_\lambda)+2\iint\log|\zeta-\eta|\,d\rho_\lambda(\eta)\,d\mu(\zeta)\\ &\le -\Sigma(\mu)-\frac{1}{2}\log\frac{\lambda}{2}+\frac{1}{4}+2\Big(-\frac{1}{\lambda}\Big(1-\frac{\lambda}{4}\Big)+\frac{1}{2}\log\frac{\lambda}{2}+\frac{1}{\lambda}-\frac{1}{2}\Big)\\ &= -\Sigma(\mu)+\frac{1}{2}\log\frac{\lambda}{2}-\frac{1}{4},\end{aligned}$$

so $\Sigma(\mu) \le \frac{1}{2}\log(\lambda/2)-1/4$, and equality occurs if and only if $\mu = \rho_\lambda$.

□

For $-1 < \tau < 1$ and $R > 0$ let $m_R^{(\tau)}$ denote the uniform distribution on the ellipse

$$E_R^{(\tau)} := \Big\{x+\mathrm{i}\,y : \frac{x^2}{(1+\tau)^2}+\frac{y^2}{(1-\tau)^2} \le R^2\Big\}, \tag{5.3.16}$$

which is sometimes called the *elliptic law*. The next proposition says that the elliptic law appears as the solution of some maximization problem for a functional containing a free entropy term.

Proposition 5.3.11 *Let $-1 < \tau < 1$ and $R > 0$. The elliptic law $m_R^{(\tau)}$ is a unique maximizer of the functional*

$$\Sigma(\mu)-\frac{1}{R^2}\int\Big(\frac{x^2}{1+\tau}+\frac{y^2}{1-\tau}\Big)\,d\mu(\zeta) \quad \text{on} \quad \mathcal{M}(\mathbb{C}) \qquad (\zeta = x+\mathrm{i}\,y).$$

Furthermore, among $\mu \in \mathcal{M}(\mathbb{C})$ with $\int(x^2/(1+\tau)+y^2/(1-\tau))\,d\mu(\zeta) \le R^2/2$ (or $= R^2/2$), $m_R^{(\tau)}$ maximizes $\Sigma(\mu)$, and $\Sigma(m_R^{(\tau)}) = \log R - 1/4$ independently of the parameter τ.

To prove the proposition we need

Lemma 5.3.12 *For* $\zeta = x + \mathrm{i}\, y$,

$$\int \log|\zeta - \eta|\, dm_R^{(\tau)}(\eta) \begin{cases} = \dfrac{1}{2R^2}\Big(\dfrac{x^2}{1+\tau} + \dfrac{y^2}{1-\tau}\Big) + \log R - \dfrac{1}{2} & \textit{if } \zeta \in E_R^{(\tau)}, \\ < \dfrac{1}{2R^2}\Big(\dfrac{x^2}{1+\tau} + \dfrac{y^2}{1-\tau}\Big) + \log R - \dfrac{1}{2} & \textit{if } \zeta \in \mathbb{C} \setminus E_R^{(\tau)}. \end{cases}$$

Proof: Since $m_R^{(\tau)} = D_R m_1^{(\tau)}$, we immediately get

$$\int \log|\zeta - \eta|\, dm_R^{(\tau)}(\eta) = \int \log|R^{-1}\zeta - \eta|\, dm_1^{(\tau)} + \log R\,.$$

So it is enough to show the case $R = 1$. Moreover we may assume $0 \le \tau < 1$ by symmetry. Set

$$\begin{cases} f_1(x,y) := \dfrac{x}{1+\tau} - \mathrm{i}\,\dfrac{y}{1-\tau} & \text{if } \zeta \in E_1^{(\tau)}, \\ f_2(x,y) := \dfrac{\zeta}{2\tau}\Big(1 - \sqrt{1 - \dfrac{4\tau}{\zeta^2}}\Big) & \text{if } \zeta \in \mathbb{C} \setminus E_1^{(\tau)}. \end{cases}$$

Then $f_2(\zeta)$ is an analytic function (a branch such that $f_2(\infty) = 0$) in $\mathbb{C} \setminus E_1^{(\tau)}$ and $f_2(x,y) = f_1(x,y)$ on the boundary of $E_1^{(\tau)}$. Since

$$\begin{cases} \dfrac{1}{2\pi}\Big(\dfrac{\partial}{\partial x} f_1(x,y) + \mathrm{i}\,\dfrac{\partial}{\partial y} f_1(x,y)\Big) = \dfrac{1}{\pi(1-\tau^2)} & \text{if } \zeta \in E_1^{(\tau)}, \\ \dfrac{1}{2\pi}\Big(\dfrac{\partial}{\partial x} f_2(x,y) + \mathrm{i}\,\dfrac{\partial}{\partial y} f_2(x,y)\Big) = 0 & \text{if } \zeta \in \mathbb{C} \setminus E_1^{(\tau)}, \end{cases}$$

one can use the Gauss integral formula ([125], Appendix IV) to obtain

$$\int \frac{dm_1^{(\tau)}(\eta)}{\zeta - \eta} = \begin{cases} f_1(x,y) & \text{if } \zeta \in E_1^{(\tau)}, \\ f_2(x,y) & \text{if } \zeta \in \mathbb{C} \setminus E_1^{(\tau)}. \end{cases}$$

Next we show that if $x + \mathrm{i}\, y \in \mathbb{C} \setminus E_1^{(\tau)}$ and $x, y \ge 0$, then

$$\begin{cases} \operatorname{Re} f_2(x,y) \le \dfrac{x}{1+\tau} & \text{(strictly if } x > 0\text{)}, \\ -\operatorname{Im} f_2(x,y) \le \dfrac{y}{1-\tau} & \text{(strictly if } y > 0\text{)}. \end{cases} \tag{5.3.17}$$

We may assume $x, y > 0$, because the case $x = 0$ or $y = 0$ is straightforward. Let $\zeta_0 = x_0 + \mathrm{i}\, y_0$ ($x_0, y_0 > 0$) be on the boundary of $E_1^{(\tau)}$. For $t \ge 1$, if we write

$x(t) - \mathrm{i}\, y(t) = f_2(t\zeta_0)/t$, then $(x(t), y(t))$ is an intersection of two hyperbolic curves:

$$\left(x - \frac{x_0}{2\tau}\right)^2 - \left(y + \frac{y_0}{2\tau}\right)^2 = \frac{x_0^2 - y_0^2 - 4\tau/t^2}{4\tau^2}, \tag{5.3.18}$$

$$\left(x - \frac{x_0}{2\tau}\right)\left(y + \frac{y_0}{2\tau}\right) = -\frac{x_0 y_0}{4\tau^2}. \tag{5.3.19}$$

Note that

$$x_0^2 - y_0^2 - 4\tau = \frac{4\tau}{(1+\tau)^2} x_0^2 - (1+\tau)^2 \le 4\tau - (1+\tau)^2 < 0.$$

Looking at the graphs of (5.3.18) and (5.3.19), one can easily see that when t increases from 1 the point $(x(t), y(t))$ moves along the left-upper half of the curve (5.3.19) in the left-lower direction from $(x(1), y(1)) = (x_0/(1+\tau), y_0/(1-\tau))$. Hence $x(t) < x_0/(1+\tau)$ and $y(t) < y_0/(1-\tau)$ for all $t > 1$, implying (5.3.17).

Let $\zeta_0 = x_0 + \mathrm{i}\, y_0$ be on the boundary of $E_1^{(\tau)}$, and for $t \ge 0$ set

$$\begin{aligned} \phi(t) &:= \int \log|t\zeta_0 - \eta|\, dm_1^{(\tau)}(\eta) \\ &= \frac{1}{2\pi(1-\tau^2)} \iint_{E_1^{(\tau)}} \log((tx_0 - x)^2 + (ty_0 - y)^2)\, dx\, dy. \end{aligned}$$

Here we may assume $x_0, y_0 \ge 0$ by symmetry. Since

$$\begin{aligned} \phi'(t) &= \frac{1}{\pi(1-\tau^2)} \iint_{E_{\tau,1}} \frac{(tx_0 - x)x_0 + (ty_0 - y)y_0}{(tx_0 - x)^2 + (ty_0 - y)^2}\, dx\, dy \\ &= x_0 \operatorname{Re} \int \frac{dm_1^{(\tau)}(\eta)}{t\zeta_0 - \eta} - y_0 \operatorname{Im} \int \frac{dm_1^{(\tau)}(\eta)}{t\zeta_0 - \eta}, \end{aligned}$$

we have

$$\phi'(t) \begin{cases} = \dfrac{tx_0^2}{1+\tau} + \dfrac{ty_0^2}{1-\tau} & \text{if } 0 < t < 1, \\ < \dfrac{tx_0^2}{1+\tau} + \dfrac{ty_0^2}{1-\tau} & \text{if } t > 1, \end{cases}$$

using (5.3.17) for the inequality. The above estimate yields

$$\phi(t) \begin{cases} = \dfrac{1}{2}\left(\dfrac{t^2x_0^2}{1+\tau} + \dfrac{t^2y_0^2}{1-\tau}\right) + \phi(0) & \text{if } 0 < t < 1, \\ < \dfrac{1}{2}\left(\dfrac{t^2x_0^2}{1+\tau} + \dfrac{t^2y_0^2}{1-\tau}\right) + \phi(0) & \text{if } t > 1. \end{cases}$$

Finally, the computation of $\phi(0) = -1/2$ is as follows:

$$\begin{aligned}\phi(0) &= \frac{1}{2\pi(1-\tau^2)} \iint_{E_{\tau,1}} \log(x^2+y^2)\,dx\,dy \\ &= \frac{1}{2\pi}\int_0^1 r\,dr \int_0^{2\pi} \log r^2((1+\tau)^2\cos^2\theta + (1-\tau)^2\sin^2\theta)\,d\theta \\ &= \int_0^1 2r\log r\,dr + \frac{1}{2\pi}\int_0^1 r\,dr\int_0^{2\pi}\log(1+\tau^2+2\tau\cos 2\theta)\,d\theta \\ &= -\frac{1}{2}.\end{aligned}$$

□

Proof of Proposition 5.3.11*:* The first assertion follows from Theorem 5.3.3 and Lemma 5.3.12. Let $Q(\zeta) := x^2/(1+\tau)+y^2/(1-\tau)$ $(\zeta = x+\mathrm{i}y)$. For any $\mu \in \mathcal{M}(\mathbb{C})$ with $\int Q(\zeta)\,d\mu(\zeta) < +\infty$, set $\lambda := \left((2/R^2)\int Q(\zeta)\,d\mu(\zeta)\right)^{1/2}$ and $\tilde{\mu} := D_{1/\lambda}\mu$. Then we have $\int Q(\zeta)\,d\tilde{\mu}(\zeta) = R^2/2$ and

$$-\Sigma(\mu) + \frac{1}{R^2}\int Q(\zeta)\,d\mu(\zeta) = -\Sigma(\tilde{\mu}) + \frac{\lambda^2}{2} - \log\lambda\,.$$

Hence the second assertion can be shown as in the proof of Proposition 5.3.4. Furthermore, by Lemma 5.3.12 we compute

$$\begin{aligned}\Sigma(m_R^{(\tau)}) &= \Sigma(m_1^{(\tau)}) + \log R \\ &= \frac{1}{2\pi(1-\tau^2)}\iint_{E_1^{(\tau)}}\left(\frac{x^2}{1+\tau}+\frac{y^2}{1-\tau}\right)dx\,dy + \log R - \frac{1}{2} \\ &= \frac{1}{2\pi}\int_0^1 r^3\,dr\int_0^{2\pi}((1+\tau)\cos^2\theta + (1-\tau)\sin^2\theta)\,d\theta \\ &\qquad + \log R - \frac{1}{2} \\ &= \log R - \frac{1}{4}.\end{aligned}$$

□

5.4 Gaussian and unitary random matrices

It is useful to give a short summary of the contents of Sections 5.1 and 5.2. We started from the level-2 large deviation theorem for the mean of independent identically distributed random variables. This result is something like the strengthening of the law of large numbers, because, beyond stating the limit distribution of the means, the rate of convergence is exactly established. Then we deduced that the differential entropy, or more generally the relative entropy, is related to the volume of atomic measures approximating the given probability measure. That was Proposition 5.1.1, giving the limit of

$$\frac{1}{n}\log \nu^n\bigl(\{x\in\mathbb{R}^n : |m_k(\kappa_n(x)) - m_k(\mu)| \le \varepsilon,\ k\le r\}\bigr), \tag{5.4.1}$$

and also the formula (5.1.8) was observed. Roughly speaking, when n is large, the Lebesgue measure of the set of those points $x \in \mathbb{R}^n$ such that the atomic measure $\frac{1}{n}(\delta(x_1)+\delta(x_2)+\cdots+\delta(x_n))$ is very close to the given measure μ is about $\exp(nS(\mu))$. Then we saw that Voiculescu's free entropy looks like the logarithm of the volume of real symmetric (or complex selfadjoint) matrices approximating the given measure in the sense of moments. This is the content of the formula (5.2.1), and it was proved in a slightly different form as Theorem 5.2.2. Since moments of a selfadjoint matrix depend only on the eigenvalues, we passed from the space of matrices to the space of eigenvalues and studied the limit of

$$\frac{1}{n^2}\log \bar{\nu}_n\bigl(\{\lambda\in\mathbb{R}^n : |m_k(\kappa_n(\lambda)) - m_k(\mu)| \le \varepsilon,\ k\le r\}\bigr). \tag{5.4.2}$$

Beyond the similarity of (5.4.1) and (5.4.2), the essential difference is not the normalization but rather the reference measure. A product measure ν^n is taken in (5.4.1), while $\bar{\nu}_n$ has the more complicated density (5.2.2).

Our aim now is to show that Voiculescu's Theorem 5.2.2 carries the essence of a large deviation result in which the free entropy (5.2.7) is the interesting part of the rate function.

Before stating the theorem let us explain more explicitly the general framework of large deviations related to random matrices. For each $n \in \mathbb{N}$ let an $n\times n$ random matrix $X(n)$ be given, and let ν_n be its distribution on $M_n(\mathbb{C})$. For instance, if $X(n)$ is real symmetric (or complex selfadjoint), then ν_n is supported on $M_n(\mathbb{R})^{sa}$ (or $M_n(\mathbb{C})^{sa}$). Assume further that ν_n is invariant under orthogonal (or unitary) conjugation. Then ν_n is in fact determined by the probability measure $\bar{\nu}_n$ induced on $\mathbb{R}^n$, the space of eigenvalues. Define a Borel measurable mapping $\mathbf{K}_n$ from $M_n(\mathbb{R})^{sa}$ (or $M_n(\mathbb{C})^{sa}$) into $\mathcal{M}(\mathbb{R})$ by

$$\mathbf{K}_n(A) := \frac{1}{n}\sum_{i=1}^{n}\delta(\lambda_i(A)), \tag{5.4.3}$$

where $\lambda_1(A),\dots,\lambda_n(A)$ are the eigenvalues of A and $\mathcal{M}(\mathbb{R})$ is endowed with the weak topology. The random measure $R_n := \mathbf{K}_n(X(n))$ is the empirical eigenvalue distribution of $X(n)$. Let P_n denote the distribution of the random measure R_n, which is a probability measure on $\mathcal{M}(\mathbb{R})$. We call P_n as well as R_n the *empirical eigenvalue distribution* of $X(n)$. For every Borel set Γ of $\mathcal{M}(\mathbb{R})$ we have

$$P_n(\Gamma) = \mathbf{Prob}(R_n\in\Gamma) = \nu_n(\mathbf{K}_n^{-1}\Gamma) = \bar{\nu}_n\bigl(\{x\in\mathbb{R}^n : \kappa_n(x)\in\Gamma\}\bigr),$$

where $\kappa_n(x) := \frac{1}{n}\sum_{i=1}^{n}\delta(x_i)$.

In the previous section many interesting distributions are realized as the maximizer of a free entropy functional of the form $\Sigma(\mu) - \int Q(x)\,d\mu(x)$. As we discussed in Sec. 4.1, some of them arise as the limit distribution μ_0 of a certain random matrix model $X(n)$. In this section we are concerned with the large deviation (in the scale n^{-2}) for (R_n) or (P_n) defined as above from $X(n)$. It is expected that a rate function I is minus a constant multiple of the free entropy functional above (up to an additive constant) and μ_0 is the unique minimizer of I with $I(\mu_0) = 0$. If this is the case, the large deviation principle for (P_n) means that the eigenvalue distribution of $X(n)$ converges exponentially fast to the limit distribution μ_0.

In the theory of large deviations, there is a standard method which is based on the concept of exponential tightness. When we are concerned with the large deviation in the scale L_n for a sequence (P_n) of measures on a Polish space $\mathcal{X}$ ($= \mathcal{M}(\mathbb{R})$ for instance), the sequence P_n is said to be *exponentially tight* if for any $\varepsilon > 0$ there exists a compact $K_\varepsilon \subset \mathcal{X}$ such that

$$\limsup_{n\to\infty} L_n \log P_n(\mathcal{X} \setminus K_\varepsilon) \le -\frac{1}{\varepsilon}\,.$$

This condition becomes trivial if the space $\mathcal{X}$ itself is compact. Let $\mathcal{A}$ be a base for the topology of $\mathcal{X}$. If for every $x \in \mathcal{X}$ we have

$$\begin{aligned} I(x) &= \sup_{G\in\mathcal{A},\, x\in G} \Big[-\limsup_{n\to\infty} L_n \log P_n(G)\Big] \\ &= \sup_{G\in\mathcal{A},\, x\in G} \Big[-\liminf_{n\to\infty} L_n \log P_n(G)\Big]\,, \end{aligned} \tag{5.4.4}$$

then (P_n) satisfies the *weak large deviation principle* with a rate function I. A standard fact ([55], Lemma 1.2.18, Sec. 4.1.2) is that when (P_n) is exponentially tight and (5.4.4) holds, (P_n) satisfies the large deviation principle and the rate function I is automatically good.

Here we note another important fact. Let (R_n) be a sequence of random probability measures on a Polish space X, and let P_n be the distribution on $\mathcal{M}(X)$ of R_n. If (P_n) satisfies the large deviation principle in the scale n^{-2} with a good rate function I having the unique minimizer μ_0, then R_n converges to μ_0 in the weak topology almost surely. The proof is easy, using the Lévy metric. Indeed, let ρ be the Lévy metric for the weak topology on $\mathcal{M}(X)$ (see (4.3.9)). For every $\varepsilon > 0$, if we take a closed set $F := \{\mu \in \mathcal{M}(X) : \rho(\mu, \mu_0) \ge \varepsilon\}$, then

$$\limsup_{n\to\infty} \frac{1}{n^2} \log P_n(F) \le -\inf\{I(\mu) : \mu \in F\} < 0,$$

because I is good and μ_0 is the unique minimizer. So we get $\sum_{n=1}^\infty P_n(F) < +\infty$. Since $P_n(F) = \mathbf{Prob}\big(\{\omega : \rho(R_n(\omega), \mu_0) \ge \varepsilon\}\big)$, the Borel-Cantelli lemma yields

$$\mathbf{Prob}\Big(\limsup_n \{\omega : \rho(R_n(\omega), \mu_0) \ge \varepsilon\}\Big) = 0\,,$$

and this gives us the desired conclusion.

Next we state the large deviation result corresponding to Theorem 5.2.2. This result is due to Ben Arous and Guionnet.

Theorem 5.4.1 *Assume that the joint distribution of the random vector $\lambda^{(n)} \in \mathbb{R}^n$ is the measure $\bar{\nu}_n$ given by the density (5.2.2). Then the sequence of random atomic measures*

$$R_n := \frac{\delta(\lambda_1^{(n)}) + \delta(\lambda_2^{(n)}) + \cdots + \delta(\lambda_n^{(n)})}{n}$$

satisfies the large deviation principle in the scale n^{-2} with the good rate function

$$I(\mu) := -\beta\Sigma(\mu) + \frac{1}{4\sigma^2}\int x^2\, d\mu(x) + \frac{\beta}{2}\log(2\beta\sigma^2) - \frac{3\beta}{4}$$

for $\mu \in \mathcal{M}(\mathbb{R})$. Furthermore, the semicircular distribution $w_{2\sqrt{2\beta\sigma^2}}$ is the unique minimizer of I with $I\big(w_{2\sqrt{2\beta\sigma^2}}\big) = 0$.

First, note that the assertion about the minimizer immediately follows from Proposition 5.3.4. The following lemma is about the rate function, though it will be shown in a more general setting in the course of proving Theorem 5.4.3.

Lemma 5.4.2 *The functional I given in the theorem is convex and lower semi-continuous in the weak topology on $\mathcal{M}(\mathbb{R})$. Moreover, $I(\mu) \geq 0$ for all $\mu \in \mathcal{M}(\mathbb{R})$, and $\{\mu \in \mathcal{M}(\mathbb{R}) : I(\mu) \leq c\}$ is compact for every $c \geq 0$. Therefore, I satisfies all requirements as a good rate function.*

Proof: Note that I is well-defined on $\mathcal{M}(\mathbb{R})$, and Proposition 5.3.4 says that it takes the minimum 0 at $w_{2\sqrt{2\beta\sigma^2}}$. We may omit the constant term from I and consider the kernel

$$\begin{aligned} F(x,y) &:= \frac{1}{8\sigma^2}(x^2+y^2) - \beta\log|x-y| \\ &= \frac{1}{8\sigma^2}\left((x^2+y^2) - 4\beta\sigma^2\log(x^2+y^2)\right) + \frac{\beta}{2}\log\frac{x^2+y^2}{(x-y)^2}\,. \end{aligned}$$

This is bounded below because the first term is and $(x^2+y^2)/(x-y)^2 \geq 1/2$ for the second term. Hence the functional

$$\iint F(x,y)\, d\mu(x)\, d\mu(y) = -\beta\Sigma(\mu) + \frac{1}{4\sigma^2}\int x^2\, d\mu(x)$$

is lower semicontinuous. The convexity of this functional can be seen immediately by Proposition 5.3.2. Choose $C > 0$ such that $F(x,y) + C \geq 0$ for all x, y. For any $\alpha > 0$, since $F(x,y) \to +\infty$ as $x^2 + y^2 \to +\infty$, let R be large enough so that $F(x,y) \geq \alpha$ if $x^2 + y^2 \geq 2R^2$. Then for every $\mu \in \mathcal{M}(\mathbb{R})$ we have

$$\begin{aligned}\mu([-R,R]^c)^2 &= (\mu \otimes \mu)\big(\{(x,y) : |x| > R,\ |y| > R\}\big) \\ &\leq (\mu \otimes \mu)\big(\{(x,y) : x^2 + y^2 \geq 2R^2\}\big) \\ &\leq \frac{1}{\alpha} \iint (F(x,y) + C)\, d\mu(x)\, d\mu(y)\,.\end{aligned}$$

This implies that for any $c \geq 0$ the set $\{\mu \in \mathcal{M}(\mathbb{R}) : \iint F(x,y)\, d\mu(x)\, d\mu(y) \leq c\}$ is tight and hence compact.

□

From now on we shall treat a more general probability measure than (5.2.2). Let $Q(x)$ be a real continuous function on $\mathbb{R}$ such that for any $\varepsilon > 0$

$$\lim_{|x| \to \infty} |x| \exp(-\varepsilon Q(x)) = 0\,. \tag{5.4.5}$$

Assume that $\bar{\nu}_n$ on $\mathbb{R}^n$ has the joint probability density

$$\frac{1}{Z_n} \exp\Big(-n \sum_{i=1}^n Q(t_i)\Big) \prod_{i<j} |t_i - t_j|^{2\beta}\,,$$

where $\beta > 0$ is fixed (independent of n) and Z_n is the normalization constant, i.e.

$$Z_n = \int \cdots \int \exp\Big(-n \sum_{i=1}^n Q(t_i)\Big) \prod_{i<j} |t_i - t_j|^{2\beta}\, dt_1 \cdots dt_n\,.$$

(The assumption (5.4.5) implies that this integral is finite.)

The large deviation theorem we are going to prove tells about the random measure $R_n := \frac{1}{n}\sum_{i=1}^n \delta(\lambda_i)$ when $\lambda \in \mathbb{R}^n$ is distributed according to $\bar{\nu}_n$, or about the probability measure

$$P_n(\Gamma) := \bar{\nu}_n\big(\{x \in \mathbb{R}^n : \kappa_n(x) \in \Gamma\}\big)$$

for Borel sets $\Gamma \subset \mathcal{M}(\mathbb{R})$.

Theorem 5.4.3 *The finite limit $B := \lim_{n\to\infty} n^{-2} \log Z_n$ exists, and (P_n) satisfies the large deviation principle in the scale n^{-2} with the good rate function*

$$I(\mu) := -\beta\Sigma(\mu) + \int Q(x)\, d\mu(x) + B$$

for $\mu \in \mathcal{M}(\mathbb{R})$. Furthermore, there exists a unique $\mu_0 \in \mathcal{M}(\mathbb{R})$ such that $I(\mu_0) = 0$.

To prove the theorem, fix a kernel

$$F(x,y) := -\beta \log|x-y| + \frac{1}{2}(Q(x)+Q(y))$$

and its cutoff

$$F_\alpha(x,y) := \min\{\phi(x,y), \alpha\} \quad \text{for} \quad \alpha > 0\,.$$

Since

$$F(x,y) \geq -\beta\bigl[\log(|x|\exp(-Q(x)/2\beta)) + \log(|y|\exp(-Q(y)/2\beta))\bigr]$$

whenever $|x|, |y| \geq 2$, it follows that $F_\alpha(x,y)$ is bounded and continuous. Therefore,

$$\mu \in \mathcal{M}(\mathbb{R}) \mapsto \iint F_\alpha(x,y)\, d\mu(x)\, d\mu(y)$$

is continuous and

$$-\beta\Sigma(\mu) + \int Q(x)\, d\mu(x) = \iint F(x,y)\, d\mu(x)\, d\mu(y)$$

is lower semicontinuous in the weak topology on $\mathcal{M}(\mathbb{R})$.

Lemma 5.4.4

$$\limsup_{n\to\infty} \frac{1}{n^2} \log Z_n \leq - \inf_{\mu \in \mathcal{M}(\mathbb{R})} \iint F(x,y)\, d\mu(x)\, d\mu(y)\,.$$

Proof: We estimate as follows:

$$\begin{aligned}
Z_n &= \int \cdots \int \exp\Bigl(-\sum_{i=1}^n Q(t_i)\Bigr) \\
&\qquad \times \exp\Bigl(-\sum_{i<j}(Q(t_i)+Q(t_j))\Bigr) \prod_{i<j} |t_i - t_j|^{2\beta}\, dt_1 \cdots dt_n \\
&= \int \cdots \int \exp\Bigl(-\sum_{i=1}^n Q(t_i)\Bigr) \exp\Bigl(-2\sum_{i<j} F(t_i,t_j)\Bigr)\, dt_1 \cdots dt_n \\
&\leq \int \cdots \int \exp\Bigl(-\sum_{i=1}^n Q(t_i)\Bigr) \\
&\qquad \times \exp\Bigl(-n^2 \iint_{\{x\neq y\}} F(x,y)\, d\mu_t(x)\, d\mu_t(y)\Bigr)\, dt_1 \cdots dt_n
\end{aligned}$$

$$\begin{aligned}
&\leq \exp\Big(-n^2 \inf_\mu \iint_{\{x\neq y\}} F(x,y)\,d\mu(x)\,d\mu(y)\Big) \\
&\qquad \times \int\cdots\int \exp\Big(-\sum_{i=1}^n Q(t_i)\Big)\,dt_1\cdots dt_n \\
&= \Big(\int e^{-Q(x)}\,dx\Big)^n \exp\Big(-n^2 \inf_\mu \iint F(x,y)\,d\mu(x)\,d\mu(y)\Big),
\end{aligned}$$

implying the lemma.

□

Lemma 5.4.5 *For every $\mu \in \mathcal{M}(\mathbb{R})$,*

$$\inf_G \Big[\limsup_{n\to\infty} \frac{1}{n^2} \log P_n(G)\Big] \leq -\iint F(x,y)\,d\mu(x)\,d\mu(y) - \liminf_{n\to\infty} \frac{1}{n^2}\log Z_n\,,$$

where G runs over a neighborhood base of μ.

Proof: For any neighborhood G of $\mu \in \mathcal{M}(\mathbb{R})$, put

$$\tilde{G} := \{t \in \mathbb{R}^n : \kappa_n(t) \in G\}\,.$$

As in the proof of Lemma 5.4.4, we get

$$\begin{aligned}
&P_n(G) = \bar{\nu}_n(\tilde{G}) \\
&= \frac{1}{Z_n}\int\cdots\int_{\tilde{G}} \exp\Big(-\sum_{i=1}^n Q(t_i)\Big)\exp\Big(-2\sum_{i<j} F(t_i,t_j)\Big)\,dt_1\cdots dt_n \\
&\leq \frac{1}{Z_n}\int\cdots\int_{\tilde{G}} \exp\Big(-\sum_{i=1}^n Q(t_i)\Big) \\
&\qquad \times \exp\Big(-n^2\iint F_\alpha(x,y)\,d\mu_t(x)\,d\mu_t(y) + n\alpha\Big)\,dt_1\cdots dt_n \\
&= \frac{1}{Z_n}\Big(\int e^{-Q(x)}\,dx\Big)^n \exp\Big(-n^2 \inf_{\mu'\in G}\iint F_\alpha(x,y)\,d\mu'(x)\,d\mu'(y) + n\alpha\Big).
\end{aligned}$$

Therefore,

$$\limsup_{n\to\infty}\frac{1}{n^2}\log P_n(G) \leq -\inf_{\mu'\in G}\iint F_\alpha(x,y)\,d\mu'(x)\,d\mu'(y) - \liminf_{n\to\infty}\frac{1}{n^2}\log Z_n\,.$$

Thanks to the weak continuity of $\mu' \mapsto \iint F_\alpha(x,y)\,d\mu'(x)\,d\mu'(y)$ we get

$$\inf_G\Big[\limsup_{n\to\infty}\frac{1}{n^2}\log P_n(G)\Big] \leq -\iint F_\alpha(x,y)\,d\mu(x)\,d\mu(y) - \liminf_{n\to\infty}\frac{1}{n^2}\log Z_n\,.$$

Letting $\alpha \to +\infty$ completes the proof of the inequality.

□

Lemma 5.4.6 *For every $\mu \in \mathcal{M}(\mathbb{R})$,*

$$\liminf_{n\to\infty} \frac{1}{n^2} \log Z_n \geq -\iint F(x,y)\, d\mu(x)\, d\mu(y)$$

and

$$\inf_G \left[\liminf_{n\to\infty} \frac{1}{n^2} \log P_n(G)\right] \geq -\iint F(x,y)\, d\mu(x)\, d\mu(y) - \limsup_{n\to\infty} \frac{1}{n^2} \log Z_n\,,$$

where G runs over a neighborhood base of μ.

Proof: It is obvious that

$$\mu \in \mathcal{M}(\mathbb{R}) \mapsto \inf\left\{\liminf_{n\to\infty} \frac{1}{n^2} \log P_n(G) : G \text{ a neighborhood of } \mu\right\}$$

is upper semicontinuous. Since $F(x,y)$ is bounded below, we get

$$\iint F(x,y)\, d\mu(x)\, d\mu(y) = \lim_{k\to\infty} \iint F(x,y)\, d\mu_k(x)\, d\mu_k(y)$$

with $\mu_k := \mu([-k,k])^{-1}\chi_{[-k,k]}\mu$. So it suffices to assume that μ has a compact support. For $\varepsilon > 0$ let ϕ_ε be a nonnegative C^∞-function supported in $[-\varepsilon,\varepsilon]$ such that $\int \phi_\varepsilon(x)\, dx = 1$, and let $\phi_\varepsilon * \mu$ be the convolution of μ with ϕ_ε. Thanks to the properties of $\Sigma(\mu)$ given in Proposition 5.3.2, it is easy to see that

$$\Sigma(\phi_\varepsilon * \mu) \geq \Sigma(\mu)\,.$$

Also

$$\lim_{\varepsilon\to+0} \int Q(x)\, d(\phi_\varepsilon * \mu)(x) = \int Q(x)\, d\mu(x)\,.$$

Hence we may assume that μ has a continuous density with compact support. Moreover, let m be the uniform distribution on an interval $[a,b]$ including $\operatorname{supp}\mu$. Then it suffices to show the required inequalities for each $(1-\delta)\mu+\delta m$ $(0 < \delta < 1)$. After all, we may assume that μ has a continuous density $f > 0$ on $\operatorname{supp}\mu = [a,b]$ so that $\delta \leq f(x) \leq \delta^{-1}$ $(a \leq x \leq b)$ for some $\delta > 0$.

For each $n \in \mathbb{N}$ let $a < a_1^{(n)} < b_1^{(n)} < a_2^{(n)} < \ldots < a_n^{(n)} < b_n^{(n)} = b$ be such that

$$\int_a^{a_i^{(n)}} f(x)\, dx = \frac{i-\frac{1}{2}}{n}\,, \qquad \int_a^{b_i^{(n)}} f(x)\, dx = \frac{i}{n} \qquad (1 \leq i \leq n).$$

Then it immediately follows that

$$\frac{\delta}{2n} \leq b_i^{(n)} - a_i^{(n)} \leq \frac{1}{2n\delta} \qquad (1 \leq i \leq n).$$

Define

$$\Delta_n := \{(t_1, \dots, t_n) \in \mathbb{R}^n : a_i^{(n)} \le t_i \le b_i^{(n)},\ 1 \le i \le n\}.$$

For any neighborhood G of μ, it is clear that

$$\Delta_n \subset \tilde{G} := \{t \in \mathbb{R}^n : \kappa_n(t) \in G\}$$

for all n large enough. Therefore, for large n we have

$$\begin{aligned}
P_n(G) &= \bar{\nu}_n(\tilde{G}) \ge \bar{\nu}_n(\Delta_n) \\
&= \frac{1}{Z_n} \int \cdots \int_{\Delta_n} \exp\Bigl(-n \sum_{i=1}^n Q(t_i)\Bigr) \prod_{i<j} |t_i - t_j|^{2\beta}\, dt_1 \cdots dt_n \\
&\ge \frac{1}{Z_n} \exp\Bigl(-n \sum_{i=1}^n \xi_i^{(n)}\Bigr) \prod_{i<j} (a_j^{(n)} - b_i^{(n)})^{2\beta} \int \cdots \int_{\Delta_n} dt_1 \cdots dt_n \\
&\ge \frac{1}{Z_n} \Bigl(\frac{\delta}{2n}\Bigr)^n \exp\Bigl(-n \sum_{i=1}^n \xi_i^{(n)}\Bigr) \prod_{i<j} (a_j^{(n)} - b_i^{(n)})^{2\beta},
\end{aligned}$$

where $\xi_i^{(n)} := \max\{Q(x) : a_i^{(n)} \le x \le b_i^{(n)}\}$. Now let $g : [0,1] \to [a,b]$ be the inverse function of $t \mapsto \int_a^t f(x)\, dx$. Since $a_i^{(n)} = g((i - \frac{1}{2})/n)$ and $b_i^{(n)} = g(i/n)$, we get

$$\lim_{n\to\infty} \frac{1}{n} \sum_{i=1}^n \xi_i^{(n)} = \int_0^1 Q(g(t))\, dt = \int_a^b Q(x) f(x)\, dt = \int Q(x)\, d\mu(x)$$

and

$$\begin{aligned}
\lim_{n\to\infty} \frac{2}{n^2} \sum_{i<j} \log(a_j^{(n)} - b_i^{(n)}) &= 2 \iint_{0\le s<t\le 1} \log(g(t) - g(s))\, ds\, dt \\
&= \int_0^1 \int_0^1 \log|g(s) - g(t)|\, ds\, dt = \iint f(x) f(y) \log|x-y|\, dx\, dy = \Sigma(\mu).
\end{aligned}$$

Therefore,

$$0 \ge \limsup_{n\to\infty} \frac{1}{n^2} \log P_n(G) \ge -\iint F(x,y)\, d\mu(x)\, d\mu(y) - \liminf_{n\to\infty} \frac{1}{n^2} \log Z_n$$

and

$$\liminf_{n\to\infty} \frac{1}{n^2} \log P_n(G) \ge -\iint F(x,y)\, d\mu(x)\, d\mu(y) - \limsup_{n\to\infty} \frac{1}{n^2} \log Z_n,$$

as desired.

□

Lemma 5.4.7 *The finite limit $B = \lim_{n\to\infty} n^{-2} \log Z_n$ exists.*

Proof: By Lemmas 5.4.4 and 5.4.6 we have

$$\limsup_{n\to\infty} \frac{1}{n^2} \log Z_n \le -\inf_{\mu} \iint F(x,y)\, d\mu(x)\, d\mu(y) \le \liminf_{n\to\infty} \frac{1}{n^2} \log Z_n \,.$$

This gives the result, because Theorem 5.3.3 says that $\mu \mapsto \iint F(x,y)\, d\mu(x)\, d\mu(y)$ attains the minimum.

□

Lemma 5.4.8 *(P_n) is exponentially tight.*

Proof: For any $\alpha > 0$ set

$$K_\alpha := \left\{ \mu \in \mathcal{M}(\mathbb{R}) : \int Q(x)\, d\mu(x) \le \alpha \right\}.$$

Since $Q(x) \to +\infty$ as $|x| \to +\infty$ by the assumption (5.4.5), it is easy to see that

$$\sup_{\mu \in K_\alpha} \mu([-R,R]^c) \to 0 \quad \text{as} \quad R \to +\infty,$$

and hence K_α is compact in the weak topology. We get

$$\begin{aligned}
P_n(K_\alpha^c) &= \tilde{\nu}_n\left(\left\{ t \in \mathbb{R}^n : \frac{1}{n} \sum_{i=1}^n Q(t_i) > \alpha \right\} \right) \\
&= \frac{1}{Z_n} \int \cdots \int_{\left\{\frac{1}{n}\sum_{i=1}^n Q(t_i) > \alpha\right\}} \exp\left(-n \sum_{i=1}^n Q(t_i) \right) \prod_{i<j} |t_i - t_j|^{2\beta}\, dt_1 \cdots dt_n \\
&\le \frac{1}{Z_n} \exp\left(-\frac{n^2 \alpha}{2} \right) \int \cdots \int \exp\left(-\frac{n}{2} \sum_{i=1}^n Q(t_i) \right) \prod_{i<j} |t_i - t_j|^{2\beta}\, dt_1 \cdots dt_n \,.
\end{aligned}$$

When $Q(x)$ is replaced by $Q(x)/2$, the finite limit

$$B_2 = \lim_{n\to\infty} \frac{1}{n^2} \log \int \cdots \int \exp\left(-\frac{n}{2} \sum_{i=1}^n Q(t_i) \right) \prod_{i<j} |t_i - t_j|^{2\beta}\, dt_1 \cdots dt_n$$

exists, and so does B by Lemma 5.4.7. Hence the above estimate gives

$$\limsup_{n\to\infty} \frac{1}{n^2} \log P_n(K_\alpha^c) \le -B + B_2 - \frac{\alpha}{2} \,.$$

Since $\alpha > 0$ is arbitrary, we have the conclusion.

□

Now we are in a position to complete the proof of Theorem 5.4.3 and its particular case, Theorem 5.4.1.

End of proof of Theorem 5.4.3*:* The proof is in the previous lammas: The weak large deviation principle holds, and the exponential tightness of (P_n) gives the large deviation principle with the good rate function $I(\mu) = \iint F(x,y)\,d\mu(x)\,d\mu(y) + B$. The existence of the unique minimizer is due to Theorem 5.3.3.

□

Assume that the minimizer μ_0 of I has support $[a,b]$ and density ψ. Differentiating the equality condition in Theorem 5.3.3 leads to the following singular integral equation:

$$\int_a^b \frac{\psi(y)}{x-y}\,dy = \frac{Q'(x)}{2\beta} \quad \text{for} \quad a < x < b, \tag{5.4.6}$$

where the integral is taken in the sense of principal value. This integral equation can be justified when $Q(x)$ satisfies a certain regularity condition (cf. [165], Sec. IV.3). For instance, it is known ([145], based on [124], Chap. 11) that if $Q(x)$ is a convex polynomial of even degree, then the solution of (5.4.6) is

$$\begin{aligned} \psi(x) &= \frac{\sqrt{(x-a)(b-x)}}{2\beta\pi^2} \int_a^b \frac{Q'(y)}{(y-x)\sqrt{(y-a)(b-y)}}\,dy \\ &= \frac{\sqrt{(x-a)(b-x)}}{2\beta\pi^2} \int_a^b \frac{Q'(x)-Q'(y)}{x-y} \frac{dy}{\sqrt{(y-a)(b-y)}}, \end{aligned} \tag{5.4.7}$$

where a, b are determined by the equations

$$\int_a^b \frac{Q'(x)}{\sqrt{(x-a)(b-x)}}\,dx = 0\,, \quad \int_a^b \frac{xQ'(x)}{\sqrt{(x-a)(b-x)}}\,dx = 2\beta\pi\,.$$

For the second equality of (5.4.7), note that one has

$$\frac{1}{\pi}\int_a^b \frac{\log|x-y|}{\sqrt{(y-a)(b-y)}}\,dy = \log\frac{b-a}{4} \qquad (a \le x \le b),$$

and hence

$$\frac{1}{\pi}\int_a^b \frac{dy}{(x-y)\sqrt{(y-a)(b-y)}} = 0 \qquad (a < x < b).$$

Also, note that if the degree of $Q(x)$ is $2p$, then $\psi(x)$ is $\sqrt{(x-a)(b-x)}$ times a polynomial of degree $2p-2$, and the convexity of $Q(x)$ is sufficient for the positivity of $\psi(x)$ on (a,b).

Next, for each $n \in \mathbb{N}$ we take a probability measure $\bar{\nu}_n$ on $\mathbb{C}^n$ having the density

$$\frac{1}{Z_n} \exp\left(-n \sum_{i=1}^{n} Q(\zeta_i)\right) \prod_{i<j} |\zeta_i - \zeta_j|^{2\beta},$$

where $\beta > 0$ is fixed and $Q(\zeta)$ is a real continuous function on $\mathbb{C}$ such that

$$\lim_{|\zeta|\to\infty} |\zeta| \exp(-\varepsilon Q(\zeta)) = 0$$

for any $\varepsilon > 0$. Then the large deviation principle is obtained for the probability measure

$$P_n(\Gamma) := \bar{\nu}_n(\{\zeta \in \mathbb{C}^n : \kappa_n(\zeta) \in \Gamma\})$$

for Borel sets $\Gamma \subset \mathcal{M}(\mathbb{C})$, where $\kappa_n(\zeta) := \frac{1}{n}\sum_{i=1}^{n} \delta(\zeta_i)$ for $\zeta = (\zeta_1, \dots, \zeta_n) \in \mathbb{C}^n$.

Theorem 5.4.9 *In the above setting, the finite limit $B := \lim_{n\to\infty} n^{-2} \log Z_n$ exists and (P_n) satisfies the large deviation principle in the scale n^{-2} with the good rate function*

$$I(\mu) := -\beta\Sigma(\mu) + \int Q(\zeta)\, d\mu(\zeta) + B \quad on \quad \mathcal{M}(\mathbb{C}).$$

Furthermore, there exists a unique $\mu_0 \in \mathcal{M}(\mathbb{C})$ such that $I(\mu_0) = 0$.

Proof: Since the theorem can be proved on the same lines as Theorem 5.4.3, we only give a sketch. Take a kernel

$$F(\zeta, \eta) := -\beta \log|\zeta - \eta| + \frac{1}{2}(Q(\zeta) + Q(\eta)) \qquad (\zeta, \eta \in \mathbb{C}).$$

Then $F(\zeta, \eta)$ is bounded below, and

$$-\beta\Sigma(\mu) + \int Q(\zeta)\, d\mu(\zeta) = \iint F(\zeta, \eta)\, d\mu(\zeta)\, d\mu(\eta)$$

is lower semicontinuous in the weak topology on $\mathcal{M}(\mathbb{C})$. One can obtain three lemmas similar to Lemmas 5.4.4–5.4.6. The proofs of the first two are the same. The third lemma is proved with a modification as follows.

A suitable regularization process can be performed as in the proof of Lemma 5.4.6, so we may assume that $\operatorname{supp}\mu = [a, b] \times [c, d]$ and μ has a continuous density f on $[a, b] \times [c, d]$ satisfying $\delta \le f \le \delta^{-1}$ for some $\delta > 0$. For each $n \in \mathbb{N}$ let $m := [\sqrt{n}]$. Let $a = x_0 < x_1 < \ldots < x_m = b$ be such that

$$\mu([x_{i-1}, x_i] \times [c, d]) = \frac{1}{m} \qquad (1 \le i \le m).$$

Noting that $m^2 \le n \le m(m+2)$, we can choose $c = y_{i,0} < y_{i,1} < \ldots < y_{i,l_i} = d$ for $1 \le i \le m$ such that $m \le l_i \le m+2$, $\sum_{i=1}^m l_i = n$ and

$$\mu([x_{i-1}, x_i] \times [y_{i,j-1}, y_{i,j}]) = \frac{1}{ml_i} \qquad (1 \le i \le m,\ 1 \le j \le l_i).$$

Arrange n pieces of rectangles $[x_{i-1}, x_i] \times [y_{i,j-1}, y_{i,j}]$ as

$$R_i^{(n)} = [a_i^{(n)}, b_i^{(n)}] \times [c_i^{(n)}, d_i^{(n)}] \qquad (1 \le i \le n),$$

and in each rectangle $R_i^{(n)}$ take a small one $S_i^{(n)}$ by dividing $R_i^{(n)}$ into 9 congruent rectangles and selecting the one in the middle. Then we get

$$\lim_{n\to\infty} \Big(\max_{1\le i\le n} \operatorname{diam}(R_i^{(n)}) \Big) \to 0\,, \tag{5.4.8}$$

$$\int_{S_i^{(n)}} d\zeta \ge \frac{\delta}{9} \int_{R_i^{(n)}} f(\zeta)\, d\zeta \ge \frac{\delta}{9m(m+2)} \ge \frac{\delta}{27n} \qquad (1 \le i \le n).$$

Define

$$\Delta_n := \{(\zeta_1, \ldots, \zeta_n) \in \mathbb{C}^n : \zeta_i \in S_i^{(n)},\ 1 \le i \le n\}\,.$$

For any neighborhood G of μ, if n is large enough, then we have

$$\Delta_n \subset \{\zeta \in \mathbb{C}^n : \kappa_n(\zeta) \in G\},$$

and so

$$\begin{aligned} P_n(G) &\ge \bar{\nu}_n(\Delta_n) \\ &\ge \frac{1}{Z_n} \exp\Big(-n \sum_{i=1}^n \max_{\zeta \in S_i^{(n)}} Q(\zeta_i) \Big) \\ &\qquad\qquad \times \prod_{i<j} \Big(\min_{\zeta \in S_i^{(n)}, \eta \in S_j^{(n)}} |\zeta - \eta| \Big)^{2\beta} \int \cdots \int_{\Delta_n} d\zeta_1 \cdots d\zeta_n \\ &\ge \frac{1}{Z_n} \Big(\frac{\delta}{27n} \Big)^n \exp\Big(-n \sum_{i=1}^n \max_{\zeta \in S_i^{(n)}} Q(\zeta_i) \Big) \prod_{i<j} \Big(\min_{\zeta \in S_i^{(n)}, \eta \in S_j^{(n)}} |\zeta - \eta| \Big)^{2\beta}. \end{aligned}$$

Therefore, to obtain the required inequalities, it suffices to show that

$$\lim_{n\to\infty} \frac{1}{n} \sum_{i=1}^n \max_{\zeta \in S_i^{(n)}} Q(\zeta_i) = \int Q(\zeta)\, d\mu(\zeta) \tag{5.4.9}$$

and

$$\liminf_{n\to\infty} \frac{2}{n^2} \sum_{i<j} \log\biggl(\min_{\zeta\in S_i^{(n)},\eta\in S_j^{(n)}} |\zeta-\eta| \biggr) \ge \Sigma(\mu)\,. \tag{5.4.10}$$

But (5.4.9) is obvious from (5.4.8). We get

$$\max_{\zeta\in R_i^{(n)},\eta\in R_j^{(n)}} |\zeta-\eta| \le \mathrm{const}\cdot \min_{\zeta\in S_i^{(n)},\eta\in S_j^{(n)}} |\zeta-\eta|,$$

and for any $\varepsilon>0$

$$\lim_{n\to\infty} \frac{2}{n^2} \#\biggl\{ (i,j) : i<j,\ \frac{\max_{\zeta\in R_i^{(n)},\eta\in R_j^{(n)}} |\zeta-\eta|}{\min_{\zeta\in S_i^{(n)},\eta\in S_j^{(n)}} |\zeta-\eta|} \le 1+\varepsilon \biggr\} = 1\,.$$

Since

$$\Sigma(\mu) \le 2 \sum_{i<j} \log\biggl(\max_{\zeta\in R_i^{(n)},\eta\in R_j^{(n)}} |\zeta-\eta| \biggr) \int_{R_i^{(n)}} f(\zeta)\,d\zeta \int_{R_j^{(n)}} f(\eta)\,d\eta\,,$$

we have

$$\begin{aligned} &\Sigma(\mu) - \liminf_{n\to\infty} \frac{2}{n^2} \sum_{i<j} \log\biggl(\min_{\zeta\in S_i^{(n)},\eta\in S_j^{(n)}} |\zeta-\eta| \biggr) \\ &\quad\le \limsup_{n\to\infty} \frac{2}{n^2} \sum_{i<j} \log\left(\frac{\max_{\zeta\in R_i^{(n)},\eta\in R_j^{(n)}} |\zeta-\eta|}{\min_{\zeta\in S_i^{(n)},\eta\in S_j^{(n)}} |\zeta-\eta|} \right) = 0\,, \end{aligned}$$

which implies (5.4.10).

Finally, Lemmas 5.4.7 and 5.4.8 are shown in the same way, and the proof is completed as before.

□

Theorem 5.4.9 includes the large deviation principle for the empirical eigenvalue distribution of an elliptic Gaussian matrix $Y(n)$ in (4.1.23) whose joint eigenvalue density was given in Lemma 4.1.10. In particular, when $Y(n) = X(n)$ is standard Gaussian, the rate function is

$$I(\mu) := -\Sigma(\mu) + \int |\zeta|^2\,d\mu(\zeta) - \frac{3}{4}$$

because $\lim_{n\to\infty} n^{-2}\log Z_n = -\frac{3}{4}$ from (4.1.31), and Proposition 5.3.6 says that the unique minimizer of I is the uniform distribution on the unit disk. The projection of the uniform distribution on the unit disk to the real or imaginary axis is the semicircle law w_1. On the other hand, the limit distribution of $X(n)$ is the distribution of a circular element c, so that the distributions of $(c+c^*)/2$ and $(c-c^*)/2\mathrm{i}$ are $w_{\sqrt{2}}$. So the limit distribution through the eigenvalue distribution is the $1/\sqrt{2}$-compression of the "real" limit. This is not strange, because $X(n)$ is non-normal and the spectral radius should be smaller than the operator norm.

For an elliptic Gaussian matrix $Y(n) = uX(n) + vX(n)^*$ $(u^2+v^2=1)$ the rate function is

$$I(\mu) := -\Sigma(\mu) + \int \Big(\frac{x^2}{1+\tau} + \frac{y^2}{1-\tau}\Big)\, d\mu(\zeta) - \frac{3}{4} \qquad (\zeta = x + \mathrm{i}\, y),$$

and the limit distribution of the empirical eigenvalue density is the elliptic law $m_1^{(\tau)}$, i.e. the uniform measure on $E_1^{(\tau)}$ in (5.3.16), where $\tau = 2uv$. Note that the constant term $-\frac{3}{4}$ of the rate function is determined from the requirement $I(m_1^{(\tau)}) = 0$ and Proposition 5.3.11.

By virtue of the fact noted above Theorem 5.4.1, the large deviations in Theorems 5.4.3 and 5.4.9 imply that the empirical eigenvalue density of the corresponding random matrix converges in the weak topology to the measure μ_0 almost surely. But the almost sure convergence in moments (slightly better than that in the weak topology) for several random matrices is already known from Theorems 4.1.5, 4.1.7 and the asymptotic freeness results in Sec. 4.3.

The large deviation theorem related to unitary random matrices is obtained similarly. Let γ_n be the Haar probability measure on the unitary group $\mathcal{U}(n)$. Let $Q(\zeta)$ be a real continuous function on the unit circle $\mathbb{T}$ and for each $n \in \mathbb{N}$ take a probability measure ν_n on $\mathcal{U}(n)$ as

$$\nu_n := \frac{1}{Z_n} \exp(-n \mathrm{Tr}\, Q(U))\, d\gamma_n(U)\,,$$

where Z_n is for normalization. The density of the measure on $\mathbb{T}^n$ induced from γ_n with respect to $d\zeta_1 \cdots d\zeta_n$ (the Haar probability measure on $\mathbb{T}^n$) was given in Lemma 4.2.1, so the above ν_n induces the measure $\bar\nu_n$ on $\mathbb{T}^n$ having the probability density

$$\frac{1}{Z_n} \exp\Big(-n \sum_{i=1}^n Q(\zeta_i)\Big) \prod_{i<j} |\zeta_i - \zeta_j|^2\,.$$

Theorem 5.4.10 *The finite limit $B := \lim_{n\to\infty} n^{-2} \log Z_n$ exists, and the empirical eigenvalue distribution P_n on $\mathcal{M}(\mathbb{T})$ of the above ν_n satisfies the large deviation principle in the scale n^{-2} with the good rate function*

$$I(\mu) := -\Sigma(\mu) + \int_{\mathbb{T}} Q(\zeta)\, d\mu(\zeta) + B \quad on \quad \mathcal{M}(\mathbb{T}).$$

Furthermore, there exists a unique $\mu_0 \in \mathcal{M}(\mathbb{T})$ such that $I(\mu_0) = 0$.

This can be proved more or less similarly to Theorems 5.4.3 and 5.4.9, so we omit the details. Below we just mention a few points.

(1) Since $\mathbb{T}$ is compact, the weak topology on $\mathcal{M}(\mathbb{T})$ is the w*-topology, and hence the exponential tightness of (P_n) is automatic.

(2) We need a regularization procedure for $\mu \in \mathcal{M}(\mathbb{T})$ as was done in the proof of Lemma 5.4.6 for $\mu \in \mathcal{M}(\mathbb{R})$. The *Poisson integral* is available for this purpose. In fact, for any $\mu \in \mathcal{M}(\mathbb{T})$ and $0 < r < 1$ define

$$f_r(e^{i\theta}) := \frac{1}{2\pi}\int_0^{2\pi} P_r(\theta - t)\,d\mu(t)\,, \quad \mu_r := \frac{1}{2\pi} f_r(e^{i\theta})\,d\theta\,,$$

where $P_r(\theta) := (1-r^2)/(1-2r\cos\theta + r^2)$, the *Poisson kernel* (used as the density in (5.3.14)). Then it is well-known that $\mu_r \to \mu$ in the w*-topology as $r \to 1$. Furthermore, a basic fact on harmonic extension and the Poisson integral ([110], Chap. I) is used to compute

$$\int_{\mathbb{T}} \log|\zeta - \eta|\,d\mu_r(\zeta) = \frac{1}{2\pi}\int_0^{2\pi} \log|re^{is} - \eta|\,d\mu(s) \qquad (\eta \in \mathbb{T})$$

and

$$\iint_{\mathbb{T}^2} \log|\zeta - \eta|\,d\mu_r(\zeta)\,d\mu_r(\eta) = \frac{1}{(2\pi)^2}\int_0^{2\pi}\int_0^{2\pi} \log|r^2e^{is} - e^{it}|\,d\mu(s)\,d\mu(t)\,.$$

Hence $\Sigma(\mu_r) \to \Sigma(\mu)$ as $r \to 1$.

(3) The large deviation in Theorem 5.4.10 has a consequence that the empirical eigenvalue density of the $n \times n$ unitary random matrix distributed according to ν_n converges to μ_0 almost surely. In particular, its mean eigenvalue distribution converges to μ_0, that is, $\int_{\mathcal{U}(n)} \mathrm{tr}_n(U^k)\,d\nu_n(U) \to m_k(\mu_0)$ for all $k \in \mathbb{Z}$. In this way, we obtain a rather general Wigner type theorem for unitary random matrices.

The following are two examples of Theorem 5.4.10 corresponding to Propositions 5.3.9 and 5.3.10. First, for $\alpha \in \mathbb{C}$, $|\alpha| < 1$, let $Q(\zeta) := \log|\zeta - \alpha|^2$ $(\zeta \in \mathbb{T})$. Then the probability measure ν_n on $\mathcal{U}(n)$ is given as

$$\nu_n = \frac{1}{Z_n}\,\frac{d\gamma_n(U)}{\det|U - \alpha I|^{2n}}\,.$$

Hence $\bar{\nu}_n$ on $\mathbb{T}^n$ is

$$\bar{\nu}_n = \frac{1}{Z_n}\,\frac{\prod_{i<j}|\zeta_i - \zeta_j|^2}{\prod_{i=1}^n |\zeta_i - \alpha|^{2n}}\,d\zeta_1 \cdots d\zeta_n\,.$$

Then the associated empirical eigenvalue distribution satisfies the large deviation principle with the rate function

$$I(\mu) := -\Sigma(\mu) + \int \log|\zeta - \alpha|^2 \, d\mu(\zeta) - \log(1 - |\alpha|^2) \quad \text{on} \quad \mathcal{M}(\mathbb{T}),$$

and the Poisson kernel measure p_α in (5.3.14) is a unique minimizer of I. Also we have

$$\lim_{n\to\infty} \frac{1}{n^2} \log \int_{\mathcal{U}(n)} \frac{d\gamma_n(U)}{\det|U - \alpha I|^{2n}} = -\log(1 - |\alpha|^2)\,.$$

It does not seem easy to directly compute the asymptotic limit of the above integral.

Second, let $\lambda > 0$ and $Q(\zeta) := -(2/\lambda)\mathrm{Re}\,\zeta$ $(\zeta \in \mathbb{T})$. Then ν_n on $\mathcal{U}(n)$ is

$$\nu_n = \frac{1}{Z_n} \exp\Big(\frac{n}{\lambda}\mathrm{Tr}\,(U + U^*)\Big)\, d\gamma_n(U)\,,$$

and $\bar{\nu}_n$ on $\mathbb{T}^n$ is

$$\bar{\nu}_n = \frac{1}{Z_n} \exp\bigg(\frac{2n}{\lambda}\sum_{i=1}^n \cos\theta_i\bigg) \prod_{i<j} |e^{\mathrm{i}\,\theta_i} - e^{\mathrm{i}\,\theta_j}|^2 \, d\theta_1 \cdots d\theta_n\,.$$

Then the empirical eigenvalue distribution satisfies the large deviation principle with the rate function

$$I(\mu) := -\Sigma(\mu) - \frac{2}{\lambda}\int \mathrm{Re}\,\zeta \, d\mu(\zeta) + B \quad \text{on} \quad \mathcal{M}(\mathbb{T}),$$

where

$$B := \begin{cases} \dfrac{1}{\lambda^2} & \text{if } \lambda \geq 2, \\[2mm] \dfrac{1}{2}\log\dfrac{\lambda}{2} + \dfrac{2}{\lambda} - \dfrac{3}{4} & \text{if } 0 < \lambda < 2, \end{cases}$$

and ρ_λ in (5.3.15) is the unique minimizer of I. Incidentally, the quantity

$$\lim_{n\to\infty} \frac{1}{n^2} \log \int_{\mathcal{U}(n)} \exp\Big(\frac{n}{\lambda}\mathrm{Tr}\,(U + U^*)\Big)\, d\gamma_n(U)$$

is equal to the above B.

5.5 The Wishart matrix

In this section we continue to study large deviation theorems related to random matrices. Our main interest here is focused on a large deviation for the empirical eigenvalue distribution of the regular and singular *Wishart matrices.* Moreover, some other random matrices are also considered, for example the complexified Wishart matrix and antisymmetric real matrices.

For each $n \in \mathbb{N}$ let a positive integer $p(n)$ be given with $p(n) \geq n$. Let $T(n)$ be a $p(n) \times n$ real random matrix all of whose entries are independent and have the identical distribution $N(0,1)$. Then the $n \times n$ random matrix $n^{-1}T(n)^tT(n)$ is the Wishart matrix with a normalization. Its distribution on the space $M_n(\mathbb{R})^+$ of $n \times n$ positive semidefinite symmetric matrices induces the eigenvalue distribution on $(\mathbb{R}^+)^n$, and according to (4.1.22) the latter distribution has the joint probability density

$$\frac{1}{Z_n}\exp\biggl(-\frac{n}{2}\sum_{i=1}^n \lambda_i\biggr)\prod_{i=1}^n \lambda_i^{(p(n)-n-1)/2}\prod_{i<j}|\lambda_i - \lambda_j| \tag{5.5.1}$$

(cf. also [4], Chap. 7).

Also, let $\hat{T}(n)$ be a $p(n) \times n$ complex random matrix such that $\mathrm{Re}\,\hat{T}_{ij}(n)$ and $\mathrm{Im}\,\hat{T}_{ij}(n)$ $(1 \leq i \leq p(n),\ 1 \leq j \leq n)$ are independent with the identical distribution $N(0,1)$. Then the $n \times n$ positive semidefinite random matrix $(2n)^{-1}\hat{T}(n)^*\hat{T}(n)$ induces the probability distribution on $M_n(\mathbb{C})^+$, and it follows from (4.4.5) that its joint eigenvalue density on $(\mathbb{R}^+)^n$ is

$$\frac{1}{Z_n}\exp\biggl(-n\sum_{i=1}^n \lambda_i\biggr)\prod_{i=1}^n \lambda_i^{p(n)-n}\prod_{i<j}(\lambda_i - \lambda_j)^2\,. \tag{5.5.2}$$

To obtain the large deviations related to $n^{-1}T(n)^tT(n)$ or $(2n)^{-1}\hat{T}(n)^*\hat{T}(n)$, it is convenient to take a joint density more general than (5.5.1) and (5.5.2). Let $Q(x)$ be a real continuous function on $\mathbb{R}^+$ such that for any $\varepsilon > 0$

$$\lim_{x\to\infty} x\exp(-\varepsilon Q(x)) = 0\,. \tag{5.5.3}$$

For each $n \in \mathbb{N}$ let $m(n) \in \mathbb{N}$ and a probability measure ν_n on $M_{m(n)}(\mathbb{R})^+$ or $M_{m(n)}(\mathbb{C})^+$ be given. Assume that ν_n induces the distribution $\bar{\nu}_n$ on $(\mathbb{R}^+)^{m(n)}$ having the density

$$\frac{1}{Z_n}\exp\biggl(-n\sum_{i=1}^{m(n)} Q(t_i)\biggr)\prod_{i=1}^{m(n)} t_i^{\gamma(n)}\prod_{1\leq i<j\leq m(n)}|t_i - t_j|^{2\beta}\,, \tag{5.5.4}$$

where $\beta > 0$ is fixed but $\gamma(n) \geq 0$ depends on n. Define the empirical eigenvalue distribution P_n on $\mathcal{M}(\mathbb{R}^+)$ by

$$P_n(\Gamma) = \bar{\nu}_n\big(\{t \in (\mathbb{R}^+)^{m(n)} : \kappa_{m(n)}(t) \in \Gamma\}\big)$$

for Borel sets $\Gamma \subset \mathcal{M}(\mathbb{R}^+)$, where $\kappa_{m(n)}(t) := \frac{1}{m(n)} \sum_{i=1}^{m(n)} \delta(t_i)$ and $\mathcal{M}(\mathbb{R}^+)$ is endowed with the weak topology as before.

Theorem 5.5.1 *Assume that $m(n)/n \to \alpha \in (0,\infty)$ and $\gamma(n)/n \to \gamma \in \mathbb{R}^+$ as $n \to \infty$. Then the finite limit $B := \lim_{n\to\infty} n^{-2} \log Z_n$ exists, and (P_n) satisfies the large deviation principle in the scale n^{-2} with the good rate function*

$$I(\mu) := -\alpha^2\beta\Sigma(\mu) + \alpha \int (Q(x) - \gamma \log x)\, d\mu(x) + B$$

for $\mu \in \mathcal{M}(\mathbb{R}^+)$. Moreover, there exists a unique $\mu_0 \in \mathcal{M}(\mathbb{R}^+)$ such that $I(\mu_0) = 0$.

To prove the theorem, let us introduce the kernel functions on $(\mathbb{R}^+)^2$ as follows:

$$\begin{aligned} F(x,y) &:= -\alpha^2\beta \log|x-y| + \frac{\alpha}{2}(Q(x)+Q(y)) - \frac{\alpha\gamma}{2}(\log x + \log y), \\ \tilde{F}_n(x,y) &:= -\frac{m(n)^2}{n^2}\beta \log|x-y| + \frac{m(n)}{2n}(Q(x)+Q(y)) \\ &\qquad -\frac{m(n)\gamma(n)}{2n^2}(\log x + \log y), \end{aligned}$$

and for $R > 0$

$$F_R(x,y) := \min\{F(x,y), R\}, \quad \tilde{F}_{n,R}(x,y) := \min\{\tilde{F}_n(x,y), R\}.$$

Since

$$F(x,y) \ge -\frac{\alpha(2\alpha\beta+\gamma)}{2}\left[\log\Big(x\exp\Big(-\frac{Q(x)}{2\alpha\beta+\gamma}\Big)\Big) + \log\Big(y\exp\Big(-\frac{Q(y)}{2\alpha\beta+\gamma}\Big)\Big)\right] \tag{5.5.5}$$

(and similarly for $\tilde{F}_n(x,y)$) whenever $x, y \ge 2$, it follows from (5.5.3) that $F_R(x,y)$ is bounded and continuous. Hence

$$-\alpha^2\beta\Sigma(\mu) + \alpha \int (Q(x) - \gamma\log x)\, d\mu(x) = \iint F(x,y)\, d\mu(x)\, d\mu(y)$$

is a well-defined and lower semicontinuous functional on $\mathcal{M}(\mathbb{R}^+)$.

Lemma 5.5.2 *For any $R > 0$, $\tilde{F}_{n,R}(x,y) \to F_R(x,y)$ uniformly as $n \to \infty$.*

Proof: Using (5.5.5) for F as well as $\tilde{F}_n$ and the assumptions on $m(n)$, $\gamma(n)$, one can see that for any $R > 0$ there exists $\delta > 0$ such that if $(x,y) \notin [\delta, \delta^{-1}] \times [\delta, \delta^{-1}]$ then $F(x,y) \geq R$ and $\tilde{F}_n(x,y) \geq R$ for all n. Furthermore, since $\log x$ and $Q(x)$ are bounded on $[\delta, \delta^{-1}]$, $\delta_1 > 0$ can be chosen so that if (x,y) does not belong to

$$\Delta := \{(x,y) : \delta \leq x \leq \delta^{-1},\ \delta \leq y \leq \delta^{-1},\ |x-y| \geq \delta_1\},$$

then $F(x,y) \geq R$ and $\tilde{F}_n(x,y) \geq R$ for all n. Obviously $\tilde{F}_n(x,y) \to F(x,y)$ as $n \to \infty$ uniformly on Δ, so the conclusion follows.

□

According to Theorem 5.3.3 there exist unique $\mu_0, \tilde{\mu}_n \in \mathcal{M}(\mathbb{R}^+)$ such that

$$\begin{aligned} \iint F(x,y)\, d\mu_0(x)\, d\mu_0(y) &= \inf_{\mu \in \mathcal{M}(\mathbb{R}^+)} \iint F(x,y)\, d\mu(x)\, d\mu(y)\,, \\ \iint \tilde{F}_n(x,y)\, d\tilde{\mu}_n(x)\, d\tilde{\mu}_n(y) &= \inf_{\mu \in \mathcal{M}(\mathbb{R}^+)} \iint \tilde{F}_n(x,y)\, d\mu(x)\, d\mu(y)\,. \end{aligned}$$

Lemma 5.5.3 *$(\tilde{\mu}_n)$ is tight, and*

$$\iint F(x,y)\, d\mu_0(x)\, d\mu_0(y) \leq \liminf_{n\to\infty} \iint \tilde{F}_n(x,y)\, d\tilde{\mu}_n(x)\, d\tilde{\mu}_n(y)\,. \tag{5.5.6}$$

Proof: It is clear that $\iint \tilde{F}_n(x,y)\, d\tilde{\mu}_n(x)\, d\tilde{\mu}_n(y) \leq c$ $(n \in \mathbb{N})$ for some $c < +\infty$. Also, by the estimate (5.5.5) for $\tilde{F}_n$, there is a $d < +\infty$ such that $\tilde{F}_n(x,y) \geq -d$ for all $x, y \in \mathbb{R}^+$ and $n \in \mathbb{N}$. For $\alpha > 0$ let

$$M_\alpha := \inf\{\tilde{F}_n(x,y) : n \in \mathbb{N},\ x, y \geq \alpha\}\,.$$

Then, by (5.5.5) for $\tilde{F}_n$ again, M_α can be arbitrarily large when $\alpha \to +\infty$. Since

$$c \geq M_\alpha \tilde{\mu}_n([\alpha,\infty))^2 - d \qquad (n \in \mathbb{N}),$$

we have $\sup_n \tilde{\mu}_n([\alpha,\infty)) \to 0$ as $\alpha \to +\infty$, which means the tightness of $(\tilde{\mu}_n)$.

Thanks to the tightness (or relative weak compactness) of $(\tilde{\mu}_n)$, one can choose a subsequence $(\tilde{\mu}_{n(m)})$ such that $\tilde{\mu}_{n(m)} \to \tilde{\mu}$ weakly for some $\tilde{\mu} \in \mathcal{M}(\mathbb{R}^+)$ and

$$\lim_{m\to\infty} \iint \tilde{F}_{n(m)}(x,y)\, d\tilde{\mu}_{n(m)}\, d\tilde{\mu}_{n(m)}(y) = \liminf_{n\to\infty} \iint \tilde{F}_n(x,y)\, d\tilde{\mu}_n(x)\, d\tilde{\mu}_n(y)\,.$$

Then

$$\begin{aligned} &\iint F(x,y)\, d\mu_0(x)\, d\mu_0(y) \\ &\quad \leq \iint F(x,y)\, d\tilde{\mu}(x)\, d\tilde{\mu}(y) \end{aligned}$$

$$
\begin{aligned}
&= \sup_{R>0} \iint F_R(x,y)\, d\tilde{\mu}(x)\, d\tilde{\mu}(y) \\
&= \sup_{R>0} \lim_{m\to\infty} \iint \tilde{F}_{n(m),R}(x,y)\, d\tilde{\mu}_{n(m)}(x)\, d\tilde{\mu}_{n(m)}(y) \quad \text{(by Lemma 5.5.2)} \\
&\le \lim_{m\to\infty} \iint \tilde{F}_{n(m)}(x,y)\, d\tilde{\mu}_{n(m)}(x)\, d\tilde{\mu}_{n(m)}(y)\,,
\end{aligned}
$$

showing (5.5.6).

□

Lemma 5.5.4

$$
\limsup_{n\to\infty} \frac{1}{n^2} \log Z_n \le -\iint F(x,y)\, d\mu_0(x)\, d\mu_0(y)\,.
$$

Proof: We estimate

$$
\begin{aligned}
Z_n &= \int\cdots\int \exp\biggl(\frac{n}{m(n)} \sum_{i=1}^{m(n)} \Bigl(-Q(t_i) + \frac{\gamma(n)}{n}\log t_i\Bigr)\biggr) \\
&\qquad \times \exp\biggl(-\frac{2n^2}{m(n)^2} \sum_{1\le i<j\le m(n)} \tilde{F}_n(t_i,t_j)\biggr)\, dt_1\cdots dt_{m(n)} \\
&\le \biggl[\int \exp\Bigl(\frac{n}{m(n)}\Bigl(-Q(x) + \frac{\gamma(n)}{n}\log x\Bigr)\Bigr)\, dx\biggr]^{m(n)} \\
&\qquad \times \exp\biggl(-n^2 \iint \tilde{F}_n(x,y)\, d\tilde{\mu}_n(x)\, d\tilde{\mu}_n(y)\biggr).
\end{aligned}
$$

Since by the assumption (5.5.3)

$$
\sup_{n\ge 1} \int \exp\Bigl(\frac{n}{m(n)}\Bigl(-Q(x) + \frac{\gamma(n)}{n}\log x\Bigr)\Bigr)\, dx < +\infty\,,
$$

the above estimate implies that

$$
\begin{aligned}
\limsup_{n\to\infty} \frac{1}{n^2}\log Z_n &\le -\liminf_{n\to\infty} \iint \tilde{F}_n(x,y)\, d\tilde{\mu}_n(x)\, d\tilde{\mu}_n(y) \\
&\le -\iint F(x,y)\, d\mu_0(x)\, d\mu_0(y)
\end{aligned}
$$

thanks to (5.5.6).

□

Lemma 5.5.5 *For every $\mu \in \mathcal{M}(\mathbb{R}^+)$,*

$$
\inf_G \biggl[\limsup_{n\to\infty} \frac{1}{n^2}\log P_n(G)\biggr] \le -\iint F(x,y)\, d\mu(x)\, d\mu(y) - \liminf_{n\to\infty} \frac{1}{n^2}\log Z_n\,,
$$

where G runs over a neighborhood base of μ.

Proof: For any neighborhood G of $\mu \in \mathcal{M}(\mathbb{R}^+)$ set $\tilde{G} := \{t \in (\mathbb{R}^+)^{m(n)} : \kappa_{m(n)}(t) \in G\}$. Then we get

$$\begin{aligned} P_n(G) &= \frac{1}{Z_n} \int \cdots \int_{\tilde{G}} \exp\Big(\frac{n}{m(n)} \sum_{i=1}^{m(n)} \Big(-Q(t_i) + \frac{\gamma(n)}{n} \log t_i \Big)\Big) \\ &\qquad \times \exp\Big(-\frac{2n^2}{m(n)^2} \sum_{1 \le i < j \le m(n)} \tilde{F}_n(t_i, t_j) \Big) \, dt_1 \cdots dt_{m(n)} \\ &\le \frac{1}{Z_n} \Big[\int \exp\Big(\frac{n}{m(n)} \Big(-Q(x) + \frac{\gamma(n)}{n} \log x \Big)\Big) \, dx \Big]^{m(n)} \\ &\qquad \times \exp\Big(-n^2 \inf_{\mu' \in G} \iint \tilde{F}_{n,R}(x,y) \, d\mu'(x) \, d\mu'(y) + nR \Big) \end{aligned}$$

for any $R > 0$. Furthermore, by Lemma 5.5.2

$$\lim_{n \to \infty} \Big(\inf_{\mu' \in G} \iint \tilde{F}_{n,R}(x,y) \, d\mu'(x) \, d\mu'(y) \Big) = \inf_{\mu' \in G} \iint F_R(x,y) \, d\mu'(x) \, d\mu'(y),$$

so that

$$\limsup_{n \to \infty} \frac{1}{n^2} \log P_n(G) \le - \inf_{\mu' \in G} \iint F_R(x,y) \, d\mu'(x) \, d\mu'(y) - \liminf_{n \to \infty} \frac{1}{n^2} \log Z_n \, .$$

Thanks to the continuity of $\mu' \mapsto \iint F_R(x,y) \, d\mu'(x) \, d\mu'(y)$, we obtain

$$\inf_G \Big[\limsup_{n \to \infty} \frac{1}{n^2} \log P_n(G) \Big] \le - \iint F_R(x,y) \, d\mu(x) \, d\mu(y) - \liminf_{n \to \infty} \frac{1}{n^2} \log Z_n \, ,$$

which yields the statement as $R \to +\infty$.

□

Lemma 5.5.6

$$\liminf_{n \to \infty} \frac{1}{n^2} \log Z_n \ge - \iint F(x,y) \, d\mu_0(x) \, d\mu_0(y) \, ,$$

and for every $\mu \in \mathcal{M}(\mathbb{R}^+)$

$$\inf_G \Big[\liminf_{n \to \infty} \frac{1}{n^2} \log P_n(G) \Big] \ge - \iint F(x,y) \, d\mu(x) \, d\mu(y) - \limsup_{n \to \infty} \frac{1}{n^2} \log Z_n \, ,$$

where G runs over a neighborhood base of μ.

Proof: By a regularization argument as in the previous section, we may assume that the support of μ is a finite interval $[0,R]$ and μ has a continuous density $f>0$ on $[0,R]$. Hence $\delta \le f(x) \le \delta^{-1}$ $(0 \le x \le R)$ for some $\delta > 0$. For each $n \in \mathbb{N}$ let $0 < a_1^{(n)} < b_1^{(n)} < a_2^{(n)} < b_2^{(n)} < \ldots < a_{m(n)}^{(n)} < b_{m(n)}^{(n)} = R$ be such that

$$\int_0^{a_i^{(n)}} f(x)\,dx = \frac{i-\frac{1}{2}}{m(n)}, \quad \int_0^{b_i^{(n)}} f(x)\,dx = \frac{i}{m(n)} \qquad (1 \le i \le m(n)).$$

We get

$$\frac{\delta}{2m(n)} \le b_i^{(n)} - a_i^{(n)} \le \frac{1}{2m(n)\delta} \qquad (1 \le i \le m(n)).$$

Define

$$\Delta_n := \{(t_1,\ldots,t_{m(n)}) \in (\mathbb{R}^+)^{m(n)} : a_i^{(n)} \le t_i \le b_i^{(n)},\ 1 \le i \le m(n)\}.$$

For any neighborhood G of μ, if n is large enough, then we have $\Delta_n \subset \{t \in (\mathbb{R}^+)^{m(n)} : \kappa_{m(n)}(t) \in G\}$, and so

$$\begin{aligned} P_n(G) &\ge \frac{1}{Z_n}\int\cdots\int_{\Delta_n} \exp\Big(-n\sum_{i=1}^{m(n)} Q(t_i)\Big)\prod_{i=1}^{m(n)} t_i^{\gamma(n)} \\ &\qquad\times \prod_{1\le i<j\le m(n)} (a_j^{(n)} - b_i^{(n)})^{2\beta} \int\cdots\int_{\Delta_n} dt_1\cdots dt_{m(n)} \\ &\ge \frac{1}{Z_n}\Big(\frac{\delta}{2m(n)}\Big)^{m(n)} \exp\Big(-n\sum_{i=1}^{m(n)} \xi_i^{(n)}\Big)\prod_{i=1}^{m(n)} (a_i^{(n)})^{\gamma(n)} \\ &\qquad\times \prod_{1\le i<j\le m(n)} (a_j^{(n)} - b_i^{(n)})^{2\beta}, \end{aligned}$$

where $\xi_i^{(n)} := \max\{Q(x) : a_i^{(n)} \le x \le b_i^{(n)}\}$. The following are easy to check:

$$\lim_{n\to\infty} \frac{1}{n}\sum_{i=1}^{m(n)} \xi_i^{(n)} = \alpha \int Q(x)\,d\mu(x),$$

$$\lim_{n\to\infty} \frac{\gamma(n)}{n^2}\sum_{i=1}^{m(n)} \log a_i^{(n)} = \alpha\gamma \int \log x\,d\mu(x),$$

$$\lim_{n\to\infty} \frac{2\beta}{n^2}\sum_{1\le i<j\le m(n)} \log(a_j^{(n)} - b_i^{(n)}) = \alpha^2\beta\Sigma(\mu).$$

Therefore,

$$0 \geq \limsup_{n\to\infty} \frac{1}{n^2} \log P_n(G) \geq -\iint F(x,y)\, d\mu(x)\, d\mu(y) - \liminf_{n\to\infty} \frac{1}{n^2} \log Z_n\,,$$

and we take the infimum for μ. Also, we obtain

$$\liminf_{n\to\infty} \frac{1}{n^2} \log P_n(G) \geq -\iint F(x,y)\, d\mu(x)\, d\mu(y) - \limsup_{n\to\infty} \frac{1}{n^2} \log Z_n\,.$$

□

End of proof of Theorem 5.5.1*:* It is immediate from Lemmas 5.5.4 and 5.5.6 that the finite limit $B := \lim_{n\to\infty} n^{-2} \log Z_n$ exists. Moreover, the exponential tightness of (P_n) can be shown similarly to the proof of Lemma 5.4.8. Therefore, Lemmas 5.5.5 and 5.5.6 imply the conclusion.

□

The joint eigenvalue distribution of the real Wishart matrix $n^{-1}T(n)^tT(n)$ with $p(n) \geq n$ has the density (5.5.1). This is a special case of (5.5.4), where $Q(x) = x/2$, $\beta = 1/2$, $m(n) = n$ and $\gamma(n) = (p(n) - n - 1)/2$. So, when $p(n)/n \to \lambda \in [1,\infty)$ as $n \to \infty$, the empirical eigenvalue distribution of $n^{-1}T(n)^tT(n)$ satisfies the large deviation principle with the good rate function

$$I(\mu) := -\frac{1}{2}\Sigma(\mu) + \frac{1}{2}\int (x - (\lambda - 1)\log x)\, d\mu(x) + B\,. \tag{5.5.7}$$

Also, since (5.5.4) becomes (5.5.2) when $Q(x) = x$, $\beta = 1$, $m(n) = n$ and $\gamma(n) = p(n) - n$, the empirical eigenvalue distribution of the complex Wishart matrix $(2n)^{-1}\hat{T}(n)^*\hat{T}(n)$ satisfies the large deviation principle as well, and the rate function is the above (5.5.7) multiplied by 2. According to Proposition 5.3.7 the Marchenko-Pastur distribution μ_λ is the minimizer of (5.5.7).

According to the Selberg integral formula of Laguerre type ([121], p. 354), the normalization constant Z_n in (5.5.2) is given as

$$\begin{aligned} Z_n &= n^{-np(n)} \int_0^\infty \cdots \int_0^\infty \exp\Bigl(-\sum_{i=1}^n x_i\Bigr) \prod_{i=1}^n x_i^{p(n)-n} \\ &\qquad\qquad \times \prod_{i<j} (x_i - x_j)^2\, dx_1 \cdots dx_n \\ &= n^{-np(n)} \prod_{j=1}^n j!\,(p(n)-j)!\,. \end{aligned}$$

Hence, by using the Stirling formula, B in (5.5.7) is computed as follows:

$$2B = \lim_{n\to\infty} \frac{1}{n^2} \log Z_n$$

$$
\begin{aligned}
&= \lim_{n\to\infty}\Bigg[\left(\frac{1}{n^2}\log\prod_{j=1}^{n} j! - \frac{1}{2}\log p(n)\right) \\
&\quad + \frac{p(n)^2}{n^2}\left(\frac{1}{p(n)^2}\log\prod_{j=0}^{p(n)-1} j! - \frac{1}{2}\log p(n)\right) \\
&\quad - \frac{(p(n)-n)^2}{n^2}\left(\frac{1}{(p(n)-n)^2}\log\prod_{j=0}^{p(n)-n-1} j! - \frac{1}{2}\log(p(n)-n)\right) \\
&\quad + \frac{p(n)^2}{2n^2}\log\frac{p(n)}{n} - \frac{(p(n)-n)^2}{2n^2}\log\frac{p(n)-n}{n}\Bigg] \\
&= -\frac{3}{4}(1+\lambda^2-(\lambda-1)^2) + \frac{\lambda^2}{2}\log\lambda - \frac{(\lambda-1)^2}{2}\log(\lambda-1) \\
&= -\frac{1}{2}(3\lambda - \lambda^2\log\lambda + (\lambda-1)^2\log(\lambda-1)).
\end{aligned}
$$

Therefore, since $I(\mu_\lambda) = 0$ for (5.5.7), we have

$$
\begin{aligned}
\Sigma(\mu_\lambda) &= \int (x - (\lambda-1)\log x)\, d\mu_\lambda(x) \\
&\quad -\frac{1}{2}(3\lambda - \lambda^2\log\lambda + (\lambda-1)^2\log(\lambda-1)).
\end{aligned} \tag{5.5.8}
$$

Combining (5.3.9) and (5.5.8) gives

$$
\Sigma(\mu_\lambda) = -1 + \frac{1}{2}(\lambda + \log\lambda + (\lambda-1)^2\log(1-\lambda^{-1})). \tag{5.5.9}
$$

Summarizing the above arguments, we have

Theorem 5.5.7 *When $p(n)/n \to \lambda \in [1,\infty)$ as $n\to\infty$, the empirical eigenvalue distribution of the Wishart matrix $n^{-1}T(n)^tT(n)$ satisfies the large deviation principle in the scale n^{-2} with the good rate function*

$$
\begin{aligned}
I(\mu) &:= -\frac{1}{2}\Sigma(\mu) + \frac{1}{2}\int (x-(\lambda-1)\log x)\, d\mu(x) \\
&\quad -\frac{1}{4}(3\lambda - \lambda^2\log\lambda + (\lambda-1)^2\log(\lambda-1))
\end{aligned}
$$

for $\mu \in \mathcal{M}(\mathbb{R}^+)$. Moreover, the Marchenko-Pastur distribution μ_λ is the unique minimizer of I, and its free entropy is given by (5.5.9).

Next, assume $p(n) < n$ and take the random atomic measure

$$
R_n := \mathbf{K}_n(n^{-1}T(n)^tT(n)),
$$

where $\mathbf{K}_n$ was introduced in (5.4.3). Note that the eigenvalues of $n^{-1}T(n)^tT(n)$ are those of $n^{-1}T(n)T(n)^t$ plus $n - p(n)$ zeros. Furthermore, it is obvious that the eigenvalue distribution of $p(n)^{-1}T(n)T(n)^t$ is given by (5.5.1) with n and $p(n)$ interchanged. Hence the eigenvalue distribution $\tilde{\nu}_n$ of $n^{-1}T(n)T(n)^t$ has the joint density

$$\frac{1}{Z_n}\exp\Bigl(-\frac{n}{2}\sum_{i=1}^{p(n)}\lambda_i\Bigr)\prod_{i=1}^{p(n)}\lambda_i^{(n-p(n)-1)/2}\prod_{1\le i<j\le p(n)}|\lambda_i-\lambda_j|\,.$$

In this way we have the following:

Lemma 5.5.8 *Let R_n and $\tilde{\nu}_n$ be as above. Then R_n is written as*

$$R_n=\frac{n-p(n)}{n}\delta(0)+\frac{p(n)}{n}\tilde{R}_n\,,$$

where the distribution of $\tilde{R}_n$ is given by

$$\tilde{P}_n(\Gamma):=\tilde{\nu}_n\bigl(\{t\in(\mathbb{R}^+)^{p(n)}:\kappa_{p(n)}(t)\in\Gamma\}\bigr)$$

for Borel sets $\Gamma\subset\mathcal{M}(\mathbb{R}^+)$.

For $0<\lambda<1$ let

$$\tilde{\mu}_\lambda:=\frac{\sqrt{4\lambda-(x-1-\lambda)^2}}{2\pi\lambda x}\chi(x)\,dx,$$

where χ denotes the characteristic function of the interval $\bigl[(1-\sqrt{\lambda})^2,(1+\sqrt{\lambda})^2\bigr]$. This probability distribution is known as a variant of the Marchenko-Pastur distribution μ_λ. They are simply related by $\tilde{\mu}_\lambda=D_\lambda\mu_{\lambda^{-1}}$.

Lemma 5.5.9 *Assume that $p(n)/n\to\lambda\in(0,1)$ as $n\to\infty$. Then the sequence $\tilde{P}_n$ in the above lemma is exponentially tight and satisfies the large deviation principle in the scale n^{-2} with the rate function*

$$\begin{aligned}\tilde{I}(\mu)\;&:=\;-\frac{\lambda^2}{2}\Sigma(\mu)+\frac{\lambda}{2}\int(x-(1-\lambda)\log x)\,d\mu(x)\\&\qquad-\frac{1}{4}(3\lambda-\lambda^2\log\lambda+(1-\lambda)^2\log(1-\lambda))\end{aligned}$$

for $\mu\in\mathcal{M}(\mathbb{R}^+)$. The distribution $\tilde{\mu}_\lambda$ is the unique minimizer of $\tilde{I}$.

Proof: By Theorem 5.5.1 for the case $m(n)=p(n)$ and $\gamma(n)=(n-p(n)-1)/2$, we know that $(\tilde{P}_n)$ satisfies the large deviation principle with the good rate function

$$\tilde{I}(\mu)=-\frac{\lambda^2}{2}\Sigma(\mu)+\frac{\lambda}{2}\int(x-(1-\lambda)\log x)\,d\mu(x)+B\,.$$

Let $I_{\lambda^{-1}}$ be the rate function given in Theorem 5.5.7 for λ^{-1} in place of λ. We get

$$\begin{aligned} \tilde{I}(D_\lambda \mu) &= -\frac{\lambda^2}{2}\Sigma(\mu) + \frac{\lambda^2}{2}\int (x - (\lambda^{-1}-1)\log x)\, d\mu(x) + B - \frac{\lambda}{2}\log\lambda\,, \\ \lambda^2 I_{\lambda^{-1}}(\mu) &= -\frac{\lambda^2}{2}\Sigma(\mu) + \frac{\lambda^2}{2}\int (x - (\lambda^{-1}-1)\log x)\, d\mu(x) \\ &\quad -\frac{1}{4}(3\lambda + \log\lambda + (1-\lambda)^2\log(\lambda^{-1}-1))\,, \end{aligned}$$

which are equal up to a constant. Hence $\tilde{I}(D_\lambda \mu) = \lambda^2 I_{\lambda^{-1}}(\mu)$ indeed holds for all $\mu \in \mathcal{M}(\mathbb{R}^+)$. This implies that $\tilde{I}$ has the unique minimizer $D_\lambda \mu_{\lambda^{-1}} = \tilde{\mu}_\lambda$, and

$$B = -\frac{1}{4}(3\lambda - \lambda^2\log\lambda + (1-\lambda)^2\log(1-\lambda))\,.$$

The exponential tightness of $(\tilde{P}_n)$ was also shown at the end of proof of Theorem 5.5.1.

□

The next theorem complements Theorem 5.5.7 about the large deviation related to the Wishart matrix. It is remarkable that the limit distribution (or the minimizer of the rate function) has an atom.

Theorem 5.5.10 *Assume that $p(n) < n$ and $p(n)/n \to \lambda \in (0,1)$ as $n \to \infty$. Then the empirical eigenvalue distribution of $n^{-1}T(n)^t T(n)$ satisfies the large deviation principle in the scale n^{-2} with the good rate function*

$$I(\mu) := \begin{cases} \tilde{I}(\tilde{\mu}) & \text{if } \mu = (1-\lambda)\delta(0) + \lambda\tilde{\mu},\ \tilde{\mu} \in \mathcal{M}(\mathbb{R}^+), \\ +\infty & \text{otherwise,} \end{cases}$$

where $\tilde{I}$ is given by Lemma 5.5.9. The minimizer of I is μ_λ $(= (1-\lambda)\delta(0) + \lambda\tilde{\mu}_\lambda)$.

By Lemmas 5.5.8 and 5.5.9 the following lemma is enough to prove the theorem.

Lemma 5.5.11 *For $n \in \mathbb{N}$ let $\tilde{R}_n$ be a random probability measure on a Polish space X, and $\tilde{P}_n$ the distribution of $\tilde{R}_n$. Let μ_0 be a fixed probability measure on X, and let $0 < \lambda_n < 1$ be such that $\lambda_n \to \lambda \in (0,1)$. If $(\tilde{P}_n)$ is exponentially tight and satisfies the large deviation principle in the scale L_n with a rate function $\tilde{I}$ on $\mathcal{M}(X)$, then the sequence of random measures $(1-\lambda_n)\mu_0 + \lambda_n\tilde{R}_n$ satisfies the same with the good rate function*

$$I(\mu) := \begin{cases} \tilde{I}(\tilde{\mu}) & \text{if } \mu = (1-\lambda)\mu_0 + \lambda\tilde{\mu},\ \tilde{\mu} \in \mathcal{M}(X), \\ +\infty & \text{otherwise.} \end{cases}$$

Proof: The distribution P_n of $(1-\lambda_n)\mu_0 + \lambda_n\tilde{R}_n$ is given by

$$P_n(\Gamma) = \tilde{P}_n(\{\tilde{\mu} \in \mathcal{M}(X) : (1-\lambda_n)\mu_0 + \lambda_n\tilde{\mu} \in \Gamma\})$$

for Borel sets $\Gamma \subset \mathcal{M}(X)$. First we show that (P_n) is exponentially tight. For any $\varepsilon > 0$ there exists a compact $\tilde{K}_\varepsilon \subset \mathcal{M}(X)$ such that

$$\limsup_{n\to\infty} L_n \log \tilde{P}_n(\tilde{K}_\varepsilon^c) \le -\frac{1}{\varepsilon}\,.$$

By noting that the weak topology on $\mathcal{M}(X)$ is metrizable, it is easy to see that the closure K_ε of $\bigcup_{n=1}^\infty \big((1-\lambda_n)\mu_0 + \lambda_n \tilde{K}_\varepsilon\big)$ is compact. Since $P_n(K_\varepsilon^c) \le \tilde{P}_n(\tilde{K}_\varepsilon^c)$, we get the conclusion.

Now it suffices to show that for every $\mu \in \mathcal{M}(X)$

$$\inf_G \Big[\liminf_{n\to\infty} L_n \log P_n(G)\Big] \ge -I(\mu)\,, \tag{5.5.10}$$

$$\inf_G \Big[\limsup_{n\to\infty} L_n \log P_n(G)\Big] \le -I(\mu)\,, \tag{5.5.11}$$

where G runs over neighborhoods of μ. Let $\mathcal{D}$ denote the set $\{(1-\lambda)\mu_0 + \lambda\tilde{\mu} : \tilde{\mu} \in \mathcal{M}(X)\}$. If $\mu \notin \mathcal{D}$, then $\mu(C) < (1-\lambda)\mu_0(C)$ for some closed $C \subset X$. Then, since $P_n(G) = 0$ for a neighborhood $G := \{\mu' \in \mathcal{M}(X) : \mu'(C) < (1-\lambda)\mu_0(C)\}$ of μ, (5.5.10) and (5.5.11) hold in this case. Next assume that $\mu \in \mathcal{D}$ and $\mu = (1-\lambda)\mu_0 + \lambda\tilde{\mu}$. For any neighborhood G of μ there exists a neighborhood $\tilde{G}$ of $\tilde{\mu}$ such that $(1-\lambda_n)\mu_0 + \lambda_n\tilde{G} \subset G$ for large n, and hence

$$\liminf_{n\to\infty} L_n \log P_n(G) \ge \liminf_{n\to\infty} L_n \log \tilde{P}_n(\tilde{G}) \ge -\tilde{I}(\tilde{\mu})\,.$$

This implies (5.5.10). On the other hand, for any neighborhood $\tilde{G}$ of $\tilde{\mu}$ there exists a neighborhood G of μ such that $\lambda_n^{-1}G - (\lambda_n^{-1} - 1)\mu_0 \subset \tilde{G}$ or

$$\{\tilde{\mu} \in \mathcal{M}(X) : (1-\lambda_n)\mu_0 + \lambda_n\tilde{\mu} \in G\} \subset \tilde{G}$$

for large n. Hence

$$\inf_G \Big[\limsup_{n\to\infty} L_n \log P_n(G)\Big] \le \inf_{\tilde{G}} \Big[\limsup_{n\to\infty} L_n \log \tilde{P}_n(\tilde{G})\Big] \le -\tilde{I}(\tilde{\mu})\,,$$

implying (5.5.11).

□

It is a consequence of Theorems 5.5.1 and 5.5.10 that the empirical eigenvalue density of the Wishart matrix $n^{-1}T(n)^t T(n)$ (or $(2n)^{-1}\hat{T}(n)^*\hat{T}(n)$) converges in the weak topology to μ_λ almost surely as $n \to \infty$, $p(n)/n \to \lambda$. Indeed, a better result is included in Proposition 4.4.11.

In the rest of this section we present three more large deviation results for random matrices.

Proposition 5.5.12 *Let $T(n)$ be an $n \times n$ standard symmetric Gaussian matrix (distributed according to $N^{(n)}(0,1)$), and let $\lambda_{n,1}, \lambda_{n,2}, \dots, \lambda_{n,n}$ be the eigenvalues of $T(n)$. Then the sequence of random atomic measures*

$$R_n := \frac{\delta(\lambda_{n,1}^2) + \delta(\lambda_{n,2}^2) + \cdots + \delta(\lambda_{n,n}^2)}{n}$$

satisfies the large deviation principle in the scale n^{-2} with the good rate function

$$I(\mu) := -\frac{1}{2}\Sigma(\mu) + \frac{1}{4}\int x\, d\mu(x) + \frac{1}{2}\log 2 - \frac{3}{4} \quad on \quad \mathcal{M}(\mathbb{R}^+).$$

The minimizer of I is the distribution $u_4^{(1)}$ given in (5.3.7).

Proof: Let P_n be the empirical eigenvalue distribution of $T(n)$ on $\mathcal{M}(\mathbb{R})$ and Q_n the distribution of R_n on $\mathcal{M}(\mathbb{R}^+)$. The relation between P_n and Q_n is simply $P_n \circ T^{-1} = Q_n$ via the transformation $T : \mathcal{M}(\mathbb{R}) \to \mathcal{M}(\mathbb{R}^+)$ defined by $T\nu := \nu \circ \sigma^{-1}$, where $\sigma(x) := x^2$. According to Theorem 5.4.1 the sequence P_n satisfies the large deviation principle in the scale n^{-2}, and the good rate function is

$$I_0(\nu) := -\Sigma(\mu) + \frac{1}{4}\int x^2\, d\nu(x) + \frac{1}{2}\log 2 - \frac{3}{4}.$$

The *contraction principle* ([55], Sec. 4.2.1) from the theory of large deviations tells us that the large deviation principle is satisfied by the sequence Q_n as well, and the rate function I on $\mathcal{M}(\mathbb{R}^+)$ is

$$I(\mu) = \inf\{I_0(\nu) : T\nu = \mu\}. \tag{5.5.12}$$

For a measure $\nu \in \mathcal{M}(\mathbb{R})$, define $\tilde{\nu} \in \mathcal{M}(\mathbb{R})$ by $\tilde{\nu}(E) := \nu(-E)$. By the convexity of I_0 (Lemma 5.4.2) we have

$$I_0\Big(\frac{\nu + \tilde{\nu}}{2}\Big) \le \frac{I_0(\nu) + I_0(\tilde{\nu})}{2} = I_0(\nu).$$

Hence the infimum of I_0 in (5.5.12) is reached at the unique symmetric measure ν_0 satisfying $T\nu_0 = \mu$. Now

$$\begin{aligned} I(\mu) &= I_0(\nu_0) \\ &= -\Sigma(\nu_0) + \frac{1}{4}\int x^2\, d\nu_0(x) + \frac{1}{2}\log 2 - \frac{3}{4} \\ &= -\frac{1}{2}\Sigma(\mu) + \frac{1}{4}\int x\, d\mu(x) + \frac{1}{2}\log 2 - \frac{3}{4} \end{aligned}$$

due to the properties (1) and (2) of the transformation T stated above (5.3.7). Finally, Proposition 5.3.5 gives the assertion about the minimizer.

□

Let $X(n)$ be an $n \times n$ standard non-selfadjoint Gaussian random matrix. The joint eigenvalue density on $(\mathbb{R}^+)^n$ of $X(n)^*X(n)$ is (5.5.2) with $p(n) = n$, and that of $|X(n)| = (X(n)^*X(n))^{1/2}$ is

$$\frac{1}{Z_n} \exp\Big(-n \sum_{i=1}^n \lambda_i^2\Big) \prod_{i<j} (\lambda_i^2 - \lambda_j^2)^2 \prod_{i=1}^n \lambda_i \,.$$

We already obtained the large deviation of the empirical eigenvalue distribution of $X(n)^*X(n)$; the good rate function is

$$I(\mu) := -\Sigma(\mu) + \int x \, d\mu(x) - \frac{3}{2} \quad \text{on} \quad \mathcal{M}(\mathbb{R}^+)$$

(i.e. the rate function in (5.5.7) multiplied by 2 with $\lambda = 1$) and the minimizer is μ_1. According to Corollary 4.3.8 the limit distribution of $|X(n)|$ is the *quarter-circular distribution*. The next proposition gives a large deviation behind this fact. The proof is a simple transformation of the above large deviation via $T : \mathcal{M}(\mathbb{R}^+) \to \mathcal{M}(\mathbb{R}^+)$ defined by $T\nu := \nu \circ \sigma$ with $\sigma(x) := x^2$ $(x \in \mathbb{R}^+)$.

Proposition 5.5.13 *Let $X(n)$ be as above. Then the empirical eigenvalue distribution of $|X(n)|$ satisfies the large deviation principle in the scale n^{-2} with the good rate function*

$$I(\mu) := -\iint \log|x^2 - y^2| \, d\mu(x) \, d\mu(y) + \int x^2 \, d\mu(x) - \frac{3}{2}$$

for $\mu \in \mathcal{M}(\mathbb{R}^+)$. Moreover, the quarter-circular distribution $\frac{1}{\pi}\sqrt{4 - x^2}\, \chi_{[0,2]}(x)\, dx$ is the unique minimizer of I.

There is a more directly defined *random matrix model of the quarter-circular element.* For $n \in \mathbb{N}$ let $T(2n)$ be an *antisymmetric* $2n \times 2n$ real random matrix such that $T_{ii}(n) = 0$ for $1 \le i \le 2n$, $\{T_{ij}(n) : 1 \le i < j \le 2n\}$ is an independent family of Gaussian random variables with the identical distribution $N(0, 1/2n)$, and $T_{ji}(n) = -T_{ij}(n)$ for $1 \le i < j \le 2n$.

Then the random eigenvalues of $\mathrm{i}\,T(2n)$ are given as $\pm\lambda_{n,1}, \pm\lambda_{n,2}, \dots, \pm\lambda_{n,n}$, where $\lambda_{n,1}, \dots, \lambda_{n,n} \ge 0$. It is known ([88], Corollary 3.2.1, or [121], Sec. 3.4) that the joint probability density of $\lambda_{n,1}, \lambda_{n,2}, \dots, \lambda_{n,n}$ is

$$\frac{1}{Z_n} \exp\Big(-n \sum_{i=1}^n \lambda_i^2\Big) \prod_{i<j} (\lambda_i^2 - \lambda_j^2)^2 \,.$$

Then the large deviation of the empirical eigenvalue distribution of $(\mathrm{i}\,T(2n))_+$ can be proved similarly to Theorem 5.4.3 or 5.5.1, and the rate function is the same as in Proposition 5.5.13. In this way, $(\mathrm{i}\,T(2n))_+$ becomes a random matrix model for the quarter-circular distribution.

Finally, let $Z(n)$ be a $p(n) \times n$ complex random matrix with $p(n) \geq n$ such that $\mathrm{Re}\, Z_{ij}(n)$ and $\mathrm{Im}\, Z_{ij}(n)$ $(1 \leq i \leq p(n),\ 1 \leq j \leq n)$ are independent with the same distribution $N(0, 1/2n)$. Let $U(n)$ be a standard $n \times n$ unitary random matrix such that $U(n)$ and $Z(n)$ are independent. According to Lemma 4.4.8 the non-selfadjoint random matrix $T(n) := U(n)(Z(n)^* Z(n))^{1/2}$ has the joint eigenvalue density

$$\frac{1}{Z_n} \exp\biggl(-n \sum_{i=1}^{n} |\zeta_i|^2\biggr) \prod_{i=1}^{n} |\zeta_i|^{2(p(n)-n)} \prod_{i<j} |\zeta_i - \zeta_j|^2 \quad \text{on} \quad \mathbb{C}^n.$$

When $p(n)/n \to \lambda \in [1, \infty)$ as $n \to \infty$, $T(n)$ constitutes a random matrix model of an R-diagonal element x such that $x^* x$ has the distribution μ_λ. The corresponding large deviation result is the following:

Proposition 5.5.14 *Let $T(n)$ be as above. Then the empirical eigenvalue distribution of $T(n)$ satisfies the large deviation principle in the scale n^{-2} with the good rate function*

$$\begin{aligned} I(\mu) \quad := \quad & -\Sigma(\mu) + \int (|\zeta|^2 - 2(\lambda - 1) \log |\zeta|) \, d\mu(\zeta) \\ & + \frac{3}{4} - \frac{1}{2}(3\lambda - \lambda^2 \log \lambda + (\lambda - 1)^2 \log(\lambda - 1)) \end{aligned}$$

for $\mu \in \mathcal{M}(\mathbb{C})$. Moreover, the annular law $m_{\sqrt{\lambda-1}, \sqrt{\lambda}}$, i.e. the uniform distribution on the annulus $\sqrt{\lambda - 1} \leq |\zeta| \leq \sqrt{\lambda}$, is the unique minimizer of I.

The scheme of the proof is similar to the previous (cf. the proofs of Theorems 5.4.9 and 5.5.1), so we omit the details. The minimizer of I is a consequence of Proposition 5.3.8, and the constant term of I is determined from $I(m_{\sqrt{\lambda-1}, \sqrt{\lambda}}) = 0$ and (5.3.12). Incidentally, it is noteworthy that the expressions of $\Sigma(\mu_\lambda)$ in (5.5.9) and of $\Sigma(m_{\sqrt{\lambda-1}, \sqrt{\lambda}})$ in (5.3.12) are the same up to an additive constant.

5.6 Entropy and large deviations revisited

This section contains some supplementary material after the many large deviation results of the previous ones. We present an example in which the rate function is the free entropy itself (up to an additive constant). This large deviation formulation yields a new characterization of free entropy.

Let $\bar{\Lambda}_n$ be the measure on $\mathbb{R}^n$ (instead of $\mathbb{R}^n_{\leq}$) induced from the Lebesgue measure Λ_n on $M_n(\mathbb{C})^{sa}$. By Lemma 4.4.6 we know that $\bar{\Lambda}_n$ has the joint density

$$C_n \prod_{i<j} (t_i - t_j)^2 \quad \text{with} \quad C_n = \frac{(2\pi)^{n(n-1)/2}}{\prod_{j=1}^{n} j!}. \tag{5.6.1}$$

The Stirling formula gives

$$\lim_{n\to\infty}\Big(\frac{1}{n^2}\log C_n+\frac{1}{2}\log n\Big)=\frac{1}{2}\log(2\pi)+\frac{3}{4}\,. \tag{5.6.2}$$

For any $R>0$, normalizing the restriction of Λ_n on $\{A\in M_n(\mathbb{C})^{sa}:\|A\|\le R\}$, we have the probability measure $\lambda_{n,R}$. Then it induces the probability measure $\bar\lambda_{n,R}$ on $[-R,R]^n$ having the density

$$\frac{1}{Z_n}\prod_{i<j}(t_i-t_j)^2\,,$$

where

$$\begin{aligned} Z_n &= \int_{-R}^{R}\cdots\int_{-R}^{R}\prod_{i<j}(t_i-t_j)^2\,dt_1\cdots dt_n \\ &= R^{n^2}\int_{-1}^{1}\cdots\int_{-1}^{1}\prod_{i<j}(t_i-t_j)^2\,dt_1\cdots dt_n \\ &= (2R)^{n^2}\prod_{j=0}^{n-1}\frac{(j+1)!\,(j!)^2}{(n+j)!}\,. \end{aligned}$$

By the Stirling formula we get

$$\lim_{n\to\infty}\frac{1}{n^2}\log Z_n=\log\frac{R}{2}\,. \tag{5.6.3}$$

The empirical distribution on $\mathcal{M}([-R,R])$ of $\lambda_{n,R}$ is given by

$$P_n(\Gamma):=\bar\lambda_{n,R}\big(\{x\in[-R,R]^n:\kappa_n(x)\in\Gamma\}\big)$$

for Borel sets $\Gamma\subset\mathcal{M}([-R,R])$.

Now, the next proposition is a version of Theorem 5.4.3 where probability measures are restricted to those supported on $[-R,R]$ with $\beta=1$ and $Q(x)=0$. In fact, Theorem 5.4.3 holds true when we take

$$Q(x)=\begin{cases}0 & \text{if } |x|\le R,\\ +\infty & \text{if } |x|>R,\end{cases}$$

though $Q(x)$ is not continuous.

Proposition 5.6.1 *The sequence P_n given above satisfies the large deviation principle in the scale n^{-2} with the rate function*

$$I(\mu):=-\Sigma(\mu)+\log\frac{R}{2}\quad on\quad \mathcal{M}([-R,R]).$$

The following theorem of Voiculescu expresses the double logarithmic integral $\Sigma(\mu)$ as a triple limit of volumes. Note that if one chooses the normalization $C_n = 1$ in (5.6.1), then the formulas become simpler: The terms $\frac{1}{2}\log n$ in (5.6.4) and $\frac{1}{2}\log(2\pi) + \frac{3}{4}$ in (5.6.5) disappear.

Theorem 5.6.2 *If μ is a probability measure supported in $[-R, R]$, then the limit*

$$\chi_R(\mu; r, \varepsilon) := \lim_{n\to\infty} \left[\frac{1}{n^2}\log\bar\Lambda_n\big(\{x \in [-R,R]^n : |m_k(\kappa_n(x)) - m_k(\mu)| \le \varepsilon,\ k \le r\}\big) + \frac{1}{2}\log n\right] \tag{5.6.4}$$

exists for every $r \in \mathbb{N}$ and $\varepsilon > 0$, and

$$\lim_{\substack{r\to\infty \\ \varepsilon\to+0}} \chi_R(\mu; r, \varepsilon) = \Sigma(\mu) + \frac{1}{2}\log(2\pi) + \frac{3}{4}. \tag{5.6.5}$$

Proof: Given μ with $\operatorname{supp}\mu \subset [-R, R]$, $r \in \mathbb{N}$ and $\varepsilon > 0$, take the closed set $F(r,\varepsilon)$ and the open set $G(r, \varepsilon)$ in $\mathcal{M}([-R, R])$ as in the proof of Proposition 5.1.1. Then the large deviation of Proposition 5.6.1 says that

$$\begin{aligned}
&\limsup_{n\to\infty} \frac{1}{n^2}\log\bar\lambda_{n,R}\big(\{x \in [-R,R]^n : \kappa_n(x) \in F(r,\varepsilon)\}\big) \\
&\qquad \le \sup\{\Sigma(\nu) : \nu \in F(r,\varepsilon)\} - \log\frac{R}{2}, \\
&\liminf_{n\to\infty} \frac{1}{n^2}\log\bar\lambda_{n,R}\big(\{x \in [-R,R]^n : \kappa_n(x) \in G(r,\varepsilon)\}\big) \\
&\qquad \ge \sup\{\Sigma(\nu) : \nu \in G(r,\varepsilon)\} - \log\frac{R}{2}.
\end{aligned}$$

It is easy to check that

$$\sup\{\Sigma(\nu) : \nu \in F(r,\varepsilon)\} = \sup\{\Sigma(\nu) : \nu \in G(r,\varepsilon)\}.$$

Hence we have

$$\begin{aligned}
&\lim_{n\to\infty} \frac{1}{n^2}\log\bar\lambda_{n,R}\big(\{x \in [-R,R]^n : \kappa_n(x) \in F(r,\varepsilon)\}\big) \\
&\qquad = \sup\{\Sigma(\nu) : \nu \in F(r,\varepsilon)\} - \log\frac{R}{2}.
\end{aligned}$$

Note that the restriction of $\bar\Lambda_n$ on $[-R, R]^n$ is equal to $C_n Z_n \bar\lambda_{n,R}$. So, thanks to (5.6.2) and (5.6.3), we have

$$\begin{aligned}
&\lim_{n\to\infty} \left[\frac{1}{n^2}\log\bar\Lambda_n\big(\{x \in [-R,R]^n : \kappa_n(x) \in F(r,\varepsilon)\}\big) + \frac{1}{2}\log n\right] \\
&\qquad = \sup\{\Sigma(\nu) : \nu \in F(r,\varepsilon)\} + \frac{1}{2}\log(2\pi) + \frac{3}{4}.
\end{aligned}$$

Letting $r \to \infty$ and $\varepsilon \to +0$ yields the conclusion.

□

The left-hand side of (5.6.5) is Voiculescu's second definiton of free entropy for a probability measure or a selfadjoint noncommutative random variable. (He used the notation $\chi(\mu)$.) The extension to N-tuples of variables will be discussed in the next chapter.

When $\bar\nu_n$ has the density (5.2.2) and $\mu \in \mathcal{M}([-R,R])$, one can apply Theorem 5.4.1 (also Lemma 5.4.2) to prove that

$$\lim_{\substack{r\to\infty\\ \varepsilon\to+0}} \lim_{n\to\infty} \frac{1}{n^2} \log \bar\nu_n\big(\{\lambda \in [-R,R]^n : |m_k(\kappa_n(\lambda)) - m_k(\mu)| \le \varepsilon,\ k \le r\}\big)$$
$$= \beta\Sigma(\mu) - \frac{1}{4\sigma^2}\int x^2\, d\mu(x) - \frac{\beta}{2}\log(2\beta\sigma^2) + \frac{3\beta}{4} \tag{5.6.6}$$

(as well as the formulas in Theorem 5.2.2), along the same lines as the proofs of Proposition 5.1.1 and the above theorem. In this way, Theorem 5.2.2 and (5.6.6) have a strong resemblance to Proposition 5.1.1, while Theorem 5.6.2 is similar to (5.1.8).

We have a variant of Theorem 5.6.2 where Λ_n is replaced by the measure $\Lambda_{+,n}$ on $M_n(\mathbb{C})^+$ in Lemma 4.4.7. Let $\bar\Lambda_{+,n}$ be the measure on $(\mathbb{R}^+)^n$ induced from $\Lambda_{+,n}$. According to (4.4.4) the joint desnity of $\bar\Lambda_{+,n}$ is

$$C_n \prod_{i<j}(t_i - t_j)^2 \quad \text{with} \quad C_n = \frac{\pi^{n^2} n!}{\left(\prod_{j=1}^n j!\right)^2},$$

and we have

$$\lim_{n\to\infty}\left(\frac{1}{n^2}\log C_n + \log n\right) = \log\pi + \frac{3}{2}. \tag{5.6.7}$$

By Lemma 4.4.7 we notice that $\bar\Lambda_{+,n}$ on $[0,R]^n$ is nothing but the translation of $\bar\Lambda_n$ restricted on $[-R/2,R/2]^n$ to $[0,R]^n$ with a different normalizing constant. So, together with (5.6.7), Theorem 5.6.2 yields the following: If μ is a probability meausre supported in $[0,R]$, then the limit

$$\chi_{+,R}(\mu; r,\varepsilon) := \lim_{n\to\infty}\bigg[\frac{1}{n^2}\log\bar\Lambda_{+,n}\big(\{x \in [0,R]^n :$$
$$|m_k(\kappa_n(x)) - m_k(\mu)| \le \varepsilon,\ k \le r\}\big) + \log n\bigg] \tag{5.6.8}$$

exists for every $r \in \mathbb{N}$ and $\varepsilon > 0$, and

$$\lim_{\substack{r\to\infty\\ \varepsilon\to+0}} \chi_{+,R}(\mu; r,\varepsilon) = \Sigma(\mu) + \log\pi + \frac{3}{2}. \tag{5.6.9}$$

Finally, for later use, we note that

$$\begin{aligned}
&\bar{\Lambda}_n\big(\{x \in [-R,R]^n : |m_k(\kappa_n(x)) - m_k(\mu)| \le \varepsilon,\ k \le r\}\big) \\
&\quad = \Lambda_n\big(\{A \in M_n(\mathbb{C})^{sa} : \|A\| \le R,\ |\mathrm{tr}_n(A^k) - m_k(\mu)| \le \varepsilon,\ k \le r\}\big), \quad (5.6.10) \\
&\bar{\Lambda}_{+,n}\big(\{x \in [0,R]^n : |m_k(\kappa_n(x)) - m_k(\mu)| \le \varepsilon,\ k \le r\}\big) \\
&\quad = \Lambda_{+,n}\big(\{A \in M_n(\mathbb{C})^{+} : \|A\| \le R,\ |\mathrm{tr}_n(A^k) - m_k(\mu)| \le \varepsilon,\ k \le r\}\big). \quad (5.6.11)
\end{aligned}$$

Notes and Remarks. The history of the Boltzmann-Gibbs entropy started at the end of the previous century. The summary on this subject in Sec. 5.1 is not at all historical. The relative entropy was introduced much later, in the 1950's, by Kullback and Leibler. Although the Boltzmann-Gibbs entropy is discussed in most books on statistical mechanics, the relative entropy is not.

The entropic central limit theorem mentioned in the introduction to this chapter is found in [14].

The general abstract framework of *large deviation* was proposed by S.R.S. Varadhan in 1966, although the topic may be traced back much earlier. Cramér's theorem for independent identically distributed variables was published in 1938, and the level-2 extension of Sanov was in 1957. One of the first systematic introductions to the theory is the book [73]. The terminology of three levels was used there. We suggest also the monographs [55] and [56] on large deviations. [53] is a standard reference in information theory; it explains basic properties of the Boltzmann-Gibbs entropy and treats its maximization under constraints. The *maximum entropy* approach to random matrices goes back to Balian [11]. Classical orthogonal polynomial matrix ensembles are discussed in [121], Sec. 19.3. If we take the choice

$$V(x) = \frac{\Gamma(p/2)\Gamma(1/2)}{\Gamma(p+1/2)}|x|^p$$

in (5.2.8), then the limiting eigenvalue density is the *Ullman distribution* $v_1^{(p)}$ discussed in Sec. 5.3.

Voiculescu's free entropy (5.2.7) was introduced in [202], based on heuristic arguments (called Voiculescu's heuristics) about the asymptotics of the Boltzmann-Gibbs entropy of random matrices. (A more rigorous derivation of Voiculescu's heuristics can be found in [15].) Theorem 5.2.2 is essentially Proposition 4.5 from [203] together with its original proof, while Theorem 5.6.2 is Voiculescu's original form. Our formulation is slightly different because we use here the joint distribution of the eigenvalues of a standard symmetric (or selfadjoint) Gaussian matrix (or more generally, (5.2.2)) as a reference measure. On the other hand, μ is assumed as in [203] to have a compact support.

The proof of Theorem 5.6.2 (also Theorem 5.2.2) via the large deviation principle has some advantages. On the one hand, it contains the existence of the limit as $n \to \infty$ in (5.6.4) (also in (5.6.6)). On the other hand, the slightly complicated Lemma 4.3.4 is not needed.

The logarithmic energy plays an important role in potential theory [114]. The generalized Frostman method for weighted potentials in Theorem 5.3.3 gives a quite useful device for solving the maximizer problem for a free entropy functional. The book [165] was published during the preparation of our manuscript and contains all that we need here about logarithmic potentials. Some of the maximization problems for free entropy treated in Sec. 5.3 were studied in [100]. Proposition 5.3.10 is taken from [94].

The remark about the Ullman distribution stated before Proposition 5.3.4 is from [195], p. 106. That reference contains more information about the Ullman measure and a section about Gaussian random matrices. We point out that the density of the roots of the Jacobi polynomials converges to the arcsine law. More generally, if the interval $[-1, 1]$ is the support of a measure, then the density of the roots of the corresponding orthogonal polynomials tends to the *arcsine law*. This follows from the Erdős-Turán theorem; see the above-mentioned [195].

The method in proving Lemmas 5.4.2, 5.4.4 and 5.4.5 is essentially the same as Ben Arous and Guionnet's in [15]. They proved the first large deviation result (Theorem 5.4.1) for random matrices, although Voiculescu's paper already had many indications that large deviation should hold. (One observes this by comparison of Theorem 5.2.2 with the conditions in (5.4.4).) The method of proving Lemma 5.4.6 (as well as Theorem 5.2.2) is similar to Voiculescu's in [203]. Theorem 5.4.9 is from [149]; a special case is the large deviation for the elliptic Gaussian matrix $Y(n)$ in (4.1.23), and its limit distribution is Girko's elliptic law [86], [90]. The paper [16] is about the large deviation related to the *real non-symmetric Gaussian matrix*, another random matrix model of the *circular law*. Theorem 5.4.10 was proved in [101] in detail. Theorem 5.5.10 has the interesting feature that the minimizer of the rate function has an atom. The approach to large deviations obtained in Sections 5.4 and 5.5 is based on the explicit form of the joint distribution of the eigenvalues, and does not extend to more general examples of random matrices.

Chapter 6

Free Entropy of Noncommutative Random Variables

Although the previous chapter focused on large deviations for random matrices, Voiculescu's free entropy emerged from the discussion. The scheme of the passage from the classical Boltzmann-Gibbs entropy to the new concept is the following. If μ is a measure supported in a finite interval $[-R, R]$, then its entropy $S(\mu)$ is a limit of volumes:

$$\lim_{\substack{r\to\infty \\ \varepsilon\to+0}} \lim_{n\to\infty} \frac{1}{n} \log \lambda^n \big(\{x \in [-R, R]^n : |m_k(\kappa_n(x)) - m_k(\mu)| \le \varepsilon,\, k \le r\}\big),$$

where λ^n is the n-dimensional Lebesgue measure, m_k denotes the kth moment and $\kappa_n(x)$ stands for the atomic measure $\big(\delta(x_1) + \delta(x_2) + \cdots + \delta(x_n)\big)/n$. When a is a selfadjoint random variable in a noncommutative probability space $(\mathcal{A}, \varphi)$, it was Voiculescu's idea to modify the classical entropy formula into

$$\lim_{\substack{r\to\infty \\ \varepsilon\to+0}} \limsup_{n\to\infty} \frac{1}{n^2} \log \nu_n \big(\{A \in M_n(\mathbb{C})^{sa} : |\mathrm{tr}_n(A^k) - \varphi(a^k)| \le \varepsilon,\ k \le r\}\big),$$

where now ν_n is an appropriate measure on the space of $n \times n$ selfadjoint matrices.

The emphasis in this chapter is on the multivariate case. The first formula admits an obvious extension to measures on $\mathbb{R}^N$ if one uses joint moments, and the classical entropy

$$-\int_{\mathbb{R}^N} f(x) \log f(x)\, dx$$

of a measure μ with density $f(x)$ is achieved. In this chapter the similar generalization of the second formula is performed: An N-tuple $(a_1, \dots, a_N)$ of noncommutative random variables is approximated in distribution by N-tuples $(A_1, \dots, A_N)$ of selfadjoint matrices, and the multivariate free entropy $\chi(a_1, \dots, a_N)$ is introduced.

One of the fundamental observations is the fact that the multivariate free entropy $\chi(a_1, \dots, a_N)$ is additive when the variables $a_1, \dots, a_N$ are in free relation: $\chi(a_1, \dots, a_N) = \chi(a_1) + \cdots + \chi(a_N)$. The converse is also true: $a_1, \dots, a_N$ are in free relation when this additivity holds with finite $\chi(a_i)$. On the one hand, this property justifies the terminology, and on the other hand it ensures that the generalization from the Boltzmann-Gibbs entropy is on the right track. The additivity of free entropy under freeness is to be compared with the additivity of Boltzmann-Gibbs entropy under stochastic independence.

Voiculescu's definition of $\chi(a_1, \dots, a_N)$ is easily modified to introduce free entropies for other types of random variables; for instance, the free entropies $\hat{\chi}(a_1, \dots, a_N)$ of non-selfadjoint $a_1, \dots, a_N$ and $\chi_u(u_1, \dots, u_N)$ of unitary $u_1, \dots, u_N$. The study of $\hat{\chi}$ and χ_u is the subject of the second half of the chapter. An important point is that the three types of free entropies $\chi, \hat{\chi}, \chi_u$ are interrelated, in particular, when their polar decompositions are considered.

6.1 Definition and basic properties

In this section the free entropy of an N-tuple of noncommutative selfadjoint random variables is defined through approximation in distribution by matrices. The subadditivity and upper semicontinuity of free entropy are proven.

The set $M_n(\mathbb{C})$ of all matrices is a noncommutative probability space with respect to the normalized trace tr_n. Consequenly, we can speak of joint moments, or more generally the joint distribution of an N-tuple of matrices. In this section we restrict ourselves to the selfadjoint ones. There is a natural linear bijection between the set $M_n(\mathbb{C})^{sa}$ of all selfadjoint matrices and $\mathbb{R}^{n^2}$ which is an isometry for the Hilbert-Schmidt and Euclidean norms. The Lebesgue measure of $\mathbb{R}^{n^2}$ induces the measure Λ_n on $M_n(\mathbb{C})^{sa}$ under the above bijection. In this chapter, for $n, N \in \mathbb{N}$ the N-fold product measure $\Lambda_n^{\otimes N}$ on the product space $(M_n(\mathbb{C})^{sa})^N$ will be denoted simply by Λ.

Let $(\mathcal{M}, \tau)$ be a tracial W^*-probability space; that is, $\mathcal{M}$ is a von Neumann algebra with a faithful normal tracial state τ. Throughout the chapter we consider such a noncommutative probability space. Let $a_1, \dots, a_N \in \mathcal{M}^{sa}$, where $\mathcal{M}^{sa}$ denotes the space of selfadjoint elements of $\mathcal{M}$. For $n, r \in \mathbb{N}$, $\varepsilon > 0$ and $R > 0$, define the following quantity:

$$
\begin{aligned}
\Gamma_R(a_1, \dots, a_N; n, r, \varepsilon) := \big\{ &(A_1, \dots, A_N) \in (M_n(\mathbb{C})^{sa})^N : \|A_i\| \le R, \\
&|\mathrm{tr}_n(A_{i_1} \cdots A_{i_k}) - \tau(a_{i_1} \cdots a_{i_k})| \le \varepsilon \text{ for all } 1 \le i_1, \dots, i_k \le N,\ 1 \le k \le r \big\}.
\end{aligned}
$$

Moreover, define

$$\chi_R(a_1,\dots,a_N;r,\varepsilon) := \limsup_{n\to\infty}\left[\frac{1}{n^2}\log\Lambda(\Gamma_R(a_1,\dots,a_N;n,r,\varepsilon)) + \frac{N}{2}\log n\right],$$

$$\chi_R(a_1,\dots,a_N) := \lim_{\substack{r\to\infty\\ \varepsilon\to+0}} \chi_R(a_1,\dots,a_N;r,\varepsilon) = \inf_{\substack{r\in\mathbb{N}\\ \varepsilon>0}} \chi_R(a_1,\dots,a_N;r,\varepsilon),$$

$$\chi(a_1,\dots,a_N) := \sup_{R>0}\chi_R(a_1,\dots,a_N).$$

Then $\chi(a_1,\dots,a_N)$ is called the *free entropy* of the N-tuple $(a_1,\dots,a_N)$.

In the case of a single $a \in \mathcal{M}^{sa}$, if μ is the distribution measure of a (with respect to τ), then we have

$$\begin{aligned}&\Gamma_R(a;n,r,\varepsilon)\\ &\quad= \left\{A\in M_n(\mathbb{C})^{sa} : \|A\|\le R,\ |\mathrm{tr}_n(A^k) - m_k(\mu)| \le \varepsilon,\ 1\le k\le r\right\}.\end{aligned}$$

Hence Theorem 5.6.2 together with (5.6.10) says that for any $R \ge \|a\|$

$$\chi_R(a;r,\varepsilon) = \lim_{n\to\infty}\left[\frac{1}{n^2}\log\Lambda(\Gamma_R(a;n,r,\varepsilon)) + \frac{1}{2}\log n\right] \tag{6.1.1}$$

(lim sup becomes lim in this case) and

$$\chi(a) = \chi_R(a) = \Sigma(\mu) + \frac{1}{2}\log(2\pi) + \frac{3}{4}. \tag{6.1.2}$$

In this way, the free entropy $\chi(a)$ defined above coincides with $\Sigma(\mu)$ treated in the previous chapter, up to an additive constant. (Note that this constant could be easily removed by an appropriate renormalization of the measure Λ_n, as was stated just before Theorem 5.6.2.) We also write $\Sigma(a)$ for $\Sigma(\mu)$.

The first result is the *subadditivity* of free entropy.

Proposition 6.1.1 *If $C := \tau(a_1^2 + \cdots + a_N^2)$, then*

$$\chi(a_1,\dots,a_N) \le \chi(a_1) + \cdots + \chi(a_N) \le \frac{N}{2}\log\frac{2\pi eC}{N}. \tag{6.1.3}$$

Proof: It is obvious that

$$\Gamma_R(a_1,\dots,a_N;n,r,\varepsilon) \subset \prod_{i=1}^N \Gamma_R(a_i;n,r,\varepsilon),$$

and hence

$$\Lambda(\Gamma_R(a_1,\dots,a_N;n,r,\varepsilon)) \le \prod_{i=1}^N \Lambda(\Gamma_R(a_i;n,r,\varepsilon)).$$

Therefore we have

$$\chi_R(a_1,\dots,a_N;r,\varepsilon) \le \sum_{i=1}^N \chi_R(a_i;r,\varepsilon),$$

and $\chi_R(a_1,\dots,a_N) \le \sum_{i=1}^N \chi_R(a_i)$ follows. This gives the first inequality in (6.1.3). According to (6.1.2) and Proposition 5.3.4, it is known that $\chi(a) \le \frac{1}{2}\log(2\pi e C_0)$ for every $a \in \mathcal{M}^{sa}$ with $\tau(a^2) \le C_0$. Therefore, putting $C_i := \tau(a_i^2)$, we have

$$\sum_{i=1}^N \chi(a_i) \le \frac{1}{2}\sum_{i=1}^N \log(2\pi e C_i) \le \frac{N}{2}\log\left(\frac{2\pi e}{N}\sum_{i=1}^N C_i\right) = \frac{N}{2}\log\frac{2\pi e C}{N}.$$

The second inequality above is due to the concavity of the logarithmic function.

□

Corollary 6.1.2 *If $\chi(a_1,\dots,a_N) > -\infty$, then the distribution of each a_i is nonatomic.*

Proof: By the above proposition we get $\chi(a_i) > -\infty$ for all i. Hence, by (6.1.2), $\Sigma(\mu_i) > -\infty$ for the distribution μ_i of X_i, and the result follows.

□

Proposition 6.1.3 *For every $1 \le L < N$,*

$$\chi(a_1,\dots,a_N) \le \chi(a_1,\dots,a_L) + \chi(a_{L+1},\dots,a_N).$$

Proof: The result follows immediately from the obvious inclusion

$$\Gamma_R(a_1,\dots,a_N;n,r,\varepsilon) \subset \Gamma_R(a_1,\dots,a_L;n,r,\varepsilon) \times \Gamma_R(a_{L+1},\dots,a_N;n,r,\varepsilon).$$

(Note that the sum of free entropies is always well-defined, due to the estimate in Proposition 6.1.1.)

□

For the single variable case we observed in (6.1.2) that $\chi(a) = \chi_R(a)$ whenever $R \ge \|a\|$. The multivariable case is quite similar.

Proposition 6.1.4 *$\chi(a_1,\dots,a_N) = \chi_R(a_1,\dots,a_N)$ whenever $R > \|a_i\|$ for every $1 \le i \le N$.*

Proof: Let $\rho := \max_i \|a_i\|$. It suffices to show that

$$\chi_R(a_1,\dots,a_N) \ge \chi_{R_1}(a_1,\dots,a_N) \quad \text{if} \quad \rho < R < R_1.$$

Let $\rho < R_0 < R < R_1 < R_2$, and define $f : [-R_2, R_2] \to [-R, R]$ by

$$f(t) := \begin{cases} \dfrac{R - R_0}{R_2 - R_0}(t + R_0) - R_0 & \text{if } -R_2 \le t \le -R_0, \\ t & \text{if } -R_0 \le t \le R_0, \\ \dfrac{R - R_0}{R_2 - R_0}(t - R_0) + R_0 & \text{if } \ \ R_0 \le t \le R_2, \end{cases}$$

Write $F(A_1, \dots, A_N) := (f(A_1), \dots, f(A_N))$ for $(A_1, \dots, A_N) \in (M_n(\mathbb{C})^{sa})^N$ with $\|A_i\| \le R_2$. Let us show that for every $r \in \mathbb{N}$ and $\varepsilon > 0$ there exist an integer $r_1 \ge r$ and an $\varepsilon_1 > 0$ such that

$$F(\Gamma_{R_1}(a_1, \dots, a_N; n, r_1, \varepsilon_1)) \subset \Gamma_R(a_1, \dots, a_N; n, r, \varepsilon) \qquad (n \in \mathbb{N}). \tag{6.1.4}$$

Let $\varepsilon_1 := \varepsilon/2$ and $0 < \delta < \varepsilon/2r(R_2 + 1)^{r-1}$, and choose an even integer $r_1 \ge r$ such that

$$\frac{\rho^{r_1} + \varepsilon_1}{R_0^{r_1}}(R_2 - R) < \delta\,.$$

For any $n \in \mathbb{N}$ and $(A_1, \dots, A_N) \in \Gamma_{R_1}(a_1, \dots, a_N; n, r_1, \varepsilon_1)$, since

$$\begin{aligned} \rho^{r_1} + \varepsilon_1 &\ge \tau(a_i^{r_1}) + \varepsilon_1 \ge \operatorname{tr}_n(A_i^{r_1}) \\ &\ge \int_{R_0 \le |t| \le R_2} t^{r_1}\, d\operatorname{tr}_n(E_{A_i}(t)) \\ &\ge R_0^{r_1} \operatorname{tr}_n(E_{A_i}([-R_2, -R_0] \cup [R_0, R_2])), \end{aligned}$$

where E_{A_i} is the spectral measure of A_i $(1 \le i \le N)$, we have

$$\begin{aligned} \operatorname{tr}_n(|f(A_i) - A_i|) &= \int |f(t) - t|\, d\operatorname{tr}_n(E_{A_i}(t)) \\ &\le (R_2 - R)\operatorname{tr}_n(E_{A_i}([-R_2, -R_0] \cup [R_0, R_2])) < \delta\,. \end{aligned} \tag{6.1.5}$$

Hence, for every $1 \le k \le r$ and $1 \le i_1, \dots, i_k \le N$, it is easy to check that

$$|\operatorname{tr}_n(f(A_{i_1}) \cdots f(A_{i_k})) - \operatorname{tr}_n(A_{i_1} \cdots A_{i_k})| \le k R_2^{k-1} \delta < \frac{\varepsilon}{2}\,.$$

This implies that

$$|\operatorname{tr}_n(f(A_{i_1}) \cdots f(A_{i_k})) - \tau(a_{i_1} \cdots a_{i_k})| < \varepsilon\,,$$

and (6.1.4) follows.

Next we compute the Radon-Nikodým derivative of the measure $\Lambda \circ f$ with respect to Λ restricted on $\{A \in M_n(\mathbb{C})^{sa} : \|A\| \le R_1\}$. We proceed as in the proof of Lemma 4.1.2 (also Lemma 4.4.6), and make the coordinate change $A \in M_n(\mathbb{C})^{sa} \leftrightarrow (U, D) \in \mathcal{U}(n)/T \times \mathbb{R}^n_\le$ by the diagonalization $A = UDU^*$, $D = \mathbf{Diag}(t_1, t_2, \dots, t_n)$ with $t_1 < t_2 < \dots < t_n$ (the other degenerate case is negligible). Since

$$dA = U(dD + D \cdot dM - dM \cdot D)U^*,$$

where $dM := -\mathrm{i}\, U^* \cdot dU$ is an infinitesimal Hermitian, we have

$$\prod_{i\le j} d(\mathrm{Re}\, A_{ij}) \prod_{i<j} d(\mathrm{Im}\, A_{ij}) = \prod_{i<j} (t_i - t_j)^2 \prod_{i=1}^n dt_i \prod_{i<j} d(\mathrm{Re}\, M_{ij}) d(\mathrm{Im}\, M_{ij}) .$$

Let $\|A\| \le R_1$. Since $f(A) = Uf(D)U^*$, we have

$$df(A) = U(f'(D)dD + f(D) \cdot dM - dM \cdot f(D))U^*,$$

so that

$$\begin{aligned}&\prod_{i\le j} d(\mathrm{Re}\, f(A)_{ij}) \prod_{i<j} d(\mathrm{Im}\, f(A)_{ij}) \\ &\quad = \prod_{i<j} (f(t_i) - f(t_j))^2 \prod_{i=1}^n f'(t_i) \prod_{i=1}^n dt_i \prod_{i<j} d(\mathrm{Re}\, M_{ij}) d(\mathrm{Im}\, M_{ij}) .\end{aligned}$$

Taking the ratio of the above two densities yields

$$\frac{d(\Lambda \circ f)}{d\Lambda}(A) = \prod_{i<j} \Big(\frac{f(t_i) - f(t_j)}{t_i - t_j} \Big)^2 \prod_{i=1}^k f'(t_i) \tag{6.1.6}$$

almost everywhere on $\{A \in M_n(\mathbb{C})^{sa} : \|A\| \le R_1\}$.

Note that

$$\Big| \frac{f(t_i) - f(t_j)}{t_i - t_j} \Big| \begin{cases} = 1 & \text{if } t_i, t_j \in (-R_0, R_0), \\ \ge \dfrac{R - R_0}{R_2 - R_0} & \text{otherwise.} \end{cases}$$

If $A \in M_n(\mathbb{C})^{sa}$ satisfies $\mathrm{tr}_n(E_A([-R_2, -R_0] \cup [R_0, R_2])) < \delta'$, then the number of eigenvalues of A contained in $[-R_2, -R_0] \cup [R_0, R_2]$ is less than $n\delta'$. Hence for every $(A_1, \dots, A_N) \in \Gamma_{R_1}(a_1, \dots, a_N; n, r_1, \varepsilon_1)$, by (6.1.5) we have

$$\frac{d(\Lambda \circ f)}{d\Lambda}(A_i) \ge \Big(\frac{R - R_0}{R_2 - R_0} \Big)^{2n^2\delta/(R_2 - R)} \qquad (1 \le i \le N).$$

Therefore, by (6.1.4) we infer that

$$\begin{aligned}&\Lambda(\Gamma_R(a_1,\dots,a_N;n,r,\varepsilon))\\&\quad\geq\left(\frac{R-R_0}{R_2-R_0}\right)^{2Nn^2\delta/(R_2-R)}\Lambda(\Gamma_{R_1}(a_1,\dots,a_N;n,r_1,\varepsilon_1))\,,\end{aligned}$$

which implies that

$$\chi_R(a_1,\dots,a_N;r,\varepsilon)\geq\chi_{R_1}(a_1,\dots,a_N)+\frac{2N\delta}{R_2-R}\log\frac{R-R_0}{R_2-R_0}\,.$$

Since $\delta>0$ is arbitrary, the desired inequality is obtained.

□

The Radon-Nikodým derivative (6.1.6) can be also obtained from the Fréchet derivative of the mapping $A\mapsto f(A)$. It is known ([22], Sec. V.3) that

$$Df(A)(H)=f^{[1]}(A)\circ H\,,$$

where $\circ$ denotes the Schur product in a basis in which A is diagonal, $\mathbf{Diag}(\lambda_1,\dots,\lambda_n)$, i.e. $A=\sum_i\lambda_i e_{ii}$ in some matrix units $(e_{ij})_{1\leq i,j\leq n}$, and moreover $f^{[1]}(A)$ is defined in these matrix units by

$$f^{[1]}(A)_{ij}:=\begin{cases}\dfrac{f(\lambda_i)-f(\lambda_j)}{\lambda_i-\lambda_j} & \text{if } \lambda_i\neq\lambda_j,\\ f'(\lambda_i) & \text{if } \lambda_i=\lambda_j.\end{cases}$$

Since the Fréchet derivative is diagonal in the basis $(e_{ij})_{1\leq i,j\leq n}$ of $M_n(\mathbb{C})$, the determinant of $Df(A)$ is easily computed, and it equals (6.1.6).

The *upper semicontinuity* of $\chi(a_1,\dots,a_N)$ is shown as follows.

Proposition 6.1.5 *Let* $(a_1,\dots,a_N)$ *and* $(a_{m,1},\dots,a_{m,N})$ *be in* $(\mathcal{M}^{sa})^N$ *for* $m\in\mathbb{N}$. *If* $(a_{m,1},\dots,a_{m,N})\to(a_1,\dots,a_N)$ *in the distribution sense and* $\sup_m\|a_{m,i}\|<+\infty$ *for* $1\leq i\leq N$, *then*

$$\chi(a_1,\dots,a_N)\geq\limsup_{m\to\infty}\chi(a_{m,1},\dots,a_{m,N})\,.$$

In particular, this is the case when $a_{m,i}\to a_i$ *strongly (i.e. in the strong operator topology) for* $1\leq i\leq N$.

Proof: Let $R>\sup_{m,i}\|a_{m,i}\|$. For any $r\in\mathbb{N}$ and $\varepsilon>0$, the assumption implies that

$$|\tau(a_{m,i_1}\cdots a_{m,i_k})-\tau(a_{i_1}\cdots a_{i_k})|<\varepsilon/2$$

for all $1 \le i_1, \dots, i_k \le N$, $1 \le k \le r$, whenever m is large enough. In this case, we have

$$\Gamma_R(a_{m,1}, \dots, a_{m,N}; n, r, \varepsilon/2) \subset \Gamma_R(a_1, \dots, a_N; n, r, \varepsilon),$$

so that

$$\chi_R(a_{m,1}, \dots, a_{m,N}) \le \chi_R(a_{m,1}, \dots, a_{m,N}; r, \varepsilon/2) \le \chi_R(a_1, \dots, a_N; r, \varepsilon).$$

Hence

$$\limsup_{m\to\infty} \chi_R(a_{m,1}, \dots, a_{m,N}) \le \chi_R(a_1, \dots, a_N),$$

which gives the result by Proposition 6.1.4.

□

We give one more basic property of free entropy which will also play a role in the next chapter. Roughly speaking, $\chi(a_1, a_2)$ is larger when a_1 and a_2 are "freer", and $\chi(a_1, a_2)$ does not change if the "degree of freeness" of the variables is left unchanged. More concretely, a particular case of the next proposition is $\chi(a_1, a_2) = \chi(a_1, a_2 + g(a_1))$ for any function g. The invariance under perturbation by polynomials of the other variable(s) is a new phenomenon compared with the Boltzmann-Gibbs entropy.

In the proof below we take polynomials of noncommuting indeterminates. The polynomial

$$P(X_1, \dots, X_N) = \sum \mathrm{coef}[i(1), i(2), \dots, i(k)] X_{i(1)} X_{i(2)} \cdots X_{i(k)}$$

is called selfadjoint if $\overline{\mathrm{coef}[i(1), i(2), \dots, i(k)]} = \mathrm{coef}[i(k), \dots, i(2), i(1)]$.

Proposition 6.1.6 *Let $a_1, \dots, a_N, b_1, \dots, b_N \in \mathcal{M}^{sa}$. If $b_1 = a_1$ and $b_i - a_i \in \{a_1, \dots, a_{i-1}\}''$ for $2 \le i \le N$, then*

$$\chi(a_1, \dots, a_N) = \chi(b_1, \dots, b_N).$$

Proof: By assumption we can choose selfadjoint noncommutative polynomials $P_{m,i}(X_1, \dots, X_{i-1})$ for $2 \le i \le N$, $m \in \mathbb{N}$, such that $P_{m,i}(a_1, \dots, a_{i-1}) \to b_i - a_i$ strongly as $m \to \infty$. Set $b_{m,1} := b_1 = a_1$ and

$$b_{m,i} := a_i + P_{m,i}(a_1, \dots, a_{i-1}) \qquad (2 \le i \le N).$$

Then $b_{m,i} \to b_i$ strongly as $m \to \infty$. Now define a map $\Psi : (M_n(\mathbb{C})^{sa})^N \to (M_n(\mathbb{C})^{sa})^N$, $\Psi(A_1, \dots, A_N) = (B_1, \dots, B_N)$, by

$$B_1 := A_1, \quad B_i := A_i + P_{m,i}(A_1, \dots, A_{i-1}) \qquad (2 \le i \le N).$$

Then the map Ψ preserves the Lebesgue measure Λ, i.e. $\Lambda \circ \Psi = \Lambda$. For any $m, r \in \mathbb{N}$, $\varepsilon > 0$ and $R > 0$ there exist $r_1 \in \mathbb{N}$ and $\varepsilon_1 > 0$ such that

$$\Psi(\Gamma_R(a_1, \dots, a_N; n, r_1, \varepsilon_1)) \subset \Gamma_R(b_{m,1}, \dots, b_{m,N}; n, r, \varepsilon) \qquad (n \in \mathbb{N}),$$

and hence

$$\chi_R(a_1, \dots, a_N; r_1, \varepsilon_1) \le \chi_R(b_{m,1}, \dots, b_{m,N}; r, \varepsilon).$$

This implies $\chi(a_1, \dots, a_N) \le \chi(b_{m,1}, \dots, b_{m,N})$, so that $\chi(a_1, \dots, a_N) \le \chi(b_1, \dots, b_N)$ by upper semicontinuity (Proposition 6.1.5). The reverse inequality follows by symmetry (we have $a_i - b_i \in \{b_1, \dots, b_{i-1}\}''$, $2 \le i \le N$, too).

□

6.2 Calculus for power series of noncommutative variables

In this section a functional calculus of operators for power series of noncommuting indeterminates is developed, and the total differential is defined. Moreover, in finite von Neumann algebras a noncommutative (free) Jacobian is introduced on the basis of the Fuglede-Kadison determinant theory.

Let $(\mathcal{M}, \tau)$ be a tracial W^*-probability space. $\mathcal{M}$ is conveniently represented on the Hilbert space $L^2(\mathcal{M}) = L^2(\mathcal{M}, \tau)$. This is the completion of $\mathcal{M}$ with respect to the inner product $\langle a, b \rangle := \tau(b^* a)$. The left multiplication $L_a b := ab$ ($b \in \mathcal{M} \subset L^2(\mathcal{M})$) is called the *standard representation* of $\mathcal{M}$. The *opposite von Neumann algebra* $\mathcal{M}^{op} = \{a^o : a \in \mathcal{M}\}$ is the same as $\mathcal{M}$ except the multiplication is reversed: $a^o b^o = (ba)^o$. The right multiplication $R_a b := ba$ is a faithful representation of the opposite von Neumann algebra $\mathcal{M}^{op}$. The cyclic vector $\xi := \mathbf{1} \in L^2(\mathcal{M})$ induces the functional τ on both $\mathcal{M}$ and $\mathcal{M}^{op}$. The von Neumann algebra tensor product $\mathcal{M} \otimes \mathcal{M}^{op}$ acts on the tensor product Hilbert space $L^2(\mathcal{M}) \otimes L^2(\mathcal{M})$. The vector $\xi \otimes \xi$ generates a functional on $\mathcal{M} \otimes \mathcal{M}^{op}$ which is a faithful normal trace on $\mathcal{M} \otimes \mathcal{M}^{op}$, denoted by $\tau \otimes \tau$. (A more detailed discussion on tensor products is found in books on operator algebras, for example [108] or [188]; however, this chapter does not require much knowledge.)

The Fuglede-Kadison determinant ([79]) is nothing else but the extension of the concept of determinant of a matrix to an arbitrary finite von Neumann algebra. For an invertible element $a \in \mathcal{M}$ the (positive-valued) *Fuglede-Kadison determinant* $\Delta(a) = \Delta(|a|)$ is defined by means of the spectral theorem. When $|a| = \int_0^\infty \lambda \, de(\lambda)$ is the spectral decomposition of $|a|$, $\Delta(a)$ is given as

$$\Delta(a) := \exp \int_0^\infty \log \lambda \, d\tau(e(\lambda)). \qquad (6.2.1)$$

The properties $\Delta(a^{-1}) = \Delta(a)^{-1}$ and $\Delta(u_1 a u_2) = \Delta(a)$ for unitaries $u_1, u_2 \in \mathcal{M}$ are straightforward consequences of the definition. The multiplicativity

$$\Delta(a_1 a_2) = \Delta(a_1)\Delta(a_2)$$

is a deeper result for invertible $a_1, a_2 \in \mathcal{M}$. In fact, the definition (6.2.1) is available for a general normal semifinite trace τ and a general element $a \in \mathcal{M}$; then $0 \leq \Delta(a) \leq \infty$. When $\mathcal{M}$ is the matrix algebra $M_n(\mathbb{C})$ and τ is the trace Tr_n, it is straightforward to see that $\Delta(a) = |\det a|$, where $\det a$ is the common determinant of a.

The von Neumann algebra on which we use the Fuglede-Kadison determinant is $\mathcal{N} := M_N(\mathbb{C}) \otimes (\mathcal{M} \otimes \mathcal{M}^{op})$. This is a finite von Neumann algebra. In our approach $\mathcal{M} \otimes \mathcal{M}^{op}$ acts on the Hilbert space $L^2(\mathcal{M}) \otimes L^2(\mathcal{M})$. The functional $\tau \otimes \tau$ is a faithful normal trace, and so is $\mathrm{Tr}_n \otimes (\tau \otimes \tau)$ on $\mathcal{N}$. When T is an arbitrary element of $\mathcal{N}$, the Fuglede-Kadison determinat $\Delta(T)$ with respect to $\mathrm{Tr}_n \otimes (\tau \otimes \tau)$ is at our disposal.

It is convenient to give a brief exposition on noncommutative derivations in the abstract setting. Let $\mathcal{A}$ be a unital algebra and $x \in \mathcal{A}$. We want to discuss a *derivation* $\partial_x : \mathcal{A} \to \mathcal{C}$. The characteristic property of any derivation is the *Leibniz rule*:

$$\partial_x(a_1 a_2) = (\partial_x a_1) a_2 + a_1 \partial_x a_2 \,.$$

Therefore we have to assume that $\mathcal{C}$ is a bimodule over $\mathcal{A}$. When $\mathcal{A}$ is a bimodule over $\mathcal{B}$ playing the role of "scalars", we can require that

$$\partial_x(b_1 a b_2) = b_1 (\partial_x a) b_2 \,.$$

In particular, $\mathcal{C}$ has to be a $\mathcal{B}$-bimodule as well. We do not make precise the assumptions on $\mathcal{A}$, $\mathcal{B}$ and $\mathcal{C}$; in the concrete situation in which we work below those will be clear. Further natural requirements are

$$\partial_x x = \mathbf{1}\,, \qquad \partial_x b = 0 \quad \text{for} \quad b \in \mathcal{B}.$$

Clearly such a derivation ∂_x exists uniquely when every element of $\mathcal{A}$ is uniquely written in the form $a = b_1 x b_2 x \cdots x b_n$, that is, $\mathcal{A}$ is algebraically freely generated by x and $\mathcal{B}$. It is time to give an example where all of our assumptions are fulfilled. Consider the polynomial algebras $\mathcal{A} := \mathbb{C}\langle X_1, X_2, \dots, X_N\rangle$, $\mathcal{B} := \mathbb{C}\langle X_2, \dots, X_N\rangle$ and $\mathcal{C} := \mathbb{C}\langle X_1, X_2, \dots, X_N\rangle$ with noncommuting indeterminates $X_1, X_2, \dots, X_n$. If we choose $x = X_1$, then $\mathcal{A}$ is algebraically freely generated by x and $\mathcal{B}$. A repeated application of the Leibniz rule yields

$$\partial_{X_1}(X_2 X_1^3 X_3 X_1) = 3X_2 X_1^2 X_3 X_1 + X_2 X_1^3 X_3 \,.$$

This example is not what we really need yet. In place of $\mathcal{C}$ we put the tensor product algebra $\mathbb{C}\langle X_1, X_2, \dots, X_N\rangle \otimes \mathbb{C}\langle X_1, X_2, \dots, X_N\rangle$, which has to be a bimodule. The $\mathbb{C}\langle X_1, X_2, \dots, X_N\rangle$-bimodule structure of $\mathcal{C}$ is determined by the rule

$$a_1(a_2 \otimes a_3)a_4 := (a_1a_2) \otimes (a_4^t a_3)\,,$$

where

$$(X_{i(1)}X_{i(2)}\cdots X_{i(k)})^t = X_{i(k)}\cdots X_{i(2)}X_{i(1)}\,.$$

Now we consider the derivation D_{X_1} with respect to the indeterminate X_1. This is uniquley defined by the requirements, and we have

$$D_{X_1}(X_2X_1^2X_3) = X_2 \otimes X_3X_1 + X_2X_1 \otimes X_3\,,$$

and generally

$$\begin{aligned} &D_{X_j}(X_{i(1)}X_{i(2)}\cdots X_{i(k)}) \\ &\quad = \sum_{i(m)=j} X_{i(1)}X_{i(2)}\cdots X_{i(m-1)} \otimes (X_{i(m+1)}X_{i(m+2)}\cdots X_{i(k)})^t\,. \end{aligned} \tag{6.2.2}$$

Next we consider formal infinite power series

$$\begin{aligned} &F(X_1, X_2, \dots, X_N) \\ &\quad = \sum \operatorname{coef}[i(1), i(2), \dots, i(k)](F)X_{i(1)}X_{i(2)}\cdots X_{i(k)} \end{aligned} \tag{6.2.3}$$

of noncommuting indeterminates $X_1, X_2, \dots, X_N$ with complex coefficients. (The summation here and below is over $1 \le i(1), i(2), \dots, i(k) \le N$, $k \in \mathbb{N}$.) The conjugate power series F^* is given by

$$F^*(X_1, X_2, \dots, X_N) := \sum \overline{\operatorname{coef}[i(1), i(2), \dots, i(k)](F)}(X_{i(1)}X_{i(2)}\cdots X_{i(k)})^t\,.$$

The N-tuple $(R_1, \dots, R_N)$ of positive numbers is called a *multi-radius of convergence* for F if the quantuty

$$M(F; R_1, R_2, \dots, R_N) := \sum |\operatorname{coef}[i(1), i(2), \dots, i(k)](F)| R_{i(1)}R_{i(2)}\cdots R_{i(k)}$$

is finite. When $a_1, a_2, \dots, a_N$ are taken from $\mathcal{M}$ and $(\|a_1\|, \dots, \|a_N\|)$ is a multi-radius of convergence for F, one can define $F(a_1, \dots, a_N) \in \mathcal{M}$ by the absolutely convergent series

$$\sum \operatorname{coef}[i(1), i(2), \dots, i(k)](F)a_{i(1)}a_{i(2)}\cdots a_{i(k)}\,.$$

In particular, if $a_1, \dots, a_N \in \mathcal{M}^{sa}$, $(\|a_1\|, \dots, \|a_N\|)$ is a multi-radius of convergence for F and $F^* = F$, then $F(a_1, \dots, a_N) \in \mathcal{M}^{sa}$.

Let $F(X_1, X_2, \dots, X_N)$ be the infinite power series (6.2.3). We set

$$D_j F(X_1, \dots, X_N) := \sum \mathrm{coef}[i(1), i(2), \dots, i(k)](F) D_{X_j}(X_{i(1)} X_{i(2)} \cdots X_{i(k)}),$$

where D_{X_j} is given by (6.2.2). Then $D_j F(a_1, \dots, a_N)$ is a well-defined element of the von Neumann algebra $\mathcal{M} \otimes \mathcal{M}^{op}$ whenever $(\|a_1\|, \dots, \|a_N\|)$ is a multi-radius of convergence for F (hence for $D_j F$ too). It is explicitly written as

$$\begin{aligned} &D_j F(a_1, \dots, a_N) \\ &\quad = \sum \mathrm{coef}[i(1), \dots, i(k)](F) \sum_{i(m)=j} (a_{i(1)} \cdots a_{i(m-1)}) \otimes (a_{i(m+1)} \cdots a_{i(k)})^o . \end{aligned}$$

Now let $(F_1, \dots, F_N)$ be an N-tuple of noncommutative power series of N variables, and let $(R_1, \dots, R_N)$ with $R_i > 0$ be a common multi-radius of convergence for F_i. The *differential* DF of $F = (F_1, \dots, F_N)$ is given by

$$DF(a_1, \dots, a_N) := \begin{bmatrix} D_1 F_1(a_1, \dots, a_N) & \cdots & D_N F_1(a_1, \dots, a_N) \\ \vdots & \ddots & \vdots \\ D_1 F_N(a_1, \dots, a_N) & \cdots & D_N F_N(a_1, \dots, a_N) \end{bmatrix}, \tag{6.2.4}$$

which is well-defined for $a_1, \dots, a_N \in \mathcal{M}$ with $\|a_i\| \le R_i$ as an element of the tensor product $\mathcal{N} = M_N(\mathbb{C}) \otimes (\mathcal{M} \otimes \mathcal{M}^{op})$.

Let $G = (G_1, \dots, G_N)$ be another N-tuple of power series. The formula

$$\begin{aligned} (G \circ F)(X_1, \dots, X_N) := \bigl(&G_1(F_1(X_1, \dots, X_N), \dots, F_N(X_1, \dots, X_N)), \\ &\dots, G_N(F_1(X_1, \dots, X_N), \dots, F_N(X_1, \dots, X_N))\bigr) \end{aligned}$$

can be used to define the composition of N-tuples of power series, since every coefficient of the composition is a finite sum. When there are $R_i > 0$ such that $(R_1, \dots, R_N)$ is a common multi-radius of convergence for F_i and $(R'_1, \dots, R'_N)$, where $R'_i := M(F_i; R_1, \dots, R_N)$, is a common multi-radius of convergence for G_i, then $(R_1, \dots, R_N)$ is a common multi-radius of convergence for all components of the composition defined above. Under this condition we have the *chain rule* as follows.

Lemma 6.2.1 *In the above situation let $a_1, \dots, a_N \in \mathcal{M}$ be such that $\|a_i\| \le R_i$. Then*

$$D(G \circ F)(a_1, \dots, a_N) = DG(F(a_1, \dots, a_N)) \cdot DF(a_1, \dots, a_N);$$

that is,

$$D_j(G\circ F)_i(a_1,\ldots,a_N)=\sum_{l=1}^{N}D_lG_i(F(a_1,\ldots,a_N))D_jF_l(a_1,\ldots,a_N)$$

for each $1\leq i,j\leq N$.

Proof: Since the differential can be calculated term by term, it suffices to show that if $G(X_1,\ldots,X_N)=X_{i_1}X_{i_2}\cdots X_{i_k}$ and $F_l(X_1,\ldots,X_N)=X_{j_l(1)}X_{j_l(2)}\cdots X_{j_l(k(l))}$, then

$$D_j(G\circ F)(a_1,\ldots,a_N)=\sum_{l=1}^{N}D_lG(F(a_1,\ldots,a_N))D_jF_l(a_1,\ldots,a_N).$$

Since

$$(G\circ F)(a_1,\ldots,a_N)=\prod_{m=1}^{k}F_{i_m}(a_1,\ldots,a_N)=\prod_{l=i_1,\ldots,i_k}(a_{j_l(1)}\cdots a_{j_l(k(l))}),$$

we get

$$\begin{aligned}
&D_j(G\circ F)(a_1,\ldots,a_N)\\
&=\sum_{m=1}^{k}\sum_{s:j_{i_m}(s)=j}(F_{i_1}(a_1,\ldots,a_N)\cdots F_{i_{m-1}}(a_1,\ldots,a_N)a_{j_{i_m}(1)}\cdots a_{j_{i_m}(s-1)})\\
&\qquad\qquad\otimes(a_{j_{i_m}(s+1)}\cdots a_{j_{i_m}(k(i_m))}F_{i_{m+1}}(a_1,\ldots,a_N)\cdots F_{i_k}(a_1,\ldots,a_N))^o\\
&=\sum_{m=1}^{k}(F_{i_1}(a_1,\ldots,a_N)\cdots F_{i_{m-1}}(a_1,\ldots,a_N))\\
&\qquad\qquad\otimes(F_{i_{m+1}}(a_1,\ldots,a_N)\cdots F_{i_k}(a_1,\ldots,a_N))^o\\
&\quad\times\sum_{s:j_{i_m}(s)=j}(a_{j_{i_m}(1)}\cdots a_{j_{i_m}(s-1)})\otimes(a_{j_{i_m}(s+1)}\cdots a_{j_{i_m}(k(i_m))})^o\\
&=\sum_{l=1}^{N}D_lG(F(a_1,\ldots,a_N))D_jF_l(a_1,\ldots,a_N).
\end{aligned}$$

□

The (positive-valued) *Jacobian* of F at $(a_1,\ldots,a_N)$ is defined by

$$|\mathcal{J}|(F)(a_1,\ldots,a_N):=\Delta(DF(a_1,\ldots,a_N)).$$

(Recall that the Fuglede-Kadison determinant is taken with respect to the trace $\mathrm{Tr}_n\otimes\tau\otimes\tau$.) The next lemma gives the Jacobian in the matrix case ($\mathcal{M}=M_n(\mathbb{C})$, $\tau=\mathrm{tr}_n$).

Lemma 6.2.2 *Let $F_1, \dots, F_N$ be noncommutative power series of N variables with $F_i = F_i^*$, and let $(R_1, \dots, R_N)$ be a common multi-radius of convergence for F_i. For the map $F := (F_1, \dots, F_N)$ on $\{(A_1, \dots, A_N) \in (M_n(\mathbb{C})^{sa})^N : \|A_i\| < R_i\}$,*

$$\begin{aligned} \frac{d(\Lambda \circ F)}{d\Lambda}(A_1, \dots, A_N) &= |\det(DF(A_1, \dots, A_N))| \\ &= \left(|\mathcal{J}|(F)(A_1, \dots, A_N)\right)^{n^2}, \end{aligned} \tag{6.2.5}$$

where det *means the usual determinant on $M_N(\mathbb{C}) \otimes M_n(\mathbb{C}) \otimes M_n(\mathbb{C})^{op} \cong M_{Nn^2}(\mathbb{C})$.*

Proof: When $\mathcal{M} = M_n(\mathbb{C})$, one can identify $\mathcal{M} \otimes \mathcal{M}^{op}$ with $M_n(\mathbb{C}) \otimes M_n(\mathbb{C}) = M_{n^2}(\mathbb{C})$ by $A \otimes B^o = A \otimes B^t$, where ^{t} stands for transpose. Identifying $(M_n(\mathbb{C})^{sa})^N$ with $\mathbb{R}^{Nn^2}$, let us show that the usual Jacobian matrix for $F(A_1, \dots, A_N)$ is unitarily conjugate to $DF(A_1, \dots, A_N) \in M_N(\mathbb{C}) \otimes M_{n^2}(\mathbb{C})$ given by (6.2.4). To show this, for $1 \le p, q \le N$ fixed, write

$$F_p(A_1, \dots, A_N) = \sum_{k=0}^{\infty} \sum_{i_1, \dots, i_k = 1}^{N} \operatorname{coef}[i(1), \dots, i(k)] A_{i(1)} \cdots A_{i(k)},$$

and put

$$\begin{aligned} &x_{ii} := (A_q)_{ii}, \quad x_{ij} := \sqrt{2}\,\mathrm{Re}\,(A_q)_{ij}, \quad x_{ji} := \sqrt{2}\,\mathrm{Im}\,(A_q)_{ij} \qquad (i < j), \\ &y_{ii} := F_p(A_1, \dots, A_N)_{ii}, \\ &y_{ij} := \sqrt{2}\,\mathrm{Re}\,F_p(A_1, \dots, A_N)_{ij}, \quad y_{ji} := \sqrt{2}\,\mathrm{Im}\,F_p(A_1, \dots, A_N)_{ij} \qquad (i < j). \end{aligned}$$

Then it is easy to check that for $i < j$ and $i' < j'$

$$\frac{\partial y_{ii}}{\partial x_{i'i'}} = (D_q F_p)_{ii,i'i'},$$

$$\begin{bmatrix} \dfrac{\partial y_{ii}}{\partial x_{i'j'}} & \dfrac{\partial y_{ii}}{\partial x_{j'i'}} \end{bmatrix} = \begin{bmatrix} (D_q F_p)_{ii,i'j'} & (D_q F_p)_{ii,j'i'} \end{bmatrix} \begin{bmatrix} 1/\sqrt{2} & \mathrm{i}/\sqrt{2} \\ 1/\sqrt{2} & -\mathrm{i}/\sqrt{2} \end{bmatrix},$$

$$\begin{bmatrix} \dfrac{\partial y_{ij}}{\partial x_{i'i'}} \\ \dfrac{\partial y_{ji}}{\partial x_{i'i'}} \end{bmatrix} = \begin{bmatrix} 1/\sqrt{2} & 1/\sqrt{2} \\ -\mathrm{i}/\sqrt{2} & \mathrm{i}/\sqrt{2} \end{bmatrix} \begin{bmatrix} (D_q F_p)_{ij,i'i'} \\ (D_q F_p)_{ji,i'i'} \end{bmatrix},$$

$$\begin{aligned} \begin{bmatrix} \dfrac{\partial y_{ij}}{\partial x_{i'j'}} & \dfrac{\partial y_{ij}}{\partial x_{j'i'}} \\ \dfrac{\partial y_{ji}}{\partial x_{i'j'}} & \dfrac{\partial y_{ji}}{\partial x_{j'i'}} \end{bmatrix} &= \begin{bmatrix} 1/\sqrt{2} & 1/\sqrt{2} \\ -\mathrm{i}/\sqrt{2} & \mathrm{i}/\sqrt{2} \end{bmatrix} \begin{bmatrix} (D_q F_p)_{ij,i'j'} & (D_q F_p)_{ij,j'i'} \\ (D_q F_p)_{ji,i'j'} & (D_q F_p)_{ji,j'i'} \end{bmatrix} \\ &\quad \times \begin{bmatrix} 1/\sqrt{2} & \mathrm{i}/\sqrt{2} \\ 1/\sqrt{2} & -\mathrm{i}/\sqrt{2} \end{bmatrix}, \end{aligned}$$

where

$$\begin{aligned}(D_qF_p)_{ij,i'j'} &= \sum_{k=1}^{\infty}\sum_{m=1}^{k}\sum_{i(m)=q} \mathrm{coef}[i(1),\dots,i(k)] \\ &\qquad \times[(A_{i(1)}\cdots A_{i(m-1)})\otimes(A_{i(m+1)}\cdots A_{i(k)})^t]_{ij,i'j'} \\ &= \sum_{k=1}^{\infty}\sum_{m=1}^{k}\sum_{i(m)=q} \mathrm{coef}[i(1),\dots,i(k)] \\ &\qquad \times(A_{i(1)}\cdots A_{i(m-1)})_{ii'}(A_{i(m+1)}\cdots A_{i(k)})_{j'j}\,.\end{aligned}$$

Arranging these formulas in the form of an $n^2 \times n^2$ matrix, we get the desired conclusion. Hence we obtain the first equality of (6.2.5), while the second equality is just the definition of $|\mathcal{J}|(F)(A_1,\dots,A_N)$, where the power n^2 is due to $\mathrm{Tr}_n \otimes \mathrm{Tr}_n = n^2 \mathrm{tr}_n \otimes \mathrm{tr}_n$.

□

6.3 Change of variable formulas for free entropy

The difference between the properties of free entropy and of Boltzmann-Gibbs entropy becomes considerable after the common subadditivity and upper semicontinuity. In this section we obtain transformation formulas for free entropy under some types of changes of the variables. Those formulas give an insight to free entropy, and will play crucial roles in the next section, where the additivity of free entropy will be discussed. When $a_1, a_2, \dots, a_n$ are selfadjoint noncommutative random variables, the separate change of variable formula means that the quantity

$$\chi(f_1(a_1),\dots,f_N(a_N)) - \sum_{i=1}^{N} \chi(f_i(a_i))$$

is independent of the continuous monotone functions f_i.

As for the Boltzmann-Gibbs entropy, when $\xi_1,\dots,\xi_N$ are random variables with joint density f on $\mathbb{R}^N$ and we transform $(\xi_1,\dots,\xi_N)$ to $(\eta_1,\dots,\eta_N)$ by a diffeomorphism F from a region including the values of $(\xi_1,\dots,\xi_N)$, the density g of $(\eta_1,\dots,\eta_N)$ is the transform of f under the coordinate change by F^{-1}. Then the following formula for Boltzmann-Gibbs entropy is easy to obtain:

$$S(\xi_1,\dots,\xi_N) = S(\eta_1,\dots,\eta_N) + \int_{\mathbb{R}^N} f(x)\log|J(x)|\,dx\,,$$

where $J(x)$ is the Jacobian of the coordinate change. Transformation formulas for free entropy are similar conceptually, but the functional calculus in noncommutative power series is involved, and the noncommutative Jacobian discussed in the previous section becomes important.

Theorem 6.3.1 *Let $F_1, \ldots, F_N$ and $G_1, \ldots, G_N$ be noncommutative power series with $F_i = F_i^*$ and $G_i = G_i^*$, such that*

$$G_i(F_1(X_1, \ldots, X_N), \ldots, F_N(X_1, \ldots, X_N)) = X_i \qquad (1 \le i \le N). \tag{6.3.1}$$

Let $(R_1, \ldots, R_N)$ and $(R_1', \ldots, R_N')$ be common multi-radii of convergence for F_i and G_i, respectively. If $a_1, \ldots, a_N \in \mathcal{M}^{sa}$ satsify

$$\|a_i\| < R_i\,, \quad M(F_i; \|a_1\|, \ldots, \|a_n\|) < R_i' \qquad (1 \le i \le N),$$

then

$$\begin{aligned}&\chi(F_1(a_1, \ldots, a_N), \ldots, F_N(a_1, \ldots, a_N)) \\ &\qquad \ge \chi(a_1, \ldots, a_N) + \log |\mathcal{J}|(F)(a_1, \ldots, a_N)\,.\end{aligned} \tag{6.3.2}$$

If in addition

$$M(G_i; \|F_1(a_1, \ldots, a_N)\|, \ldots, \|F_N(a_1, \ldots, a_N)\|) < R_i \qquad (1 \le i \le N),$$

then equality holds in (6.3.2).

Proof: For $\rho_1, \ldots, \rho_N > 0$, we can define $\Gamma_{\rho_1,\ldots,\rho_N}(a_1, \ldots, a_N; n, r, \varepsilon)$ similarly to $\Gamma_R(a_1, \ldots, a_N; n, r, \varepsilon)$ by taking $\|A_i\| \le \rho_i$ instead of $\|A_i\| \le R$, and we have $\chi_{\rho_1,\ldots,\rho_N}(a_1, \ldots, a_N)$ like $\chi_R(a_1, \ldots, a_N)$. Then the proof of Proposition 6.1.4 can be slightly modified to show that

$$\chi(a_1, \ldots, a_N) = \chi_{\rho_1,\ldots,\rho_N}(a_1, \ldots, a_N)$$

whenever $\|a_i\| < \rho_i$ for $1 \le i \le N$.

Choose $\|a_i\| < \rho_i < R_i$ and $M(F_i; \rho_1, \ldots, \rho_N) < \rho_i' < R_i'$ for $1 \le i \le N$. (This is possible because F_i is not constant, due to (6.3.1).) Given $r \in \mathbb{N}$ and $\varepsilon > 0$, it is not difficult to see that there exist $r_1 \in \mathbb{N}$ and $\varepsilon_1 > 0$ such that

$$\begin{aligned}&F(\Gamma_{\rho_1,\ldots,\rho_N}(a_1, \ldots, a_N; n, r_1, \varepsilon_1)) \\ &\qquad \subset \Gamma_{\rho_1',\ldots,\rho_N'}(F_1(a_1, \ldots, a_N), \ldots, F_N(a_1, \ldots, a_N); n, r, \varepsilon) \qquad (n \in \mathbb{N}),\end{aligned}$$

where $F := (F_1, \ldots, F_N)$ on $\{(A_1, \ldots, A_N) \in (M_n(\mathbb{C})^{sa})^N : \|A_i\| < R_i\}$. By Lemma 6.2.2 this yields

$$\begin{aligned}&\log \Lambda(\Gamma_{\rho_1',\ldots,\rho_N'}(F_1(a_1, \ldots, a_N), \ldots, F_N(a_1, \ldots, a_N); n, r, \varepsilon)) \\ &\qquad \ge \log \Lambda(\Gamma_{\rho_1,\ldots,\rho_N}(a_1, \ldots, a_N; n, r_1, \varepsilon_1)) \\ &\qquad\qquad + n^2 \inf\{\log |\mathcal{J}|(F)(A_1, \ldots, A_N)\} \qquad (n \in \mathbb{N}),\end{aligned} \tag{6.3.3}$$

where the infimum is taken over $(A_1, \dots, A_N) \in \Gamma_{\rho_1,\dots,\rho_N}(a_1, \dots, a_N; n, r_1, \varepsilon_1)$.

In the following let $b_1, \dots, b_N \in \mathcal{M}^{sa}$. By the assumption on the multi-radius of convergence, there exists $C \geq 1$ such that

$$\begin{aligned} \|DF(b_1, \dots, b_N)\| \leq C \quad &\text{if} \quad \|b_i\| \leq \rho_i, \\ \|DG(b_1, \dots, b_N)\| \leq C \quad &\text{if} \quad \|b_i\| \leq \rho_i'. \end{aligned}$$

By Lemma 6.2.1 applied to (6.3.1) we know that

$$DG(F(b_1, \dots, b_N)) \cdot DF(b_1, \dots, b_N) = \mathbf{1},$$

and hence $C^2 DF(b_1, \dots, b_N)^* DF(b_1, \dots, b_N) \geq \mathbf{1}$ in $M_N(\mathbb{C}) \otimes \mathcal{M} \otimes \mathcal{M}^{op}$. So we get

$$C^{-2}\mathbf{1} \leq DF(b_1, \dots, b_N)^* DF(b_1, \dots, b_N) \leq C^2 \mathbf{1} \quad \text{if} \quad \|b_i\| \leq \rho_i.$$

For any $\delta > 0$ with $\delta(2C + \delta) < C^{-2}$, put $\delta_1 := \delta(2C + \delta)$ and $\delta_2 := \delta_1/(C^{-2} - \delta_1)$. One can choose an $N \times N$ matrix P whose entries are noncommutative polynomials of $2n$ variables and which satisfies

$$\|DF(b_1, \dots, b_N) - P(b_1 \otimes \mathbf{1}, \dots, b_N \otimes \mathbf{1}, \mathbf{1} \otimes b_1^o, \dots, \mathbf{1} \otimes b_N^o)\| < \delta,$$

and hence

$$\begin{aligned} \|DF(b_1, \dots, b_N)^* DF(b_1, \dots, b_N) - P(b_1 \otimes \mathbf{1}, \dots, b_N \otimes \mathbf{1}, \\ \mathbf{1} \otimes b_1^o, \dots, \mathbf{1} \otimes b_N^o)^* P(b_1 \otimes \mathbf{1}, \dots, b_N \otimes \mathbf{1}, \mathbf{1} \otimes b_1^o, \dots, \mathbf{1} \otimes b_N^o)\| \leq \delta_1 \end{aligned}$$

whenever $\|b_i\| \leq \rho_i$. Furthermore, choose a polynomial $Q(t)$ such that

$$|Q(t) - \log t| \leq \delta \quad \text{for} \quad C^{-2} - \delta_1 \leq t \leq C^2 + \delta_1.$$

Here note that if $a, b \geq \alpha \mathbf{1}$ and $\|a - b\| < \alpha$, then

$$\|\log a - \log b\| \leq \frac{\|a - b\|}{\alpha - \|a - b\|}.$$

Indeed, from the operator monotonicity of log we have

$$\begin{aligned} -\|a - b\| b^{-1} &\leq \log(b - \|a - b\|\mathbf{1}) - \log b \leq \log a - \log b \\ &\leq \log(b + \|a - b\|\mathbf{1}) - \log b \leq \|a - b\| b^{-1} \end{aligned}$$

and $b \geq (\alpha - \|a - b\|)\mathbf{1}$.

From the above estimates we have

$$\begin{aligned}
&\| \log DF(a_1, \dots, a_N)^* DF(a_1, \dots, a_N) - Q(P(a_1 \otimes \mathbf{1}, \dots, a_N \otimes \mathbf{1}, \\
&\quad \mathbf{1} \otimes a_1^o, \dots, \mathbf{1} \otimes a_N^o)^* P(a_1 \otimes \mathbf{1}, \dots, a_N \otimes \mathbf{1}, \mathbf{1} \otimes a_1^o, \dots, \mathbf{1} \otimes a_N^o)) \| \\
&\quad \le \delta + \delta_2 \,. \qquad (6.3.4)
\end{aligned}$$

In the same way, in $M_N(\mathbb{C}) \otimes M_n(\mathbb{C}) \otimes M_n(\mathbb{C})^{op} \cong M_{Nn^2}(\mathbb{C})$ we have

$$\begin{aligned}
&\| \log DF(A_1, \dots, A_N)^* DF(A_1, \dots, A_N) - Q(P(A_1 \otimes I, \dots, A_N \otimes I, \\
&\quad I \otimes A_1^t, \dots, I \otimes A_N^t)^* P(A_1 \otimes I, \dots, A_N \otimes I, I \otimes A_1^t, \dots, I \otimes A_N^t)) \| \\
&\quad \le \delta + \delta_2 \qquad (6.3.5)
\end{aligned}$$

whenever $A_i \in M_n(\mathbb{C})^{sa}$, $\|A_i\| \le \rho_i$. Since

$$\begin{aligned}
(\mathrm{Tr}_N \otimes \mathrm{tr}_n \otimes \mathrm{tr}_n)\big(Q(P(A_1 \otimes I, \dots, A_N \otimes I, I \otimes A_1^t, \dots, I \otimes A_N^t)^* \\
\times P(A_1 \otimes I, \dots, A_N \otimes I, I \otimes A_1^t, \dots, I \otimes A_N^t))\big)
\end{aligned}$$

is a polynomial (of degree ≤ 2) in the multiple moments $\mathrm{tr}_n(A_{i_1} \cdots A_{i_k})$ and the same is true for $a_1, \dots, a_N$ as well, it follows from (6.3.4) and (6.3.5) that if r_1 is large enough and ε_1 is small enough, then

$$\begin{aligned}
&\big|(\mathrm{Tr}_N \otimes \mathrm{tr}_n \otimes \mathrm{tr}_n)\big(\log(DF(A_1, \dots, A_N)^* DF(A_1, \dots, A_N))\big) \\
&\quad -(\mathrm{Tr}_N \otimes \tau \otimes \tau)\big(\log(DF(a_1, \dots, a_N)^* DF(a_1, \dots, a_N))\big)\big| \\
&\quad \le \delta + 2N(\delta + \delta_2)
\end{aligned}$$

for all $(A_1, \dots, A_N) \in \Gamma_{\rho_1, \dots, \rho_N}(a_1, \dots, a_N; n, r_1, \varepsilon_1)$. The above inequality means that

$$| \log |\mathcal{J}|(F)(A_1, \dots, A_N) - \log |\mathcal{J}|(F)(a_1, \dots, a_N)| \le \frac{1}{2}\delta + N(\delta + \delta_2) \ (=: \delta_3).$$

This and (6.3.3) imply that

$$\begin{aligned}
&\chi_{\rho_1', \dots, \rho_N'}(F_1(a_1, \dots, a_N), \dots, F_N(a_1, \dots, a_N); r, \varepsilon) \\
&\quad \ge \chi_{\rho_1, \dots, \rho_N}(a_1, \dots, a_N; r_1, \varepsilon_1) + \log |\mathcal{J}|(F)(a_1, \dots, a_N) - \delta_3,
\end{aligned}$$

so that

$$\begin{aligned}
&\chi_{\rho_1', \dots, \rho_N'}(F_1(a_1, \dots, a_N), \dots, F_N(a_1, \dots, a_N)) \\
&\quad \ge \chi_{\rho_1, \dots, \rho_N}(a_1, \dots, a_N) + \log |\mathcal{J}|(F)(a_1, \dots, a_N) - \delta_3 \,.
\end{aligned}$$

Since δ_3 can be arbitrarily small, we obtain (6.3.2) from the fact noted at the beginning of the proof.

For the second assertion we need to show that

$$F_i(G_1(X_1,\dots,X_N),\dots,G_N(X_1,\dots,X_N)) = X_i \qquad (1 \le i \le N).$$

Consider F on a neighborhood of 0 in $(M_n(\mathbb{C})^{sa})^N$ ($\cong \mathbb{R}^{Nn^2}$). Since $DF(0)$ is invertible by Lemma 6.2.1 and (6.3.1), it follows from the fact noticed in the proof of Lemma 6.2.2 that F is invertible in a neighborhood of 0, so $F \circ G = \mathrm{id}$ ($G := (G_1,\dots,G_N)$) in a neighborhood of $F(0)$. By analyticity we get $F \circ G = \mathrm{id}$ in a neighborhood of 0 in $(M_n(\mathbb{C})^{sa})^N$. Since n is arbitrary, one can easily see that $F \circ G = \mathrm{id}$ as noncommutative power series, as desired. Thus the first assertion gives

$$\begin{aligned}\chi(a_1,\dots,a_N) \ &\ge\ \chi(F_1(a_1,\dots,a_N),\dots,F_N(a_1,\dots,a_N)) \\ &\qquad + \log|\mathcal{J}|(G)(F(a_1,\dots,a_N)).\end{aligned}$$

Moreover, the multiplicativity of the determinant Δ gives

$$|\mathcal{J}|(G)(F(a_1,\dots,a_N)) \cdot |\mathcal{J}|(F)(a_1,\dots,a_N) = 1,$$

so that equality is valid in (6.3.2).

□

The following is a simple application of the above theorem. The free entropy behaves in exactly the same way as the Boltzmann-Gibbs entropy under linear transformations of the variables.

Corollary 6.3.2 *Let $a_1,\dots,a_N \in \mathcal{M}^{sa}$.*

(1) *If $A = [\alpha_{ij}]_{i,j=1}^N \in M_N(\mathbb{R})$ and $\beta_1,\dots,\beta_N \in \mathbb{R}$, then*

$$\chi\Big(\sum_j \alpha_{1j}a_j + \beta_1\mathbf{1},\dots,\sum_j \alpha_{Nj}a_j + \beta_N\mathbf{1}\Big) = \chi(a_1,\dots,a_N) + \log|\det A|\,.$$

(2) *If $a_1,\dots,a_N$ are linearly dependent, then $\chi(a_1,\dots,a_N) = -\infty$.*

Proof: (1) It is enough to treat the case $\beta_i = 0$ ($1 \le i \le N$). If $F(X_1,\dots,X_N) := \big(\sum_j \alpha_{1j}X_j,\dots,\sum_j \alpha_{Nj}X_j\big)$, then

$$DF(a_1,\dots,a_N) = A \otimes \mathbf{1}_{\mathcal{M}\otimes\mathcal{M}^{op}},$$

so that $|\mathcal{J}|(F)(a_1,\dots,a_N) = |\det A|$. Hence the result is obvious from Theorem 6.3.1 if A is invertible. When A is singular, both sides are $-\infty$ from (2), below.

(2) We may assume that $a_1 = \alpha_2 a_2 + \cdots + \alpha_N a_N$ ($\alpha_i \in \mathbb{R}$). For every $\varepsilon > 0$, since $a_1 = \varepsilon a_1 + (1-\varepsilon)\alpha_2 a_2 + \cdots + (1-\varepsilon)\alpha_N a_N$, it follows from (1) that

$$\chi(a_1,\dots,a_N) = \chi(a_1,\dots,a_N) + \log\varepsilon\,,$$

which means $\chi(a_1,\dots,a_N) = -\infty$.

An essential point in the following special case of Theorem 6.3.1 is that the assumptions there are automatically fulfilled here.

Proposition 6.3.3 *Let* $a_1, \dots, a_N \in \mathcal{M}^{sa}$, *and let* $P_1, \dots, P_N$ *be noncommutative polynomials of* N *variables with* $P_i = P_i^*$. *Then for any* $\rho \in \mathbb{R}$ *sufficiently near* 0, *if* $F_{\rho,i}(X_1, \dots, X_n) := X_i + \rho P_i(X_1, \dots, X_n)$, *then*

$$\begin{aligned}&\chi(F_{\rho,1}(a_1, \dots, a_N), \dots, F_{\rho,N}(a_1, \dots, a_N)) \\ &\qquad = \chi(a_1, \dots, a_N) + \log |\mathcal{J}|(F_\rho)(a_1, \dots, a_N).\end{aligned}$$

Proof: Let us apply Theorem 6.3.1. For this we have to show that for any $\rho \in \mathbb{R}$ near 0 there are noncommutative power series $G_{\rho,i}$ $(1 \le i \le n)$ with $G_{\rho,i} = G_{\rho,i}^*$ such that

(i) $G_{\rho,i}(F_{\rho,1}(X_1, \dots, X_N), \dots, F_{\rho,n}(X_1, \dots, X_N)) = X_i \quad (1 \le i \le N)$,

(ii) $(M(F_{\rho,1}; \|a_1\| + \varepsilon, \dots, \|a_n\| + \varepsilon), \dots, M(F_{\rho,n}; \|a_1\| + \varepsilon, \dots, \|a_n\| + \varepsilon))$ is a common multi-radius of convergence for the $G_{\rho,i}$ for some $\varepsilon > 0$.

Note that the other two conditions about the multi-radius of convergence for the $F_{\rho,i}$ are trivial because the $F_{\rho,i}$ are polynomials.

Now choose $R > 0$ and $\varepsilon > 0$ so that $\|a_i\| + 2\varepsilon < R$. Let $\mathcal{A}_R$ denote the space of all noncommutative power series $F = F(X_1, \dots, X_N)$ such that $F = F^*$ and $M(F; R, \dots, R) < +\infty$. It is straightforward to see that $\mathcal{A}_R$ is an involutive Banach algebra with the norm $\|F\|_R := M(F; R, \dots, R)$. Consider a Banach space $(\mathcal{A}_R)^N$ with the norm $\|(F_1, \dots, F_N)\|_R := \max_i \|F_i\|_R$. Letting

$$\mathcal{D}^N := \{(F_i) \in (\mathcal{A}_R)^N : \|(F_i)\|_R \le R + \varepsilon,\ F_i = F_i^*,\ 1 \le i \le N\},$$

one can define $\Psi_\rho : \mathcal{D}^N \to (\mathcal{A}_R)^N$ for $\rho \in \mathbb{R}$ by

$$\Psi_\rho(F_1, \dots, F_N) := \left(\tfrac{1}{2}(X_i + F_i - \rho P_i(F_1, \dots, F_N))\right)_{1 \le i \le N},$$

because $P_i(F_1, \dots, F_N)^* = P_i(F_1, \dots, F_N)$. Then

$$\|\Psi_\rho(F_1, \dots, F_N)\|_R \le \tfrac{1}{2}(2R + \varepsilon + |\rho| \|(P_i)\|_{R+\varepsilon}) \quad \text{for} \quad (F_i) \in \mathcal{D}^N, \tag{6.3.6}$$

and since $\|F_{i_1} \cdots F_{i_k} - F'_{i_1} \cdots F'_{i_k}\|_R \le k(R+\varepsilon)^{k-1} \|(F_i - F'_i)\|_R$, there exists $K > 0$ such that

$$\|\Psi_\rho(F_1, \dots, F_n) - \Psi_\rho(F'_1, \dots, F'_n)\|_R \le (\tfrac{1}{2} + K|\rho|) \|(F_i - F'_i)\|_R \tag{6.3.7}$$

for $(F_i),(F_i') \in \mathcal{D}^N$. Now assume $|\rho| < 1/(\varepsilon^{-1}\|(P_i)\|_{R+\varepsilon} + 2K)$. Then it follows from (6.3.6) and (6.3.7) that Ψ_ρ is a contraction (with Lipschitz constant < 1) of $\mathcal{D}^N$ into itself. Hence, by Banach's contraction lemma, it has a fixed point $(G_{\rho,i}) \in \mathcal{D}^N$ so that $G_{\rho,i} + \rho P_i(G_{\rho,1},\dots,G_{\rho,n}) = X_i$, i.e.

(i′) $F_{\rho,i}(G_{\rho,1}(X_1,\dots,X_n),\dots,G_{\rho,n}(X_1,\dots,X_n)) = X_i \quad (1 \le i \le N)$.

Furthermore, since

$$M(F_{\rho,i};\|a_1\|+\varepsilon,\dots,\|a_N\|+\varepsilon) \le \|F_{\rho,i}\|_{R-\varepsilon} \le R-\varepsilon+|\rho|\|P_i\|_{R-\varepsilon} \le R,$$

the condition (ii) is satisfied. Finally, (i) can be obtained from (i′) as in the last part of the proof of Theorem 6.3.1.

□

It is quite natural to expect that the free entropy $\chi(a_1,\dots,a_N)$ is $-\infty$ whenever some algebraic relation is satisfied among $a_1,\dots,a_N$. Although it seems difficult to show this directly from the definition, one can benefit from the above proposition. We illustrate it in a simple example.

Example 6.3.4 Let $a_1,a_2 \in \mathcal{M}^{sa}$. If $a_1a_2 = a_2a_1$, then $\chi(a_1,a_2) = -\infty$.

In order to apply Proposition 6.3.3 we choose $P_1(X_1,X_2) := \mathrm{i}\,(X_1X_2 - X_2X_1)$ and $P_2(X_1,X_2) := 0$. Since $F_{\rho,1}(a_1,a_2) = a_1 + \rho P_1(a_1,a_2) = a_1$ and $F_{\rho,2}(a_1,a_2) = a_2$, we have

$$\chi(a_1,a_2) = \chi(a_1,a_2) + \log|\mathcal{J}|(F_\rho)(a_1,a_2)\,. \tag{6.3.8}$$

We compute

$$DF_\rho(a_1,a_2) = \begin{bmatrix} \mathbf{1}\otimes\mathbf{1} + \rho\mathrm{i}\,(\mathbf{1}\otimes a_2^o - a_2\otimes\mathbf{1}) & \rho\mathrm{i}\,(a_1\otimes\mathbf{1} - \mathbf{1}\otimes a_1^o) \\ 0 & \mathbf{1}\otimes\mathbf{1} \end{bmatrix}$$

and

$$DF_\rho(a_1,a_2)DF_\rho(a_1,a_2)^* = \mathbf{1} + \rho x + \rho^2 y\,,$$

where

$$x := \begin{bmatrix} 0 & \mathrm{i}\,(a_1\otimes\mathbf{1} - \mathbf{1}\otimes a_1^o) \\ -\mathrm{i}\,(a_1\otimes\mathbf{1} - \mathbf{1}\otimes a_1^o) & 0 \end{bmatrix},$$

$$y := \begin{bmatrix} (a_1\otimes\mathbf{1} - \mathbf{1}\otimes a_1^o)^2 + (\mathbf{1}\otimes a_2^o - a_2\otimes\mathbf{1})^2 & 0 \\ 0 & 0 \end{bmatrix}.$$

Now we perform a direct computation by means of the Taylor series of the logarithm:

$$\begin{aligned}\log|\mathcal{J}|(F_\rho)(a_1,a_2) &= \frac{1}{2}(\mathrm{Tr}_2\otimes\tau\otimes\tau)(\log(DF_\rho(a_1,a_2)DF_\rho(a_1,a_2)^*)) \\ &= \frac{1}{2}\sum_{m=1}^{\infty}\frac{(-1)^{m-1}}{m}(\mathrm{Tr}_2\otimes\tau\otimes\tau)((\rho x+\rho^2 y)^m) \\ &= \frac{1}{2}\rho^2(\tau\otimes\tau)((\mathbf{1}\otimes a_2^o - a_2\otimes\mathbf{1})^2)+o(\rho^2) \\ &= \rho^2(\tau(a_2^2)-\tau(a_2)^2)+o(\rho^2)\,.\end{aligned}$$

If $\tau(a_2^2) > \tau(a_2)^2$, then we get $\log|\mathcal{J}|(F_\rho)(a_1,a_2) > 0$ for small ρ. This forces $\chi(a_1,a_2) = -\infty$, thanks to (6.3.8). Otherwise, a_2 must be a scalar and $\chi(a_1,a_2) = -\infty$.

□

From the subadditivity and the previous example, we know that $\chi(a_1,\dots,a_N) = -\infty$ if there are two commuting elements among $a_1,\dots,a_N$. The above argument should be available in principle when some algebraic relation is satisfied by the elements $a_1,\dots,a_N$. For instance, in the case $a_1a_2a_3 + a_3a_2a_1 = 0$, it is enough to apply Proposition 6.3.3 to the polynomials $P_2(X_1,X_2,X_3) = X_1X_2X_3 + X_3X_2X_1$ and $P_1 = P_3 = 0$, and the argument of the example works.

Another kind of transformation formula will be also useful in the next section. It is restricted to separate change of each variable, but a larger class of continuous functions are available. We first give a lemma which provides a smoothing technique for the free entropy under the functional calculus.

Lemma 6.3.5 *Let $a \in \mathcal{M}^{sa}$ with $\chi(a) > -\infty$, and let f be a continuous and increasing function on $\mathbb{R}$. Then there exists a sequence (f_m) of C^∞-functions on $\mathbb{R}$ such that $f_m' > 0$ on $\mathbb{R}$, $\|f_m(a) - f(a)\| \to 0$ and $\chi(f_m(a)) \to \chi(f(a))$.*

Proof: Let μ be the distribution of a. Since $\Sigma(\mu) > -\infty$, there are $0 < \delta(m) < 1/m$ ($m \in \mathbb{N}$) such that

$$\iint_{|s-t|<\delta(m)} \log|s-t|\,d\mu(s)\,d\mu(t) \ge -\frac{1}{m},$$

$$\iint_{|s-t|<\delta(m)} d\mu(s)\,d\mu(t) \le \frac{1}{m\log m}.$$

For each m choose a C^∞-function $\phi_m \ge 0$ supported in a neighborhood of 0 with $\int \phi_m\,dt = 1$ such that

$$|(f*\phi_m)(t) - f(t)| \le \frac{\delta(m)}{2m} \quad \text{for} \quad t \in \operatorname{supp}\mu.$$

Define C^∞-functions f_m by

$$f_m(t) := \frac{t}{m} + (f * \phi_m)(t) \qquad (t \in \mathbb{R}).$$

Then it is obvious that $f'_m(t) \geq 1/m$ and $\|f_m(a) - f(a)\| \to 0$. If $s, t \in \operatorname{supp}\mu$ and $|s-t| \geq \delta(m)$, then, assuming $s > t$, we estimate

$$\begin{aligned}
&f_m(s) - f_m(t) \\
&\quad = \frac{s-t}{m} + (f * \phi_m)(s) - (f * \phi_m)(t) \\
&\quad \geq \frac{\delta(m)}{m} + f(s) - f(t) - |(f * \phi_m)(s) - f(s)| - |(f * \phi_m)(t) - f(t)| \\
&\quad \geq f(s) - f(t)\,.
\end{aligned}$$

Hence we have

$$\begin{aligned}
\Sigma(f_m(a)) &= \iint \log|f_m(s) - f_m(t)|\,d\mu(s)\,d\mu(t) \\
&\geq \iint_{|s-t|\geq\delta(m)} \log|f(s) - f(t)|\,d\mu(s)\,d\mu(t) \\
&\qquad + \iint_{|s-t|<\delta(m)} \log\frac{|s-t|}{m}\,d\mu(s)\,d\mu(t) \\
&\geq \iint \log|f(s) - f(t)|\,d\mu(s)\,d\mu(t) \\
&\qquad + \iint_{|s-t|<\delta(m)} \log|s-t|\,d\mu(s)\,d\mu(t) \\
&\qquad - \log m \iint_{|s-t|<\delta(m)} d\mu(s)\,d\mu(t) \\
&\geq \Sigma(f(a)) - \frac{2}{m}
\end{aligned}$$

whenever m is large, so that $|f(s) - f(t)| < 1$ if $s, t \in \operatorname{supp}\mu$ and $|s-t| < \delta(m)$. Therefore,

$$\Sigma(f(a)) \leq \liminf_{m\to\infty} \Sigma(f_m(a))\,.$$

This together with Proposition 5.3.2 shows that $\Sigma(f_m(a)) \to \Sigma(f(a))$; that is, $\chi(f_m(a)) \to \chi(f(a))$.

□

Proposition 6.3.6 *Let $a_1, \dots, a_N \in \mathcal{M}^{sa}$ with $\chi(a_i) > -\infty$, and let $f_1, \dots, f_N$ be continuous increasing functions on $\mathbb{R}$. Then*

$$\chi(f_1(a_1), \dots, f_N(a_N)) \geq \chi(a_1, \dots, a_N) + \sum_{i=1}^N \bigl[\chi(f_i(a_i)) - \chi(a_i)\bigr]\,.$$

Moreover, equality holds here if $f_1, \dots, f_N$ are strictly increasing,

Proof: If $\chi(f_i(a_i)) = -\infty$ for some i, then both sides of the required inequality are $-\infty$ thanks to Proposition 6.1.1. Hence we may assume that $\chi(f_i(a_i)) > -\infty$ for all i. Now it suffices to show that

$$\chi(f_1(a_1), a_2, \dots, a_N) \geq \chi(a_1, \dots, a_N) + \chi(f_1(a_1)) - \chi(a_1). \tag{6.3.9}$$

In fact, repeated use of this inequality yields the conclusion. To show this, by Lemma 6.3.5 and Proposition 6.1.5 we may assume that f_1 is a C^∞-function f on $\mathbb{R}$ such that $f' > 0$ on $\mathbb{R}$. Put

$$K(s,t) := \begin{cases} \dfrac{f(s) - f(t)}{s - t} & \text{if } s \neq t, \\ f'(t) & \text{if } s = t, \end{cases}$$

and $L(s,t) := \log K(s,t)$, which is continuous on $\mathbb{R}^2$. Then, thanks to (6.1.2),

$$\begin{aligned} \chi(f(a_1)) - \chi(a_1) &= \iint \log\Bigl(\frac{f(s) - f(t)}{s - t}\Bigr)\, d\mu_1(s)\, d\mu_1(t) \\ &= (\tau \otimes \tau)(L(a_1 \otimes \mathbf{1}, \mathbf{1} \otimes a_1)), \end{aligned} \tag{6.3.10}$$

where μ_1 is the distribution of a_1 and the latter is defined in $(\mathcal{M} \otimes \mathcal{M}, \tau \otimes \tau)$.

Write $F(A_1, \dots, A_N) := (f(A_1), A_2, \dots, A_N)$ on $(M_n(\mathbb{C})^{sa})^N$. Let $R_1 \geq R > \max_i \|a_i\|$, for which $f([-R, R]) \subset [-R_1, R_1]$. For each $r \in \mathbb{N}$ and $\varepsilon > 0$, by a polynomial approximation of f on $[-R, R]$, one can choose $r_1 \in \mathbb{N}$ and $\varepsilon_1 > 0$ such that

$$\begin{aligned} &F(\Gamma_R(a_1, \dots, a_N; n, r_1, \varepsilon_1)) \\ &\qquad \subset \Gamma_{R_1}(f(a_1), a_2, \dots, a_N; n, r, \varepsilon) \qquad (n \in \mathbb{N}). \end{aligned} \tag{6.3.11}$$

The Jacobian of $F(A_1, \dots, A_N)$ depends on A_1 only, and as in the proof of Proposition 6.1.4 we have

$$\frac{d(\Lambda \circ f)}{d\Lambda}(A_1) = \prod_{i<j} \Bigl(\frac{f(t_i) - f(t_j)}{t_i - t_j}\Bigr)^2 \prod_{i=1}^n f'(t_i)$$

under the coordinate change $M_n(\mathbb{C})^{sa} \leftrightarrow \mathcal{U}(n)/T \times \mathbb{R}^n_{\leq}$. Since the eigenvalues of $K(A_1 \otimes I, I \otimes A_1)$ are $K(t_i, t_j)$ $(1 \leq i, j \leq n)$, this means that

$$\begin{aligned} \frac{d(\Lambda \circ f)}{d\Lambda}(A_1) &= \det K(A_1 \otimes I, I \otimes A_1) \\ &= \exp(\mathrm{Tr}_n \otimes \mathrm{Tr}_n)(L(A_1 \otimes I, I \otimes A_1)). \end{aligned} \tag{6.3.12}$$

For any $\delta > 0$ we choose a real polynomial $P(s,t)$ of two variables such that $|L(s,t) - P(s,t)| < \delta$ on $[-R,R]^2$. Then by (6.3.10) and (6.3.12), we have

$$\begin{aligned}
&\left|\frac{1}{n^2}\log\frac{d(\Lambda\circ f)}{d\Lambda}(A_1) - \big[\chi(f(a_1)) - \chi(a_1)\big]\right| \\
&\quad = |(\mathrm{tr}_n\otimes\mathrm{tr}_n)(L(A_1\otimes I, I\otimes A_1)) - (\tau\otimes\tau)(L(a_1\otimes\mathbf{1},\mathbf{1}\otimes a_1))| \\
&\quad \le 2\delta + |(\mathrm{tr}_n\otimes\mathrm{tr}_n)(P(A_1\otimes I, I\otimes A_1)) - (\tau\otimes\tau)(P(a_1\otimes\mathbf{1},\mathbf{1}\otimes a_1))| \\
&\quad \le 3\delta
\end{aligned}$$

for all $(A_1,\dots,A_N) \in \Gamma_R(a_1,\dots,a_N;n,r_1,\varepsilon_1)$, $n\in\mathbb{N}$, whenever r_1 is large enough and ε_1 is small enough. This and (6.3.11) imply that

$$\begin{aligned}
&\chi_{R_1}(f(a_1),a_2,\dots,a_N;r,\varepsilon) \\
&\quad \ge \chi_R(a_1,\dots,a_N;r_1,\varepsilon_1) - 3\delta + \chi(f(a_1)) - \chi(a_1)\,,
\end{aligned}$$

which gives (6.3.9) by Proposition 6.1.4.

Finally, when $f_1,\dots,f_N$ are strictly increasing, we can apply the already known inequality to the inverse functions, so the reverse inequality is obtained.

□

6.4 Additivity of free entropy

The subadditivity of free entropy, $\chi(a_1,\dots,a_N) \le \chi(a_1) + \dots + \chi(a_N)$, holds for any N-tuple of selfadjoint random variables. A crucial problem is to characterize when equality holds here. This problem was settled by Voiculescu: The free relation of noncommutative selfadjoint random variables is equivalent to the additivity of their free entropy.

The highlight of the section is the following *additivity theorem.*

Theorem 6.4.1 *Let $a_1,\dots,a_N \in \mathcal{M}^{sa}$. If $a_1,\dots,a_N$ are in free relation, then*

$$\chi(a_1,\dots,a_N) = \chi(a_1) + \dots + \chi(a_N)\,.$$

Conversely, if $\chi(a_i) > -\infty$ for $1 \le i \le N$ and the above equality holds, then $a_1,\dots,a_N$ are in free relation.

The proof will be divided into two parts. First, we prepare an approximate freeness property of Haar distributed unitary matrices, which will be a key technique in proving that the free relation implies additivity.

The notion of *approximate freeness* for matrices is introduced as follows. Let $(M_n(\mathbb{C})^{\star N}, \mathrm{tr}_n^{\star N})$ be the free product of N copies of $(M_n(\mathbb{C}), \mathrm{tr}_n)$, and let j_i be the injection of $M_n(\mathbb{C})$ into its ith copy in $M_n(\mathbb{C})^{\star N}$. When $\Omega_i \subset M_n(\mathbb{C})$ $(1 \le i \le N)$, $r \in \mathbb{N}$ and $\varepsilon > 0$ are given, the subsets $\Omega_1, \dots, \Omega_N$ are said to be (r, ε)-*free* if

$$|\mathrm{tr}_n(A_1 A_2 \dots A_k) - \mathrm{tr}_n^{\star N}(\tilde{A}_1 \tilde{A}_2 \cdots \tilde{A}_k)| \le \varepsilon$$

for all $A_1, A_2 \dots, A_k \in \bigsqcup_{i=1}^N \Omega_i$, $1 \le k \le r$, where $\tilde{A} := j_i(A)$ for $A \in \Omega_i$.

Lemma 6.4.2 *Let $r \in \mathbb{N}$, $\varepsilon > 0$, $R > 0$ and $\theta > 0$ be given. Then there exists $n_0 \in \mathbb{N}$ such that*

$$\begin{aligned} \gamma_n^{\otimes N}\big(\{(U_1, \dots, U_N) \in \mathcal{U}(n)^N : \{A_1^{(0)}, \dots, A_{k_0}^{(0)}\}, \{U_1 A_1^{(1)} U_1^*, \dots, U_1 A_{k_1}^{(1)} U_1^*\}, \\ \dots, \{U_N A_1^{(N)} U_N^*, \dots, U_N A_{k_N}^{(N)} U_N^*\} \textit{ are } (r, \varepsilon)\textit{-free}\}\big) \ge 1 - \theta \end{aligned}$$

for all $n \ge n_0$ and for any choice of $1 \le N \le r$, $1 \le k_0, k_1, \dots, k_N \le r$ and $A_j^{(i)} \in M_n(\mathbb{C})$ with $\|A_j^{(i)}\| \le R$ $(0 \le i \le N,\ 1 \le j \le k_i)$.

Proof: We may fix N and $k_0, k_1, \dots, k_N$. When $(U_1, \dots, U_N) \in \mathcal{U}(n)^N$ and $A_j^{(i)} \in M_n(\mathbb{C})$ are given as above, let $\Omega_i := \{U_i A_1^{(i)} U_i^*, \dots, U_i A_{k_i}^{(i)} U_i^*\}$ for $0 \le i \le N$, with $U_0 := I_n$. Moreover, let $B^{(1)}, \dots, B^{(p)}$ be an enumeration of

$$T_1 T_2 \cdots T_s - \mathrm{tr}_n(T_1 T_2 \cdots T_s) I_n\,,$$

where $1 \le s \le r$ and $T_1, \dots, T_s$ are from $\{A_1^{(i)}, \dots, A_{k_i}^{(i)}\}$ for some $0 \le i \le N$. Then $\mathrm{tr}_n(B^{(t)}) = 0$ and $\|B^{(t)}\| \le 2(1+R)^r$ for $1 \le t \le p$. It can be easily seen that every joint moment $\mathrm{tr}_n(A_1 A_2 \cdots A_k)$, where $1 \le k \le r$ and $A_1, \dots, A_k \in \bigsqcup_{i=0}^N \Omega_i$, is a polynomial of $\mathrm{tr}_n(T_1 T_2 \cdots T_s)$ as above and

$$\mathrm{tr}_n(U_{i_1}^{m_1} B_1 U_{i_2}^{m_2} B_2 \cdots U_{i_l}^{m_l} B_l)\,, \tag{6.4.1}$$

where $l \le 2r$, $1 \le i_1, \dots, i_l \le N$, $m_1, \dots, m_l \in \mathbb{Z} \setminus \{0\}$ and for $1 \le j \le l$ either $B_j \in \{B^{(1)}, \dots, B^{(p)}\}$ or $B_j = I_n$ and $i_j \ne i_{j+1}$ (with $i_{l+1} = i_1$). Here note that the polynomial is determined only by the pattern of $(A_1, \dots, A_k)$ from $\bigsqcup_{i=0}^N \Omega_i$, independently of n and the choice of U_i and $A_j^{(i)}$. According to Lemma 4.3.2, the term (6.4.1) converges to 0 as $n \to \infty$ in L^2-mean and hence in probability with respect to $\gamma_n^{\otimes N}$, and the convergence is uniform for the choice of $A_j^{(i)}$ with $\|A_j^{(i)}\| \le R$. Since $\mathrm{tr}_n^{\star N}(\tilde{A}_1 \tilde{A}_2 \cdots \tilde{A}_k)$ is given by putting 0 into all terms (6.4.1) in the polynomial, we observe that

$$|\mathrm{tr}_n(A_1 A_2 \cdots A_k) - \mathrm{tr}_n^{\star N}(\tilde{A}_1 \tilde{A}_2 \cdots \tilde{A}_k)| \to 0 \quad \text{as} \quad n \to \infty$$

in probability, and the convergence is uniform as above. Thus the conclusion follows because a finite number of $(A_1, \dots, A_k)$ are involved.

□

Lemma 6.4.3 *Let $a_1, \dots, a_N \in \mathcal{M}^{sa}$ and $1 \le L < N$. Assume that $\{a_1, \dots, a_L\}$ and $\{a_{L+1}, \dots, a_N\}$ are free, $\chi(a_1, \dots, a_L) > -\infty$ and $\chi(a_{L+1}, \dots, a_N) > -\infty$. Then, for every $r \in \mathbb{N}$, $\varepsilon > 0$ and $R > \max_i \|a_i\|$, there exists $\varepsilon_1 > 0$ such that*

$$\lim_{n\to\infty} \frac{\Lambda(\Xi_n(r, \varepsilon_1) \cap \Theta_n(r, \varepsilon))}{\Lambda(\Xi_n(r, \varepsilon_1))} = 1 ,$$

where

$$\begin{aligned} \Xi_n(r, \varepsilon_1) &:= \Gamma_R(a_1, \dots, a_L; n, r, \varepsilon_1) \times \Gamma_R(a_{L+1}, \dots, a_N; n, r, \varepsilon_1) , \\ \Theta_n(r, \varepsilon) &:= \Gamma_R(a_1, \dots, a_N; n, r, \varepsilon) . \end{aligned}$$

Proof: From the above definition of approximate freeness as well as the freeness of $\{a_1, \dots, a_L\}$ and $\{a_{L+1}, \dots, a_N\}$, for given r, ε and R we can choose $\varepsilon_1 > 0$ such that if $(A_1, \dots, A_L; A_{L+1}, \dots, A_N) \in \Xi_n(r, \varepsilon_1)$ and if $\{A_1, \dots, A_L\}$ and $\{A_{L+1}, \dots, A_N\}$ are (r, ε_1)-free, then $(A_1, \dots, A_L; A_{L+1}, \dots, A_N) \in \Theta_n(r, \varepsilon)$. For every $\theta > 0$, by the previous lemma there is an $n_0 \in \mathbb{N}$ such that

$$\begin{aligned} &\gamma_n(\{U \in \mathcal{U}(n) : \\ &\qquad \{A_1, \dots, A_L\} \text{ and } \{UA_{L+1}U^*, \dots, UA_NU^*\} \text{ are } (r, \varepsilon_1)\text{-free}\}) \ge 1 - \theta \end{aligned} \tag{6.4.2}$$

for all $n \ge n_0$ and for all $A_i \in M_n(\mathbb{C})^{sa}$ with $\|A_i\| \le R$ $(1 \le i \le N)$. According to the assumption of finite free entropies and Proposition 6.1.4, the Λ-measure of $\Xi_n(r, \varepsilon_1)$ is positive whenever n is large. For such n the probability measure σ_n on $\Xi_n(r, \varepsilon_1)$ can be defined by normalizing the restriction of Λ on $\Xi_n(r, \varepsilon_1)$. Then, since σ_n as well as the set $\Xi_n(r, \varepsilon_1)$ is invariant under the action of $\mathcal{U}(n)$ given by $(A_1, \dots, A_L; A_{L+1}, \dots, A_N) \mapsto (A_1, \dots, A_L; UA_{L+1}U^*, \dots, UA_NU^*)$ for $U \in \mathcal{U}(n)$, we have

$$\begin{aligned} &\frac{\Lambda(\Xi_n(r, \varepsilon_1) \cap \Theta_n(r, \varepsilon))}{\Lambda(\Xi_n(r, \varepsilon_1))} \\ &\qquad = \int_{\Xi_n(r,\varepsilon_1)} \left(\int_{\mathcal{U}(n)} \psi(A_1, \dots, A_L; UA_{L+1}U^*, \dots, UA_NU^*) \, d\gamma_n(U) \right) d\sigma_n , \end{aligned}$$

where ψ is the characteristic function of $\Xi_n(r, \varepsilon_1) \cap \Theta_n(r, \varepsilon)$. The choice of ε_1 and (6.4.2) show that

$$\int_{\mathcal{U}(n)} \psi(A_1, \dots, A_L; UA_{L+1}U^*, \dots, UA_NU^*) \, d\gamma_n(U) \ge 1 - \theta$$

for all $(A_1, \dots, A_L; A_{L+1}, \dots, A_N) \in \Xi_n(r, \varepsilon_1)$. Therefore, we infer that

$$\frac{\Lambda(\Xi_n(r, \varepsilon_1) \cap \Theta_n(r, \varepsilon))}{\Lambda(\Xi_n(r, \varepsilon_1))} \ge 1 - \theta$$

whenever n is large enough, and the conclusion is obtained.

□

Proof of Theorem 6.4.1 (*the first part*): We can prove a slightly stronger result, namely,

$$\chi(a_1, a_2, \dots, a_N) = \chi(a_1) + \chi(a_2, \dots, a_N) \tag{6.4.3}$$

when a_1 and $\{a_2, \dots, a_N\}$ are in free relation. By Proposition 6.1.3 it suffices to show that $\chi(a_1, a_2, \dots, a_N) \ge \chi(a_1) + \chi(a_2, \dots, a_N)$, so we may assume that $\chi(a_1) > -\infty$ and $\chi(a_2, \dots, a_N) > -\infty$. Hence for any $r \in \mathbb{N}$, $\varepsilon > 0$ and $R > \max_i \|a_i\|$, there is an $\varepsilon_1 > 0$ for which the conclusion of Lemma 6.4.3 is satisfied when applied to a_1 and $\{a_2, \dots, a_N\}$. Then, in the notation of Lemma 6.4.3,

$$\begin{aligned}
&\chi_R(a_1, \dots, a_N; r, \varepsilon) \\
&\quad = \limsup_{n\to\infty} \left[\frac{1}{n^2} \log \Lambda(\Theta_n(r, \varepsilon)) + \frac{N}{2} \log n\right] \\
&\quad \ge \limsup_{n\to\infty} \left[\frac{1}{n^2} \log \Lambda(\Xi_n(r, \varepsilon_1)) + \frac{N}{2} \log n\right] \\
&\quad = \limsup_{n\to\infty} \left[\frac{1}{n^2} \log(\Gamma_R(a_1; n, r, \varepsilon_1)) + \frac{1}{2} \log n\right. \\
&\qquad\qquad \left. + \frac{1}{n^2} \log \Lambda(\Gamma_R(a_2, \dots, a_N; n, r, \varepsilon_1) + \frac{N-1}{2} \log n\right] \\
&\quad = \chi_R(a_1; r, \varepsilon_1) + \chi_R(a_2, \dots, a_N; r, \varepsilon_1) \\
&\quad \ge \chi(a_1) + \chi(a_2, \dots, a_N)\,.
\end{aligned}$$

The above third equality is due to the existence of the limit in (6.1.1). Therefore, (6.4.3) is obtained.

□

The other part of the additivity theorem states that the additivity of free entropy implies the free relation. The strategy consists in proving this first for semicircular random variables. The discussion of the semicircular case is related to a maximization problem, more precisely, to the uniqueness of the maximizer.

Let $a_1, \dots, a_N \in \mathcal{M}^{sa}$ be noncommutative random variables with $\tau(a_i^2) = C_i$ $(1 \le i \le N)$. Proposition 6.1.1 tells us that

$$\chi(a_1, \dots, a_N) \le \sum_{i=1}^N \chi(a_i) \le \frac{1}{2} \sum_{i=1}^N \log(2\pi e C_i) \le \frac{N}{2} \log\left(\frac{2\pi e}{N} \sum_{i=1}^N C_i\right).$$

Equality holds in the first inequality above when $a_1, \dots, a_N$ are in free relation. This is provided by the already proven part of the additivity theorem. The equality case in the second inequality is charaterized thanks to Proposition 5.3.4; that is, equality takes place if and only if a_i has the semicircle distribution $w_{2\sqrt{C_i}}$ for each i. The equality in the third inequality is equivalent to $C_1 = \dots = C_N$. In this way, we infer that under the constraint $\tau(a_1^2+\cdots+a_N^2)\le C$ the free entropy $\chi(a_1,\dots,a_N)$ attains the maximum $\frac{N}{2}\log(2\pi eC/N)$ if $a_1,\dots,a_N$ are in free relation and the distribution of each a_i is w_r, $r=2\sqrt{C/N}$. The content of the next theorem is the uniqueness of the maximizer.

Theorem 6.4.4 *If $a_1,\dots,a_N\in\mathcal{M}^{sa}$ satisfy $\tau(a_1^2+\cdots+a_N^2)\le C$, then the free entropy $\chi(a_1,\dots,a_N)$ attains the maximum $\frac{N}{2}\log(2\pi eC/N)$ if and only if $a_1,\dots,a_N$ are in free relation and the distribution of each a_i is w_r, $r=2\sqrt{C/N}$.*

Proof: By the above argument it suffices to show that the joint distribution of $(a_1,\dots,a_N)$ for which the maximal value is attained is uniquely determined. We may assume that $C=N$ and the distribution of each a_i is w_2 (this is necessary to attain the maximum). In particular, $\tau(a_i)=0$ and $\tau(a_i^2)=1$.

We are going to show that any local maximizer $(a_1,\dots,a_N)$ satisfying the above conditions must be free. We perturb a_i by a noncommutative polynomial $P=P^*$. According to Proposition 6.3.3,

$$\begin{aligned}&\chi(a_1,\dots,a_{i-1},a_i+\rho P(a_1,\dots,a_N),a_{i+1},\dots,a_N)\\&\quad=\chi(a_1,\dots,a_N)+\log|\mathcal{J}|(F_\rho)(a_1,\dots,a_N)\end{aligned}$$

for all $\rho\in\mathbb{R}$ near 0. Since

$$b_i(\rho):=\tau\big((a_i+\rho P(a_1,\dots,a_N))^2\big)^{-1/2}(a_i+\rho P(a_1,\dots,a_N))$$

belongs to $\mathcal{M}^{sa}$ and $\tau(b_i(\rho)^2)=1$, we have

$$\begin{aligned}\chi(a_1,\dots,a_N)&\ge\chi(a_1,\dots,a_{i-1},b_i(\rho),a_{i+1},\dots,a_N)\\&=\chi(a_1,\dots,a_{i-1},a_i+\rho P(a_1,\dots,a_N),a_{i+1},\dots,a_N)\\&\qquad-\frac{1}{2}\log\tau\big((a_i+\rho P(a_1,\dots,a_N))^2\big)\\&=\chi(a_1,\dots,a_N)+\log|\mathcal{J}|(F_\rho)(a_1,\dots,a_N)\\&\qquad-\frac{1}{2}\log\tau\big((a_i+\rho P(a_1,\dots,a_N))^2\big)\end{aligned}$$

by assumption and Corollary 6.3.2. Therefore,

$$\log|\mathcal{J}|(F_\rho)(a_1,\dots,a_N)-\frac{1}{2}\log\tau((a_i+\rho P(a_1,\dots,a_N))^2)\le 0 \tag{6.4.4}$$

when $|\rho|$ is small. Furthermore, the following is immediate to check (cf. Example 6.3.4):

$$\log|\mathcal{J}|(F_\rho)(a_1,\dots,a_N)=\frac{1}{2}(\mathrm{Tr}_N\otimes\tau\otimes\tau)(\log(\mathbf{1}+\rho x+\rho^2y)),$$

where x and y are selfadjoint elements of $M_N(\mathbb{C}) \otimes (\mathcal{M} \otimes \mathcal{M}^{op})$ and

$$(\tau \otimes \tau)(x) = 2(\tau \otimes \tau)(D_i P(a_1, \dots, a_N)).$$

Therefore, the Taylor expansion of the logarithm yields

$$\log |\mathcal{J}|(F_\rho)(a_1, \dots, a_N) = \rho(\tau \otimes \tau)(D_i P(a_1, \dots, a_N)) + O(\rho^2),$$

so that

$$\frac{d}{d\rho}\bigg|_{\rho=0} \log |\mathcal{J}|(F_\rho)(a_1, \dots, a_N) = (\tau \otimes \tau)(D_i P(a_1, \dots, a_N)).$$

On the other hand, since $\tau(a_i^2) = 1$, it is obvious that

$$\frac{d}{d\rho}\bigg|_{\rho=0} \frac{1}{2} \log \tau\big((a_i + \rho P(a_1, \dots, a_N))^2\big) = \tau(a_i P(a_1, \dots, a_N)).$$

By (6.4.4) these imply that

$$\tau(a_i P(a_1, \dots, a_N)) = (\tau \otimes \tau)(D_i P(a_1, \dots, a_N)). \tag{6.4.5}$$

Since both sides of (6.4.5) are linear in P, this relation must hold for any polynomial, not only for the selfadjoint ones.

Let $1 \le i_1, \dots, i_k \le N$ and choose $i = i_1$ and $P(X_1, \dots, X_N) = X_{i_2} \cdots X_{i_k}$. Then from (6.4.5) we have

$$\tau(a_{i_1} a_{i_2} \cdots a_{i_k}) = \sum_{s : i_s = i_1} \tau(a_{i_2} \cdots a_{i_{s-1}}) \tau(a_{i_{s+1}} \cdots a_{i_k}).$$

This recursion formula uniquely determines the joint distribution of $(a_1, \dots, a_N)$, and it gives the conclusion.

□

We are now in a position to complete the proof of the additivity theorem. For standard semicircular variables the full statement is already shown; additivity implying free relation is contained in the previous theorem. The strategy is to treat the general random variables as functions of semicircular ones.

Proof of Theorem 6.4.1 (*the second part*): For each i, since the distribution of a_i is nonatomic by the assumption $\chi(a_i) > -\infty$, one can choose a continuous and increasing function $f_i : \mathbb{R} \to \mathbb{R}$ such that the distribution of $f_i(a_i)$ is the standard semicircle law w_2. Then, by Proposition 6.3.6 and the additivity assumption, we have

$$\chi(f_1(a_1), \dots, f_N(a_N)) \ge \sum_{i=1}^N \chi(f_i(a_i)) = \frac{N}{2} \log(2\pi e),$$

so that $\chi(f_1(a_1),\dots,f_N(a_n))$ takes the maximal value $\frac{N}{2}\log(2\pi e)$. Hence Theorem 6.4.4 implies that $f_1(a_1),\dots,f_N(a_N)$ are in free relation, and so are $a_1,\dots,a_N$ because $a_i \in \{f_i(a_i)\}''$.

□

In the proof of the additivity theorem we obtained the equality (6.4.3) under the condition that a_1 and $\{a_2,\dots,a_N\}$ are in free relation. This is slightly stronger than the statement of the first part of Theorem 6.4.1, so we may call it the *strong additivity theorem.* The converse is not known in this case, and it is an interesting question whether the equality

$$\chi(a_1,\dots,a_L,a_{L+1},\dots,a_N) = \chi(a_1,\dots,a_L) + \chi(a_{L+1},\dots,a_N) \tag{6.4.6}$$

holds when $\{a_1,\dots,a_L\}$ and $\{a_{L+1},\dots,a_N\}$ are in free relation. The question is open, but one can see from the above proof that the block additivity (6.4.6) holds if the first block $(a_1,\dots,a_L)$ is *regular*, that is, replacing lim sup in the definition $\chi(a_1,\dots,a_L)$ by lim inf yields the same quantity.

6.5 Free entropies of unitary and non-selfadjoint random variables

The distribution measure of a unitary random variable is supported on the unit circle, and the large deviation theorem for random unitaries suggests that the free entropy must be the double logarithmic integral of the distribution. Given an N-tuple of unitaries, one should follow a different approach: The N-tuple can be approximated in distribution by unitary matrices; the measure of the approximating unitaries can be taken, and one can imitate the definition of the free entropy of an N-tuple of selfadjoint random variables. The free entropy of an N-tuple of non-selfadjoint random variables is similar. Simple non-selfadjoint matrices are used in the matricial approximation.

Let $u_1,\dots,u_N \in \mathcal{M}$ be unitaries. For $n,r \in \mathbb{N}$ and $\varepsilon > 0$ define

$$\begin{aligned}\Gamma_u(u_1,\dots,u_N;n,r,\varepsilon) &:= \big\{(U_1,\dots,U_N) \in \mathcal{U}(n)^N : \\ &|\mathrm{tr}_n(U'_{i_1}\cdots U'_{i_k}) - \tau(u'_{i_1}\cdots u'_{i_k})| \le \varepsilon \text{ for all } 1 \le i_1,\dots,i_k \le 2N,\ 1 \le k \le r\big\},\end{aligned}$$

where

$$\begin{aligned}(U'_1,\dots,U'_{2N}) &:= (U_1,\dots,U_N,U_1^*,\dots,U_N^*),\\ (u'_1,\dots,u'_{2N}) &:= (u_1,\dots,u_N,u_1^*,\dots,u_N^*).\end{aligned}$$

The *free entropy* of the N-tuple $(u_1,\dots,u_N)$ of unitaries is defined as follows:

$$\chi_u(u_1,\dots,u_N;r,\varepsilon) := \limsup_{n\to\infty} \frac{1}{n^2}\log\gamma(\Gamma_u(u_1,\dots,u_N;n,r,\varepsilon))$$

(here γ is the Haar probability measure on $\mathcal{U}(n)^N$, i.e. $\gamma = \gamma_n^{\otimes N}$),

$$\chi_u(u_1,\ldots,u_N) := \lim_{\substack{r\to\infty\\ \varepsilon\to+0}} \chi_u(u_1,\ldots,u_N;r,\varepsilon).$$

For the case of a single unitary $u \in \mathcal{M}$ we have

$$\Gamma_u(u;n,r,\varepsilon) = \bigl\{U \in \mathcal{U}(n) : |\mathrm{tr}_n(U^k) - m_k(\mu)| \le \varepsilon,\ -r \le k \le r\bigr\},$$
$$\gamma_n(\Gamma_u(u;n,r,\varepsilon)) = \bar{\gamma}_n\bigl(\{\zeta \in \mathbb{T}^n : \kappa_n(\zeta) \in F(r,\varepsilon)\}\bigr),$$

where μ is the distribution on $\mathbb{T}$ of u, $\bar{\gamma}_n$ is the measure on $\mathbb{T}^n$ induced by γ_n (having the density given in Lemma 4.2.1), and

$$F(r,\varepsilon) := \bigl\{\nu \in \mathcal{M}(\mathbb{T}) : |m_k(\nu) - m_k(\mu)| \le \varepsilon,\ -r \le k \le r\bigr\}.$$

To show that the one variable case reduces to the double logarithmic integral, one can proceed as in the proof of Theorem 5.6.2. The large deviation result in Theorem 5.4.10 with $Q(\zeta) = 0$ (then $B = 0$) implies that

$$\begin{aligned}
&\limsup_{n\to\infty} \frac{1}{n^2} \log \bar{\gamma}_n\bigl(\{\zeta \in \mathbb{T}^n : \kappa_n(\zeta) \in F(r,\varepsilon)\}\bigr)\\
&\quad \le \sup\{\Sigma(\nu) : \nu \in F(r,\varepsilon)\} = \sup\{\Sigma(\nu) : \nu \in G(r,\varepsilon)\}\\
&\quad \le \liminf_{n\to\infty} \frac{1}{n^2} \log \bar{\gamma}_n\bigl(\{\zeta \in \mathbb{T}^n : \kappa_n(\zeta) \in G(r,\varepsilon)\}\bigr),
\end{aligned}$$

where $G(r,\varepsilon)$ is gotten by replacing $\le \varepsilon$ by $< \varepsilon$ in the definition of $F(r,\varepsilon)$. Therefore,

$$\begin{aligned}
\chi_u(u;r,\varepsilon) &= \lim_{n\to\infty} \frac{1}{n^2} \log \gamma_n(\Gamma_u(u;n,r,\varepsilon)) \\
&= \sup\{\Sigma(\nu) : \nu \in F(r,\varepsilon)\}.
\end{aligned} \tag{6.5.1}$$

Since the latter tends to $\Sigma(\mu)$ as $r \to \infty$ and $\varepsilon \to +0$, we observe that

$$\chi_u(u) = \Sigma(\mu) = \iint_{\mathbb{T}^2} \log|\zeta - \eta|\, d\mu(\zeta)\, d\mu(\eta). \tag{6.5.2}$$

Several properties of $\chi_u(u_1,\ldots,u_N)$ are completely similar to those of the previous $\chi(a_1,\ldots,a_N)$. For example, one readily verifies

Proposition 6.5.1 *The following hold for unitaries $u_1,\ldots,u_N \in \mathcal{M}$.*

(1) *Negativity:* $\chi_u(u_1,\ldots,u_N) \le 0$.

(2) *Subadditivity: For every $1 \le L < N$,*

$$\chi_u(u_1,\ldots,u_N) \le \chi_u(u_1,\ldots,u_L) + \chi_u(u_{L+1},\ldots,u_N).$$

(3) *Upper semicontinuity: Let* $(u_1,\dots,u_N)$ *and* $(u_{m,1},\dots,u_{m,N})$ *be* N*-tuples of unitaries in* $\mathcal{M}$ *for* $m \in \mathbb{N}$*. If* $(u_{m,1},\dots,u_{m,N}) \to (u_1,\dots,u_N)$ *in distribution, then*

$$\chi_u(u_1,\dots,u_N) \ge \limsup_{m\to\infty} \chi_u(u_{m,1},\dots,u_{m,N}).$$

In particular, this is the case if $u_{m,i} \to u_i$ *weakly for* $1 \le i \le N$*.*

The next result is a unitary counterpart of Proposition 6.1.6.

Proposition 6.5.2 *Let* $u_1,\dots,u_N,v_1,\dots,v_N \in \mathcal{M}$ *be unitaries. If* $v_1 = u_1$ *and* $v_i u_i^* \in \{u_1,\dots,u_{i-1}\}''$ *for* $2 \le i \le N$*, then*

$$\chi_u(u_1,\dots,u_N) = \chi_u(v_1,\dots,v_N).$$

Proof: Since the assumption implies also that $u_i v_i^* \in \{v_1,\dots,v_{i-1}\}''$ for $2 \le i \le N$, it suffices to show that

$$\chi_u(u_1,\dots,u_N) \le \chi_u(v_1,\dots,v_N).$$

One can choose selfadjoint noncommutative polynomials $P_{m,i}(X_1,X_2,\dots,X_{2i-2})$ for $2 \le i \le N$, $m \in \mathbb{N}$ such that

$$\exp\Big(\mathrm{i}\,P_{m,i}\Big(\frac{u_1+u_1^*}{2},\frac{u_1-u_1^*}{2\mathrm{i}},\dots,\frac{u_{i-1}+u_{i-1}^*}{2},\frac{u_{i-1}-u_{i-1}^*}{2\mathrm{i}}\Big)\Big) \to v_i u_i^*$$

strongly* as $m \to \infty$. Set $v_{m,1} := v_1 = u_1$ and, for $2 \le i \le N$,

$$v_{m,i} := \exp\Big(\mathrm{i}\,P_{m,i}\Big(\frac{u_1+u_1^*}{2},\frac{u_1-u_1^*}{2\mathrm{i}},\dots,\frac{u_{i-1}+u_{i-1}^*}{2},\frac{u_{i-1}-u_{i-1}^*}{2\mathrm{i}}\Big)\Big)u_i.$$

Then $v_{m,i} \to v_i$ strongly* as $m \to \infty$. If a map $\Psi : \mathcal{U}(n)^N \to \mathcal{U}(n)^N$, $\Psi(U_1,\dots,U_N) = (V_1,\dots,V_N)$, is defined by $V_1 := U_1$ and, for $2 \le i \le N$,

$$V_i := \exp\Big(\mathrm{i}\,P_{m,i}\Big(\frac{U_1+U_1^*}{2},\frac{U_1-U_1^*}{2\mathrm{i}},\dots,\frac{U_{i-1}+U_{i-1}^*}{2},\frac{U_{i-1}-U_{i-1}^*}{2\mathrm{i}}\Big)\Big)U_i,$$

then it is obvious that $\gamma \circ \Psi = \gamma$ holds due to the multiplication invariance of γ. For any $m, r \in \mathbb{N}$ and $\varepsilon > 0$ one can easily see that there exist $r_1 \in \mathbb{N}$ and $\varepsilon_1 > 0$ such that

$$\Psi(\Gamma_u(u_1,\dots,u_N;n,r_1,\varepsilon_1)) \subset \Gamma_u(v_{m,1},\dots,v_{m,N};n,r,\varepsilon) \qquad (n \in \mathbb{N}).$$

This yields

$$\chi_u(u_1,\dots,u_N;r_1,\varepsilon_1) \le \chi_u(v_{m,1},\dots,v_{m,N};r,\varepsilon),$$

so that $\chi_u(u_1,\dots,u_N) \le \chi_u(v_{m,1},\dots,v_{m,N})$. Hence the required inequality follows as $m \to \infty$, thanks to the upper semicontinuity.

□

The additivity theorem for $\chi_u(u_1,\dots,u_n)$ is completely analogous to the selfadjoint case in Theorem 6.4.1.

Theorem 6.5.3 *Let $u_1,\dots,u_N \in \mathcal{M}$ be unitaries. If $u_1,\dots,u_N$ are in *-free relation (i.e. $\{u_1,u_1^*\},\dots,\{u_N,u_N^*\}$ are free), then*

$$\chi_u(u_1,\dots,u_N) = \chi_u(u_1) + \cdots + \chi_u(u_N).$$

*Conversely, if $\chi_u(u_i) > -\infty$ for $1 \le i \le N$ and the above equality holds, then $u_1,\dots,u_N$ are in *-free relation.*

The proof of the theorem will be given in Sec. 6.6.

Now let us turn to the free entropy of an N-tuple $(a_1,\dots,a_N)$ of non-selfadjoint random variables. One possibility is to use the Descartes decomposition $a_i = b_i + \mathrm{i}\, c_i$ and to pass to the $2N$-tuple $(b_1,b_2,\dots,b_N,c_1,c_2,\dots,c_N)$ of selfadjoint variables. The other possibility is to approximate the non-selfadjoint variables with non-selfadjoint matrices. We are first going to follow the latter idea, but it turns out soon that the two possible ways are actually equivalent.

The Lebesgue measure $\hat\Lambda_n$ on $M_n(\mathbb{C})$ given in (4.1.25) is induced by the natural isometry $M_n(\mathbb{C}) \cong \mathbb{R}^{2n^2}$. Consider the map $A \in M_n(\mathbb{C}) \mapsto (B,C) \in (M_n(\mathbb{C})^{sa})^2$ which is given by the Descartes decomposition $A = B + \mathrm{i}\,C$. Since

$$\|A\|_{HS}^2 = \mathrm{Tr}(A^*A) = \mathrm{Tr}(B^*B + C^*C) = \|B\|_{HS}^2 + \|C\|_{HS}^2$$

($\|\cdot\|_{HS}$ denotes the Hilbert-Schmidt norm), this map induces a linear isometry on $\mathbb{R}^{2n^2}$ via $M_n(\mathbb{C}) \cong \mathbb{R}^{2n^2} \cong (M_n(\mathbb{C})^{sa})^2$, so we have

Lemma 6.5.4 *Under the map $A \in M_n(\mathbb{C}) \mapsto (B,C) \in (M_n(\mathbb{C})^{sa})^2$ above, $\hat\Lambda_n$ on $M_n(\mathbb{C})$ corresponds to $\Lambda_n \otimes \Lambda_n$ on $(M_n(\mathbb{C})^{sa})^2$.*

Let $a_1,\dots,a_N$ be (non-selfadjoint) elements of $\mathcal{M}$. The definition of the *free entropy* of the N-tuple $(a_1,\dots,a_N)$ is a slight modification of the selfadjoint case. For $n,r \in \mathbb{N}$, $\varepsilon > 0$ and $R > 0$ we define

$$\begin{aligned}\hat\Gamma_R(a_1,\dots,a_N;n,r,\varepsilon) := \big\{&(A_1,\dots,A_N) \in M_n(\mathbb{C})^N : \|A_i\| \le R,\\ &|\mathrm{tr}_n(A'_{i_1}\cdots A'_{i_k}) - \tau(a'_{i_1}\cdots a'_{i_k})| \le \varepsilon \text{ for all } 1 \le i_1,\dots,i_k \le 2N,\ 1 \le k \le r\big\},\end{aligned}$$

where

$$\begin{aligned}(A'_1,\dots,A'_{2N}) &:= (A_1,\dots,A_N,A_1^*,\dots,A_N^*),\\ (a'_1,\dots,a'_{2N}) &:= (a_1,\dots,a_N,a_1^*,\dots,a_N^*).\end{aligned}$$

Moreover,

$$\hat{\chi}_R(a_1,\dots,a_N;r,\varepsilon) := \limsup_{n\to\infty}\left[\frac{1}{n^2}\log\hat{\Lambda}(\hat{\Gamma}_R(a_1,\dots,a_N;n,r,\varepsilon)) + N\log n\right],$$

where $\hat{\Lambda}$ is short for $\hat{\Lambda}_n^{\otimes N}$. Then the definition of $\hat{\chi}(a_1,\dots,a_N)$ follows the selfadjoint case.

Proposition 6.5.5 *Let $a_1,\dots,a_N \in \mathcal{M}$. If $b_i := (a_i+a_i^*)/2$ and $c_i := (a_i-a_i^*)/2\mathrm{i}$, then*

$$\begin{aligned}&\chi_R(b_1,c_1,\dots,b_N,c_N) \ge \hat{\chi}_R(a_1,\dots,a_N) \ge \chi_{R/2}(b_1,c_1,\dots,b_N,c_N),\\ &\hat{\chi}(a_1,\dots,a_N) = \chi(b_1,c_1,\dots,b_N,c_N).\end{aligned}$$

Proof: It is easy to check that

$$\begin{aligned}&\Gamma_R(b_1,c_1,\dots,b_N,c_N;n,r,\varepsilon)\\ &\qquad\supset\{(B_1,C_1,\dots,B_N,C_N)\in(M_n(\mathbb{C})^{sa})^{2N}:\\ &\qquad\qquad\qquad(B_1+\mathrm{i}C_1,\dots,B_N+\mathrm{i}C_N)\in\hat{\Gamma}_R(a_1,\dots,a_N;n,r,\varepsilon)\},\\ &\hat{\Gamma}_{2R}(a_1,\dots,a_N;n,r,\varepsilon)\\ &\qquad\supset\{(B_1+\mathrm{i}C_1,\dots,B_N+\mathrm{i}C_N)\in M_n(\mathbb{C})^N:\\ &\qquad\qquad\qquad(B_1,C_1,\dots,B_N,C_N)\in\Gamma_R(b_1,c_1,\dots,b_N,c_N;n,r,\varepsilon/2^r)\}.\end{aligned}$$

By Lemma 6.5.4 these imply that

$$\begin{aligned}&\chi_R(b_1,c_1,\dots,b_N,c_N;r,\varepsilon) \ge \hat{\chi}_R(a_1,\dots,a_N;r,\varepsilon),\\ &\hat{\chi}_{2R}(a_1,\dots,a_N;r,\varepsilon) \ge \chi_R(b_1,c_1,\dots,b_N,c_N;r,\varepsilon/2^r).\end{aligned}$$

Hence we get the conclusions.

□

The above proposition enables us to extend some properties of free entropy of selfadjoint random variables to the non-selfadjoint case. For instance, we have

Proposition 6.5.6 $\hat{\chi}(a_1,\dots,a_N) = \hat{\chi}_R(a_1,\dots,a_N)$ *whenever* $R > 2\max_i \|a_i\|$.

Proposition 6.5.7 *For $a_1,\dots,a_N \in \mathcal{M}$,*

1. $\hat{\chi}(\sum_j \alpha_{1j}a_j + \beta_1\mathbf{1},\dots,\sum_j \alpha_{Nj}a_j + \beta_N\mathbf{1}) = \hat{\chi}(a_1,\dots,a_N) + 2\log|\det A|$ *for every $A = [\alpha_{ij}] \in M_n(\mathbb{C})$ and $\beta_1,\dots,\beta_N \in \mathbb{C}$,*
2. $\hat{\chi}(a_1,\dots,a_N) = -\infty$ *when $a_1,\dots,a_N$ are linearly dependent, and*
3. $\hat{\chi}(a_1,\dots,a_N)$ *does not change when any a_i is replaced by a_i^*.*

The non-selfadjoint extension $\hat{\chi}(a_1,\dots,a_N)$ has the subadditivity and upper semicontinuity properties as well. Furthermore, from Example 6.3.4 and the subadditivity we know that $\hat{\chi}(a_1,\dots,a_N) = -\infty$ if there is a normal element among $a_1,\dots,a_N$. This means that the free entropy $\hat{\chi}$ is of no use for normal random variables.

It is evident that $a_1,\dots,a_N$ are *-free circular elements with the same variance C if and only if $b_1,c_1,\dots,b_N,c_N$ are free semicircular elements with the same variance $C/2$ (i.e. with radius $\sqrt{2C}$). So the following is the translation of Theorem 6.4.4.

Proposition 6.5.8 *If $a_1,\dots,a_N \in \mathcal{M}$ satisfy $\tau(a_1^*a_1 + \cdots + a_N^*a_N) \le C$, then $\hat{\chi}(a_1,\dots,a_N) \le N\log(\pi eC/N)$, and equality is attained if and only if $a_1,\dots,a_N$ are *-free circular elements of the same variance C/N.*

Example 6.5.9 Let a be an *elliptic element* in Example 2.6.7, that is,

$$a = \sqrt{\frac{1+\alpha}{2}}b_0 + \mathrm{i}\sqrt{\frac{1-\alpha}{2}}c_0\,,$$

where b_0, c_0 are free standard semicircular elements and $-1 < \alpha < 1$. Then by Proposition 6.5.5 and Theorem 6.4.1 we compute

$$\hat{\chi}(a) = \chi\Big(\sqrt{\frac{1+\alpha}{2}}b_0\Big) + \chi\Big(\sqrt{\frac{1-\alpha}{2}}c_0\Big) = \log(\pi e\sqrt{1-\alpha^2})\,.$$

Consider the functional

$$\hat{\chi}(a) - \Big(\frac{\tau(b^2)}{1+\alpha} + \frac{\tau(c^2)}{1-\alpha}\Big) \quad \text{for} \quad a = b + \mathrm{i}c \ (b,c \in \mathcal{M}^{sa}).$$

By Theorem 6.4.1 and Proposition 5.3.4 we observe that this functional attains its maximal value if and only if a is the above elliptic element.

□

6.6 Relation between different free entropies

The aim of this section is to make a bridge between free entropy of unitary random variables and that of non-selfadjoint ones via the polar decomposition. A key idea is as follows. Let $u_1,\dots,u_N$ be unitaries and $h_1,\dots,h_N$ be positive variables. On the one hand, we have the free entropy $\hat{\chi}(u_1h_1,\dots,u_Nh_N)$ of the N-tuple of non-selfadjoint variables, and on the other hand we may consider the $2N$-tuple $(u_1,\dots,u_N,h_1,\dots,h_N)$, or rather $(u_1,\dots,u_N,h_1^2,\dots,h_N^2)$, in which unitary and positive random variables are mixed. We define the *free entropy of* such *mixed tuples*, and next obtain its connection with $\hat{\chi}(u_1h_1,\dots,u_Nh_N)$.

The measure $\Lambda_{+,n}$ on $M_n(\mathbb{C})^+$ introduced in Sec. 4.4 will be used below. According to Lemma 4.4.7 this measure is the restriction of the Lebesgue measure Λ_n on $M_n(\mathbb{C})^+$ with a different normalizing constant. Let $u_1, \dots, u_N \in \mathcal{M}$ be unitaries and $h_1, \dots, h_L \in \mathcal{M}^+$. For $n, r \in \mathbb{N}$, $\varepsilon > 0$ and $R > 0$ we define

$$\begin{aligned}
&\Gamma_{(u,+),R}(u_1, \dots, u_N; h_1, \dots, h_L; n, r, \varepsilon) \\
&\quad := \big\{(U_1, \dots, U_N; H_1, \dots, H_L) \in \mathcal{U}(n)^N \times (M_n(\mathbb{C})^+)^L : \|H_i\| \le R, \\
&\qquad\qquad |\mathrm{tr}_n(B_{i_1} \cdots B_{i_k}) - \tau(b_{i_1} \cdots b_{i_k})| \le \varepsilon \\
&\qquad\qquad \text{for all } 1 \le i_1, \dots, i_k \le 2N + L,\ 1 \le k \le r\big\},
\end{aligned}$$

where

$$\begin{aligned}
(B_1, \dots, B_{2N+L}) &:= (U_1, \dots, U_N, U_1^*, \dots, U_N^*, H_1, \dots, H_L), \\
(b_1, \dots, b_{2N+L}) &:= (u_1, \dots, u_N, u_1^*, \dots, u_N^*, h_1, \dots, h_L),
\end{aligned}$$

and further define

$$\begin{aligned}
&\chi_{(u,+),R}(u_1, \dots, u_N; h_1, \dots, h_L; r, \varepsilon) \\
&\quad := \limsup_{n \to \infty} \bigg[\frac{1}{n^2} \log(\gamma \otimes \Lambda_+)(\Gamma_{(u,+),R}(u_1, \dots, u_N; h_1, \dots, h_L; n, r, \varepsilon)) \\
&\qquad\qquad + L \log n\bigg],
\end{aligned}$$

where $\gamma \otimes \Lambda_+$ is used for $\gamma_n^{\otimes N} \otimes \Lambda_{+,n}^{\otimes L}$ on $\mathcal{U}(n)^N \times (M_n(\mathbb{C})^+)^L$. Now the *free entropy* $\chi_{(u,+)}(u_1, \dots, u_N; h_1, \dots, h_L)$ *of mixed type* is defined as earlier; we let $r \to \infty$, $\varepsilon \to +0$ and then $R \to \infty$. When the set $\{h_1, \dots, h_L\}$ of positive elements is empty, our definition reduces to $\chi_u(u_1, \dots, u_N)$ given in the previous section. On the other hand, we write $\Gamma_{+,R}(h_1, \dots, h_L; n, r, \varepsilon)$, $\chi_+(h_1, \dots, h_L)$, etc. when no unitaries are present.

It is obvious that the free entropy $\chi_{(u,+)}$ has the subadditivity property, like χ and χ_u. Moreover, the following upper semicontinuity of $\chi_{(u,+)}$ can be shown in a way similar to Propositions 6.1.4 and 6.1.5.

Proposition 6.6.1 *Let $u_1, \dots, u_N$ and $u_{m,1}, \dots, u_{m,N}$ $(m \in \mathbb{N})$ be unitaries in $\mathcal{M}$. Let $h_1, \dots, h_L$ and $h_{m,1}, \dots, h_{m,L}$ $(m \in \mathbb{N})$ be in $\mathcal{M}^+$. If*

$$(u_{m,1}, \dots, u_{m,N}, h_{m,1}, \dots, h_{m,L}) \to (u_1, \dots, u_N, h_1, \dots, h_L)$$

in the distribution sense and $\sup_m \|h_{m,i}\| < +\infty$ $(1 \le i \le L)$, then

$$\begin{aligned}
&\chi_{(u,+)}(u_1, \dots, u_N; h_1, \dots, h_L) \\
&\quad \ge \limsup_{m \to \infty} \chi_{(u,+)}(u_{m,1}, \dots, u_{m,N}; h_{m,1}, \dots, h_{m,L}).
\end{aligned}$$

For a single $h \in \mathcal{M}^+$ let μ be the distribution on $\mathbb{R}^+$ of h and $R \geq \|h\|$. Since

$$\Gamma_{+,R}(h; n, r, \varepsilon) = \left\{H \in M_n(\mathbb{C})^+ : \|H\| \leq R, \, |\mathrm{tr}_n(H^k) - m_k(\mu)| \leq \varepsilon, \, k \leq r\right\},$$

it follows from (5.6.8), (5.6.9) and (5.6.11) that the limit

$$\chi_{+,R}(h; r, \varepsilon) = \lim_{n\to\infty} \left[\frac{1}{n^2} \log \Lambda_{+,n}(\Gamma_{+,R}(h; n, r, \varepsilon)) + \log n\right] \tag{6.6.1}$$

exists for every $r \in \mathbb{N}$ and $\varepsilon > 0$, and

$$\begin{aligned} \chi_+(h) = \chi_{+,R}(h) &= \Sigma(\mu) + \log \pi + \frac{3}{2} \\ &= \chi(h) + \frac{1}{2} \log \frac{\pi}{2} + \frac{3}{4}. \end{aligned} \tag{6.6.2}$$

Furthermore, from Proposition 5.3.5 we know that, among $h \in \mathcal{M}^+$ with $\tau(h) \leq C$, the free entropy $\chi_+(h)$ attains the maximal value $\log(\pi e C)$ if (and only if) h has the distribution

$$\frac{\sqrt{4Ct - t^2}}{2\pi Ct} \chi_{[0,4C]}(t) \, dt$$

or equivalently $h^{1/2}$ is a quarter-circular element of radius $2\sqrt{C}$. From the subadditivity we have

$$\chi_+(h_1, \dots, h_N) \leq \chi_+(h_1) + \cdots + \chi_+(h_N) \leq N \log \frac{\pi e C}{N}$$

whenever $h_1, \dots, h_N \in \mathcal{M}^+$ satisfy $\tau(h_1^2 + \cdots + h_N^2) \leq C$.

The next proposition says that the free entropy χ_+ is nothing but χ restricted on positive random variables up to additive constants, extending the relation between $\chi_+(h)$ and $\chi(h)$ in (6.6.2).

Proposition 6.6.2 *The equality*

$$\chi_+(h_1, \dots, h_N) = \chi(h_1, \dots, h_N) + \frac{N}{2}\left(\log \frac{\pi}{2} + \frac{3}{2}\right)$$

holds for every $N \in \mathbb{N}$ and $h_1, \dots, h_N \in \mathcal{M}^+$.

Proof: For $r \in \mathbb{N}$, $\varepsilon > 0$ and $R > 0$, it is obvious that

$$\Gamma_{+,R}(h_1, \dots, h_N; n, r, \varepsilon) \subset \Gamma_R(h_1, \dots, h_N; n, r, \varepsilon)$$

(the right-hand side is taken in $(M_n(\mathbb{C})^{sa})^N \supset (M_n(\mathbb{C})^+)^N$). Hence by Lemma 4.4.7 we get

$$\begin{aligned}
&\chi_{+,R}(h_1,\dots,h_N;r,\varepsilon)\\
&\quad\le \limsup_{n\to\infty}\left[\frac{1}{n^2}\log\Lambda_+(\Gamma_R(h_1,\dots,h_N;n,r,\varepsilon))+N\log n\right]\\
&\quad= \limsup_{n\to\infty}\left[\frac{1}{n^2}\log\Lambda(\Gamma_R(h_1,\dots,h_N;n,r,\varepsilon))+\frac{N}{2}\log n\right.\\
&\qquad\qquad\left.+N\Big(\frac{1}{n^2}\log C_n+\frac{1}{2}\log n\Big)\right]\\
&\quad= \chi_R(h_1,\dots,h_N;r,\varepsilon)+\frac{N}{2}\Big(\log\frac{\pi}{2}+\frac{3}{2}\Big),
\end{aligned}$$

where C_n is the constant in Lemma 4.4.7 and satisfies

$$\lim_{n\to\infty}\Big(\frac{1}{n^2}\log C_n+\frac{1}{2}\log n\Big)=\frac{1}{2}\Big(\log\frac{\pi}{2}+\frac{3}{2}\Big)$$

by the Stirling formula. Therefore, we have

$$\chi_+(h_1,\dots,h_N)\le\chi(h_1,\dots,h_N)+\frac{N}{2}\Big(\log\frac{\pi}{2}+\frac{3}{2}\Big).$$

To show the reverse inequality, for $\delta>0$ we take $(h_1+\delta\mathbf{1},\dots,h_N+\delta\mathbf{1})$ instead of $(h_1,\dots,h_N)$, and also $R>\max_i\|h_i\|+\delta$. Thanks to Corollary 6.3.2 (1), Proposition 6.1.4 and the translation invariance of Λ_n, we can estimate

$$\begin{aligned}
&\chi(h_1,\dots,h_N)=\chi(h_1+\delta\mathbf{1},\dots,h_N+\delta\mathbf{1})\\
&\quad=\lim_{\substack{r\to\infty\\ \varepsilon\to+0}}\limsup_{n\to\infty}\left[\frac{1}{n^2}\log\Lambda(\Gamma_{+,R}(h_1+\delta\mathbf{1},\dots,h_N+\delta\mathbf{1};n,r,\varepsilon))+\frac{N}{2}\log n\right]\\
&\quad=\lim_{\substack{r\to\infty\\ \varepsilon\to+0}}\limsup_{n\to\infty}\left[\frac{1}{n^2}\log\Lambda_+(\Gamma_{+,R}(h_1+\delta\mathbf{1},\dots,h_N+\delta\mathbf{1};n,r,\varepsilon))\right.\\
&\qquad\qquad\left.+N\log n-N\Big(\frac{1}{n^2}\log C_n+\frac{1}{2}\log n\Big)\right]\\
&\quad\le\chi_+(h_1+\delta\mathbf{1},\dots,h_N+\delta\mathbf{1})-\frac{N}{2}\Big(\log\frac{\pi}{2}+\frac{3}{2}\Big).
\end{aligned}$$

Using the upper semicontinuity (Proposition 6.6.1) as $\delta\to+0$, we obtain the result. □

The following relation between two free entropies $\hat{\chi}$ and $\chi_{(u,+)}$ might be expected from the definitions in the light of Lemma 4.4.7.

Theorem 6.6.3 *If $u_1,\dots,u_N\in\mathcal{M}$ are unitaries and $h_1,\dots,h_N\in\mathcal{M}^+$, then*

$$\begin{aligned}
\hat{\chi}(u_1h_1,\dots,u_Nh_N) &= \chi_{(u,+)}(u_1,\dots,u_N;h_1^2,\dots,h_N^2)\\
&\le \chi_u(u_1,\dots,u_N)+\chi(h_1^2,\dots,h_N^2)+\frac{N}{2}\Big(\log\frac{\pi}{2}+\frac{3}{2}\Big).
\end{aligned}$$

To prove the theorem, we need to approximate the unitary part of A by polynomials of A, A^*. The approximation here must be uniform for $A \in M_n(\mathbb{C})$ with $\|A\| \le R$ in some sense. The next lemma provides the right approximation procedure for our purpose.

Let $a \in \mathcal{M}$ and assume that the distribution of $|a|$ is nonatomic. Let $a = u|a|$ be the polar decomposition. Note that $\ker a = \{0\}$ by assumption, and hence $u \in \mathcal{M}$ must be a unitary (since $\mathcal{M}$ is a finite von Neumann algebra). Let $\|\cdot\|_p$ denote the Schatten p-norm with respect to τ as well as tr_n.

Lemma 6.6.4 *With the above assumption and notation, for every $p \ge 1$, $\varepsilon > 0$ and $R \ge \|a\|$, there exist $n_0, r \in \mathbb{N}$, $\delta > 0$ and a real polynomial $P(t)$ such that $\|u - aP(a^*a)\|_p \le \varepsilon$, and such that, for each $n \ge n_0$, if $A \in M_n(\mathbb{C})$ with $\|A\| \le R$ is non-singular and U is the unitary part of A, and if*

$$|\mathrm{tr}_n((A^*A)^k) - \tau((a^*a)^k)| \le \delta \qquad (1 \le k \le r), \tag{6.6.3}$$

then $\|U - AP(A^*A)\|_p \le \varepsilon$.

Proof: Let μ be the distribution of $|a|$. For every $\alpha, \beta > 0$, since

$$u - a(|a| + \alpha\mathbf{1})^{-1} = u(\mathbf{1} - |a|(|a| + \alpha\mathbf{1})^{-1}) = \alpha u(|a| + \alpha\mathbf{1})^{-1},$$

we have

$$\begin{aligned} \|u - a(|a| + \alpha\mathbf{1})^{-1}\|_p^p &= \|\alpha(|a| + \alpha\mathbf{1})^{-1}\|_p^p \\ &= \int_0^\infty \left(\frac{\alpha}{t+\alpha}\right)^p d\mu(t) \le \mu([0,\beta]) + \left(\frac{\alpha}{\beta}\right)^p. \end{aligned}$$

Similarly, for any non-singular $A \in M_n(\mathbb{C})$ with $A = U|A|$, we have

$$\|U - A(|A| + \alpha I)^{-1}\|_p^p = \frac{1}{n}\sum_{i=1}^n \left(\frac{\alpha}{\lambda_i + \alpha}\right)^p \le \frac{1}{n}\#\{i : \lambda_i \le \beta\} + \left(\frac{\alpha}{\beta}\right)^p,$$

where $(0 <)\, \lambda_1 \le \lambda_2 \le \ldots \le \lambda_n$ are the eigenvalues of $|A|$.

Now for each $n \in \mathbb{N}$, since μ is nonatomic, one can choose $0 < \xi_1^{(n)} < \xi_2^{(n)} < \ldots < \xi_n^{(n)} = \|a\|$ such that $\mu([0, \xi_i^{(n)}]) = i/n$ $(1 \le i \le n)$. Then it immediately follows that

$$\tau((a^*a)^k) = \int_0^\infty t^{2k}\, d\mu(t) = \lim_{n\to\infty} \frac{1}{n}\sum_{i=1}^n (\xi_i^{(n)})^{2k} \qquad (k \in \mathbb{N}).$$

Let $\beta > 0$ be fixed so that $\mu([0, 2\beta]) < \varepsilon^p/2$. By Lemma 4.3.4 there are $r \in \mathbb{N}$ and $\delta > 0$ such that, for every $n \in \mathbb{N}$, if $(\lambda_1, \ldots, \lambda_n) \in (\mathbb{R}^+)^n_\le$ satisfies

$$\left|\frac{1}{n}\sum_{i=1}^n \lambda_i^{2k} - \frac{1}{n}\sum_{i=1}^n (\xi_i^{(n)})^{2k}\right| \le 2\delta \qquad (1 \le k \le r),$$

then

$$\frac{1}{n}\sum_{i=1}^{n}(\lambda_i^2 - (\xi_i^{(n)})^2)^2 \le \beta^4\varepsilon^p . \tag{6.6.4}$$

Next, choose $n_0 \in \mathbb{N}$ such that

$$\left|\frac{1}{n}\sum_{i=1}^{n}(\xi_i^{(n)})^{2k} - \tau((a^*a)^k)\right| \le \delta \qquad (1 \le k \le r)$$

whenever $n \ge n_0$. Then, for any $n \ge n_0$, (6.6.4) is valid if $A \in M_n(\mathbb{C})$ satisfies (6.6.3). Furthermore, when (6.6.3) is satisfied, we have

$$\frac{1}{n}\#\{i : \lambda_i \le \beta\} \le \frac{11}{18}\varepsilon^p . \tag{6.6.5}$$

Indeed, put $l := \#\{i : \lambda_i \le \beta\}$ and $m := \#\{i : \xi_i^{(n)} \le 2\beta\}$. If $m < i \le l$, then $\lambda_i \le \lambda_l \le \beta$ but $2\beta < \xi_{m+1}^{(n)} \le \xi_i^{(n)}$, so $(\lambda_i^2 - (\xi_i^{(n)})^2)^2 \ge (4\beta^2 - \beta^2)^2 = 9\beta^4$. Hence (6.6.4) implies $\frac{1}{n}(l-m)\cdot 9\beta^4 \le \beta^4\varepsilon^p$, so that $l/n \le m/n + \varepsilon^p/9$. Since

$$\frac{m}{n} = \mu([0,\xi_m^{(n)}]) \le \mu([0,2\beta]) \le \frac{\varepsilon^p}{2},$$

we have $l/n \le \varepsilon^p/2 + \varepsilon^p/9$, showing (6.6.5).

From all the above estimates, we infer that, for each $\alpha > 0$ and $n \ge n_0$, if $A \in M_n(\mathbb{C})$ with $\|A\| \le R$ is non-singular and satisfies (6.6.3), then

$$\|U - A(|A| + \alpha I)^{-1}\|_p^p \le \frac{11}{18}\varepsilon^p + \left(\frac{\alpha}{\beta}\right)^p$$

and

$$\|u - a(|a| + \alpha\mathbf{1})^{-1}\|_p^p \le \frac{\varepsilon^p}{2} + \left(\frac{\alpha}{\beta}\right)^p .$$

Choose $\alpha > 0$ such that $(\alpha/\beta)^p \le \varepsilon^p/18$, and next choose a polynomial $P(t)$ such that

$$|P(t) - (\sqrt{t} + \alpha)^{-1}| \le \frac{1}{R}\left(1 - \left(\frac{2}{3}\right)^{1/p}\right)\varepsilon \quad \text{for} \quad 0 \le t \le R^2.$$

Then for each $n \ge n_0$ and A as above, we obtain

$$\begin{aligned}
&\|U - A(|A| + \alpha I)^{-1}\|_p \le \left(\frac{2}{3}\right)^{1/p}\varepsilon , \\
&\|AP(A^*A) - A(|A| + \alpha I)^{-1}\|_p \\
&\qquad \le \|A\|\,\|P(A^*A) - (|A| + \alpha I)^{-1}\| \le \left(1 - \left(\frac{2}{3}\right)^{1/p}\right)\varepsilon ,
\end{aligned}$$

so $\|U - AP(A^*A)\|_p \le \varepsilon$ holds, and similary $\|u - aP(a^*a)\|_p \le \varepsilon$.

□

Proof of Theorem 6.6.3: First, the inequality in the theorem is a consequence of the subadditivity of $\chi_{(u,+)}$ and Proposition 6.6.2. Define $\Psi : M_n(\mathbb{C}) \to \mathcal{U}(n) \times M_n(\mathbb{C})^+$ by $\Psi(A) := (U, A^*A)$, where U is the unitary part of A. This is bijective except for the negligible singular elements (in $M_n(\mathbb{C})$ and $M_n(\mathbb{C})^+$). Put $a_i := u_i h_i$, so that $h_i^2 = a_i^* a_i$. Let $n, r \in \mathbb{N}$, $\varepsilon > 0$ and $R > \max\{1, \|h_1\|, \dots, \|h_N\|\}$. It is straightforward to see that there are $r_1 \in \mathbb{N}$ and $\varepsilon_1 > 0$ such that

$$\Psi(\hat{\Gamma}_R(a_1, \dots, a_N; n, r_1, \varepsilon_1)) \subset \mathcal{U}(n) \times \Gamma_{+,R^2}(h_1^2, \dots, h_N^2; n, r, \varepsilon),$$

and by Lemma 4.4.7

$$\hat{\Lambda}(\hat{\Gamma}_R(a_1, \dots, a_N; n, r_1, \varepsilon_1)) \le \Lambda_+(\Gamma_{+,R^2}(h_1^2, \dots, h_N^2; n, r, \varepsilon))$$

for all $n \in \mathbb{N}$. This yields $\hat{\chi}(a_1, \dots, a_N) \le \chi_+(h_1^2, \dots, h_N^2)$. Hence we may assume that $\chi(h_1^2, \dots, h_N^2) > -\infty$, so the distribution of each h_i is nonatomic.

Let $\varepsilon_0 > 0$ be such that $r\varepsilon_0(R^2 + \varepsilon_0)^{r-1} \le \varepsilon/3$. By Lemma 6.6.4 there exist $n_0, r_0 \in \mathbb{N}$, $\delta > 0$ and real polynomials $P_i(t)$ $(1 \le i \le N)$ such that $\|u_i - a_i P_i(a_i^* a_i)\|_r \le \varepsilon_0$, and such that, for each $1 \le i \le N$ and $n \ge n_0$, if $A_i \in M_n(\mathbb{C})$ is non-singular with $A_i = U_i |A_i|$, $\|A_i\| \le R$, and

$$|\mathrm{tr}_n((A_i^* A_i)^k) - \tau((a_i^* a_i)^k)| \le \delta \qquad (1 \le k \le r_0),$$

then $\|U_i - A_i P_i(A_i^* A_i)\|_r \le \varepsilon_0$. For $A_i \in M_n(\mathbb{C})$ $(1 \le i \le N)$ satisfying the above conditions, we set

$$\begin{aligned}
(B_1, \dots, B_{3N}) &:= (U_1, \dots, U_N, U_1^*, \dots, U_N^*, A_1^* A_1, \dots, A_N^* A_N), \\
(B_1', \dots, B_{3N}') &:= (A_1 P_1(A_1^* A_1), \dots, A_N P_N(A_N^* A_N), \\
&\qquad P_1(A_1^* A_1) A_1^*, \dots, P_N(A_N^* A_N) A_N^*, A_1^* A_1, \dots, A_N^* A_N),
\end{aligned}$$

as well as

$$\begin{aligned}
(b_1, \dots, b_{3N}) &:= (u_1, \dots, u_N, u_1^*, \dots, u_N^*, a_1^* a_1, \dots, a_N^* a_N), \\
(b_1', \dots, b_{3N}') &:= (a_1 P_1(a_1^* a_1), \dots, a_N P_N(a_N^* a_N), \\
&\qquad P_1(a_1^* a_1) a_1^*, \dots, P_N(a_N^* a_N) a_N^*, a_1^* a_1, \dots, a_N^* a_N).
\end{aligned}$$

Then for any $n \ge n_0$ and $1 \le i_1, \dots, i_k \le 3N$ $(1 \le k \le r)$, by using the Hölder inequality, it can be verified that

$$\begin{aligned}
&|\mathrm{tr}_n(B_{i_1} \cdots B_{i_k}) - \mathrm{tr}_n(B_{i_1}' \cdots B_{i_k}')| \\
&\quad \le \|B_{i_1} \cdots B_{i_k} - B_{i_1}' \cdots B_{i_k}'\|_1 \le k\varepsilon_0(R^2 + \varepsilon_0)^{k-1} \le \frac{\varepsilon}{3},
\end{aligned}$$

and similarly $|\tau(b_{i_1}\cdots b_{i_k}) - \tau(b'_{i_1}\cdots b'_{i_k})| \le \varepsilon/3$. Now choose r_1 ($\ge 2r_0$) large enough and ε_1 ($\le \delta$) small enough so that if $(A_1,\dots,A_N) \in \hat\Gamma_R(a_1,\dots,a_N;n,r_1,\varepsilon_1)$ then $|\mathrm{tr}_n(B'_{i_1}\cdots B'_{i_k}) - \tau(b'_{i_1}\cdots b'_{i_k})| \le \varepsilon/3$ for all $1 \le i_1,\dots,i_k \le 3N$ $(1 \le k \le r)$. Therefore, for $n \ge n_0$ we obtain

$$\Psi(\hat\Gamma_R(a_1,\dots,a_N;n,r_1,\varepsilon_1)) \subset \Gamma_{(u,+),R^2}(u_1,\dots,u_N;h_1^2,\dots,h_N^2;n,r,\varepsilon)$$

(up to negligible sets), and hence by Lemma 4.4.7

$$\begin{aligned}&\hat\Lambda(\hat\Gamma_R(a_1,\dots,a_N;n,r_1,\varepsilon_1))\\ &\qquad \le (\gamma\otimes\Lambda_+)(\Gamma_{(u,+),R^2}(u_1,\dots,u_N;h_1^2,\dots,h_N^2;n,r,\varepsilon)).\end{aligned}$$

This implies that

$$\hat\chi(a_1,\dots,a_N) \le \chi_{(u,+)}(u_1,\dots,u_N;h_1^2,\dots,h_N^2).$$

Conversely, given $r \in \mathbb{N}$, $\varepsilon > 0$ and $R > 0$, by approximating $\sqrt{t}$ on $[0,R^2]$ by a polynomial, it is seen that there are $r_1 \in \mathbb{N}$ and $\varepsilon_1 > 0$ such that

$$\Gamma_{(u,+),R^2}(u_1,\dots,u_N;h_1^2,\dots,h_N^2;n,r_1,\varepsilon_1) \subset \Psi(\hat\Gamma_R(a_1,\dots,a_N;n,r,\varepsilon))$$

(up to negligible sets) for all $n \in \mathbb{N}$. This gives the reverse inequality.

□

Theorem 6.6.3 gives

$$\hat\chi(a_1,\dots,a_N) \le \chi_u(u_1,\dots,u_N) + \chi(a_1^*a_1,\dots,a_N^*a_N) + \frac{N}{2}\Bigl(\log\frac{\pi}{2}+\frac{3}{2}\Bigr)$$

for every $a_1,\dots,a_N \in \mathcal{M}$ and all unitaries $u_1,\dots,u_N \in \mathcal{M}$ satisfying $a_i = u_i|a_i|$. In particular, we have the following corollary. Its proof was indeed included in the first paragraph of the proof of Theorem 6.6.3.

Corollary 6.6.5 *Let $a_1,\dots,a_N \in \mathcal{M}$. If $\hat\chi(a_1,\dots,a_N) > -\infty$, then the distribution of $a_i^*a_i$ is nonatomic (hence $\ker a_i = \{0\}$) for every $1 \le i \le N$.*

Next, we show that the inequality in Theorem 6.6.3 can be replaced by equality in some cases of free relation. First, we take a free family $\{h_1,\dots,h_N\}$ which is also free from $\{u_1,\dots,u_N,u_1^*,\dots,u_N^*\}$. Then an exact relation among $\hat\chi$, χ_u and χ is obtained as follows. Thus we have a formula for χ_u in terms of $\hat\chi$ (hence χ).

Theorem 6.6.6 *Let $u_1,\dots,u_N \in \mathcal{M}$ be unitaries and $h_1,\dots,h_N \in \mathcal{M}^+$. If $\{u_1,\dots,u_N,u_1^*,\dots,u_N^*\}$, $h_1,\dots,h_N$ are free, then*

$$\hat\chi(u_1h_1,\dots,u_Nh_N) = \chi_u(u_1,\dots,u_N) + \sum_{i=1}^N \chi(h_i^2) + \frac{N}{2}\Bigl(\log\frac{\pi}{2}+\frac{3}{2}\Bigr).$$

In particular, if $h_1,\dots,h_N$ are free standard (i.e. of radius 2) quarter-circular elements and they are free from $\{u_1,\dots,u_N,u_1^*,\dots,u_N^*\}$, then

$$\begin{aligned}\chi_u(u_1,\dots,u_N) &= \hat{\chi}(u_1h_1,\dots,u_Nh_N) - N\log(\pi e)\\ &= \chi(b_1,c_1,\dots,b_N,c_N) - N\log(\pi e),\end{aligned}$$

where $u_ih_i = b_i + \mathrm{i}\,c_i$ with selfadjoint b_i, c_i.

To prove the theorm we show the next lemma, which will play in the present situation the same role as Lemma 6.4.3 did in the proof of Theorem 6.4.1.

Lemma 6.6.7 *Let $u_1,\dots,u_N,h_1,\dots,h_N$ be as in Theorem 6.6.6, and assume that $\chi_u(u_1,\dots,u_N) > -\infty$ and $\chi_+(h_i^2) > -\infty$ $(1 \le i \le N)$. Then, for every $r \in \mathbb{N}$, $\varepsilon > 0$ and $R > \max_i \|h_i\|^2$, there exists $\varepsilon_1 > 0$ such that*

$$\lim_{n\to\infty} \frac{(\gamma\otimes\Lambda_+)(\Xi_n(r,\varepsilon_1)\cap\Theta_n(r,\varepsilon))}{(\gamma\otimes\Lambda_+)(\Xi_n(r,\varepsilon_1))} = 1\,,$$

where

$$\begin{aligned}\Xi_n(r,\varepsilon_1) &:= \Gamma_u(u_1,\dots,u_N;n,r,\varepsilon_1)\times\prod_{i=1}^N \Gamma_{+,R}(h_i^2;n,r,\varepsilon_1)\,,\\ \Theta_n(r,\varepsilon) &:= \Gamma_{(u,+),R}(u_1,\dots,u_N;h_1^2,\dots,h_N^2;n,r,\varepsilon)\,.\end{aligned}$$

Proof: Thanks to the freeness of $\{u_1,\dots,u_N,u_1^*,\dots,u_N^*\}$, $h_1^2,\dots,h_N^2$, one can choose $\varepsilon_1 > 0$ such that if $(U_1,\dots,U_N;H_1,\dots,H_N) \in \Xi_n(r,\varepsilon_1)$ and $\{U_1,\dots,U_N, U_1^*,\dots,U_N^*\}$, $\{H_1\},\dots,\{H_N\}$ are (r,ε_1)-free, then $(U_1,\dots,U_N;H_1,\dots,H_N) \in \Theta_n(r,\varepsilon)$. For every $\theta > 0$, according to Lemma 6.4.2 there exists $n_0 \in \mathbb{N}$ such that

$$\begin{aligned}\gamma(\{(V_1,\dots,V_N)\in\mathcal{U}(n)^N : \{U_1,\dots,U_N,U_1^*,\dots,U_N^*\},\{V_1H_1V_1^*\},\dots,\\ \{V_NH_NV_N^*\} \text{ are } (r,\varepsilon_1)\text{-free}\}) \ge 1-\theta \qquad (6.6.6)\end{aligned}$$

for all $n \ge n_0$ and for any choice of $U_i \in \mathcal{U}(n)$ and $H_i \in M_n(\mathbb{C})^+$ with $\|H_i\| \le R$ $(1 \le i \le N)$. Since $\chi_u(u_1,\dots,u_N) > -\infty$ and $\chi_+(h_i^2) > -\infty$, the $\gamma\otimes\Lambda_+$-measure of $\Xi_n(r,\varepsilon_1)$ is positive for large n, so we define the probability measure σ_n on $\Xi_n(r,\varepsilon_1)$ by normalizing the restriction of $\gamma\otimes\Lambda_+$ on $\Xi_n(r,\varepsilon_1)$. Since σ_n is invariant under the action of $\mathcal{U}(n)^N$ on $\Xi_n(r,\varepsilon_1)$ given by $(U_1,\dots,U_N;H_1,\dots,H_N)\mapsto (U_1,\dots,U_N;V_1H_1V_1^*,\dots,V_NH_NV_N^*)$ for $(V_1,\dots,V_N)\in\mathcal{U}(n)^N$, we have

$$\begin{aligned}\frac{(\gamma\otimes\Lambda_+)(\Xi_n(r,\varepsilon_1)\cap\Theta_n(r,\varepsilon))}{(\gamma\otimes\Lambda_+)(\Xi_n(r,\varepsilon_1))} &= \int_{\Xi_n(r,\varepsilon_1)}\biggl(\int_{\mathcal{U}(n)^N}\psi(U_1,\dots,U_N;\\ &\qquad V_1H_1V_1^*,\dots,V_NH_NV_N^*)\,d\gamma(V_1,\dots,V_N)\biggr)\,d\sigma_n\,,\end{aligned}$$

where ψ is the characteristic function of $\Xi_n(r,\varepsilon_1)\cap\Theta_n(r,\varepsilon)$. From (6.6.6) we get

$$\int_{\mathcal{U}(n)^N} \psi(U_1, \dots, U_N; V_1 H_1 V_1^*, \dots, V_N H_N V_N^*)\, d\gamma(V_1, \dots, V_N) \geq 1 - \theta$$

for all $(U_1, \dots, U_N; H_1, \dots, H_N) \in \Xi_n(r, \varepsilon_1)$. Therefore,

$$\frac{(\gamma \otimes \Lambda_+)(\Xi_n(r, \varepsilon_1) \cap \Theta_n(r, \varepsilon))}{(\gamma \otimes \Lambda_+)(\Xi_n(r, \varepsilon_1))} \geq 1 - \theta$$

whenever n is large, and we have the result.

□

Proof of Theorem 6.6.6: By Theorem 6.6.3 and (6.6.2) it suffices to show that

$$\chi_{(u,+)}(u_1, \dots, u_N; h_1^2, \dots, h_N^2) \geq \chi_u(u_1, \dots, u_N) + \sum_{i=1}^N \chi_+(h_i^2), \tag{6.6.7}$$

so we may assume that $\chi_u(u_1, \dots, u_N) > -\infty$ and $\chi_+(h_i^2) > -\infty$ $(1 \leq i \leq N)$. For any $r \in \mathbb{N}$, $\varepsilon > 0$ and $R > \max_i \|h_i\|^2$, let $\varepsilon_1 > 0$ be as in Lemma 6.6.7. Then we have

$$\begin{aligned}
&\chi_{(u,+),R}(u_1, \dots, u_N; h_1^2, \dots, h_N^2; r, \varepsilon) \\
&\quad \geq \limsup_{n\to\infty} \left[\frac{1}{n^2} \log(\gamma \otimes \Lambda_+)(\Xi_n(r, \varepsilon_1)) + N \log n \right] \\
&\quad = \limsup_{n\to\infty} \left[\frac{1}{n^2} \log \gamma(\Gamma_u(u_1, \dots, u_N; n, r, \varepsilon_1)) \right. \\
&\qquad\qquad \left. + \sum_{i=1}^N \left(\frac{1}{n^2} \log \Lambda_{+,n}(\Gamma_{+,R}(h_i^2; n, r, \varepsilon_1)) + \log n \right) \right] \\
&\quad = \chi_u(u_1, \dots, u_N; r, \varepsilon_1) + \sum_{i=1}^N \chi_{+,R}(h_i^2; r, \varepsilon_1).
\end{aligned}$$

Above we used the fact that lim sup becomes lim in (6.6.1). Thus (6.6.7) is shown. The second part is clear from the fact mentioned before Proposition 6.6.2 and Proposition 6.5.5.

□

When the roles of $u_1, \dots, u_N$ and $h_1, \dots, h_N$ are exchanged in Theorem 6.6.6, we have

Theorem 6.6.8 *Let $u_1, \dots, u_N \in \mathcal{M}$ be unitaries and $h_1, \dots, h_N \in \mathcal{M}^+$. If $\{u_1, u_1^*\}, \dots, \{u_N, u_N^*\}$, $\{h_1, \dots, h_N\}$ are free, then*

$$\hat{\chi}(u_1 h_1, \dots, u_N h_N) = \sum_{i=1}^N \chi_u(u_i) + \chi(h_1^2, \dots, h_N^2) + \frac{N}{2}\left(\log \frac{\pi}{2} + \frac{3}{2}\right).$$

If $u_1, \dots, u_N$ are Haar unitaries in addition, then

$$\hat{\chi}(u_1h_1,\dots,u_Nh_N)=\chi(h_1^2,\dots,h_N^2)+\frac{N}{2}\left(\log\frac{\pi}{2}+\frac{3}{2}\right).$$

Proof: By Theorem 6.6.3 and Proposition 6.6.2 it suffices to show that

$$\chi_{(u,+)}(u_1,\dots,u_N;h_1^2,\dots,h_N^2)\ge\sum_{i=1}^N\chi_u(u_i)+\chi_+(h_1^2,\dots,h_N^2),\tag{6.6.8}$$

and we may assume $\chi_u(u_i)>-\infty$ and $\chi_+(h_1^2,\dots,h_N^2)>-\infty$. For $n,r\in\mathbb{N}$, $\varepsilon>0$ and $R>0$ we set

$$\Xi_n(r,\varepsilon):=\prod_{i=1}^N\Gamma_u(u_i;n,r,\varepsilon)\times\Gamma_{+,R}(h_1^2,\dots,h_N^2;n,r,\varepsilon),$$

and $\Theta_n(r,\varepsilon)$ is the same as in Lemma 6.6.7. By the freeness assumption there is $\varepsilon_1>0$ such that if $(U_1,\dots,U_N;H_1,\dots,H_N)\in\Xi_n(r,\varepsilon_1)$ and $\{U_1,U_1^*\},\dots,\{U_N,U_N^*\}$, $\{H_1,\dots,H_N\}$ are (r,ε_1)-free, then $(U_1,\dots,U_N;H_1,\dots,H_N)\in\Theta_n(r,\varepsilon)$. For every $\theta>0$, by Lemma 6.4.2 there exists $n_0\in\mathbb{N}$ such that

$$\begin{aligned}\gamma(\{(V_1,\dots,V_N)\in(\mathcal{U}(n))^N:\{V_1U_1V_1^*,V_1U_1^*V_1^*\},\dots,\{V_NU_NV_N^*,V_NU_N^*V_N^*\},\\ \{H_1,\dots,H_N\}\text{ are }(r,\varepsilon_1)\text{-free}\})\ge1-\theta\end{aligned}$$

for all $n\ge n_0$ and for all $U_i\in\mathcal{U}(n)$ and $H_i\in M_n(\mathbb{C})^+$ with $\|H_i\|\le R$ $(1\le i\le N)$. Then, as in the proof of Lemma 6.6.7, we have

$$\frac{(\gamma\otimes\Lambda_+)(\Xi_n(r,\varepsilon_1)\cap\Theta_n(r,\varepsilon))}{(\gamma\otimes\Lambda_+)(\Xi_n(r,\varepsilon_1))}\ge1-\theta$$

for large n. Therefore,

$$\lim_{n\to\infty}\frac{(\gamma\otimes\Lambda_+)(\Xi_n(r,\varepsilon_1)\cap\Theta_n(r,\varepsilon))}{(\gamma\otimes\Lambda_+)(\Xi_n(r,\varepsilon_1))}=1.$$

This implies that

$$\begin{aligned}&\chi_{(u,+),R}(u_1,\dots,u_N;h_1^2,\dots,h_N^2;r,\varepsilon)\\ &\quad\ge\limsup_{n\to\infty}\left[\frac{1}{n^2}\log(\gamma\otimes\Lambda_+)(\Xi_n(r,\varepsilon_1))+N\log n\right]\\ &\quad=\limsup_{n\to\infty}\left[\frac{1}{n^2}\sum_{i=1}^N\log\gamma_n(\Gamma_u(u_i;n,r,\varepsilon_1))\right.\\ &\qquad\qquad\left.+\frac{1}{n^2}\log\Lambda_+(\Gamma_{+,R}(h_1^2,\dots,h_N^2;n,r,\varepsilon_1))+N\log n\right]\\ &\quad=\sum_{i=1}^N\chi_u(u_i;r,\varepsilon_1)+\chi_{+,R}(h_1^2,\dots,h_N^2;r,\varepsilon_1),\end{aligned}$$

thanks to (6.5.1), and we obtain (6.6.8).

□

Next, we apply the relation shown above to get the additivity properties of the free entropies χ_u and $\hat{\chi}$. We first give the change of variable formula similar to Proposition 6.3.6 for $\chi_{(u,+)}$. To do so, we need a smoothing technique like Lemma 6.3.5. We denote by $\mathcal{F}_{\mathbb{T}}$ the set of all functions $f : \mathbb{T} \to \mathbb{T}$ which are given as $f(e^{\mathrm{i}\,t}) = e^{\mathrm{i}\,\phi(t)}$ by a continuous increasing function ϕ on $[0, 2\pi]$ with $\phi(0) = 0$, $\phi(2\pi) = 2\pi$. An $f \in \mathcal{F}_{\mathbb{T}}$ is said to be C^∞ if ϕ is. Note that if ϕ is differentiable at $t \in [0, 2\pi]$, then

$$\lim_{\eta\to\zeta} \left| \frac{f(\eta) - f(\zeta)}{\eta - \zeta} \right| = \phi'(t) \quad \text{for} \quad \zeta = e^{\mathrm{i}\,t}.$$

In this case we write $|f'(e^{\mathrm{i}\,t})|$ instead of $\phi'(t)$. For each unitary $u \in \mathcal{M}$ and $f \in \mathcal{F}_{\mathbb{T}}$ one can define the unitary $f(u)$ by functional calculus, that is, $f(u) := \int_{\mathbb{T}} f(\zeta)\, de(\zeta)$ for the spectral decomposition $u = \int_{\mathbb{T}} \zeta\, de(\zeta)$.

Lemma 6.6.9 *Let $u \in \mathcal{M}$ be a unitary with $\chi_u(u) > -\infty$, and let $f \in \mathcal{F}_{\mathbb{T}}$. Then there exists a sequence (f_m) of C^∞-functions in $\mathcal{F}_{\mathbb{T}}$ such that $|f'_m| > 0$ on $\mathbb{T}$, $\|f_m(u) - f(u)\| \to 0$ and $\chi_u(f_m(u)) \to \chi_u(f(u))$.*

On the other hand, we denote by $\mathcal{F}_{\mathbb{R}^+}$ the set of all continuous increasing functions $g : \mathbb{R}^+ \to \mathbb{R}^+$ with $g(0) = 0$.

Lemma 6.6.10 *Let $h \in \mathcal{M}^+$, $\chi(h) > -\infty$, and $g \in \mathcal{F}_{\mathbb{R}^+}$. Then there exists a sequence (g_m) of C^∞-functions in $\mathcal{F}_{\mathbb{R}^+}$ such that $g'_m > 0$ on $\mathbb{R}^+$, $\|g_m(h)-g(h)\| \to 0$ and $\chi(g_m(h)) \to \chi(g(h))$.*

Lemma 6.6.10 is essentially included in Lemma 6.3.5, so we omit the proof. The proof of Lemma 6.6.9 is similar, with some modifications as follows.

Proof of Lemma 6.6.9*:* Let μ be the distribution of u. Write $f(e^{\mathrm{i}\,t}) = e^{\mathrm{i}\,\phi(t)}$ with $\phi : [0, 2\pi] \to [0, 2\pi]$, and extend ϕ to $\mathbb{R}$ periodically, that is, $\phi(-t) := \phi(2\pi - t) - 2\pi$ and $\phi(2\pi + t) := \phi(t) + 2\pi$ for $t \in [0, 2\pi]$, and so on. Since $\Sigma(\mu) = \chi_u(u) > -\infty$ by (6.5.2), there are $0 < \delta(m) < 1/m$ ($m \in \mathbb{N}$) such that

$$\iint_{|\zeta-\eta|<\delta(m)} \log|\zeta - \eta|\, \mu(\zeta)\, d\mu(\eta) \geq -\frac{1}{m}, \tag{6.6.9}$$

$$\iint_{|\zeta-\eta|<\delta(m)} d\mu(\zeta)\, d\mu(\eta) \leq \frac{1}{m \log m}. \tag{6.6.10}$$

For each m choose a C^∞-function $\psi_m \geq 0$ with compact support in $(-\pi, \pi)$ and $\int \psi_m\, dt = 1$ such that

$$|\tilde{\phi}_m(t) - \phi(t)| \leq \frac{\delta(m)}{2m} \quad \text{for} \quad t \in \mathbb{R}, \tag{6.6.11}$$

where $\tilde{\phi}_m$ is defined by

$$\tilde{\phi}_m(t) := (\phi * \psi_m)(t) - (\phi * \psi_m)(0).$$

Note that $\tilde{\phi}_m(0) = 0$, $\tilde{\phi}_m(2\pi) = 2\pi$ and $\tilde{\phi}_m$ is periodic with period 2π. Now define

$$\phi_m(t) := \frac{1}{m}t + \left(1 - \frac{1}{m}\right)\tilde{\phi}_m(t) \qquad (t \in \mathbb{R}),$$

and $f_m(e^{\mathrm{i}\,t}) := e^{\mathrm{i}\,\phi_m(t)}$ for $e^{\mathrm{i}\,t} \in \mathbb{T}$. Then f_m is a C^∞-function in $\mathcal{F}_\mathbb{T}$ and $|f'_m| \geq 1/m$ on $\mathbb{T}$. Moreover, we get $\|f_m - f\|$ (the sup norm on $\mathbb{T}$) $\to 0$, and hence $\|f_m(u) - f(u)\| \to 0$.

Let m be so large that $\|f_m - f\| \leq (\sqrt{2} - 1)/2$ and $|f(\zeta) - f(\eta)| < 1$ if $|\zeta - \eta| < \delta(m)$. Note that if $|f(\zeta) - f(\eta)| < 1$, then

$$|f_m(\zeta) - f_m(\eta)| \leq |f(\zeta) - f(\eta)| + 2\|f_m - f\| < \sqrt{2}.$$

Write

$$\begin{aligned} \chi_u(f_m(u)) &= \iint \log|f_m(\zeta) - f_m(\eta)|\, d\mu(\zeta)\, d\mu(\eta) \\ &= \left(\iint_{|\zeta-\eta|<\delta(m)} + \iint_{|f(\zeta)-f(\eta)|\geq 1} + \iint_{\substack{|\zeta-\eta|\geq\delta(m) \\ |f(\zeta)-f(\eta)|<1}}\right) \\ &\qquad \log|f_m(\zeta) - f_m(\eta)|\, d\mu(\zeta)\, d\mu(\eta). \end{aligned}$$

When $|\zeta - \eta| < \delta(m)$, since $|f_m(\zeta) - f_m(\eta)| < \sqrt{2}$, we can write $f_m(\zeta) = e^{\mathrm{i}\,\phi_m(t)}$ ($\zeta = e^{\mathrm{i}\,t}$) and $f_m(\eta) = e^{\mathrm{i}\,\phi_m(s)}$ ($\eta = e^{\mathrm{i}\,s}$) with $|\phi_m(t) - \phi_m(s)| < \pi/2$. Since

$$\begin{aligned} |f_m(\zeta) - f_m(\eta)| &\geq \frac{2\sqrt{2}}{\pi}|\phi_m(t) - \phi_m(s)| \\ &\geq \frac{2\sqrt{2}}{\pi} \cdot \frac{|t-s|}{m} \geq \frac{2\sqrt{2}}{\pi} \cdot \frac{|\zeta - \eta|}{m}, \end{aligned}$$

we get by (6.6.9) and (6.6.10)

$$\begin{aligned} &\iint_{|\zeta-\eta|<\delta(m)} \log|f_m(\zeta) - f_m(\eta)|\, d\mu(\zeta)\, d\mu(\eta) \\ &\quad \geq \left(\log\frac{2\sqrt{2}}{\pi} - \log m\right)\frac{1}{m\log m} - \frac{1}{m} = \left(\log\frac{2\sqrt{2}}{\pi}\right)\frac{1}{m\log m} - \frac{2}{m}, \end{aligned}$$

implying

$$\liminf_{m\to\infty} \iint_{|\zeta-\eta|<\delta(m)} \log|f_m(\zeta) - f_m(\eta)|\, d\mu(\zeta)\, d\mu(\eta) \geq 0. \tag{6.6.12}$$

When $|f(\zeta) - f(\eta)| \geq 1$, since

$$\begin{aligned} |f_m(\zeta) - f_m(\eta)| &\geq |f(\zeta) - f(\eta)| - |f_m(\zeta) - f(\zeta)| - |f_m(\eta) - f(\eta)| \\ &> 1 - (\sqrt{2} - 1) > \frac{1}{2}, \end{aligned}$$

we get $|\log |f_m(\zeta) - f_m(\eta)|| \leq \log 2$ and

$$\begin{aligned} &\lim_{m\to\infty} \iint_{|f(\zeta)-f(\eta)|\geq 1} \log |f_m(\zeta) - f_m(\eta)| \, d\mu(\zeta) \, d\mu(\eta) \\ &\quad = \iint_{|f(\zeta)-f(\eta)|\geq 1} \log |f(\zeta) - f(\eta)| \, d\mu(\zeta) \, d\mu(\eta) \end{aligned} \tag{6.6.13}$$

by the bounded convergence theorem. Next, when $|\zeta-\eta| \geq \delta(m)$ and $|f(\zeta)-f(\eta)| < 1$, we can write $f_m(\zeta) = e^{\mathrm{i}\,\phi_m(t)}$ $(\zeta = e^{\mathrm{i}\,t})$ and $f_m(\eta) = e^{\mathrm{i}\,\phi_m(s)}$ $(\eta = e^{\mathrm{i}\,s})$ with $|\phi_m(t) - \phi_m(s)| < \pi/2$. Letting $t > s$ and noting that $t - s \geq |\zeta - \eta| \geq \delta(m)$, we get

$$\begin{aligned} \frac{\pi}{2} &> \phi_m(t) - \phi_m(s) \\ &= \frac{1}{m}(t-s) + \Big(1 - \frac{1}{m}\Big)(\tilde{\phi}_m(t) - \tilde{\phi}_m(s)) \\ &\geq \frac{\delta(m)}{m} + \Big(1 - \frac{1}{m}\Big)(\phi(t) - \phi(s) - |\tilde{\phi}_m(t) - \phi(t)| - |\tilde{\phi}_m(s) - \phi(s)|) \\ &\geq \Big(1 - \frac{1}{m}\Big)(\phi(t) - \phi(s)) \end{aligned}$$

by (6.6.11), and hence

$$\begin{aligned} |f_m(\zeta) - f_m(\eta)| &\geq |e^{\mathrm{i}\,(1-\frac{1}{m})(\phi(t)-\phi(s))} - 1| \\ &\geq \Big(1 - \frac{1}{m}\Big)|e^{\mathrm{i}\,(\phi(t)-\phi(s))} - 1| \\ &= \Big(1 - \frac{1}{m}\Big)|f(\zeta) - f(\eta)|. \end{aligned}$$

(For the second inequality above, note that $|e^{\mathrm{i}\,\alpha\theta} - 1| \geq \alpha|e^{\mathrm{i}\,\theta} - 1|$ for $0 \leq \alpha \leq 1$ and $0 \leq \theta \leq \pi/2$.) Therefore we have

$$\begin{aligned} &\iint_{\substack{|\zeta-\eta|\geq\delta(m) \\ |f(\zeta)-f(\eta)|<1}} \log |f_m(\zeta) - f_m(\eta)| \, d\mu(\zeta) \, d\mu(\eta) \\ &\quad \geq \log\Big(1 - \frac{1}{m}\Big) \iint_{|\zeta-\eta|\geq\delta(m)} d\mu(\zeta) \, d\mu(\eta) \\ &\qquad\quad + \iint_{\substack{|\zeta-\eta|\geq\delta(m) \\ |f(\zeta)-f(\eta)|<1}} \log |f(\zeta) - f(\eta)| \, d\mu(\zeta) \, d\mu(\eta) \\ &\quad \geq \log\Big(1 - \frac{1}{m}\Big) \cdot \frac{1}{m \log m} + \iint_{|f(\zeta)-f(\eta)|<1} \log |f(\zeta) - f(\eta)| \, d\mu(\zeta) \, d\mu(\eta), \end{aligned}$$

implying

$$\liminf_{m\to\infty} \iint_{\substack{|\zeta-\eta|\ge\delta(m)\\ |f(\zeta)-f(\eta)|<1}} \log|f_m(\zeta)-f_m(\eta)|\,d\mu(\zeta)\,d\mu(\eta)$$
$$\ge \iint_{|f(\zeta)-f(\eta)|<1} \log|f(\zeta)-f(\eta)|\,d\mu(\zeta)\,d\mu(\eta)\,. \tag{6.6.14}$$

The above estimates (6.6.12)–(6.6.14) together imply that

$$\liminf_{m\to\infty} \chi_u(f_m(u)) \ge \chi_u(f(u))\,.$$

This and the upper semicontinuity give the conclusion.

□

Lemma 6.6.11 *Let $u_1,\dots,u_N \in \mathcal{M}$ be unitaries with $\chi_u(u_i) > -\infty$ and $h_1,\dots,$ $h_L \in \mathcal{M}^+$ with $\chi_+(h_j) > -\infty$. Then*

$$\begin{aligned}&\chi_{(u,+)}(f_1(u_1),\dots,f_N(u_N);g_1(h_1),\dots,g_L(h_L))\\ &\quad\ge \chi_{(u,+)}(u_1,\dots,u_N;h_1,\dots,h_L)\\ &\qquad+\sum_{i=1}^N\big[\chi_u(f_i(u_i))-\chi_u(u_i)\big]+\sum_{j=1}^L\big[\chi(g_j(h_j))-\chi(h_j)\big]\end{aligned}$$

for every $f_1,\dots,f_N \in \mathcal{F}_\mathbb{T}$ and $g_1,\dots,g_L \in \mathcal{F}_{\mathbb{R}^+}$.

Proof: By Lemmas 6.6.9 and 6.6.10 together with Proposition 6.6.1 we may show the following two cases:

(a) If f is a C^∞-function in $\mathcal{F}_\mathbb{T}$ with $|f'|>0$ on $\mathbb{T}$, then

$$\begin{aligned}&\chi_{(u,+)}(f(u_1),u_2,\dots,u_N;h_1,\dots,h_N)\\ &\quad\ge \chi_{(u,+)}(u_1,\dots,u_N;h_1,\dots,h_N)+\chi_u(f(u_1))-\chi_u(u_1)\,.\end{aligned}$$

(b) If g is a C^∞-function in $\mathcal{F}_{\mathbb{R}^+}$ with $g'>0$ on $\mathbb{R}^+$, then

$$\begin{aligned}&\chi_{(u,+)}(u_1,\dots,u_N;g(h_1),h_2,\dots,h_N)\\ &\quad\ge \chi_{(u,+)}(u_1,\dots,u_N;h_1,\dots,h_N)+\chi(g(h_1))-\chi(h_1)\,.\end{aligned}$$

The proof of (b) is the same as Proposition 6.3.6. We sketch the similar proof of (a). For $\zeta,\eta\in\mathbb{T}$ define

$$K(\zeta,\eta) := \begin{cases} \left|\dfrac{f(\zeta)-f(\eta)}{\zeta-\eta}\right| & \text{if } \zeta\ne\eta,\\ |f'(\zeta)| & \text{if } \zeta=\eta. \end{cases}$$

Then $L(\zeta,\eta) := \log K(\zeta,\eta)$ is continuous on $\mathbb{T}^2$, and

$$\chi_u(f(u_1)) - \chi_u(u_1) = (\tau\otimes\tau)(L(u_1\otimes \mathbf{1}, \mathbf{1}\otimes u_1)).$$

Write $F(U_1,\dots,U_N;H_1,\dots,H_L) := (f(U_1),U_2,\dots,U_N;H_1,\dots,H_L)$ on $\mathcal{U}(n)^N \times (M_n(\mathbb{C})^+)^L$. For every $r\in\mathbb{N}$ and $\varepsilon>0$, by approximating f by a trigonometric polynomial, we see that

$$\begin{aligned} &F(\Gamma_{(u,+),R}(u_1,\dots,u_N;h_1,\dots,h_L;n,r_1,\varepsilon_1)) \\ &\qquad \subset \Gamma_{(u,+),R}(f(u_1),u_2,\dots,u_N;h_1,\dots,h_L;n,r,\varepsilon) \qquad (n\in\mathbb{N}) \end{aligned}$$

for some $r_1\in\mathbb{N}$ and $\varepsilon_1>0$. Since

$$\begin{aligned} \frac{d(\gamma_n\circ f)}{d\gamma_n}(U_1) &= \prod_{i<j}\left|\frac{f(\zeta_i)-f(\zeta_j)}{\zeta_i-\zeta_j}\right|^2 \prod_{i=1}^n |f'(\zeta_i)| \\ &= \exp(\mathrm{Tr}_n\otimes\mathrm{Tr}_n)(L(U_1\otimes I, I\otimes U_1)) \end{aligned}$$

($\zeta_1,\dots,\zeta_N$ are the eigenvalues of U_1), we can show as in the proof of Proposition 6.3.6 that for any $\delta>0$ there are $r_1\in\mathbb{N}$ and $\varepsilon_1>0$ such that

$$\left|\frac{1}{n^2}\log\frac{d(\gamma_n\circ f)}{d\gamma_n}(U_1) - \big[\chi_u(f(u_1)) - \chi_u(u_1)\big]\right| \le 3\delta$$

for all $(U_1,\dots,U_N;H_1,\dots,H_L)\in\Gamma_{(u,+),R}(u_1,\dots,u_N;n,r_1,\varepsilon_1)$, $n\in\mathbb{N}$, and the inequality in (a) is obtained.

□

If $f_1,\dots,f_N\in\mathcal{F}_\mathbb{T}$ and $g_1,\dots,g_L\in\mathcal{F}_{\mathbb{R}^+}$ are strictly increasing (in terms of angle for f_i), then the inequality in Lemma 6.6.11 can be replaced by equality.

Proposition 6.6.12 *If $u_1,\dots,u_N\in\mathcal{M}$ are unitaries, then $\chi_u(u_1,\dots,u_N)=0$ if and only if $u_1,\dots,u_N$ are *-free Haar unitaries.*

Proof: Choose free standard quarter-circular elements $h_1,\dots,h_N$ which are free from $\{u_1,\dots,u_N,u_1^*,\dots,u_N^*\}$. Theorem 6.6.6 says that $\chi_u(u_1,\dots,u_N)=0$ if and only if $\hat\chi(u_1h_1,\dots,u_Nh_N) = N\log(\pi e)$. According to Proposition 6.5.8 the latter equality holds if and only if $u_1h_1,\dots,u_Nh_N$ are *-free circular elements, which is equivalent to $u_1,\dots,u_N$ being *-free Haar unitaries.

□

Now we are in a position to supply the promised proof of Theorem 6.5.3.

Proof of Theorem 6.5.3*:* The proof of the first part is essentially included in the proof of Theorem 6.6.8. In fact, when $(h_1,\dots,h_N)$ is empty in the proof of (6.6.8), it can read as a proof of the first part here. So the details are left to the reader. To prove the second, assume that $\chi_u(u_i)>-\infty$ for $1\le i\le N$ and the additivity holds. For each i, since the distribution of u_i is nonatomic, there is a (unique) $f_i \in \mathcal{F}_{\mathbb{T}}$ such that the distribution of $f_i(u_i)$ is the Haar probability measure on $\mathbb{T}$, so $\chi_u(f_i(u_i)) = 0$. Then, by Lemma 6.6.11 (in the case of the set $\{h_j\}$ being empty) and the additivity assumption, we get

$$\chi_u(f_1(u_1), \dots, f_N(u_N)) \geq \sum_{i=1}^{N} \chi_u(f_i(u_i)) = 0\,.$$

Hence Proposition 6.6.12 implies that $f_1(u_1), \dots, f_N(u_N)$ are *-free, and so are $u_1, \dots, u_N$ because $u_i \in \{f_i(u_i)\}''$.

□

Theorem 6.6.13 *Let $a_1, \dots, a_N \in \mathcal{M}$ be such that $a_i = u_i h_i$ with a *-free pair of a unitary $u_i \in \mathcal{M}$ and $h_i \in \mathcal{M}^+$. If $a_1, \dots, a_N$ are *-free, then*

$$\hat{\chi}(a_1, \dots, a_N) = \hat{\chi}(a_1) + \cdots + \hat{\chi}(a_N)\,.$$

*Conversely, if $\hat{\chi}(a_i) > -\infty$ for $1 \leq i \leq N$ and the above equality holds, then $a_1, \dots, a_N$ are *-free.*

Proof: If $a_1, \dots, a_N$ are *-free, then $u_1, \dots, u_N, h_1, \dots, h_N$ are *-free due to the *-freeness of u_i, h_i. Hence Theorems 6.6.6 and 6.5.3 imply that

$$\hat{\chi}(a_1, \dots, a_N) = \sum_{i=1}^{N} \chi_u(u_i) + \sum_{i=1}^{N} \chi(h_i^2) + \frac{N}{2}\left(\log\frac{\pi}{2} + \frac{3}{2}\right) = \sum_{i=1}^{N} \hat{\chi}(a_i)\,.$$

Conversely, assume that $\hat{\chi}(a_i) > -\infty$ for $1 \leq i \leq N$ and the additivity holds. Since $\chi_u(u_i) > -\infty$ and $\chi(h_i^2) > -\infty$, one can choose $f_i \in \mathcal{F}_{\mathbb{T}}$ and $g_i \in \mathcal{F}_{\mathbb{R}^+}$ such that $f_i(u_i)$ is a Haar unitary and $g_i(h_i)^2$ is a standard quarter-circular. Then, letting $b_i := f_i(u_i)g_i(h_i)$ and using Theorem 6.6.3, Lemma 6.6.11 (applied to f_i, $g_i(t^{1/2})^2$) and Theorem 6.6.6, we get

$$\begin{aligned}
\hat{\chi}(b_1, \dots, b_N) &= \chi_{(u,+)}(f_1(u_1), \dots, f_N(u_N); g_1(h_1)^2, \dots, g_N(h_N)^2) \\
&\geq \hat{\chi}_{(u,+)}(u_1, \dots, u_N; h_1^2, \dots, h_N^2) \\
&\quad + \sum_{i=1}^{N} [\chi_u(f_i(u)) - \chi_u(u_i)] + \sum_{i=1}^{N} [\chi(g_i(h_i)^2) - \chi(h_i^2)] \\
&= \hat{\chi}(a_1, \dots, a_N) - \sum_{i=1}^{N} \hat{\chi}(a_i) + N\log(\pi e) = N\log(\pi e)\,.
\end{aligned}$$

Hence Proposition 6.5.8 implies that $b_1, \dots, b_N$ are *-free standard circulars, so that $a_1, \dots, a_N$ are *-free because $a_i \in \{b_i, b_i^*\}''$.

□

Theorem 6.6.13 can be applied in particular when $a_1, \dots, a_N$ are R-diagonal elements. Specializing to the case $\hat{\chi}(a)$ of a single non-selfadjoint $a \in \mathcal{M}$, we state

Proposition 6.6.14 *Let $a \in \mathcal{M}$ with $\hat{\chi}(a) > -\infty$, and let $a = uh$ be the polar decomposition. Then*

$$\hat{\chi}(a) \le \chi_u(u) + \chi(a^*a) + \frac{1}{2}\log\frac{\pi}{2} + \frac{3}{4},$$

*and equality is attained if and only if u, h are *-free. Moreover, the equality*

$$\hat{\chi}(a) = \chi(a^*a) + \frac{1}{2}\log\frac{\pi}{2} + \frac{3}{4}$$

holds if and only if a is R-diagonal.

Proof: Theorem 6.6.6 contains the "if" part of the first assertion. To see the "only if", choose $f \in \mathcal{F}_{\mathbb{T}}$ and $g \in \mathcal{F}_{\mathbb{R}^+}$ such that $f(u)$ is a Haar unitary and $g(h)^2$ is a standard quarter-circular. Then the equality $\hat{\chi}(uh) = \chi_u(u) + \chi_+(h^2)$ implies $\hat{\chi}(f(u)g(h)) = \log(\pi e)$ as in the proof of Theorem 6.6.13, and this means that $f(u)g(h)$ is a standard circular and hence u, h are *-free. The second assertion is immediate from the first.

□

By (6.6.2) we restate the above proposition as follows.

Corollary 6.6.15 *Let $\mu \in \mathcal{M}(\mathbb{R}^+)$ have compact support and $\Sigma(\mu) > -\infty$. If $a \in \mathcal{M}$ is such that a^*a has the distribution μ, then*

$$\hat{\chi}(a) \le \Sigma(\mu) + \log\pi + \frac{3}{2},$$

and equality is attained if and only if a is R-diagonal.

Example 6.6.16 Let μ_λ be the free Poisson distribution (i.e. the Marchenko-Pastur distribution) for $\lambda \ge 1$. If a is an R-diagonal element such that a^*a has the distribution μ_λ, then Corollary 6.6.15 and (5.5.9) give

$$\hat{\chi}(a) = \log\pi + \frac{1}{2}(1 + \lambda + \log\lambda + (\lambda-1)^2\log(1-\lambda^{-1}))\,.$$

Consider the functional

$$\hat{\chi}(a) - \tau(a^*a) + 2(\lambda-1)\Delta(a) \quad \text{for} \quad a \in \mathcal{M},$$

where $\Delta(a)$ is the Fuglede-Kadison determinant of a. If μ is the distribution of a^*a, then by Corollary 6.6.15 we have

$$\begin{aligned}
&\hat{\chi}(a) - \tau(a^*a) + 2(\lambda-1)\Delta(a) \\
&\quad \le \Sigma(\mu) - \int (t - (\lambda-1)\log t)\,d\mu(t) + \log\pi + \frac{3}{2},
\end{aligned}$$

because $2\Delta(a) = \int \log t \, d\mu(t)$. According to Proposition 5.3.7, the above functional attains the maximal value if and only if a is an R-diagonal element a such that a^*a has the distribution μ_λ.

□

Notes and Remarks. After the case of a single selfadjoint random variable in [202], the (multivariate) free entropy $\chi(a_1, \dots, a_N)$ was extensively developed in a series of papers by Voiculescu [203], [205]–[208].

Basic properties such as subadditivity and upper semicontinuity in Sec. 6.1 and noncommutative functional calculus for power series in Sec. 6.2 were presented in [203]. As for the change of variable formulas, Theorem 6.3.1 was proved in [203], while Propositions 6.3.3 and 6.3.6 are taken from [205].

The first part of Theorem 6.4.1 is contained in [203]. The proof based on the approximate freeness property (Lemma 6.4.2) is taken from [208], and the strong additivity (6.4.3) was given there in a slightly more general form. On the other hand, the converse part together with Theorem 6.4.4 was in [205]. [208] contains several results like Lemma 6.4.2 on the approximate freeness of Haar distributed unitary matrices; the concentration technique of Gromov and Milman is essential in that paper. The proof here is a direct application of the L^2-convergence in Lemma 4.3.2; thus the result of Gromov and Milman is not used.

In the classical case, when a random variable X has a continuously differentiable density f, the Fisher information of X is defined as

$$I(X) := \int \left(\frac{d}{dt} \log f(t) \right)^2 f(t) \, dt = \int \frac{f'(t)^2}{f(f)} \, dt.$$

For any random variable X with finite variance, one can reformulate the differential formula in [14] as follows:

$$S(X + \sqrt{s}Z) - \frac{1}{2} \int_0^s I(X + \sqrt{t}Z) \, dt = S(X) \qquad (s \geq 0),$$

where Z is a standard Gaussian random variable such that X and Z are independent. The free analogue of the Fisher information for a single selfadjoint variable a was introduced in [202] as $\Phi(a) := \frac{4}{3}\pi^2 \int f(t)^3 \, dt$ if the distribution of a has the density f (otherwise $\Phi(a) := +\infty$). The free analogue of the above differential formula was also obtained:

$$\Sigma(a + \sqrt{s}S) - \frac{1}{2} \int_0^s \Phi(a + \sqrt{t}S) \, dt = \Sigma(a) \qquad (s \geq 0),$$

where S is a standard semicircular variable free from a. (Constant coefficients in this formula and in the definition of Φ are properly changed from those in [202];

see also [207].) The above formula yields

$$\chi(a) = \frac{1}{2}\int_0^\infty \Bigl(\frac{1}{1+t} - \Phi(a+\sqrt{t}S)\Bigr)\,dt + \frac{1}{2}\log(2\pi e)\,.$$

By using the notion of noncommutative Hilbert transform, Voiculescu introduced the (relative) free information $\Phi^*(a_1,\dots,a_N : B)$ of selfadjoint variables $a_1,\dots,a_N$ with respect to a subalgebra B in [207]. Furthermore, he defined the (relative) free entropy of $(a_1,\dots,a_N)$ with respect to B as

$$\begin{aligned}&\chi^*(a_1,\dots,a_N : B)\\ &\quad := \frac{1}{2}\int_0^\infty \Bigl(\frac{N}{1+t} - \Phi^*(a_1+\sqrt{t}S_1,\dots,a_N+\sqrt{t}S_N : B)\Bigr)\,dt + \frac{N}{2}\log(2\pi e)\,,\end{aligned}$$

where $S_1,\dots,S_N$ are standard semicircular variables such that $B, S_1,\dots,S_N$ are free. For a single selfadjoint a, it was shown that $\Phi^*(a : \mathbb{C})$ coincides with $\Phi(a)$ whenever the distribution of a has a density, and hence $\chi^*(a : \mathbb{C}) = \chi(a)$. It is not known whether χ and χ^* coincide in the general multivariable case. The matricial entropy χ is also called the "microstates" free entropy, while χ^* is often called the "microstates-free" free entropy. The free informations Φ and Φ^* and the free entropy χ^* are not treated in this book.

There are some important inequalities known for free entropy and for free information. For instance, the free Cramér-Rao inequality shown in [202] says that $\Phi(a)\tau(a^2) \ge 1$, and equality holds if and only if a is centered semicircular. The free entropy power inequality in [186] says that $e^{2\chi(a+b)} \ge e^{2\chi(a)} + e^{2\chi(b)}$ holds for selfadjoint variables a, b; its extension to the multivariable case is not known. The similar inequalities for Φ^* and χ^* are in [207].

The contents of Sections 6.5 and 6.6 are taken from [103]. The interrelation among different types of free entropies given in Theorems 6.6.3 and 6.6.6 makes it easy for us to transform properties of the selfadjoint free entropy into properties of the unitary free entropy, and vice versa. The maximization problems at the end of this chapter were treated in [135] too. The same kind of maximization problems for Φ^* and χ^* were discussed in [134].

Brown's spectral distribution measure of a noncommutative random variable in a tracial W^*-probability space is based on the *Fuglede-Kadison determinant.* The function $\lambda \mapsto (2\pi)^{-1}\log\Delta(a-\lambda\mathbf{1})$ is subharmonic on $\mathbb{C}$, and the representing Riesz measure μ_a is called the *Brown measure* for a. For an *R-diagonal element* the Brown measure is supported on an annulus or on a disk, see Haagerup and Larsen [97]. In some examples the Brown measure coincides with the limiting eigenvalue density of the corresponding random matrix model; the complexified Wishart matrix is such a model.

Chapter 7

Relation to Operator Algebras

This chapter is mostly concerned with relations and applications of free probability theory to von Neumann algebras, in particular, to free group factors. We take a noncommutative probability space $(\mathcal{M}, \tau)$ of a von Neumann algebra $\mathcal{M}$ with a faithful normal tracial state τ. A selfadjoint element a in $\mathcal{M}$ with nonatomic distribution generates a von Neumann algebra isomorphic to $\mathcal{L}(\mathbb{Z})$. If $a_1, \dots, a_n$ are such elements in free relation, then the generated von Neumann algebra $\{a_1, \dots, a_n\}''$ is isomorphic to $\star_{i=1}^{n} \mathcal{L}(\mathbb{Z}) = \mathcal{L}(\mathbf{F}_n)$, the free group factor associated with the free group with n generators. In this way, the free group factors are naturally realized via free families of noncommutative random variables, and it turns out that three concepts of freeness, free product and free group factors have the same essence.

The long-standing and famous isomorphism problem of free group factors is whether it is possible to have $\mathcal{L}(\mathbf{F}_n) \cong \mathcal{L}(\mathbf{F}_m)$ if $n \neq m$. Great progress was recently made on this problem by F. Rădulescu and by K. Dykema as a continuation of Voiculescu's work; however, the final solution has not yet been achieved. Random matrix models provide a powerful machinery in their studies. Rădulescu and Dykema independently discovered a one-parameter family of type II_1 factors interpolating free group factors, which is central in the recent development of theory of free group factors.

There are remarkable applications of free entropy to the theory of factors akin to free group factors. Several open questions on factors were recently answered by using the free entropy technique. For instance, the non-existence of Cartan subalgebras in free group factors was shown by Voiculescu, and L. Ge proved that free group factors are *prime* in the sense that they are not isomorphic to the tensor product of two factors of type II_1.

Voiculescu introduced the notion of free entropy dimension by differentiating free entropy in a certain way. The advantage is that the free entropy dimension is more sensitive to freeness than the free entropy itself. There is another related notion of free dimension defined by Dykema for some class of von Neumann algebras.

7.1 Free group factors and semicircular systems

First of all, recall some basic terminology on von Neumann algebras, though we already used it in previous chapters. A *von Neumann algebra* (or W^**-algebra*) $\mathcal{M}$ on a Hilbert space $\mathcal{H}$ is a $*$-subalgebra of $B(\mathcal{H})$ which is closed in the strong (or equivalently weak) operator topology and contains the identity operator $\mathbf{1}$. The theory of von Neumann algebras was initiated by the monumental work of Murray and von Neumann [122]. The *double commutant theorem* of von Neumann says that a $*$-subalgebra $\mathcal{M}$ of $B(\mathcal{H})$ is a von Neumann algebra if and only if $\mathcal{M} = \mathcal{M}''$, where $\mathcal{M}' := \{a \in B(\mathcal{H}) : ab = ba,\ b \in \mathcal{M}\}$, the *commutant* of $\mathcal{M}$, and $\mathcal{M}'' := (\mathcal{M}')'$. A von Neumann algebra $\mathcal{M}$ is called a *factor* if its *center* is trivial, i.e. $\mathcal{M} \cap \mathcal{M}' = \mathbb{C}\mathbf{1}$. Factors are classified into *types* I_n $(n = 1, 2, \dots, \infty)$, II_1, II_∞, III_λ $(0 \le \lambda \le 1)$. We do not enter into the details on the classification of factors. But we give a few remarks about type I and type II factors here. A type I_n $(n < \infty)$ factor is just the matrix algebra $M_n(\mathbb{C})$, and a type I_∞ factor is $B(\mathcal{H})$ with $\dim \mathcal{H} = \infty$. Type II_1 factors are characterized as factors which have a faithful normal tracial state (unique and faithful automatically). This type of factors will be the main topic of our subsequent discussions. Any type II_∞ factor is represented as the tensor product $\mathcal{M} \otimes B(\mathcal{H})$ of a type II_1 factor $\mathcal{M}$ and $B(\mathcal{H})$ of type I_∞.

A von Neumann algebra $\mathcal{M}$ (acting on a separable Hilbert space) is said to be *hyperfinite* or *AFD* if there is an increasing sequence $(\mathcal{M}_n)$ of finite-dimensional $*$-subalgebras of $\mathcal{M}$ which generates $\mathcal{M}$, i.e. $\mathcal{M} = \left(\bigcup_n \mathcal{M}_n\right)''$. A remarkable result due to Murray and von Neumann [123] is that a hyperfinite type II_1 factor is unique (up to isomorphism). It is constructed, for example, by the closure of the infinite tensor $M_2(\mathbb{C}) \otimes M_2(\mathbb{C}) \otimes \cdots$ via the GNS representation by $\mathrm{tr}_2 \otimes \mathrm{tr}_2 \otimes \cdots$ (tr_2 being the tracial state on $M_2(\mathbb{C})$). The hyperfinite II_1 factor is usually denoted by R. As was finally proved in epoch-making work of Connes [52], several conditions equivalent to AFD for von Neumann algebras are known. Among others, a von Neumann algebra $\mathcal{M} \subset B(\mathcal{H})$ is AFD if and only if it is *injective*, that is, there exists a conditional expectation (or norm one projection) from $B(\mathcal{H})$ onto $\mathcal{M}$. Furthermore, Connes (and Haagerup for type III_1) proved that an AFD factor of each type except type III_0 is unique.

There is another important notion for II_1 factors. Let $\mathcal{M}$ be a type II_1 factor with a normal tracial state τ, and $B(\mathcal{H})$ a I_∞ factor with the usual trace Tr. For every $0 < t < \infty$ choose a projection p in $\mathcal{M} \otimes B(\mathcal{H})$ such that $(\tau \otimes \mathrm{Tr})(p) = t$. Then a II_1 factor $\mathcal{M}_t$ is defined as $p(\mathcal{M} \otimes B(\mathcal{H}))p$. The isomorphism class of $\mathcal{M}_t$ is independent of the choice of p having a given trace value t. When $0 < t \le 1$, we may define $\mathcal{M}_t = p\mathcal{M}p$ with a projection $p \in \mathcal{M}$ such that $\tau(p) = t$. One can readily see that

$$\mathcal{M}_{t_1 t_2} \cong (\mathcal{M}_{t_1})_{t_2} \qquad (t_1, t_2 > 0), \tag{7.1.1}$$

and hence

$$\{t \in (0, \infty) : \mathcal{M}_t \cong \mathcal{M}\}$$

forms a subgroup of the multiplicative group $(0, \infty)$. This is the so-called *funda-*

mental group of $\mathcal{M}$. One can equivalently define it as the set of $t \in (0, \infty)$ such that there exists an automorphism α of $\mathcal{M} \otimes B(\mathcal{H})$ satisfying $(\tau \otimes \mathrm{Tr}) \circ \alpha = t(\tau \otimes \mathrm{Tr})$. Concerning the hyperfinite (or injective) II_1 factor R, its uniqueness implies that the fundamental group of R is the whole $(0, \infty)$.

The subjects of this chapter are more or less related to free group factors. So let us first recall the construction of free group factors for the reader's convenience. Free group factors are typical examples of the group von Nuemann algebra associated with a discrete group.

Let G be a general discrete group. The Hilbert space $\ell^2(G)$ consists of $\xi : G \to \mathbb{C}$ such that $\sum_{g \in G} |\xi(g)|^2 < +\infty$, whose inner product is $\langle \xi, \eta \rangle := \sum_{g \in G} \xi(g)\overline{\eta(g)}$. The left and right regular representations of G on $\ell^2(G)$ are defined by

$$(L_g\xi)(h) := \xi(g^{-1}h), \quad (R_g\xi)(h) := \xi(hg) \qquad (\xi \in \ell^2(G),\ g, h \in G).$$

Then the (left) *group von Neumann algebra* $\mathcal{L}(G)$ of G is generated by $L_G := \{L_g : g \in G\}$, i.e. $\mathcal{L}(G) := (L_G)''$, and the right group von Neumann algebra is $\mathcal{R}(G) := (R_G)''$. It is obvious that $\mathcal{L}(G)$ and $\mathcal{R}(G)$ are commuting, namely,

$$\mathcal{L}(G) \subset \mathcal{R}(G)'. \tag{7.1.2}$$

Define a vector state τ on $\mathcal{L}(G)$ by

$$\tau(a) := \langle a\delta_e, \delta_e \rangle \qquad (a \in \mathcal{L}(G)),$$

where e is the identity of G and $\delta_g(h) = \delta_{gh}$ for $g, h \in G$. It is clear that δ_e is cyclic for both $\mathcal{L}(G)$ and $\mathcal{R}(G)$. Hence, thanks to (7.1.2), δ_e is a cyclic and separating vector for $\mathcal{L}(G)$, so τ is faithful on $\mathcal{L}(G)$. Since

$$\tau(L_gL_h) = \delta_e(gh) = \delta_e(hg) = \tau(L_hL_g),$$

we see that τ is a faithful normal tracial state on $\mathcal{L}(G)$. A conjugate-linear isometry J on $\ell^2(G)$ with $J^2 = \mathbf{1}$ is given by

$$(J\xi)(g) := \overline{\xi(g^{-1})} \qquad (\xi \in \ell^2(G),\ g \in G),$$

which is a particular case of the so-called *modular conjugation*. In fact, it is easy to check that

$$\begin{aligned} &Ja\delta_e = a^*\delta_e \qquad (a \in \mathcal{L}(G)),\\ &\mathcal{L}(G)' = J\mathcal{L}(G)J = \mathcal{R}(G). \end{aligned}$$

This means that equality indeed holds in (7.1.2).

Now let G be an infinite discrete group, and we determine when $\mathcal{L}(G)$ becomes a factor. For any $a \in \mathcal{L}(G)$, since

$$(L_h^* a L_h \delta_e)(g) = \tau(L_g^* L_h^* a L_h) = \tau(L_{hgh^{-1}}^* a) = (a\delta_e)(hgh^{-1})$$

and δ_e is separating for $\mathcal{L}(G)$, it is seen that a belongs to $\mathcal{L}(G)'$ if and only if $(a\delta_e)(hgh^{-1}) = (a\delta_e)(g)$ for all $g, h \in G$, that is, $a\delta_e \in \ell^2(G)$ is constant on each conjugacy class of G. This says that if the conjugacy class of each $g \in G \setminus \{e\}$ is infinite, then $a\delta_e \in \mathbb{C}\delta_e$ or $a \in \mathbb{C}\mathbf{1}$ for any a in the center of $\mathcal{L}(G)$. Hence $\mathcal{L}(G)$ becomes a factor if G is an *ICC group*, and vice versa. The free group $\mathbf{F}_n$ with n generators ($n = 2, 3, \ldots, \infty$) is a typical example of ICC groups. In this way, we have the type II_1 factors $\mathcal{L}(\mathbf{F}_n)$, called the *free group factors*. It is worth noting that the free group factors are not hyperfinite; indeed, $\mathcal{L}(\mathbf{F}_2)$ was the first example of non-hyperfinite von Neumann algebras discovered by Murray and von Neumann.

Next, for convenience, let us recall and fix terms about free families of some special kinds. Let $(\mathcal{A}, \varphi)$ be a C^*-probability space. Let $(a_i)_{i\in I}$ be a finite or countable family of noncommutative random variables in $\mathcal{A}$. We say that $(a_i)_{i\in I}$ is a *semicircular system* if it is a free family of standard semicircular elements (of distribution w_2). A semicircular system is sometimes assumed to have the distribution w_1, but we prefer w_2 in accordance with the limit distribution of the standard Gaussian random matrix model (cf. Theorem 4.1.7). Also, we say that $(a_i)_{i\in I}$ is a *circular system* if it is a *-free family of standard circular elements, that is, $\big((a_i + a_i^*)/\sqrt{2}\big)_{i\in I} \cup \big((a_i - a_i^*)/\sqrt{2}\mathrm{i}\big)_{i\in I}$ is a semicircular system (cf. Example 2.6.2).

The next proposition tells us that the von Neumann algebra generated by a semicircular or circular system is isomorphic to a free group factor.

Proposition 7.1.1 *Let $(\mathcal{M}, \varphi)$ be a W^*-probability space with a faithful normal state φ on $\mathcal{M}$.*

(1) *If $(a_i)_{i\in I}$ is a semicircular system in $\mathcal{M}$, then the generated von Neumann algebra $\{a_i : i \in I\}''$ is isomorphic to $\mathcal{L}(\mathbf{F}_{|I|})$.*

(2) *If $(a_i)_{i\in I}$ is a circular system in $\mathcal{M}$, then the von Neumann algebra generated by $(a_i)_{i\in I}$ is isomorphic to $\mathcal{L}(\mathbf{F}_{2|I|})$.*

(3) *If $(u_i)_{i\in I}$ is a finite or countable free family of Haar unitaries in $\mathcal{M}$, then $\{u_i : i \in I\}''$ is isomorphic to $\mathcal{L}(\mathbf{F}_{|I|})$.*

Proof: (2) is an immediate consequence of (1). If $a = a^*$ has the distribution w_2 and $f : [-2, 2] \to [0, 1]$ is defined by $f'(t) = \frac{1}{2\pi}\sqrt{4 - t^2}$, then $u := \exp(2\pi\mathrm{i}\, f(a))$ is a Haar unitary. So, it is enough to show (3) only. Let g_i ($i \in I$) be the generators of $\mathbf{F}_{|I|}$, and let L be the left regular representation of $\mathbf{F}_{|I|}$ on $\ell^2(\mathbf{F}_{|I|})$, as explained above. Let $\mathcal{A}$ and $\mathcal{B}$ denote the (algebraic) *-algebras generated by $(L_{g_i})_{i\in I}$ and $(u_i)_{i\in I}$, respectively. A *-homomorphism $\rho : \mathcal{A} \to \mathcal{B}$ is defined by

$$\rho(L_{g_{i_1}}^{k_1} L_{g_{i_2}}^{k_2} \cdots L_{g_{i_n}}^{k_n}) = u_{i_1}^{k_1} u_{i_2}^{k_2} \cdots u_{i_n}^{k_n}$$

for $i_1 \neq i_2 \neq \ldots \neq i_n$ and $k_j \in \mathbb{Z} \setminus \{0\}$ for $1 \leq j \leq n$. Since

$$\tau\big(L_{g_{i_1}}^{k_1} \cdots L_{g_{i_n}}^{k_n}\big) = \tau\big(L_{g_{i_1}^{k_1} \cdots g_{i_n}^{k_n}}\big) = 0 = \varphi\big(u_{i_1}^{k_1} \cdots u_{i_n}^{k_n}\big)$$

by the assumption of free Haar unitaries, we have

$$\tau(a) = \varphi \circ \rho(a) \qquad (a \in \mathcal{A}).$$

Thanks to the faithfulness of φ, this implies that ρ is a *-isomorphism between $\mathcal{A}$ and $\mathcal{B}$. When $\pi_{(\mathcal{A},\tau)}$ and $\pi_{(\mathcal{B},\varphi)}$ denote the GNS representations of $(\mathcal{A}, \tau)$ and $(\mathcal{B}, \varphi)$, we have

$$\{u_i : i \in I\}'' \cong \pi_{(\mathcal{B},\varphi)}(\mathcal{B})'' \cong \pi_{(\mathcal{A},\tau)}(\mathcal{A})'' = \mathcal{L}(\mathbf{F}_{|I|})\,.$$

□

The above proof shows that if $(a_i)_{i\in I}$ is a semicircular system in $(\mathcal{M}, \varphi)$, then φ is automatically tracial on $\{a_i : i \in I\}''$. This was indeed shown in Proposition 2.2.6.

The free product of noncommutative probability spaces was explained in Sec. 2.1 at the algebraic level. In the following let us discuss the free product at a more analytic level to complement the contents of Sec. 2.1. To do so, we first need to define the free product of Hilbert spaces. Let $(\mathcal{H}_i, \xi_i)_{i\in I}$ be a family of Hilbert spaces with distiguished vectors $\xi_i \in \mathcal{H}_i$ ($\|\xi_i\| = 1$). Set $\mathcal{H}_i^0 := \mathcal{H}_i \ominus \mathbb{C}\xi_i$. The *Hilbert space free product* $\mathcal{H}$ with a distinguished vector ξ ($\|\xi\| = 1$) is the direct sum of $\mathbb{C}\xi$ and the tensor products $\mathcal{H}_{i_1}^0 \otimes \mathcal{H}_{i_2}^0 \otimes \cdots \otimes \mathcal{H}_{i_n}^0$ for all $i_1 \neq i_2 \neq \ldots \neq i_n$, that is,

$$\mathcal{H} := \mathbb{C}\xi \oplus \bigoplus_{n=1}^{\infty} \left(\bigoplus_{i_1 \neq i_2 \neq \ldots \neq i_n} \mathcal{H}_{i_1}^0 \otimes \mathcal{H}_{i_2}^0 \otimes \cdots \otimes \mathcal{H}_{i_n}^0 \right).$$

We write $(\mathcal{H}, \xi) = \star_{i\in I}(\mathcal{H}_i, \xi_i)$. For each $i \in I$, take a subspace $\mathcal{H}(i)$ of $\mathcal{H}$ as

$$\mathcal{H}(i) := \mathbb{C}\xi \oplus \bigoplus_{n=1}^{\infty} \left(\bigoplus_{i \neq i_1 \neq i_2 \neq \ldots \neq i_n} \mathcal{H}_{i_1}^0 \otimes \mathcal{H}_{i_2}^0 \otimes \cdots \otimes \mathcal{H}_{i_n}^0 \right),$$

and define a unitary operator $V_i : \mathcal{H}_i \otimes \mathcal{H}(i) \to \mathcal{H}$ as follows:

$$V_i : \begin{cases} \xi_i \otimes \xi \mapsto \xi, & \\ \eta \otimes \xi \mapsto \eta & \text{for } \eta \in \mathcal{H}_i^0, \\ \xi_i \otimes \zeta \mapsto \zeta & \text{for } \zeta \in \mathcal{H}_{i_1}^0 \otimes \cdots \otimes \mathcal{H}_{i_n}^0, \\ \eta \otimes \zeta \mapsto \eta \otimes \zeta & \text{for } \eta \in \mathcal{H}_i^0,\ \zeta \in \mathcal{H}_{i_1}^0 \otimes \cdots \otimes \mathcal{H}_{i_n}^0. \end{cases} \tag{7.1.3}$$

Then a *-representation $\lambda_i : B(\mathcal{H}_i) \to B(\mathcal{H})$ can be defined by

$$\lambda_i(a) := V_i(a \otimes \mathbf{1}_{\mathcal{H}(i)})V_i^* \qquad (a \in B(\mathcal{H}_i)). \tag{7.1.4}$$

When $(\mathcal{M}_i)_{i\in I}$ is a family of von Neumann algebras $\mathcal{M}_i$ on $(\mathcal{H}_i, \xi_i)$, where ξ_i is cyclic and separating for $\mathcal{M}_i$, the *von Neumann algebra free product* $\star_{i\in I}\mathcal{M}_i$ is generated by $\lambda_i(\mathcal{M}_i)$, $i \in I$:

$$\star_{i\in I}\mathcal{M}_i := \left(\bigcup_{i\in I}\lambda_i(\mathcal{M}_i)\right)'' .$$

We can also introduce the free product of *-representations using the above setting. For $i \in I$ let $\mathcal{A}_i$ be a *-algebra and π_i a *-representation of $\mathcal{A}_i$ on $(\mathcal{H}_i, \xi_i)$. Define the algebraic free product $\star_{i\in I}\mathcal{A}_i$ as in Sec. 2.1:

$$\star_{i\in I}\mathcal{A}_i := \mathbb{C}\mathbf{1} \oplus \bigoplus_{n=1}^{\infty}\left(\bigoplus_{i_1\neq i_2\neq\ldots\neq i_n} \mathcal{A}_{i_1}^0\mathcal{A}_{i_2}^0\cdots\mathcal{A}_{i_n}^0\right)$$

with the product given in (2.1.1) and (2.1.2). Using the notation in (7.1.4), we define a *-representation $\lambda_i : \mathcal{A}_i \to B(\mathcal{H})$ by

$$\lambda_i(a) := V_i(\pi_i(a)\otimes \mathbf{1}_{\mathcal{H}(i)})V_i^* \qquad (a \in \mathcal{A}_i).$$

Then the *free product representation* $\pi = \star_{i\in I}\pi_i$ of $\mathcal{A} = \star_{i\in i}\mathcal{A}_i$ is defined by

$$\pi(a_1a_2\cdots a_n) := \lambda_{i_1}(a_1)\lambda_{i_2}(a_2)\cdots\lambda_{i_n}(a_n)$$

for $a_j \in \mathcal{A}_{i_j}$, $i_1 \neq i_2 \neq \ldots \neq i_n$. For instance, let φ_i be a state on $\mathcal{A}_i$ and $(\pi_i, \mathcal{H}_i, \xi_i)$ the GNS cyclic representation of $(\mathcal{A}_i, \varphi_i)$, so that $\varphi_i(a) = \langle\pi_i(a)\xi_i, \xi_i\rangle$ $(a \in \mathcal{A}_i)$ and $\mathcal{H}_i = \overline{\pi_i(\mathcal{A}_i)\xi_i}$. Then the *free product state* φ on $\mathcal{A} = \star_{i\in I}\mathcal{A}_i$ is given as $\varphi(a) := \langle\pi(a)\xi, \xi\rangle$, where $(\mathcal{H}, \xi) = \star_{i\in I}(\mathcal{H}_i, \xi_i)$ and $\pi = \star_{i\in I}\pi_i$. Then one can easily show that $(\pi, \mathcal{H}, \xi)$ is the GNS representation of $(\mathcal{A}, \varphi)$ and $\pi(\mathcal{A})'' = \star_{i\in I}\pi_i(\mathcal{A}_i)''$.

The following is a von Neumann algebra version of Example 2.1.2.

Example 7.1.2 Let $(G_i)_{i\in I}$ be a family of discrete groups and $G = \star_{i\in I}G_i$ the free product group. With the identities e_i of G_i and e of G we have

$$(\ell^2(G), \delta_e) = \mathop{\star}_{i\in I}(\ell^2(G_i), \delta_{e_i})$$

under the identification

$$\delta_{g_1\cdots g_n} = \delta_{g_1}\otimes\cdots\otimes\delta_{g_n} \in \ell^2(G_{i_1})^0\otimes\cdots\otimes\ell^2(G_{i_n})^0$$

for $g_j \in G_{i_j}\setminus\{e_{i_j}\}$, $i_1 \neq i_2 \neq \ldots \neq i_n$. Let $L^{(i)}$ denote the regular representation of G_i on $\ell^2(G_i)$ and L that of G on $\ell^2(G)$. Then it is straightforward to check that $\lambda_i(L_g^{(i)}) = L_g$ for all $g \in G_i$. Hence

$$\mathcal{L}(G) = \left(\bigcup_{i\in I}\lambda_i(\mathcal{L}(G_i))\right)'' = \mathop{\star}_{i\in I}\mathcal{L}(G_i).$$

In particular, since $\mathbf{F}_{|I|} = \star_{i\in I}\mathbb{Z}$, we have

$$\mathcal{L}(\mathbf{F}_{|I|}) = \mathop{\star}_{i\in I} \mathcal{L}(\mathbb{Z})\,. \tag{7.1.5}$$

□

The definition of free relation modeled the free product construction, so that we have the following property as a matter of course.

Proposition 7.1.3 *Let $\mathcal{M}_i$ be a von Neumann algebra on $(\mathcal{H}_i, \xi_i)$ for $i \in I$, and let $\mathcal{M} = \star_{i\in I}\mathcal{M}_i$ be the von Neumann algebra free product on $(\mathcal{H}, \xi) = \star_{i\in I}(\mathcal{H}_i, \xi_i)$. Then $(\mathcal{M}_i)_{i\in I}$ is a free family in $(\mathcal{M}, \varphi)$, where $\varphi = \langle \cdot\,\xi, \xi\rangle$ and $\mathcal{M}_i$ is identified with $\lambda_i(\mathcal{M}_i)$.*

Proof: Let $i_1 \neq i_2 \neq \ldots \neq i_n$ and $a_j \in \mathcal{M}_{i_j}$, and assume that $\varphi(\lambda_{i_j}(a_j)) = \langle a_j\xi_{i_j}, \xi_{i_j}\rangle = 0$ or v$a_j\xi_{i_j} \in \mathcal{H}^0_{i_j}$ for $1 \leq j \leq n$. Then it is readily seen that

$$\lambda_{i_1}(a_1)\cdots\lambda_{i_n}(a_n)\xi = a_1\xi_{i_1} \otimes \cdots \otimes a_n\xi_{i_n} \in \mathcal{H}^0_{i_1} \otimes \cdots \otimes \mathcal{H}^0_{i_n}\,,$$

and hence $\varphi(\lambda_{i_1}(a_1)\cdots\lambda_{i_n}(a_n)) = 0$.

□

We explained in Sec. 4.2 that the standard selfadjoint (or real symmetric) Gaussian random matrices form an asymptotic model of a semicircular system. On the other hand, the full Fock space picture provides a direct model of a semicircular system, as can be seen more or less from Theorem 1.1.5 and Example 2.2.1. This is fully summarized in the next theorem.

Theorem 7.1.4 *Let $\mathcal{H}$ be a Hilbert space with an orthonormal basis $(f_i)_{i\in I}$. Set $s(f_i) := \ell(f_i)^* + \ell(f_i)$ and $\mathcal{M} := \{s(f_i) : i \in I\}''$ on the full Fock space $\mathcal{F}(\mathcal{H})$. Then:*

(1) *The vacuum vector Φ is cyclic and separating for $\mathcal{M}$.*

(2) *$\tau := \langle \cdot\,\Phi, \Phi\rangle$ is a faithful normal tracial state on $\mathcal{M}$.*

(3) *$(s(f_i))_{i\in I}$ is a semicircular system in $(\mathcal{M}, \tau)$.*

(4) *$\mathcal{M} \cong \mathcal{L}(\mathbf{F}_{|I|})$.*

The assertion (3) is already known from Theorem 1.1.5 and Example 2.2.1 (the argument of Example 2.2.1 is valid for many $\ell(f_i)$'s). The proof below is based on a general property of free products (Proposition 7.1.3). We give two lemmas to prove the theorem. The first lemma treats the particular case where $\dim\mathcal{H} = 1$.

Lemma 7.1.5 *Let $\mathcal{H} = \mathbb{C}f$ be one-dimensional and $\mathcal{M} := \{s(f)\}''$ on $\mathcal{F}(\mathcal{H})$. Then $(\mathcal{M}, \mathcal{F}(\mathcal{H}), \Phi)$ is unitarily conjugate to $(\mathcal{L}(\mathbb{Z}), \ell^2(\mathbb{Z}), \delta_0)$.*

Proof: Since $\mathcal{F}(\mathcal{H}) = \bigoplus_{n=0}^{\infty} \mathbb{C}f^{\otimes n}$ with $f^{\otimes 0} = \Phi$ and

$$(\ell(f)^* + \ell(f))^n \Phi - f^{\otimes n} \in \bigoplus_{k=0}^{n-1} \mathbb{C}f^{\otimes k} \qquad (n \geq 1),$$

it follows that Φ is cyclic for $\mathcal{M}$. Hence Φ is cyclic and separating for $\mathcal{M}$, because $\mathcal{M}$ is commutative. Since the distribution of $s(f)$ with respect to $\tau = \langle \cdot \Phi, \Phi \rangle$ is w_2 by Theorem 1.1.5, $\mathcal{M}$ is generated by a Haar unitary U (see the proof of Proposition 7.1.1). Thus we see that

$$(\mathcal{M}, \mathcal{F}(\mathcal{H}), \Phi) \cong (L^\infty(\mathbb{T}), L^2(\mathbb{T}), 1),$$

and the latter is unitarily conjugate to $(\mathcal{L}(\mathbb{Z}), \ell^2(\mathbb{Z}), \delta_0)$ via the Fourier transform. □

Lemma 7.1.6 *Let $\mathcal{M}_i$ be a von Neumann algebra on $(\mathcal{H}_i, \xi_i)$ for $i \in I$, and let $(\mathcal{H}, \xi) = \star_{i \in I}(\mathcal{H}_i, \xi_i)$. If ξ_i is cyclic and separating for $\mathcal{M}_i$ for each $i \in I$, then ξ is cyclic and separating for $\star_{i \in I}\mathcal{M}_i$ too.*

Proof: Besides the "left" free product $\mathcal{M} = \star_{i \in I}\mathcal{M}_i$ one can define the "right" free product as follows: Set

$$\mathcal{H}_r(i) := \mathbb{C}\xi \oplus \bigoplus_{n=1}^{\infty} \left(\bigoplus_{i_1 \neq i_2 \neq \ldots \neq i_n \neq i} \mathcal{H}_{i_1}^0 \otimes \mathcal{H}_{i_2}^0 \otimes \cdots \otimes \mathcal{H}_{i_n}^0 \right)$$

and define a unitary operator $W_i : \mathcal{H}_r(i) \otimes \mathcal{H}_i \to \mathcal{H}$ analogously to (7.1.3). Let $\rho_i : B(\mathcal{H}_i) \to B(\mathcal{H})$ be a *-representation, given as

$$\rho_i(a) := W_i(\mathbf{1}_{\mathcal{H}_r(i)} \otimes a)W_i^* \qquad (a \in B(\mathcal{H}_i)).$$

Now let $\mathcal{N}_i := \mathcal{M}_i'$ on $\mathcal{H}_i$ and define the "right" free product

$$\mathcal{N} := \left(\bigcup_{i \in I} \rho_i(\mathcal{N}_i) \right)''.$$

Since we immediately see that $\lambda_i(a)\rho_j(b) = \rho_j(b)\lambda_i(a)$ for all $i, j \in I$ and $a \in \mathcal{M}_i$, $b \in \mathcal{N}_j$, we have $\mathcal{M} \subset \mathcal{N}'$. It is clear by construction that ξ is cyclic for $\mathcal{N}$ as well as for $\mathcal{M}$. So ξ is also separating for $\mathcal{M}$. □

It is worth noting that $\mathcal{M} = \mathcal{N}'$ indeed holds under the assumption of the above lemma.

Proof of Theorem 7.1.4: For $i \in I$ set $\mathcal{H}_i := \mathbb{C} f_i$. Since $\mathcal{H} = \bigoplus_{i\in I} \mathcal{H}_i$ and $\mathcal{F}(\mathcal{H}_i) = \mathbb{C}\Phi_i \oplus \bigoplus_{n=1}^\infty \mathbb{C} f_i^{\otimes n}$, we have

$$\begin{aligned}
\bigoplus_{n=1}^\infty \mathcal{H}^{\otimes n} &= \bigoplus_{n=1}^\infty \bigoplus_{i_1,\dots,i_n\in I} \mathcal{H}_{i_1}\otimes\cdots\otimes\mathcal{H}_{i_n} \\
&= \bigoplus_{n,k=1}^\infty \bigoplus_{\substack{i_1\neq i_2\neq\dots\neq i_k \\ n_j\in\mathbb{N},\, n_1+\cdots+n_k=n}} \mathbb{C} f_{i_1}^{\otimes n_1}\otimes\cdots\otimes \mathbb{C} f_{i_k}^{\otimes n_k} \\
&= \bigoplus_{k=1}^\infty \bigoplus_{i_1\neq i_2\neq\dots\neq i_k} \mathcal{F}(\mathcal{H}_{i_1})^0\otimes\cdots\otimes\mathcal{F}(\mathcal{H}_{i_k})^0 ,
\end{aligned}$$

which shows that

$$(\mathcal{F}(\mathcal{H}),\Phi) = \mathop{\star}_{i\in I} (\mathcal{F}(\mathcal{H}_i),\Phi_i) .$$

It is readily verified that for each $i \in I$

$$\lambda_i(\ell_i(f_i)) = V_i(\ell_i(f_i)\otimes \mathbf{1}_{\mathcal{F}(\mathcal{H})(i)})V_i^* = \ell(f_i),$$

and hence $\lambda_i(s_i(f_i)) = s(f_i)$, where $\ell_i(\cdot)$ and $s_i(\cdot)$ mean $\ell(\cdot)$ and $s(\cdot)$ defined on $\mathcal{F}(\mathcal{H}_i)$. Therefore, we have

$$\mathcal{M} = \{\lambda_i(s_i(f_i)) : i\in I\}'' = \mathop{\star}_{i\in I} \mathcal{M}_i ,$$

where $\mathcal{M}_i = \{s_i(f_i)\}''$ on $\mathcal{F}(\mathcal{H}_i)$. Now (1) follows from Lemmas 7.1.5 and 7.1.6. Proposition 7.1.3 together with Theorem 1.1.5 implies (3), so Proposition 7.1.1 and its remark yield (4) and (2). Also, (4) can be more directly shown as

$$\mathcal{M} = \mathop{\star}_{i\in I} \mathcal{M}_i \cong \mathop{\star}_{i\in I} \mathcal{L}(\mathbb{Z}) = \mathcal{L}(\mathbf{F}_{|I|})$$

by Lemma 7.1.5 and (7.1.5).

□

Furthermore, a circular system can be canonically realized on the full Fock space. Set $c_i := \ell(f_i)^* + \ell(g_i)$ and $\tilde{c}_i := (s(f_i) + \mathrm{i}\, s(g_i))/\sqrt{2}$ on $\mathcal{F}(\mathcal{H})$, where $(f_i)_{i\in I}\cup(g_i)_{i\in I}$ is an orthonormal basis of $\mathcal{H}$. Theorem 7.1.4 shows that $(\tilde{c}_i)_{i\in I}$ is a circular system in $(\mathcal{M} = \{s(f_i), s(g_i) : i\in I\}'', \tau = \langle\cdot\Phi,\Phi\rangle)$, and $\mathcal{M}$ is isomorphic to $\mathcal{L}(\mathbf{F}_{2|I|})$. Also, $(c_i)_{i\in I}$ is a circular system, as we remarked before Corollary 4.3.8 (see also Example 2.6.2). The unitary operator $\mathcal{F}(U)$ on $\mathcal{F}(\mathcal{H})$ preserving the vacuum vector is induced from the unitary operator U on $\mathcal{H}$ given by $Uf_i := (f_i - \mathrm{i}\, g_i)/\sqrt{2}$, $Ug_i := (f_i + \mathrm{i}\, g_i)/\sqrt{2}$ $(i\in I)$. Then one gets

$$\mathcal{F}(U)\ell(f_i)\mathcal{F}(U)^* = \frac{\ell(f_i) - \mathrm{i}\,\ell(g_i)}{\sqrt{2}}, \quad \mathcal{F}(U)\ell(g_i)\mathcal{F}(U)^* = \frac{\ell(f_i) + \mathrm{i}\,\ell(g_i)}{\sqrt{2}},$$

so $\mathcal{F}(U)c_i\mathcal{F}(U)^* = \tilde{c}_i$ and $\{c_i : i\in I\}''$ is unitarily conjugate to $\{\tilde{c}_i : i\in I\}''$.

7.2 Interpolated free group factors

The discussions in the previous section tell us that semicircular and circular systems are essentially the same as free group factors from the viewpoint of von Neumann algebras. Indeed, in this section it will turn out that the technique using (semi)circular systems and their random matrix models is quite useful in the analysis of free group factors.

From now on we always consider a tracial W^*-probability space $(\mathcal{M}, \tau)$ of a von Neumann algebra $\mathcal{M}$ and a faithful normal tracial state τ. We first give some lemmas whose proofs are typical applications of Gaussian random matrix models.

Lemma 7.2.1 *In $(\mathcal{M}, \tau)$ let $(a_s)_{s\in S}$ be a semicircular system and $(e_{ij})_{i,j=1}^N$ a system of matrix units with $\tau(e_{ii}) = 1/N$ such that $\{a_s : s \in S\}$ and $\{e_{ij} : 1 \le i, j \le N\}$ are in free relation. Define*

$$\begin{aligned} \Omega_1 &:= \{N^{1/2} e_{1i} a_s e_{i1} : 1 \le i \le N,\ s \in S\}, \\ \Omega_2 &:= \{N^{1/2} e_{1i} a_s e_{j1} : 1 \le i < j \le N,\ s \in S\}. \end{aligned}$$

Then the following hold in $(e_{11}\mathcal{M}e_{11}, N\tau|_{e_{11}\mathcal{M}e_{11}})$:

(1) *Ω_1 is a semicircular system.*

(2) *Ω_2 is a circular system.*

(3) *$\Omega_1 \cup \Omega_2$ is a *-free family.*

Proof: For $n \in N\mathbb{N}$ let $(H(s,n))_{s\in S}$ be an independent family of $n \times n$ standard selfadjoint Gaussian matrices. Also, let $(E_{ij})_{i,j=1}^N$ be the usual matrix units of $M_N(\mathbb{C})$, and let $E_{ij}(n)$ ($n \in N\mathbb{N}$) be the $n \times n$ block-diagonal matrix all of whose diagonals are E_{ij}. Then Corollary 4.3.6 says that

$$\Big((\{H(s,n)\})_{s\in S},\ \{E_{ij}(n) : 1 \le i, j \le N\}\Big)$$

is asymptotically free as $n \to \infty$ (through multiples of N), and its limit distribution is equal to the distribution of

$$\Big((\{a_s\})_{s\in S},\ \{e_{ij} : 1 \le i, j \le N\}\Big)$$

due to the assumptions on (a_s) and (e_{ij}). So, for any $1 \le i, j \le N$ the distribution of $e_{1i} a_s e_{j1}$ is equal to the limit distribution of $E_{1i}(n) H(s,n) E_{j1}(n)$. Note that $(N^{1/2} E_{1i}(n) H(s,n) E_{j1}(n))_{1 \le i \le j \le N,\, s\in S}$ is an independent family of $n/N \times n/N$ standard selfadjoint (for $i = j$) or non-selfadjoint (for $i < j$) Gaussian matrices (if the trivial summand of size $n - n/N$ is neglected). Hence the result follows from Corollaries 4.3.6 and 4.3.8 together.

□

In previous chapters we sometimes discussed compressions of noncommutative random variables, cf. Examples 2.4.7, 2.6.9 and Lemma 4.4.10. The following is another result of the same kind.

Lemma 7.2.2 *In $(\mathcal{M}, \tau)$ let R be a copy of the hyperfinite II_1 factor, and let $(a_s)_{s \in S}$ be a semicircular system such that R and $\{a_s : s \in S\}$ are in free relation. Let $p \in R$ be a nonzero projection. Then in $(p\mathcal{M}p, \tau(p)^{-1}\tau|_{p\mathcal{M}p})$, $(\tau(p)^{-1/2} p a_s p)_{s \in S}$ is a semicircular system which is free from pRp.*

Proof: First assume that $\tau(p) = l/2^k$, a dyadic rational number. Note that $\big(R, (\{u a_s u^*\})_{s \in S}\big)$ is free again if $u \in R$ is a unitary. Furthermore, note that if $p, q \in R$ are projections with $\tau(p) = \tau(q)$, then $u^* p u = q$ for some unitary $u \in R$. So we may show the result for a specific p having the above trace value. Represent $R = M_{2^k}(\mathbb{C}) \otimes M_2(\mathbb{C}) \otimes M_2(\mathbb{C}) \otimes \cdots$ and choose a diagonal $p \in M_{2^k}(\mathbb{C})$ $(\subset R)$ with $\tau(p) = l/2^k$. Let $N \in \mathbb{N}$ be given, and for $n = 2^{k+N}, 2^{k+N+1}, \ldots$ let $H(s,n)$ $(s \in S)$ be independent $n \times n$ standard Gaussian matrices. Also, for such n and $A \in M_{2^{k+N}}(\mathbb{C})$ let $D(A,n)$ be the block-diagonal matrix all of whose diagonals are A. Then it is obvious that the distribution of $(D(A,n))_{A \in M_{2^{k+N}}(\mathbb{C})}$ is equal to that of $(A)_{A \in M_{2^{k+N}}(\mathbb{C})}$ as a family in R. Hence Corollary 4.3.6 shows that

$$\Big((\{H(s,n)\})_{s \in S}, \{D(A,n) : A \in M_{2^{k+N}}(\mathbb{C})\}\Big)$$

is asymptotically free as $n = 2^{k+N+j}$, $j \to \infty$, and its limit distribution is the distribution of

$$\Big((\{a_s\})_{s \in S}, M_{2^{k+N}}(\mathbb{C})\Big)$$

(where $M_{2^{k+N}}(\mathbb{C}) \subset R$). Note that $\tau(p)^{-1/2} D(p,n) H(s,n) D(p,n)$ $(s \in S)$ are the same kind of random matrices of size $nl/2^k$ together with block-diagonal matrices $D(p,n)D(A,n)D(p,n) = D(pAp,n)$ (if the trivial summand of size $n - nl/2^k$ is neglected). Hence Corollary 4.3.6 can be used again to see that $(\tau(p)^{-1/2} p a_s p)_{s \in S}$ is a semicircular system free from $pM_{2^{k+N}}(\mathbb{C})p$ $(\subset pRp)$. Letting $N \to \infty$ gives the result.

For general p, choose a sequence (p_m) of nonzero projections in R such that $\tau(p_m)$ are dyadic rationals and $p_m \to p$ strongly. Then

$$\big((\tau(p_m)^{-1/2} p_m a_s p_m)_{s \in S}, (p_m x p_m)_{x \in R}\big) \to \big((\tau(p)^{-1/2} p a_s p)_{s \in S}, (pxp)_{x \in R}\big)$$

in distribution as $m \to \infty$. Since the conclusion holds for each p_m, we can pass to the limit.

□

Below, for a family Ω $(\subset \mathcal{M})$ containing non-selfadjoint elements, we shall often say simply that Ω is free when it is *-free.

Lemma 7.2.3 *In $(\mathcal{M}, \tau)$ let (a_1, a_2) be a semicircular pair, h a quarter-circular element and b_1, b_2 normal elements. Assume that (a_1, a_2, h, b_1, b_2) is a free family.In*

$(\mathcal{M}\otimes M_2(\mathbb{C}), \tau\otimes \mathrm{tr}_2)$ set

$$y := \frac{1}{\sqrt{2}}\begin{bmatrix} a_1 & h \\ h & a_2 \end{bmatrix}, \quad b := \begin{bmatrix} b_1 & 0 \\ 0 & b_2 \end{bmatrix}.$$

Then y is semicircular and y, b are free.

Proof: Let u, v be Haar unitaries in $\mathcal{M}$ such that (a_1, a_2, h, u, v) is free (we may enlarge $\mathcal{M}$ if necessary to choose such u, v). Since (a_1, a_2, h, v, u^*vu) is free, we may assume by functional calculus that $b_1 = f_1(v)$ and $b_2 = f_2(u^*vu)$ with some measurable functions $f_1, f_2 : \mathbb{T} \to \mathbb{C}$. Then b_1 and $ub_2u^* = f_2(v)$ are commuting. It is enough to show the required conclusion for

$$\tilde{y} := \begin{bmatrix} \mathbf{1} & 0 \\ 0 & u \end{bmatrix} y \begin{bmatrix} \mathbf{1} & 0 \\ 0 & u^* \end{bmatrix} = \frac{1}{\sqrt{2}}\begin{bmatrix} a_1 & hu^* \\ uh & ua_2u^* \end{bmatrix},$$

$$\tilde{b} := \begin{bmatrix} \mathbf{1} & 0 \\ 0 & u \end{bmatrix} b \begin{bmatrix} \mathbf{1} & 0 \\ 0 & u^* \end{bmatrix} = \begin{bmatrix} b_1 & 0 \\ 0 & ub_2u^* \end{bmatrix}.$$

It is easy to see that (a_1, ua_2u^*, h, u, v) is a free family. Moreover, Proposition 4.4.2 implies that uh is circular. Now the proof is an application of the random matrix model. For $n \in \mathbb{N}$ let

$$H(2n) = \begin{bmatrix} H_1(n) & X(n)^* \\ X(n) & H_2(n) \end{bmatrix}$$

be a $2n \times 2n$ standard selfadjoint Gaussian matrix with $n \times n$ blocks $H_1(n), H_2(n), X(n)$. Also, let $D(2n)$ $(n \in \mathbb{N})$ be constant diagonal matrices with $n\times n$ diagonal blocks $D_1(n), D_2(n)$ such that the limit distribution of $(D_1(n), D_2(n))$ is the distribution of (b_1, ub_2u^*). Then, applying Corollaries 4.3.6 and 4.3.8 to $n \times n$ blocks shows that

$$\Big(\{\sqrt{2}\,H_1(n)\}, \{\sqrt{2}\,H_2(n)\}, \{\sqrt{2}\,X(n), \sqrt{2}\,X(n)^*\}, \{D_1(n), D_2(n)\}\Big)$$

is asymptotically free and its limit distribution is equal to the distribution of $(\{a_1\}, \{ua_2u^*\}, \{uh, hu^*\}, \{b_1, ub_2u^*\})$. (Here the multiple constant $\sqrt{2}$ is just to adjust variances of matrix entries to n.) In this way, it is shown that the distribution of $(\tilde{y}, \tilde{b})$ coincides with the limit distribution of $(H(2n), D(2n))$. Now the desired conclusion follows from Corollary 4.3.6.

□

Theorem 7.2.4 *Let R be a hyperfinite II_1 factor. For $n = 1, 2, \ldots, \infty$,*

$$R \star \mathcal{L}(\mathbf{F}_n) \cong \mathcal{L}(\mathbf{F}_{n+1}),$$

where $\mathbf{F}_1$ means $\mathbb{Z}$.

Proof: According to (7.1.5) it suffices to show that

$$R \star \mathcal{L}(\mathbb{Z}) \cong \mathcal{L}(\mathbf{F}_2).$$

Write $\mathcal{M} := R \star \mathcal{L}(\mathbb{Z})$ with the canonical (free product) trace τ. Let a be a semicircular element generating $\mathcal{L}(\mathbb{Z})$. Represent $R = M_2(\mathbb{C}) \otimes M_2(\mathbb{C}) \otimes \cdots$ with matrix units $(e_{ij}^m)_{i,j=0}^1$ of the mth $M_2(\mathbb{C})$, and for each $k \in \mathbb{N}$ write

$$e_k(i_1 \dots i_k, j_1 \dots j_k) := e_{i_1 j_1}^1 \otimes \cdots \otimes e_{i_k j_k}^k \qquad (i_m, j_m = 0, 1),$$

which are matrix units of $\bigotimes_1^k M_2(\mathbb{C})$ ($\subset R$). In

$$\mathcal{A}_k := e_k(0 \dots 0, 0 \dots 0) \mathcal{M} e_k(0 \dots 0, 0 \dots 0),$$

set

$$\begin{aligned} a_k(i_1 \dots i_k) &:= 2^{k/2} e_k(0 \dots 0, i_1 \dots i_k) a e_k(i_1 \dots i_k, 0 \dots 0), \\ c_k(i_1 \dots i_k, j_1 \dots j_k) &:= 2^{k/2} e_k(0 \dots 0, i_1 \dots i_k) a e_k(j_1 \dots j_k, 0 \dots 0) \end{aligned}$$

for $i_1 \dots i_k < j_1 \dots j_k$ (lexicographic order). Then Lemma 7.2.1 implies that in $(\mathcal{A}_k, 2^k \tau|_{\mathcal{A}_k})$

$$\Omega_1 := \{a_k(i_1 \dots i_k) : i_1, \dots, i_k = 0, 1\}$$

and

$$\Omega_2 := \{c_k(i_1 \dots i_k, j_1 \dots j_k) : i_m, j_m = 0, 1 \text{ with } i_1 \dots i_k < j_1 \dots j_k\}$$

are a semicircular system and a circular system, respectively, and $\Omega_1 \cup \Omega_2$ is a free family. Let $c_k(0 \dots 0, 0 \dots 01) = u_k h_k$ be the polar decomposition. Then u_k is a Haar unitary and h_k is quarter-circular by Proposition 4.4.2. Now define a selfadjoint y and a normal b in $\mathcal{M}$ by

$$\begin{aligned} y := \sum_{k=1}^{\infty} \big[& e_k(0 \dots 01, 0 \dots 01) a e_k(0 \dots 01, 0 \dots 01) \\ & + 2^{-k/2} e_k(0 \dots 0, 0 \dots 0) h_k e_k(0 \dots 0, 0 \dots 01) \\ & + 2^{-k/2} e_k(0 \dots 01, 0 \dots 0) h_k e_k(0 \dots 0, 0 \dots 0) \big], \\ b := \sum_{k=1}^{\infty} & \frac{1}{k} e_k(0 \dots 01, 0 \dots 0) u_k e_k(0 \dots 0, 0 \dots 01). \end{aligned}$$

Let us see that y and b generate $\mathcal{M}$. Take spectral projections of b to get $e_k(0 \dots 01, 0 \dots 01)$ $(k \geq 1)$. These extract $e_k(0 \dots 0, 0 \dots 0) h_k e_k(0 \dots 0, 0 \dots 01)$ $(k \geq 1)$

from y, whose polar parts are $e_k(0\dots 0, 0\dots 01)$. These generate all matrix units $e_k(i_1\dots i_k, j_1\dots j_k)$. Hence we get all $e_k(0\dots 01, 0\dots 01)ae_k(0\dots 01, 0\dots 01)$ and $c_k(0\dots 0, 0\dots 01)$, generating a.

Note that the spectral measure of b has no atoms, because the u_k are Haar unitaries. So it suffices by Proposition 7.1.1 to show that y is semicircular and y, b are free. To do so, set

$$\begin{aligned} y_n := \sum_{k=1}^{n} \big[& e_k(0\dots 01, 0\dots 01)ae_k(0\dots 01, 0\dots 01) \\ & +2^{-k/2}e_k(0\dots 0, 0\dots 0)h_k e_k(0\dots 0, 0\dots 01) \\ & +2^{-k/2}e_k(0\dots 01, 0\dots 0)h_k e_k(0\dots 0, 0\dots 0)\big] \\ & +e_n(0\dots 0, 0\dots 0)ae_n(0\dots 0, 0\dots 0), \\ b_n := \sum_{k=1}^{n} & \frac{1}{k}e_k(0\dots 01, 0\dots 0)u_k e_k(0\dots 0, 0\dots 01) + \frac{1}{n+1}e_n(0\dots 0, 0\dots 0). \end{aligned}$$

Since Ω_1 and Ω_2 are uniformly bounded, it is clear that $\|y_n - y\| \to 0$. Also $\|b_n - b\| \to 0$ is clear. Hence it is enough to show the same conclusion for y_n and b_n for each n. We proceed by induction. For $n = 1$ we may write

$$y_1 = \frac{1}{\sqrt{2}}\begin{bmatrix} a_1(0) & h_1 \\ h_1 & a_1(1) \end{bmatrix}, \quad b_1 = \begin{bmatrix} \frac{1}{2}\mathbf{1} & 0 \\ 0 & u_1 \end{bmatrix} \quad \text{in} \quad \mathcal{A}_1 \otimes M_2(\mathbb{C}),$$

so y_1 is semicircular and y_1, b_1 are free by Lemma 7.2.3. Suppose the conclusion holds for $n - 1$, and set

$$\begin{aligned} y_n' &:= \sqrt{2}(y_n - e_1(1,1)ae_1(1,1) - e_1(0,0)h_1e_1(0,1) - e_1(1,0)h_1e_1(0,0)), \\ b_n' &:= b_n - e_1(1,0)u_1e_1(0,1). \end{aligned}$$

Note that y_n' and b_n' correspond to y_{n-1} and b_{n-1} taken in $(\mathcal{A}_1, 2\tau|_{\mathcal{A}_1})$ for $a_1(0) = \sqrt{2}e_1(0,0)ae_1(0,0)$ instead of a. Hence, applying the induction hypothesis to $\mathcal{A}_1$, we see that y_n' is semicircular and y_n', b_n' are free. Since

$$y_n = \frac{1}{\sqrt{2}}\begin{bmatrix} y_n' & h_1 \\ h_1 & a_1(1) \end{bmatrix}, \quad b_n = \begin{bmatrix} b_n' & 0 \\ 0 & u_1 \end{bmatrix} \quad \text{in} \quad \mathcal{A}_1 \otimes M_2(\mathbb{C}),$$

it remains, thanks to Lemma 7.2.3, to show that $(y_n', a_1(1), h_1, b_n', u_1)$ is a free family. For this, note that

$$\Big(\{e_n(0i_2\dots i_n, 0j_2\dots j_n) : i_m, j_m = 0,1\},\ \{a_1(0)\},\ \{a_1(1)\},\ \{c_1(0,1)\}\Big)$$

is free in $(\mathcal{A}_1, 2\tau|_{\mathcal{A}_1})$, which can be shown like Lemma 7.2.1 by using the random matrix model. Since y_n' and b_n' are generated by $\{e_n(0i_2\dots i_n, 0j_2\dots j_n)\}$ and $a_1(0)$, it follows that $(\{y_n', b_n'\}, \{a_1(1)\}, \{h_1, u_1\})$ is free. So we have the required assertion because both (y'_n, b'_n) and (h_1, u_1) are free pairs.

□

Besides the above theorem there are several known relations involving free group factors and free products. For instance, Voiculescu proved that for every $N \in \mathbb{N}$ and $n \in \mathbb{N} \cup \{\infty\}$

$$\mathcal{L}(\mathbf{F}_{n+1})_{1/N} \cong \mathcal{L}(\mathbf{F}_{N^2n+1}) \quad \text{or} \quad \mathcal{L}(\mathbf{F}_{n+1}) \cong \mathcal{L}(\mathbf{F}_{N^2n+1}) \otimes M_N(\mathbb{C}) \tag{7.2.1}$$

and

$$\mathcal{L}(\mathbf{F}_n \star \mathbb{Z}/N\mathbb{Z})_{1/N} \cong \mathcal{L}(\mathbf{F}_{N^2n-N+1}) . \tag{7.2.2}$$

The following, as well as Theorem 7.2.4, is due to Dykema: For every $N \in \mathbb{N}$ and $n \in \mathbb{N} \cup \{\infty\}$,

$$\begin{aligned} &(\mathcal{L}(\mathbf{F}_n) \otimes M_N(\mathbb{C}))_{1/N} \cong \mathcal{L}(\mathbf{F}_{N^2n}) \\ &\qquad \text{or} \quad \mathcal{L}(\mathbf{F}_n) \otimes M_N(\mathbb{C}) \cong \mathcal{L}(\mathbf{F}_{N^2n}) \otimes M_N(\mathbb{C}) . \end{aligned} \tag{7.2.3}$$

All of these formulas can be proved by manipulating semicircular and circular systems approximated by Gaussian random matrix models.

It follows from (7.2.1) for $n = \infty$ that the fundamental group of $\mathcal{L}(\mathbf{F}_\infty)$ contains all positive rational numbers. In fact, Rădulescu proved more strongly that the fundamental group of $\mathcal{L}(\mathbf{F}_\infty)$ is the whole $(0, \infty)$. This fact will be seen below as a by-product of the construction of interpolated free group factors. It is remarkable that we still have almost no exact information about fundamental groups of II_1 factors except for R and $\mathcal{L}(\mathbf{F}_\infty)$.

Next let us show that $R \star R \cong \mathcal{L}(\mathbf{F}_2)$. To prove this, the following lemma is useful.

Lemma 7.2.5 *Let (p, q) be a free pair of projections in $(\mathcal{M}, \tau)$ such that $\tau(p) = \tau(q) = 1/2$. Let $\mathcal{N} := \{p, q\}''$. Then*

$$\mathcal{N} \cong L^\infty([0, \pi/2], 2\, d\theta/\pi) \otimes M_2(\mathbb{C}),$$

where $\tau|_\mathcal{N} = \left(\frac{2}{\pi}\int_0^{\pi/2} \cdot\, d\theta\right) \otimes \mathrm{tr}_2$ and p, q are represented as

$$p = \begin{bmatrix} 1 & 0 \\ 0 & 0 \end{bmatrix}, \quad q = \begin{bmatrix} \cos^2\theta & \cos\theta\sin\theta \\ \cos\theta\sin\theta & \sin^2\theta \end{bmatrix}. \tag{7.2.4}$$

Moreover, the distribution of pqp in $(p\mathcal{N}p, 2\tau|_{p\mathcal{N}p})$ is

$$\frac{1}{\pi\sqrt{t(1-t)}}\chi_{(0,1)}(t)\, dt .$$

Proof: Since p and q are free, the distribution of $p+q$ (also $p+(\mathbf{1}-q)$) is $\frac{1}{2}(\delta(0)+\delta(1)) \boxplus \frac{1}{2}(\delta(0)+\delta(1))$. Hence, thanks to Example 3.2.2, $p\wedge q$, $p\wedge(\mathbf{1}-q)$, $(\mathbf{1}-p)\wedge q$ and $(\mathbf{1}-p)\wedge(\mathbf{1}-q)$ are all $\{0\}$. So the structure theorem for two projections says (cf. [188], pp. 306–308) that

$$\mathcal{N} \cong L^\infty([0,\pi/2],\nu)\otimes M_2(\mathbb{C})$$

with identifications $\tau|_{\mathcal{N}} = \left(\int_0^{\pi/2}\cdot\, d\nu\right)\otimes \mathrm{tr}_2$ and (7.2.4), where ν is a probability measure on $[0,\pi/2]$. Then we have

$$(2p-1)(2q-1) = \begin{bmatrix} \cos 2\theta & \sin 2\theta \\ -\sin 2\theta & \cos 2\theta \end{bmatrix},$$

which is a Haar unitary. Therefore,

$$\int_0^{\pi/2} \cos 2n\theta\, d\nu(\theta) = 0 \qquad (n\in\mathbb{Z},\, n\neq 0),$$

which forces ν to equal $2\,d\theta/\pi$. Moreover, the latter assertion is true because

$$\tau((pqp)^n) = \frac{1}{\pi}\int_0^{\pi/2} \cos^{2n}\theta\, d\theta = \frac{1}{2\pi}\int_0^1 t^n \frac{dt}{\sqrt{t(1-t)}} \qquad (n\in\mathbb{N}).$$

□

Theorem 7.2.6 *Let R and $\tilde{R}$ be hyperfinite II_1 factors. Then*

$$R\star\tilde{R}\cong R\star\mathcal{L}(\mathbb{Z})$$

by an isomorphism mapping R identically to itself.

Proof: The proof will be divided into several steps.

Step 1. First, let $\mathcal{M} := R\star\tilde{R}$, and let τ be the canonical (free product) trace. Choose projections p in R and q in $\tilde{R}$ such that $\tau(p)=\tau(q)=1/2$. Let $u\in R$ and $v\in\tilde{R}$ be partial isometries such that

$$u^*u = p,\quad uu^* = \mathbf{1}-p,\quad v^*v = q,\quad vv^* = \mathbf{1}-q.$$

Represent $\mathcal{N} := \{p,q\}''$ as in the above lemma, and set

$$x := \begin{bmatrix} 0 & 0 \\ 1 & 0 \end{bmatrix},\quad y := \begin{bmatrix} -\cos\theta\sin\theta & -\sin^2\theta \\ \cos^2\theta & \cos\theta\sin\theta \end{bmatrix},$$

$$w := \begin{bmatrix} \cos\theta & -\sin\theta \\ \sin\theta & \cos\theta \end{bmatrix}.$$

Then x, y are partial isometries and w is a unitary such that

$$x^*x = p\,, \quad xx^* = \mathbf{1} - p\,, \quad y^*y = q\,, \quad yy^* = \mathbf{1} - q\,,$$
$$p = w^*qw\,, \quad x = w^*yw\,.$$

Step 2. Show that $p\mathcal{M}p$ is generated by

$$\Omega_0 := pRp \cup w^*q\tilde{R}qw \cup \{pqp,\ x^*u,\ w^*y^*vw\}\,.$$

Set

$$\Omega := pRp \cup w^*q\tilde{R}qw \cup \{q,\ u,\ w^*y^*vw\}\,.$$

Since Ω contains p and q, it is obvious that Ω generates $\mathcal{M}$. Let $e_{11} = p$, $e_{12} = x^*$, $e_{21} = x$, and $e_{22} = \mathbf{1} - p$, which form matrix units. For $\Sigma \subset \mathcal{M}$ the linear span of $\bigcup_{j,k=1}^2 e_{1j}\Sigma e_{k1}$ is denoted by $\Theta(\Sigma)$. Note that $\Theta(\Sigma^*) = \Theta(\Sigma)^*$ and $\Theta(\Sigma_1\Sigma_2) \subset \Theta(\Sigma_1)\Theta(\Sigma_2)$. Hence one can easily see that $\Theta(\Omega)$ generates $e_{11}\mathcal{M}e_{11} = p\mathcal{M}p$. It remains to show that $\Theta(\Omega)$ is generated by Ω_0. But this holds because pqx and x^*qx are generated by pqp, and $pu = ux = 0$, $up = u$, etc.

Step 3. For $\Sigma_1, \Sigma_2 \subset \mathcal{M}$ we use the notation $\Xi(\Sigma_1, \Sigma_2)$ for the set of alternating products from Σ_1 and Σ_2, that is, the set of $a_1a_2\cdots a_n$ where $a_j \in \Sigma_{i_j}$ and $i_1 \neq i_2 \neq \ldots \neq i_n$, containing $\mathbf{1}$ (the trivial product). Let $a := 2p - \mathbf{1}$ and $b := 2q - \mathbf{1}$. Show that if $z \in (\tilde{R} \cup \mathcal{N})''$ $(= (\tilde{R} \cup \{a\})'')$ and $\tau(z) = \tau(pz) = 0$, then z is the strong limit of a bounded sequence in the linear span of $\Xi(\tilde{R}^0, \{a\}) \backslash \{\mathbf{1}, a\}$, where $\tilde{R}^0$ denotes the set of elements of $\tilde{R}$ with trace zero. Since the linear span of $\Xi(\tilde{R}^0, \{a\})$ is a dense *-subalgebra of $(\tilde{R} \cup \mathcal{N})''$, z is the strong limit of a bounded sequence (z_n) in span $\Xi(\tilde{R}^0, \{a\})$ by the Kaplansky density theorem (see [188], II.4.8). Write $z_n = \alpha_n\mathbf{1} + \beta_n a + z_n'$, where $z_n' \in \mathrm{span}(\Xi(\tilde{R}^0, \{a\}) \setminus \{\mathbf{1}, a\})$. From the freeness of p and $\tilde{R}$ it is easily checked that $\tau(z') = \tau(pz') = 0$ for all $z' \in \Xi(\tilde{R}^0, \{a\}) \setminus \{\mathbf{1}, a\}$. Hence $\alpha_n = \tau(z_n) \to 0$ and $\alpha_n + \beta_n = 2\tau(pz_n) \to 0$, implying $z_n' \to z$ strongly. Similarly, if $z \in (R \cup \mathcal{N})''$ and $\tau(z) = \tau(qz) = 0$, then z is the strong limit of a bounded sequence in the linear span of $\Xi(R^0, \{b\}) \setminus \{\mathbf{1}, b\}$.

Step 4. Show that x^*u is a Haar unitary in $p\mathcal{M}p$. Note that $\tau(x^*) = 0$ and $px^* = 0$. So by Step 3, to prove that $\tau((x^*u)^n) = 0$ for $n \geq 1$, it suffices to show that $\tau(z_1uz_2u\cdots z_nu) = 0$ for every $z_i \in \Xi(\tilde{R}^0, \{a\}) \setminus \{\mathbf{1}, a\}$. But since $ua = u$ and $au = -u$, the array inside the trace is an alternating product from $\{a, u\}$ and $\tilde{R}^0$, so zero trace follows from freeness and $\tau(a) = \tau(u) = 0$. Hence x^*u is a Haar unitary in $p\mathcal{M}p$. Similarly, by using the second assertion in Step 3, it follows that y^*v is a Haar unitary in $q\mathcal{M}q$, so w^*y^*vw is one in $p\mathcal{M}p$.

Step 5. Show that $\big(pRp,\ \{pqp\},\ \{x^*u\}\big)$ is free in $p\mathcal{M}p$. Set

$$g_k := (pqp)^k - 2\tau((pqp)^k) \qquad (k \geq 1).$$

Let $S_1 := \{z \in pRp : \tau(z) = 0\}$, $S_2 := \{g_k : k \geq 1\}$ and $S_3 := \{(x^*u)^j : j \in \mathbb{Z},\ j \neq 0\}$. We need to show that $\tau(a_1a_2\cdots a_n) = 0$ for every $a_j \in S_{i_j}$, $i_1 \neq i_2 \neq \ldots \neq i_n$.

It is clear that such $a_1a_2\cdots a_n$ can be written as an alternating product from

$$\Phi := \{u, u^*\} \cup \{z,\, uz,\, zu^*,\, uzu^* : z \in pRp,\, \tau(z) = 0\}$$

and

$$\Psi := \{x, x^*\} \cup \{g_k,\, xg_k,\, g_kx^*,\, xg_kx^* : k \geq 1\}\,.$$

Here $\Psi \subset \mathcal{N}$. Since $px = 0$ and $px^* = x^*$, $\tau(z) = \tau(pz) = 0$ for all $z \in \Psi$. So by Step 3 we may show that an alternatling product from Φ and $\Xi(\tilde{R}^0, \{a\}) \setminus \{\mathbf{1}, a\}$ has trace zero. Note that $\tau(uz) = \tau(zu^*) = 0$, $\tau(uzu^*) = \tau(z)$ and $za = az = z$ for all $z \in pRp$. Also, note that $ua = u$, $au^* = u^*$, $au = -u$ and $u^*a = -u^*$. Therefore, an alternating product from Φ and $\Xi(\tilde{R}^0, \{a\}) \setminus \{\mathbf{1}, a\}$ becomes an alternating product from R^0 and $\tilde{R}^0$, which has trace zero by freeness.

Step 6. Show that

$$\big(pRp,\, w^*q\tilde{R}qw,\, \{pqp\},\, \{x^*u\},\, \{w^*y^*vw\}\big)$$

is free in $p\mathcal{M}p$. Set

$$\tilde{\mathcal{N}_0} := \big(pRp \cup \{pqp,\, x^*u\}\big)''\ (\subset p\mathcal{M}p)\,, \quad \tilde{\mathcal{N}_1} := w\tilde{\mathcal{N}_0}w^*\ (\subset q\mathcal{M}q)\,.$$

Note that $\tilde{\mathcal{N}_1} \subset (R \cup \mathcal{N})''$. We may show that $\big(\tilde{\mathcal{N}_1},\, q\tilde{R}q,\, \{y^*v\}\big)$ is free in $q\mathcal{M}q$. Let $\tilde{S}_1 := \{z \in \tilde{\mathcal{N}_1} : \tau(z) = 0\}$, $\tilde{S}_2 := \{z \in q\tilde{R}q : \tau(z) = 0\}$ and $\tilde{S}_3 := \{(y^*v)^j : j \in \mathbb{Z},\, j \neq 0\}$. A product $a_1a_2\cdots a_n$ of $a_j \in \tilde{S}_{i_j}$, $i_1 \neq i_2 \neq \ldots \neq i_n$, is written as an alternating product from

$$\tilde{\Phi} := \{v, v^*\} \cup \{z,\, vz,\, zv^*,\, vzv^* : z \in q\tilde{R}q,\, \tau(z) = 0\}$$

and

$$\tilde{\Psi} := \{y, y^*\} \cup \tilde{S}_1 \cup y\tilde{S}_1 \cup \tilde{S}_1y^* \cup y\tilde{S}_1y^*\,.$$

Since $qy = 0$ and $qy^* = y^*$, $\tau(z) = \tau(qz) = 0$ for all $z \in \tilde{\Psi}$. Moreover, $\tilde{\Psi} \subset (R \cup \mathcal{N})''$. Hence it suffices by Step 3 to show that an alternating product from $\tilde{\Phi}$ and $\Xi(R^0, \{b\}) \setminus \{\mathbf{1}, b\}$ has trace zero. As in Step 5, such a product is an alternating product from $\tilde{R}^0$ and R^0, so the conclusion is obtained.

Step 7. Since the distribution of pqp has no atoms by Lemma 7.2.5, $\{pqp\}''$ is isomorphic to $\mathcal{L}(\mathbb{Z})$. Hence Steps 2, 4 and 6 imply by Proposition 7.1.1 that

$$p(R \star \tilde{R})p \cong pRp \star q\tilde{R}q \star \mathcal{L}(\mathbf{F}_3)\,,$$

so by Theorem 7.2.4

$$p(R \star \tilde{R})p \cong pRp \star \mathcal{L}(\mathbf{F}_4)\,,$$

where the isomorphism maps pRp ($\subset p(R \star \tilde{R})p$) identically to itself.

Step 8. Next, let $\mathcal{M} := R \star \mathcal{L}(\mathbb{Z})$, and let τ be the free product trace on $\mathcal{M}$. Let $p, u \in R$ be as in Step 1, and let q be a projection in $\mathcal{A} := \mathcal{L}(\mathbb{Z})$ such that $\tau(q) = 1/2$. Also, let $\mathcal{N} := \{p, q\}''$, x, y and w be as in Step 1. It is seen as in Step 2 that $p\mathcal{M}p$ is generated by

$$pRp \cup w^*q\mathcal{A}w \cup w^*y^*\mathcal{A}yw \cup \{pqp,\, x^*u\}\,.$$

The assertions in Step 3 hold in the present case where $\tilde{R}$ is replaced by $\mathcal{A}$. Then the proofs of Step 4 (for x^*u) and Step 5 are the same with $\mathcal{A}$ in place of $\tilde{R}$. Furthermore, it can be shown that

$$\bigl(pRp\,, w^*q\mathcal{A}w,\; w^*y^*\mathcal{A}yw,\; \{pqp\},\; \{x^*u\}\bigr)$$

is free in $p\mathcal{M}p$. The proof here is a slight modification of Step 6, so the full details are left to the reader. Since $w^*q\mathcal{A}w$ and $w^*y^*\mathcal{A}yw$ are isomorphic to $\mathcal{L}(\mathbb{Z})$, we have

$$p(R \star \mathcal{L}(\mathbb{Z}))p \cong pRp \star \mathcal{L}(\mathbf{F}_4)$$

by an isomorphism mapping pRp ($\subset p(R \star \mathcal{L}(\mathbb{Z}))p$) onto itself. Combining this with the isomorphism in Step 7, we have

$$p(R \star \tilde{R})p \cong p(R \star \mathcal{L}(\mathbb{Z}))p,$$

and the result follows by tensoring with $M_2(\mathbb{C})$.

□

Now we are in a good position to introduce the *interpolated free group factors.* The one-parameter family $\mathcal{L}(\mathbf{F}_r)$ $(1 < r \le \infty)$ of II_1 factors possesses the following properties:

(i) When $r = n \in \{2, 3, \ldots, \infty\}$, $\mathcal{L}(\mathbf{F}_r)$ is isomorphic to the free group factor $\mathcal{L}(\mathbf{F}_n)$.

(ii) *Compression formula*: For every $1 < r \le \infty$ and $0 < t < \infty$,

$$\mathcal{L}(\mathbf{F}_r)_t \cong \mathcal{L}(\mathbf{F}_{1+(r-1)/t^2})\,.$$

(iii) *Addition formula*: For every $1 < r, r' \le \infty$,

$$\mathcal{L}(\mathbf{F}_r) \star \mathcal{L}(\mathbf{F}_{r'}) \cong \mathcal{L}(\mathbf{F}_{r+r'})\,.$$

The properties (i) and (ii) may be used as a definition of $\mathcal{L}(\mathbf{F}_r)$ for non-integer $r > 1$, namely,

$$\mathcal{L}(\mathbf{F}_r) = \mathcal{L}(\mathbf{F}_n)_t \quad \text{with} \quad t = \left(\frac{n-1}{r-1}\right)^{1/2}.$$

The formula (7.2.1) is a special case of (ii). (Needless to say, the notation $\mathbf{F}_r$ itself makes no sense.)

In the following we give two equivalent definitions of interpolated free group factors. The first is Rădulescu's definition and the second is due to Dykema. The two definitions are a bit different, but a semicircular system plays an essential role in both.

In $(\mathcal{M}, \tau)$ let $(a_s)_{s\in S}$ be an infinite semicircular system, and let R be a copy of the hyperfinite II_1 factor such that R and $\{a_s : s \in S\}$ are in free relation. (In the first definition below R is of no use.)

1° Let two distinct $\sigma_0, \sigma_1 \in S$ be fixed, and let h be a nonzero projection in $\{a_{\sigma_1}\}''$. Let p_s, q_s $(s \in S \setminus \{\sigma_0, \sigma_1\})$ be projections in $\{a_{\sigma_1}\}''$ such that $p_s = q_s$ or $p_s q_s = 0$. Let

$$r := 1 + 2\tau(h)\tau(\mathbf{1} - h) + \sum_{s\in S\setminus\{\sigma_0,\sigma_1\}} k_s \tau(p_s)\tau(q_s),$$

where $k_s := 1$ if $p_s = q_s$ and $k_s := 2$ if $p_s q_s = 0$. Then $\mathcal{L}(\mathbf{F}_r)$ is a type II_1 factor isomorphic to

$$\left(\{a_{\sigma_1}, h a_{\sigma_0}(\mathbf{1} - h)\} \cup \{p_s a_s q_s : s \in S \setminus \{\sigma_0, \sigma_1\}\}\right)''.$$

(Note that $\{a_{\sigma_1}, h a_{\sigma_0}(\mathbf{1} - h)\}''$ and also the above double commutant become factors.)

2° Let $1 < r \le \infty$, and let the p_s $(s \in S)$ be projections in R such that

$$r = 1 + \sum_{s\in S} \tau(p_s)^2.$$

Then $\mathcal{L}(\mathbf{F}_r)$ is a type II_1 factor isomorphic to

$$\left(R \cup \{p_s a_s p_s : s \in S\}\right)''.$$

Our first nontrivial task is to show that the above definition of $\mathcal{L}(\mathbf{F}_r)$ depends only on r, and is independent of the choice of the projections. From now on let us follow Dykema's approach.

Lemma 7.2.7 *In the definition 2° let q_s $(s \in S)$ be another choice of projections in R such that $r = 1 + \sum_{s\in S} \tau(q_s)^2$. Then*

$$\left(R \cup \{p_s a_s p_s : s \in S\}\right)'' \cong \left(R \cup \{q_s a_s q_s : s \in S\}\right)''$$

by an isomorphism mapping R identically to itself.

Proof: We consider a certain standard form. Represent $R = M_2(\mathbb{C}) \otimes M_2(\mathbb{C}) \otimes \cdots$, and set $f_0 := 1$ and $f_k := e_k(0 \ldots 1, 0 \ldots 1)$ for $k \in \mathbb{N}$ with the notation in the proof of Theorem 7.2.4. Then $(f_k)_{k \geq 1}$ is an orthogonal sequence of projections with $\tau(f_k) = 2^{-k}$. When $1 < r < \infty$, expand $r - 1$ on the base 4: $r - 1 = N_0 + N_1 4^{-1} + N_2 4^{-2} + \cdots$, where $N_0 \in \{0\} \cup \mathbb{N}$ and $N_j \in \{0, 1, 2, 3\}$ for $j \geq 1$. When $r = \infty$, let $N_0 = \infty$ and $N_j = 0$ for $j \geq 1$. Choose mutually disjoint $S_0, S_1, \ldots \subset S$ such that $\#S_j = N_j$ $(j \geq 0)$, and let $k(s) = j$ for $s \in S_j$. Obviously,

$$r = 1 + \sum_{s \in \tilde{S}} \tau(f_{k(s)})^2, \quad \text{where} \quad \tilde{S} := \bigcup_{j=0}^{\infty} S_j \,.$$

Then it is enough to show that

$$\mathcal{N}_1 := \big(R \cup \{p_s a_s p_s : s \in \hat{S}\}\big)'' \cong \mathcal{N}_0 := \big(R \cup \{f_{k(s)} a_s f_{k(s)} : s \in \tilde{S}\}\big)'',$$

where $\hat{S} := \{s \in S : p_s \neq 0\}$ (the above right-hand side may be considered as a standard form). Here and below, an isomorphism is always supposed to map R identically to itself.

Note that if $u_s \in R$ $(s \in S)$ are unitaries, then $\big(R, (\{u_s a_s u_s^*\})_{s \in S}\big)$ is free again. This can be seen by a simple induction argument. Indeed, to see that $(R, \{u_1 a_{s(1)} u_1^*\}, \ldots, \{u_n a_{s(n)} u_n^*\})$ is free, we may show the freeness of $(R, \{a_{s(1)}\}, \{u_2 a_{s(2)} u_2^*\}, \ldots, \{u_n a_{s(n)} u_n^*\})$ with u_k for $u_1^* u_k$. But this reduces to the freeness of $(R, \{u_2 a_{s(2)} u_2^*\}, \ldots, \{u_n a_{s(n)} u_n^*\})$, because $a_{s(1)}$ is free from $R \cup \{a_{s(2)}, \ldots, a_{s(n)}\}$. Thus we may assume that each p_s is a (possibly infinite) sum of projections from $(f_k)_{k \geq 0}$. So write $p_s = \sum_{k \in K_s} f_k$, where $K_s \subset \mathbb{N}$ if $p_s \neq 1$ and $K_s = \{0\}$ if $p_s = 1$. Then we have

$$\mathcal{N}_1 = \big(R \cup \{f_k a_s f_{k'} : k, k' \in K_s,\, s \in \hat{S}\}\big)''.$$

The rest of the proof will be divided into three parts.

Step 1. Show that

$$\begin{aligned} \mathcal{N}_1 \cong \mathcal{N}_2 := \big(R &\cup \{f_k a_{\beta(k,k,s)} f_k : k \in K_s,\, s \in \hat{S}\} \\ &\cup \{f_k a_{\beta(k,k',s)} f_{k'},\, f_{k'} a_{\beta(k,k',s)} f_k : k, k' \in K_s,\, k' < k,\, s \in \hat{S}\}\big)'', \end{aligned} \tag{7.2.5}$$

where $\beta : \{(k, k', s) : k, k' \in K_s,\, k' \leq k,\, s \in \hat{S}\} \to S$ is an injection. First assume that $\hat{S}$ and all the K_s are finite, and let $K := \max\{k : k \in K_s,\, s \in \hat{S}\}$. For $n = 2^{K+N}, 2^{K+N+1}, \ldots$ let $H(s, n)$ $(s \in S)$ and $D(A, n)$ $(A \in M_{2^{K+N}}(\mathbb{C}))$ be as in the proof of Lemma 7.2.2. Then the distribution of

$$\Big(M_{2^{K+N}}(\mathbb{C}),\, (f_k a_s f_k)_{k \in K_s,\, s \in \hat{S}},\, (f_k a_s f_{k'},\, f_{k'} a_s f_k)_{k, k' \in K_s,\, k' < k,\, s \in \hat{S}}\Big) \tag{7.2.6}$$

(where $M_{2^{K+N}}(\mathbb{C}) \subset R$) is the limit distribution of

$$\Big((D(A,n))_{A\in M_{2K+N}(\mathbb{C})},\ \big(D(f_k,n)H(s,n)D(f_k,n)\big)_{k\in K_s,\,s\in\hat{S}},$$
$$\big(D(f_k,n)H(s,n)D(f_{k'},n),\ D(f_{k'},n)H(s,n)D(f_k,n)\big)_{k,k'\in K_s,\,k'<k,\,s\in\hat{S}}\Big).$$

Here, due to the orthogonality of (f_k), one can replace $H(s,n)$ against $D(f_k,n)$ and $D(f_{k'},n)$ by $H(\beta(k,k',s),n)$ for each $k,k'\in K_s$, $k'\le k$. Therefore, the distribution of (7.2.6) is equal to that of

$$\Big(M_{2K+N}(\mathbb{C}),\ (f_k a_{\beta(k,k,s)} f_k)_{k\in K_s,\,s\in\hat{S}},$$
$$(f_k a_{\beta(k,k',s)} f_{k'},\ f_{k'} a_{\beta(k,k',s)} f_k)_{k,k'\in K_s,\,k'<k,\,s\in\hat{S}}\Big),$$

and this implies that (7.2.5) is satisfied when the R's in $\mathcal{N}_1$ and in $\mathcal{N}_2$ are replaced by $M_{2K+N}(\mathbb{C})$. The limit as $N\to\infty$ gives the assertion in the case of finite $\hat{S}$ and K_s. The general case then follows by taking a suitable inductive limit.

Step 2. Show that $\mathcal{N}_2\cong\mathcal{N}_3:=\big(R\cup\{f_{l(s)}a_s f_{l(s)}:s\in T\}\big)''$ for some $T\subset S$ and some $l(s)\in\{0\}\cup\mathbb{N}$ for $s\in T$. As in Step 1 we may show the assertion when $\hat{S}$ and all the K_s are finite. For each $k'<k$, $f_{k'}$ is the sum of $2^{k-k'}$ orthogonal projections which are equivalent in $M_{2^k}(\mathbb{C})$ ($\subset R$) to f_k (by permutation matrices), so one can write

$$f_{k'}=\sum_{j=1}^{2^{k-k'}}\Pi_{kk'j}f_k\Pi_{kk'j}^{-1},$$

where the $\Pi_{kk'j}$ are permutation matrices in $M_{2^k}(\mathbb{C})$. Then the right-hand side of (7.2.5) is

$$\begin{aligned}
&\big(R\cup\{f_k a_{\beta(k,k,s)} f_k : k\in K_s,\ s\in\hat{S}\}\\
&\quad\cup\{f_k a_{\beta(k,k',s)}\Pi_{kk'j}f_k,\ f_k\Pi_{kk'j}^{-1}a_{\beta(k,k',s)}f_k :\\
&\qquad\qquad k,k'\in K_s,\ k'<k,\ 1\le j\le 2^{k-k'},\ s\in\hat{S}\}\big)''\\
&=\big(R\cup\{f_k a_{\beta(k,k,s)} f_k : k\in K_s,\ s\in\hat{S}\}\\
&\quad\cup\{f_k a^{(1)}_{\beta(k,k',s),j}f_k,\ f_k a^{(2)}_{\beta(k,k',s),j}f_k :\\
&\qquad\qquad k,k'\in K_s,\ k'<k,\ 1\le j\le 2^{k-k'},\ s\in\hat{S}\}\big)'',
\end{aligned}$$

where

$$a^{(1)}_{\beta(k,k',s),j}:=\frac{a_{\beta(k,k',s)}\Pi_{kk'j}+\Pi_{kk'j}^{-1}a_{\beta(k,k',s)}}{\sqrt{2}},$$
$$a^{(2)}_{\beta(k,k',s),j}:=\frac{a_{\beta(k,k',s)}\Pi_{kk'j}-\Pi_{kk'j}^{-1}a_{\beta(k,k',s)}}{\sqrt{2}\mathrm{i}}.$$

The distribution of $(f_k a^{(i)}_{\beta(k,k',s),j} f_k)_{k,k'\in K_s,\, k'<k,\, 1\le j\le 2^{k-k'},\, i=1,2,\, s\in\hat S}$ is the limit distribution of $\big(D(f_k,n)H^{(i)}_{kk'sj}(n)D(f_k,n)\big)_{k,k'\in K_s,\, k'<k,\, 1\le j\le 2^{k-k'},\, i=1,2,\, s\in\hat S}$, where

$$H^{(1)}_{kk'sj}(n) := \frac{H(\beta(k,k',s),n)D(\Pi_{kk'j},n) + D(\Pi^{-1}_{kk'j},n)H(\beta(k,k',s),n)}{\sqrt{2}},$$

$$H^{(2)}_{kk'sj}(n) := \frac{H(\beta(k,k',s),n)D(\Pi_{kk'j},n) - D(\Pi^{-1}_{kk'j},n)H(\beta(k,k',s),n)}{\sqrt{2}\,\mathrm{i}}.$$

Since $D(f_k,n)H^{(i)}_{kk'sj}(n)D(f_k,n)$ $(i=1,2)$ are regarded as submatrices of the real and imaginary parts of a standard non-selfadjoint Gaussian matrix, one can replace $H^{(i)}_{kk'sj}(n)$ by $H(t^{(i)}_{kk'sj},n)$, where $t^{(1)}_{kk'sj}, t^{(2)}_{kk'sj} \in S$ $(k,k'\in K_s,\ k'<k,\ 1\le j\le 2^{k-k'},$ $s\in\hat S)$ are all distinct. Now the proof proceeds as in Step 1. Note that the value r is preserved in the procedure of Steps 1 and 2:

$$r = 1 + \sum_{s\in T} \tau(f_{l(s)})^2 .$$

Step 3. Finally, show that $\mathcal{N}_3 \cong \mathcal{N}_0$. One can apply the random matrix model as in Step 2 to get

$$\big(R \cup \{f_k a_{s_i} f_k : 1 \le i \le 4\}\big)'' \cong \big(R \cup \{f_{k-1} a_s f_{k-1}\}\big)'', \tag{7.2.7}$$

whenever $s_1,\dots,s_4 \in S$ are distinct, $s\in S$ and $k\ge 1$. When T is finite, $\mathcal{N}_3 \cong \mathcal{N}_0$ follows from a finite number of applications of (7.2.7). Assume that T is infinite. If $r<\infty$ and r is not rational on the basis 4, then there exists an increasing sequence T_m $(m\in\mathbb{N})$ of finite subsets of T such that

$$\sum_{s\in T_m} 4^{-l(s)} = \sum_{j=0}^{m} N_j 4^{-j}, \qquad \bigcup_{m=1}^{\infty} T_m = T .$$

By (7.2.7) we have a compatible sequence of isomorphisms

$$\big(R \cup \{f_{l(s)} a_s f_{l(s)} : s \in T_m\}\big)'' \cong \left(R \cup \left\{ f_{k(s)} a_s f_{k(s)} : s \in \bigcup_{j=0}^{m} S_j \right\}\right)'' .$$

Taking the inductive limit gives $\mathcal{N}_3 \cong \mathcal{N}_0$. If r is rational on the base 4 and $m_0 := \max\{j : N_j \ne 0\}$, then there exists an increasing sequence T_m $(m>m_0)$ of subsets of T such that

$$\sum_{s\in T_m} 4^{-l(s)} = \sum_{j=0}^{m_0} N_j 4^{-j} - 4^{-m}, \qquad \bigcup_{m>m_0} T_m = T .$$

Choose $\sigma \in S_{m_0}$ (so $k(\sigma) = m_0$). Application of (7.2.7) gives

$$\begin{aligned}&\left(R \cup \{f_{l(s)} a_s f_{l(s)} : s \in T_m\}\right)'' &(7.2.8)\\ &\cong \left(R \cup \{f_{k(s)} a_s f_{k(s)} : s \in \tilde{S} \setminus \{\sigma\}\} \cup \{f_j a_{t_{ji}} f_j : m_0 < j \le m,\ 1 \le i \le 3\}\right)'',\end{aligned}$$

where the $t_{ji} \in S \setminus \tilde{S}$ ($j > m_0$, $1 \le i \le 3$) are all distinct. Moreover, similarly to the arguments in Steps 1 and 2 we can see that the right-hand side of (7.2.8) is isomorphic to

$$\left(R \cup \{f_{k(s)} a_s f_{k(s)} : s \in \tilde{S} \setminus \{\sigma\}\} \cup \{f_{m_0} a_\sigma f_{m_0} - g_m a_\sigma g_m\}\right)'' \ (\subset \mathcal{N}_0),$$

where $g_m \in R$ are projections such that $g_m \le f_{m_0}$ and $\tau(g_m) = 2^{-m}$. Then $\mathcal{N}_3 \cong \mathcal{N}_0$ follows by taking an inductive limit. If $r = \infty$ then, since $\sum_{s \in T_m} 4^{-l(s)} \ge 1$ for disjoint finite subsets T_m of T ($m \in \mathbb{N}$), the situation can be transformed to the case $\#T(0) = \infty$ by an isomorphism using (7.2.7) infinitely many times, where $T(k) := \{s \in T : l(s) = k\}$. This can be further transformed to the case $\#T(k) = \infty$ for all $k \ge 0$, and finally to the case $\#T(0) = \infty$ and $\#T(k) = 0$ for all $k \ge 1$, so $\mathcal{N}_3 \cong \mathcal{N}_0 = \mathcal{L}(\mathbf{F}_\infty)$.

□

Now property (i) is shown as follows. When $r \in \{2, 3, \ldots, \infty\}$ and $n = r - 1$, we may choose a semicircular system $(a_1, \ldots, a_n)$ and $p_i = 1$ ($1 \le i \le n$) in the definition 2°. Then by Proposition 7.1.1 and Theorem 7.2.4 we have

$$\mathcal{L}(\mathbf{F}_r) = R \star \{a_i : 1 \le i \le n\}'' = R \star \mathcal{L}(\mathbf{F}_n) \cong \mathcal{L}(\mathbf{F}_{n+1}).$$

Property (ii) and the fact that $\mathcal{L}(\mathbf{F}_r)$ is a type II_1 factor are proved as follows.

Theorem 7.2.8 *For every $1 < r \le \infty$, $\mathcal{L}(\mathbf{F}_r)$ is a type II_1 factor and*

$$\mathcal{L}(\mathbf{F}_r)_t \cong \mathcal{L}(\mathbf{F}_{1+(r-1)/t^2}) \qquad (0 < t < \infty).$$

Proof: Let $\mathcal{L}(\mathbf{F}_r) = \left(R \cup \{p_s a_s p_s : s \in S\}\right)''$ as in the definition 2° with $r = 1 + \sum_{s \in S} \tau(p_s)^2$. Let $p \in R$ be a nonzero projection with $\tau(p) = t$. By Lemma 7.2.7 we may assume that $p_s \le p$ for all $s \in S$. Since

$$\begin{aligned}p\mathcal{L}(\mathbf{F}_r)p &= \left(pRp \cup \{p_s a_s p_s : s \in S\}\right)''\\ &= \left(pRp \cup \{p_s (\tau(p)^{-1/2} p a_s p) p_s : s \in S\}\right)'',\end{aligned}$$

we have, thanks to Lemma 7.2.2,

$$p\mathcal{L}(\mathbf{F}_r)p \cong \mathcal{L}(\mathbf{F}_{r'}),$$

where

$$r' = 1 + \sum_{s \in S} \left(\frac{\tau(p_s)}{\tau(p)}\right)^2 = 1 + \frac{r-1}{t^2}.$$

Let $t = ((r-1)/(n-1))^{1/2}$ in the above, where $r \le n \in \mathbb{N}$. Then, by property (i), $p\mathcal{L}(\mathbf{F}_r)p$ is a factor for any projection $p \in R$ with $\tau(p) = t$. This means that $\mathcal{L}(\mathbf{F}_r)$ is a factor, and the compression formula holds for $0 < t \le 1$. The case $t > 1$ follows from $\mathcal{L}(\mathbf{F}_r) \cong \mathcal{L}(\mathbf{F}_{1+(r-1)/t^2})_{1/t}$ and (7.1.1).

□

Property (iii) is shown as follows.

Theorem 7.2.9 *For every* $1 < r, r' \le \infty$,

$$\mathcal{L}(\mathbf{F}_r) \star \mathcal{L}(\mathbf{F}_{r'}) \cong \mathcal{L}(\mathbf{F}_{r+r'}) .$$

Proof: In $(\mathcal{M}, \tau)$ let $(a_s)_{s \in S \cup T}$ be a semicircular system where S, T are disjoint sets, and let $R, \tilde{R}$ be copies of the hyperfinite II_1 factors such that R, $\tilde{R}$ and $\{a_s : s \in S \cup T\}$ are in free relation. Write

$$\begin{aligned} \mathcal{L}(\mathbf{F}_r) &\cong \left(R \cup \{p_s a_s p_s\}_{s \in S}\right)'' , \\ \mathcal{L}(\mathbf{F}_{r'}) &\cong \left(\tilde{R} \cup \{\tilde{p}_t a_t \tilde{p}_t\}_{t \in T}\right)'' , \end{aligned}$$

where $p_s \in R$ $(s \in S)$ and $\tilde{p}_t \in \tilde{R}$ $(t \in T)$ are projections such that $r = 1 + \sum_{s \in S} \tau(p_s)^2$ and $r' = 1 + \sum_{t \in T} \tau(\tilde{p}_t)^2$. Since $R \cup \{p_s a_s p_s : s \in S\}$ and $\tilde{R} \cup \{\tilde{p}_t a_t \tilde{p}_t : t \in T\}$ are in free relation, we have

$$\mathcal{L}(\mathbf{F}_r) \star \mathcal{L}(\mathbf{F}_{r'}) \cong \left(R \cup \tilde{R} \cup \{p_s a_s p_s : s \in S\} \cup \{\tilde{p}_t a_t \tilde{p}_t : t \in T\}\right)'' .$$

Since $(R \cup \tilde{R})'' \cong R \star \tilde{R} \cong R \star \mathcal{L}(\mathbb{Z})$ by Theorem 7.2.6, we can choose a semicircular element a in $(R \cup \tilde{R})''$ such that R and a are in free relation and $R \cup \{a\}$ generates $(R \cup \tilde{R})''$. Furthermore, choose unitaries $u_t \in (R \cup \tilde{R})''$ $(t \in T)$ such that $q_t := u_t \tilde{p}_t u_t^* \in R$. Then

$$\begin{aligned} &\left(R \cup \tilde{R} \cup \{p_s a_s p_s : s \in S\} \cup \{\tilde{p}_t a_t \tilde{p}_t : t \in T\}\right)'' \\ &\quad = \left(R \cup \{a\} \cup \{p_s a_s p_s : s \in S\} \cup \{q_t u_t a_t u_t^* q_t : t \in T\}\right)'' \end{aligned} \tag{7.2.9}$$

and $(a, a_s, u_t a_t u_t^*)_{s \in S,\, t \in T}$ is a semicircular system free from R. Since

$$r + r' = 1 + 1 + \sum_{s \in S} \tau(p_s)^2 + \sum_{t \in T} \tau(q_t)^2 ,$$

it follows that $\mathcal{L}(\mathbf{F}_{r+r'})$ is given as (7.2.9), and the addition formula is proved.

□

The properties (ii) and (iii) supply important knowledge concerning ismorphism between the (interpolated) free group factors. The following is a direct consequence of (ii) in the case of $r = \infty$.

Corollary 7.2.10 *The fundamental group of* $\mathcal{L}(\mathbf{F}_\infty)$ *is* $(0, \infty)$.

Corollary 7.2.11 $\mathcal{L}(\mathbf{F}_r)$ $(1 < r < \infty)$ *are all stably isomorphic, that is,*

$$\mathcal{L}(\mathbf{F}_r) \otimes B(\mathcal{H}) \cong \mathcal{L}(\mathbf{F}_s) \otimes B(\mathcal{H})$$

for every $1 < r, s < \infty$, *where* $B(\mathcal{H})$ *is of type* I_∞.

Proof: Note that $\mathcal{M}_t \otimes B(\mathcal{H}) \cong \mathcal{M} \otimes B(\mathcal{H})$ for any II_1 factor $\mathcal{M}$ and any $t > 0$. So the corollary is obvious from property (ii).

□

Corollary 7.2.12 *For every* $1 < r, s \leq \infty$ *the isomorphism class of* $\mathcal{L}(\mathbf{F}_r) \otimes \mathcal{L}(\mathbf{F}_s)$ *depends only on* $(r-1)(s-1)$. *In particular,*

$$\mathcal{L}(\mathbf{F}_r) \otimes \mathcal{L}(\mathbf{F}_\infty) \cong \mathcal{L}(\mathbf{F}_s) \otimes \mathcal{L}(\mathbf{F}_\infty)$$

for every $r, s > 1$.

Proof: Note that if $\mathcal{M}$ and $\mathcal{N}$ are general II_1 factors, then $\mathcal{M}_t \otimes \mathcal{N}_{1/t} \cong \mathcal{M} \otimes \mathcal{N}$ for any $t > 0$. This can be easily seen by taking a tensor product of projections $p \in \mathcal{M}$ and $q \in \mathcal{N} \otimes B(\mathcal{H})$ such that $\tau_\mathcal{M}(p) = t$ (< 1) and $(\tau_\mathcal{N} \otimes \mathrm{Tr})(q) = 1/t$. Hence the proof is a simple application of (ii).

□

Corollary 7.2.13 *One (and only one) of the following two statements is true:*

(1) $\mathcal{L}(\mathbf{F}_r)$ $(1 < r < \infty)$ *are all isomorphic, and the fundamental group of* $\mathcal{L}(\mathbf{F}_r)$ *is* $(0, \infty)$ *for any* $1 < r < \infty$.

(2) $\mathcal{L}(\mathbf{F}_r)$ $(1 < r < \infty)$ *are mutually non-isomorphic, and the fundamental group of* $\mathcal{L}(\mathbf{F}_r)$ *is* $\{1\}$ *for any* $1 < r < \infty$.

Proof: Note that, by property (ii), if $\mathcal{L}(\mathbf{F}_r) \cong \mathcal{L}(\mathbf{F}_{r'})$ for some $r \neq r'$, then the fundamental group of $\mathcal{L}(\mathbf{F}_r)$ is non-trivial. So, suppose that the fundamental group of $\mathcal{L}(\mathbf{F}_r)$ contains $\alpha \neq 1$ for some $1 < r < \infty$. Choose $1 < s, s' < \infty$ such that $((s-1)/(s'-1))^{1/2} = \alpha$. Since (ii) implies that

$$\mathcal{L}(\mathbf{F}_s) \cong \mathcal{L}(\mathbf{F}_r)_\beta \cong \mathcal{L}(\mathbf{F}_r)_{\alpha\beta} \cong \mathcal{L}(\mathbf{F}_{s'}),$$

where $\beta = ((r-1)/(s-1))^{1/2}$, we have $\mathcal{L}(\mathbf{F}_{s+t}) \cong \mathcal{L}(\mathbf{F}_{s'+t})$ for all $t > 1$ by property (iii). One can choose s, s' arbitrarily near 1 and conclude that the statement (1) holds true.

□

Moreover, it is known that if (1) holds true in the above, then we also have $\mathcal{L}(\mathbf{F}_r) \cong \mathcal{L}(\mathbf{F}_\infty)$ for every $r > 1$.

Concerning the long-standing isomorphism problem of free group factors, the last corollary says that we have one of two extreme cases, according to whether the fundamental group is trivial or not. The problem is still open.

7.3 Free entropy dimension

In this section let $(\mathcal{M},\tau)$ be a tracial W^*-probability space as before. Based on the free entropy studied in the previous chapter, Voiculescu further introduced the notion of free entropy dimension for N-tuples of selfadjoint elements of $\mathcal{M}$. This dimension $\delta(a_1,\dots,a_N)$ is defined as a certain kind of differential of the free entropy $\chi(a_1,\dots,a_N)$ when perturbed by a semicircular system. It will turn out that the value of $\delta(a_1,\dots,a_N)$ is closely related to the freeness of $a_1,\dots,a_N$, and so is the additivity of $\chi(a_1,\dots,a_N)$.

Let $a_1,\dots,a_N \in \mathcal{M}^{sa}$, and let $(S_1,\dots,S_N)$ be a semicircular system in $\mathcal{M}$ such that $\{a_1,\dots,a_N\}$ and $\{S_1,\dots,S_N\}$ are in free relation. Then the *free entropy dimension* $\delta(a_1,\dots,a_N)$ is defined by

$$\delta(a_1,\dots,a_N) := N + \limsup_{\varepsilon\to+0} \frac{\chi(a_1+\varepsilon S_1,\dots,a_N+\varepsilon S_N)}{|\log\varepsilon|}\,.$$

(The above $S_1,\dots,S_N$ always exist when we enlarge $\mathcal{M}$ by taking a free product with another von Neumann algebra.) Note that the joint distribution of $a_1 + \varepsilon S_1,\dots,a_N+\varepsilon S_N$ is independent of the choice of $S_1,\dots,S_N$, and so $\delta(a_1,\dots,a_N)$ is uniquely determined.

The following is obvious from the definition and Proposition 6.1.3.

Proposition 7.3.1 *For every* $1 \le L < N$,

$$\delta(a_1,\dots,a_N) \le \delta(a_1,\dots,a_L) + \delta(a_{L+1},\dots,a_N)\,.$$

For the case of a single $a \in \mathcal{M}^{sa}$ we can exactly compute $\delta(a)$ as follows.

Theorem 7.3.2 *Let $a, S \in \mathcal{M}^{sa}$, where S is standard semicircular and free from a. If μ is the distribution of a, then*

$$\lim_{\varepsilon\to+0} \frac{\chi(a+\varepsilon S)}{|\log\varepsilon|} = -\sum_{t\in\mathbb{R}} \mu(\{t\})^2 = -(\mu\otimes\mu)(\Delta),$$

and hence

$$\delta(a) = 1 - \sum_{t\in\mathbb{R}} \mu(\{t\})^2\,,$$

where $\Delta := \{(s,t)\in\mathbb{R}^2 : s = t\}$.

Before proving the theorem let us recall some monotonicity (or subordination) properties of additive free convolution without proof. Let μ_1,μ_2 be compactly supported probability measures on $\mathbb{R}$ and $\mu_3 := \mu_1 \boxplus \mu_2$. If μ_1 has density $f_1 = d\mu_1/dt$ belonging to $L^p(\mathbb{R})$ for some $1 < p \le \infty$, then μ_3 has density f_3 satisfying

$$\|f_3\|_p \le \|f_1\|_p . \tag{7.3.1}$$

The proof is based on an analytic subordination principle in the upper half-plane $\mathbb{C}^+$, which says that if $F(x + \mathrm{i}\, y)$ is a harmonic function on $\mathbb{C}^+$ and $\omega : \mathbb{C}^+ \to \mathbb{C}^+$ is an analytic function satisfying $\lim_{z\to\infty} |\omega(z)| = \infty$ and $\operatorname{Im} \omega(z) \ge \operatorname{Im} z$, then

$$\int |F(\omega(x + \mathrm{i}\, y))|^p \, dx \le \int |F(x + \mathrm{i}\, y)|^p \, dx$$

for every $1 \le p \le \infty$ and every $y > 0$. This is an analogue of the known subordination principle in the disk ([60], Chap. 6). In addition, although it will not be needed here, it is noteworthy that the following monotonicity can be proved by applying the above subordination principle:

$$\Sigma(\mu_1 \boxplus \mu_2) \ge \Sigma(\mu_1) \tag{7.3.2}$$

for any compactly supported probability measures μ_1, μ_2 on $\mathbb{R}$.

Proof of Theorem 7.3.2: First, note that the limit in question does not change when S is replaced by tS ($t > 0$). So in the proof below we use S having the distribution w_1 (instead of w_2) for notational convenience. For $\varepsilon > 0$ let μ_ε be the distribution of $a + \varepsilon S$. By (6.1.2) we need to show that

$$\lim_{\varepsilon \to +0} \frac{\Sigma(\mu_\varepsilon)}{|\log \varepsilon|} = -(\mu \otimes \mu)(\Delta) .$$

It suffices to prove the following two estimtes:

$$\lim_{\varepsilon \to +0} \frac{1}{|\log \varepsilon|} \left| \Sigma(\mu_\varepsilon) - \iint \log |s - t + \mathrm{i}\, \varepsilon| \, d\mu_\varepsilon(s)\, d\mu_\varepsilon(t) \right| = 0 , \tag{7.3.3}$$

$$\lim_{\varepsilon \to +0} \frac{1}{|\log \varepsilon|} \iint \log |s - t + \mathrm{i}\, \varepsilon| \, d\mu_\varepsilon(s)\, d\mu_\varepsilon(t) = -(\mu \otimes \mu)(\Delta) . \tag{7.3.4}$$

By the monotonicity property mentioned above, μ_ε has the density $f_\varepsilon = d\mu_\varepsilon/dt$ and

$$\begin{aligned} \|f_\varepsilon\|_2 \le \|w_\varepsilon\|_2 &= \frac{2}{\pi \varepsilon^2} \left[\int_{-\varepsilon}^{\varepsilon} (\varepsilon^2 - t^2) \, dt \right]^{1/2} \\ &= \frac{2}{\pi \varepsilon^{1/2}} \left[\int_{-1}^{1} (1 - t^2) \, dt \right]^{1/2} = \mathrm{const} \cdot \varepsilon^{-1/2} . \end{aligned}$$

Put $g_\varepsilon(x) := \log|1 + \mathrm{i}\,\varepsilon/x|$. Then

$$\begin{aligned}
\|g_\varepsilon\|_2 &= \left[2\int_0^\infty \left(\log\left(1+\frac{\varepsilon^2}{x^2}\right)^{1/2}\right)^2 dx\right]^{1/2} \\
&= \left[\frac{\varepsilon}{2}\int_0^\infty \left(\log\left(1+\frac{1}{x^2}\right)\right)^2 dx\right]^{1/2} = \mathrm{const}\cdot\varepsilon^{1/2}.
\end{aligned}$$

Hence we have

$$\begin{aligned}
0 &\le \iint \log|s-t+\mathrm{i}\,\varepsilon|\,d\mu_\varepsilon(s)\,d\mu_\varepsilon(t) - \Sigma(\mu_\varepsilon) \\
&= \iint f_\varepsilon(s)f_\varepsilon(t)\log\left|1+\mathrm{i}\,\frac{\varepsilon}{s-t}\right| ds\,dt \\
&\le \int f_\varepsilon(s)\|f_\varepsilon\|_2\|g_\varepsilon\|_2\,ds \le \mathrm{const}.
\end{aligned}$$

This implies (7.3.3).

Next, let us prove (7.3.4). Concerning the spectral measures e_a and $e_{a+\varepsilon S}$ of a and $a+\varepsilon S$, since $\|S\| = 1$, we get

$$\begin{aligned}
&e_a((-\infty, t-\varepsilon)) \wedge e_{a+\varepsilon S}([t,\infty)) = 0, \\
&e_{a+\varepsilon S}((-\infty, t)) \wedge e_a([t+\varepsilon,\infty)) = 0,
\end{aligned}$$

which imply that $\tau\big(e_a((-\infty,t-\varepsilon))\big) \le \tau\big(e_{a+\varepsilon S}((-\infty,t))\big) \le \tau\big(e_a((-\infty,t+\varepsilon))\big)$, that is, $\mu((-\infty,t-\varepsilon)) \le \mu_\varepsilon((-\infty,t)) \le \mu((-\infty,t+\varepsilon))$, and hence, for any $s<t$,

$$\mu_\varepsilon((s,t)) \le \mu((s-\varepsilon,t+\varepsilon)), \quad \mu((s,t)) \le \mu_\varepsilon((s-\varepsilon,t+\varepsilon)).$$

For $0 \le s \le t$ write

$$\Delta(s,t) := \{(x,y)\in\mathbb{R}^2 : s \le |x-y| < t\}.$$

Then for any $r>0$ we can estimate

$$\begin{aligned}
&(\mu_\varepsilon\otimes\mu_\varepsilon)(\Delta(0,r)) \\
&\quad\le (\mu_\varepsilon\otimes\mu_\varepsilon)\Big(\bigcup_{n\in\mathbb{Z}} (n\varepsilon - r, (n+1)\varepsilon + r)\times\big[n\varepsilon,(n+1)\varepsilon\big)\Big) \\
&\quad\le (\mu_\varepsilon\otimes\mu_\varepsilon)\Big(\bigcup_{n\in\mathbb{Z}} (n\varepsilon - (r+\varepsilon), (n+1)\varepsilon + (r+\varepsilon))\times\big[n\varepsilon,(n+1)\varepsilon\big)\Big) \\
&\quad\le (\mu\otimes\mu_\varepsilon)(\Delta(0,r+2\varepsilon)) \\
&\quad\le (\mu\otimes\mu_\varepsilon)\Big(\bigcup_{n\in\mathbb{Z}} \big[n\varepsilon,(n+1)\varepsilon\big)\times\big(n\varepsilon - (r+2\varepsilon), (n+1)\varepsilon + (r+2\varepsilon)\big)\Big) \\
&\quad\le (\mu\otimes\mu)\Big(\bigcup_{n\in\mathbb{Z}} \big[n\varepsilon,(n+1)\varepsilon\big)\times\big(n\varepsilon - (r+3\varepsilon), (n+1)\varepsilon + (r+3\varepsilon)\big)\Big) \\
&\quad\le (\mu\otimes\mu)(\Delta(0,r+4\varepsilon)),
\end{aligned}$$

and similarly

$$(\mu\otimes\mu)(\Delta(0,r))\le(\mu_\varepsilon\otimes\mu_\varepsilon)(\Delta(0,r+4\varepsilon))\,.$$

Now, for any $0<\delta<1$, let $0<\varepsilon<1$ be such that $5\varepsilon<\varepsilon^\delta$, and let $R:=\|a\|+1$ be such that $\operatorname{supp}\mu_\varepsilon\subset[-R,R]$. If $s,t\in[-R,R]$ and $|s-t|\ge\varepsilon^\delta$, then

$$\delta\log\varepsilon\le\log|s-t+\mathrm{i}\,\varepsilon|\le\log(2R+\varepsilon)\,.$$

Hence we have

$$\biggl|\iint_{|s-t|\ge\varepsilon^\delta}\log|s-t+\mathrm{i}\,\varepsilon|\,d\mu_\varepsilon(s)\,d\mu_\varepsilon(t)\biggr|\le\log(2R+\varepsilon)+\delta|\log\varepsilon|,$$

so that

$$\limsup_{\varepsilon\to+0}\frac{1}{|\log\varepsilon|}\biggl|\iint_{|s-t|\ge\varepsilon^\delta}\log|s-t+\mathrm{i}\,\varepsilon|\,d\mu_\varepsilon(s)\,d\mu_\varepsilon(t)\biggr|\le\delta\,. \tag{7.3.5}$$

Since

$$\log\varepsilon\le\log|s-t+\mathrm{i}\,\varepsilon|\le\log(1+\varepsilon)\le-\log\varepsilon\quad\text{if}\quad|s-t|<1\,,$$

we have

$$\begin{aligned}\frac{1}{|\log\varepsilon|}&\biggl|\iint_{\Delta(5\varepsilon,\varepsilon^\delta)}\log|s-t+\mathrm{i}\,\varepsilon|\,d\mu_\varepsilon(s)\,d\mu_\varepsilon(t)\biggr|\\&\le(\mu_\varepsilon\otimes\mu_\varepsilon)(\Delta(5\varepsilon,\varepsilon^\delta))\le(\mu\otimes\mu)(\Delta(\varepsilon,\varepsilon^\delta+4\varepsilon)),\end{aligned}$$

so that

$$\lim_{\varepsilon\to+0}\frac{1}{|\log\varepsilon|}\biggl|\iint_{\Delta(5\varepsilon,\varepsilon^\delta)}\log|s-t+\mathrm{i}\,\varepsilon|\,d\mu_\varepsilon(s)\,d\mu_\varepsilon(t)\biggr|=0\,. \tag{7.3.6}$$

Finally, since $\log\varepsilon\le\log|s-t+\mathrm{i}\,\varepsilon|\le\log(6\varepsilon)$ if $(s,t)\in\Delta(0,5\varepsilon)$, we have

$$\begin{aligned}-(\mu\otimes\mu)(\Delta(0,9\varepsilon))&\le-(\mu_\varepsilon\otimes\mu_\varepsilon)(\Delta(0,5\varepsilon))\\&\le\frac{1}{|\log\varepsilon|}\iint_{\Delta(0,5\varepsilon)}\log|s-t+\mathrm{i}\,\varepsilon|\,d\mu_\varepsilon(s)\,d\mu_\varepsilon(t)\\&\le\Bigl(-1+\frac{6}{|\log\varepsilon|}\Bigr)(\mu_\varepsilon\otimes\mu_\varepsilon)(\Delta(0,5\varepsilon))\\&\le\Bigl(-1+\frac{6}{|\log\varepsilon|}\Bigr)(\mu\otimes\mu)(\Delta(0,\varepsilon))\,.\end{aligned}$$

Therefore,

$$\lim_{\varepsilon\to+0}\frac{1}{|\log\varepsilon|}\iint_{\Delta(0,5\varepsilon)}\log|s-t+\mathrm{i}\,\varepsilon|\,d\mu_\varepsilon(s)\,d\mu_\varepsilon(t)=-(\mu\otimes\mu)(\Delta)\,. \tag{7.3.7}$$

Combining (7.3.5)–(7.3.7), we obtain

$$\limsup_{\varepsilon\to+0}\left|\frac{1}{|\log\varepsilon|}\iint\log|s-t+\mathrm{i}\,\varepsilon|\,d\mu_\varepsilon(s)\,d\mu_\varepsilon(t)+(\mu\otimes\mu)(\Delta)\right|\le\delta\,,$$

which implies (7.3.4).

□

The above theroem implies the lower semicontinuity of $\delta(a)$ as follows.

Corollary 7.3.3 *Let $a, a_m \in \mathcal{M}^{sa}$ for $m \in \mathbb{N}$. If $\sup_m \|a_m\| < +\infty$ and $a_m \to a$ in distribution, then*

$$\delta(a)\le\liminf_{m\to\infty}\delta(a_m)\,.$$

In particular, this is the case if $a_m \to a$ strongly.

Proof: The assumption implies that $(\mu_m\otimes\mu_m)(f)\to(\mu\otimes\mu)(f)$ for every continuous function f on $\mathbb{R}^2$, where μ, μ_m are the distributions of a, a_m. Then it is straightforward to see that

$$(\mu\otimes\mu)(\Delta)\ge\limsup_{m\to\infty}(\mu_m\otimes\mu_m)(\Delta)\,.$$

This gives the result by Theorem 7.3.2.

□

Corollary 7.3.4 *Let $a_1,\dots,a_N \in \mathcal{M}^{sa}$, and let μ_i be the distribution of a_i. Then:*

(1) $\delta(a_1,\dots,a_N)\le N$.

(2) *If $\delta(a_1,\dots,a_N)=N$, then all μ_i are nonatomic.*

(3) *If $a_1,\dots,a_N$ are in free relation, then*

$$\delta(a_1,\dots,a_N)=\delta(a_1)+\dots+\delta(a_N)=N-\sum_{i=1}^{N}\sum_{t\in\mathbb{R}}\mu_i(\{t\})^2\,.$$

Proof: (1) and (2) immediately follow from Proposition 7.3.1 and Theorem 7.3.2. (3) follows from Theorems 6.4.1 and 7.3.2.

□

In view of its name, one may expect that the free entropy dimension admits a nonnegative value. Although its nonnegativity will be intrinsically characterized in a future theorem, we first prove

Proposition 7.3.5 *If $a_1, \dots, a_N$ belong to a von Neumann subalgebra of $\mathcal{M}$ isomorphic to $\mathcal{L}(\mathbf{F}_M)$ for some $M \in \mathbb{N}$, then*

$$\delta(a_1, \dots, a_N) \geq 0\,.$$

Proof: The assumption means that there exists a semicircular system $(b_1, \dots, b_M)$ in $\mathcal{M}$ such that $a_i \in \{b_1, \dots, b_M\}''$ for $1 \leq i \leq N$. Choose another semicircular system $(S_1, \dots, S_N)$ which is in free relation to $\{b_1, \dots, b_M\}$. By Propositions 6.1.3, 6.1.6 and Theorem 6.4.1 we have

$$\begin{aligned} &\chi(a_1 + \varepsilon S_1, \dots, a_N + \varepsilon S_N) \\ &\qquad \geq \chi(a_1 + \varepsilon S_1, \dots, a_N + \varepsilon S_N, b_1, \dots, b_M) - \sum_{j=1}^{M} \chi(b_j) \\ &\qquad = \chi(\varepsilon S_1, \dots, \varepsilon S_N, b_1, \dots, b_M) - \sum_{j=1}^{M} \chi(b_j) \\ &\qquad = \sum_{i=1}^{N} \chi(\varepsilon S_i) = \sum_{i=1}^{N} \chi(S_i) + N \log \varepsilon\,, \end{aligned}$$

which implies that

$$\limsup_{\varepsilon \to +0} \frac{\chi(a_1 + \varepsilon S_1, \dots, a_N + \varepsilon S_N)}{|\log \varepsilon|} \geq -N,$$

and hence the result follows.

□

Next we prepare a technical result on restricted Minkowski sums. For a while, all subsets of $\mathbb{R}^n$ are assumed to be measurable, and the same λ denotes the Lebesgue measure on $\mathbb{R}^n$ of different dimensions. The *Minkowski sum* of $A, B \subset \mathbb{R}^n$ is $A + B := \{x + y : a \in A, b \in B\}$. The *Brunn-Minkowski inequality*

$$\lambda(A + B)^{1/n} \geq \lambda(A)^{1/n} + \lambda(B)^{1/n}$$

is known (see [151]). Given a subset $\Theta \subset A \times B$, the *restricted Minkowski sum* is defined as

$$A +_\Theta B := \{a + b : (a, b) \in \Theta\}\,.$$

The following is a modified Brunn-Minkowski inequality in which the exponent $1/n$ is replaced by $2/n$.

Lemma 7.3.6 *There exists a universal constant $c > 0$ such that, for any $0 < \rho < 1$, $n \in \mathbb{N}$ and $A, B \subset \mathbb{R}^n$, if*

$$\rho \leq \left(\frac{\lambda(A)}{\lambda(B)}\right)^{1/n} \leq \rho^{-1}$$

and if $\Theta \subset A \times B$ $(\subset \mathbb{R}^{2n})$ *satisfies*

$$\lambda(\Theta) \geq \big(1 - c\min\{\rho\sqrt{n}, 1\}\big)\lambda(A \times B),$$

then

$$\lambda(A +_\Theta B)^{2/n} \geq \lambda(A)^{2/n} + \lambda(B)^{2/n}.$$

Proof: We divide the proof into two parts.

Step 1. Let $0 < \rho < 1$ and

$$\Theta := \big\{(a, b) \in \mathbb{B}^n \times \rho\mathbb{B}^n : \|a + b\| \leq \sqrt{1 + \rho^2}\big\},$$

where $\mathbb{B}^n$ denotes the unit ball of $\mathbb{R}^n$. Then we show that

$$\lambda(\Theta) \leq \big(1 - c\min\{\rho\sqrt{n}, 1\}\big)\lambda(\mathbb{B}^n \times \rho\mathbb{B}^n) \tag{7.3.8}$$

for some constant $c > 0$ independent of n and ρ. To do so, it suffices to show that, for a constant $c' > 0$, if $1 \geq \|a\| \geq 1 - s/n$ where $s := \frac{1}{2}\min\{\rho\sqrt{n}, 1\}$, then one gets

$$\lambda\big(\big\{b \in \rho\mathbb{B}^n : \|a + b\| > \sqrt{1 + \rho^2}\big\}\big) \geq c'\lambda(\rho\mathbb{B}^n) \tag{7.3.9}$$

uniformly for n and ρ. Indeed, we then have

$$\begin{aligned} \lambda(\Theta) &= \lambda(\mathbb{B}^n \times \rho\mathbb{B}^n) - \lambda((\mathbb{B}^n \times \rho\mathbb{B}^n) \setminus \Theta) \\ &\leq \lambda(\mathbb{B}^n \times \rho\mathbb{B}^n) \\ &\qquad - \int_{1 \geq \|a\| \geq 1 - s/n} \lambda\big(\big\{b \in \rho\mathbb{B}^n : \|a + b\| > \sqrt{1 + \rho^2}\big\}\big)\, d\lambda(a) \\ &\leq \left(1 - \left[1 - \left(1 - \frac{s}{n}\right)^n\right]c'\right)\lambda(\mathbb{B}^n \times \rho\mathbb{B}^n), \end{aligned}$$

which implies (7.3.8) because

$$1 - \left(1 - \frac{s}{n}\right)^n \geq \frac{s}{2} \qquad (0 \leq s \leq 1,\ n \in \mathbb{N}).$$

To prove (7.3.9), we may assume that $a = (r, 0, \ldots, 0)$, where $1 \geq r \geq 1 - s/n$ and $n \geq 2$ (the case $n = 1$ is easily treated). For $b = (t, b')$ $(b' \in \mathbb{R}^{n-1})$, the conditions $b \in \rho\mathbb{B}^n$ and $\|a + b\| \leq \sqrt{1 + \rho^2}$ mean that

$$t^2 + \|b'\|^2 \leq \rho^2 \quad \text{and} \quad (r + t)^2 + \|b'\|^2 \leq 1 + \rho^2,$$

that is,

$$\|b'\|^2 \leq \begin{cases} \rho^2 - t^2 & \text{for } -\rho \leq t \leq u_1, \\ 1 + \rho^2 - (r + t)^2 & \text{for } u_1 \leq t \leq u_2, \end{cases}$$

where

$$u_1 := \frac{1-r^2}{2r}, \quad u_2 := \sqrt{1+\rho^2} - r\,.$$

Note that $u_2 < \rho$ and

$$u_1 \leq \frac{1-\left(1-\frac{\rho}{2\sqrt{n}}\right)^2}{2\left(1-\frac{\rho}{2\sqrt{n}}\right)} \leq \frac{\rho}{2\sqrt{n}-\rho} \leq \frac{\rho}{\sqrt{n}}\,.$$

Therefore, we have

$$\begin{aligned}
&\lambda\left(\{b \in \rho\mathbb{B}^n : \|a+b\| \leq \sqrt{1+\rho^2}\}\right)\\
&\leq \lambda(\mathbb{B}^{n-1})\left[\int_{-\rho}^{u_1} (\rho^2-t^2)^{\frac{n-1}{2}}\,dt + \int_{u_1}^{u_2} (1+\rho^2-(r+t)^2)^{\frac{n-1}{2}}\,dt\right]\\
&\leq \lambda(\mathbb{B}^{n-1})\left[\int_{-\rho}^{\rho/\sqrt{n}} (\rho^2-t^2)^{\frac{n-1}{2}}\,dt + \int_{\rho/\sqrt{n}}^{u_2} (1+\rho^2-(r+t)^2)^{\frac{n-1}{2}}\,dt\right]. \quad (7.3.10)
\end{aligned}$$

The two integrals in (7.3.10) can be compared with $\int_{-\rho}^{\rho}(\rho^2-t^2)^{\frac{n-1}{2}}\,dt$. For the first integral, one has

$$\frac{\int_{-\rho}^{\rho/\sqrt{n}}(\rho^2-t^2)^{\frac{n-1}{2}}\,dt}{\int_{-\rho}^{\rho}(\rho^2-t^2)^{\frac{n-1}{2}}\,dt} = \frac{\int_{-\sqrt{n}}^{1}\left(1-\frac{t^2}{n}\right)^{\frac{n-1}{2}}\,dt}{\int_{-\sqrt{n}}^{\sqrt{n}}\left(1-\frac{t^2}{n}\right)^{\frac{n-1}{2}}\,dt},$$

which is stricly smaller than 1 uniformly for n. When $\rho/\sqrt{n} \leq t \leq u_2$, since $(1+\rho^2-(r+t)^2)/(\rho^2-t^2)$ is decreasing in t and $r \geq 1-\rho/2\sqrt{n}$, one gets

$$\begin{aligned}
\frac{1+\rho^2-(r+t)^2}{\rho^2-t^2} &\leq \frac{1+\rho^2-\left(r+\frac{\rho}{\sqrt{n}}\right)^2}{\rho^2-\frac{\rho^2}{n}}\\
&\leq \frac{1+\rho^2-\left(1+\frac{\rho}{2\sqrt{n}}\right)^2}{\rho^2-\frac{\rho^2}{n}} \leq \frac{n-\sqrt{n}-\frac{1}{4}}{n-1}\,.
\end{aligned}$$

Therefore, the second integral in (7.3.10) is dominated by

$$\left(\frac{n-\sqrt{n}-\frac{1}{4}}{n-1}\right)^{\frac{n-1}{2}} \int_{-\rho}^{\rho} (\rho^2-t^2)^{\frac{n-1}{2}}\,dt\,.$$

Since

$$\left(\frac{n-\sqrt{n}-\frac{1}{4}}{n-1}\right)^{\frac{n-1}{2}} \to 0 \quad \text{as} \quad n \to \infty,$$

the estimates of two integrals imply that

$$\lambda\bigl(\{b \in \rho\mathbb{B}^n : \|a+b\| \le \sqrt{1+\rho^2}\}\bigr) \le c''\lambda(\rho\mathbb{B}^n)$$

for some $0 < c'' < 1$ uniformly for n and ρ, yielding (7.3.9).

Step 2. Step 1 shows that the conclusion of the lemma holds for the particular case

$$A = \rho_1\mathbb{B}^n\,, \quad B = \rho_2\mathbb{B}^n\,, \quad \Theta = \{(a,b) \in A \times B : a+b \in \rho_3\mathbb{B}^n\}, \tag{7.3.11}$$

where $\rho_1, \rho_2, \rho_3 > 0$. In fact, the case $\rho_1 = 1 > \rho_2 = \rho$ is a direct consequence, and the above case follows by symmetry and homogeneity. Now let $A, B \subset \mathbb{R}^n$ and $\Theta \subset A \times B$ be general as stated in the lemma. It suffices to show that there are A_0, B_0 and Θ_0 of the form (7.3.11) such that

$$\begin{aligned}&\lambda(A_0) = \lambda(A)\,, \quad \lambda(B_0) = \lambda(B)\,, \\ &\lambda(\Theta_0) \ge \lambda(\Theta)\,, \quad \lambda(A_0 +_{\Theta_0} B_0) \le \lambda(A +_\Theta B)\,.\end{aligned}$$

Set

$$C := A +_\Theta B\,, \quad \Theta_1 := \{(a,b) \in A \times B : a+b \in C\}\,,$$

and take $\rho_1, \rho_2, \rho_3 > 0$ so that

$$\lambda(A) = \lambda(\rho_1\mathbb{B}^n)\,, \quad \lambda(B) = \lambda(\rho_2\mathbb{B}^n)\,, \quad \lambda(C) = \lambda(\rho_3\mathbb{B}^n)\,.$$

Then $\Theta \subset \Theta_1$, and one can estimate

$$\begin{aligned}\lambda(\Theta_1) &= \int_{\mathbb{R}^n}\int_{\mathbb{R}^n} \chi_A(a)\chi_B(b)\chi_C(a+b)\,da\,db \\ &\le \int_{\mathbb{R}^n}\int_{\mathbb{R}^n} \chi_{\rho_1\mathbb{B}^n}(a)\chi_{\rho_2\mathbb{B}^n}(b)\chi_{\rho_3\mathbb{B}^n}(a+b)\,da\,db \\ &= \lambda\bigl(\{(a,b) \in \rho_1\mathbb{B}^n \times \rho_2\mathbb{B}^n : a+b \in \rho_3\mathbb{B}^n\}\bigr)\,.\end{aligned} \tag{7.3.12}$$

This is due to a rearrangement inequality for an integral of a product of nonnegative functions by spherical symmetrizations (see [43], Theorem 3.4). Thus the required conditions are satisfied for $A_0 := \rho_1\mathbb{B}^n$, $B_0 := \rho_2\mathbb{B}^n$ and $\Theta_0 := \{(a,b) \in A_0 \times B_0 : a+b \in \rho_3\mathbb{B}^n\}$, so the proof is completed.

□

Here it is convenient to introduce the following terminology of Voiculescu. For $a_1, \dots, a_N \in \mathcal{M}^{sa}$, it is said that the N-tuple $(a_1, \dots, a_N)$ has *finite-dimensional approximants* (or *f.d.a.*, for short) if, for every $r \in \mathbb{N}$, $\varepsilon > 0$ and $R > \max_i \|a_i\|$, $\Gamma_R(a_1, \dots, a_N; n, r, \varepsilon) \ne \emptyset$ for some n (equivalently for any sufficiently large n). Evidently, this condition is necessary for $\chi(a_1, \dots, a_N) > -\infty$.

The next proposition gives the free entropy power inequality under perturbation by a semicircular system.

Proposition 7.3.7 *Let $a_1, \dots, a_N \in \mathcal{M}^{sa}$ and assume that $(a_1, \dots, a_N)$ has finite-dimensional approximants. If $(S_1, \dots, S_N)$ is a semicircular system free from $\{a_1, \dots, a_N\}$, then for every $t_1, \dots, t_N > 0$*

$$\begin{aligned}
&\exp\Bigl(\frac{2}{N}\chi(a_1 + t_1S_1, \dots, a_N + t_NS_N)\Bigr) \\
&\quad \geq \exp\Bigl(\frac{2}{N}\chi(a_1, \dots, a_N)\Bigr) + \exp\Bigl(\frac{2}{N}\chi(t_1S_1, \dots, t_NS_N)\Bigr) \\
&\quad = \exp\Bigl(\frac{2}{N}\chi(a_1, \dots, a_N)\Bigr) + 2\pi e\biggl(\prod_{i=1}^N t_i\biggr)^{2/N}.
\end{aligned}$$

Proof: Let $R > \max_i \|a_i\| + 2\max_i t_i$. For $n, r \in \mathbb{N}$ and $\varepsilon > 0$ set

$$\begin{aligned}
A_n(r,\varepsilon) &:= \Gamma_R(a_1, \dots, a_N; n, r, \varepsilon), \\
B_n(r,\varepsilon) &:= \prod_{i=1}^N \Gamma_R(t_iS_i; n, r, \varepsilon), \\
\Omega_n(r,\varepsilon) &:= \Gamma_R(a_1, \dots, a_N, t_1S_1, \dots, t_NS_N; n, r, \varepsilon), \\
C_n(r,\varepsilon) &:= \Gamma_{2R}(a_1 + t_1S_1, \dots, a_N + t_NS_N; n, r, \varepsilon).
\end{aligned}$$

For any given $r \in \mathbb{N}$ and $\varepsilon > 0$, choose $\varepsilon' > 0$ such that

$$\begin{aligned}
&\{(A_1 + B_1, \dots, A_N + B_N) : (A_1, \dots, A_N, B_1, \dots, B_N) \in \Omega_n(r, \varepsilon')\} \\
&\quad \subset C_n(r,\varepsilon).
\end{aligned} \tag{7.3.13}$$

Moreover, by Lemma 6.4.3 there exists $\varepsilon_1 > 0$ such that

$$\lim_{n\to\infty} \frac{\Lambda\bigl((A_n(r,\varepsilon_1) \times B_n(r,\varepsilon_1)) \cap \Omega_n(r,\varepsilon')\bigr)}{\Lambda(A_n(r,\varepsilon_1) \times B_n(r,\varepsilon_1))} = 1. \tag{7.3.14}$$

(Note that Lemma 6.4.3 is valid under the assumption of f.d.a. in place of finite free entropy, as is clear from the proof.) Set

$$\Theta_n := (A_n(r,\varepsilon_1) \times B_n(r,\varepsilon_1)) \cap \Omega_n(r,\varepsilon').$$

Then (7.3.13) and (7.3.14) imply that

$$A_n(r,\varepsilon_1) +_{\Theta_n} B_n(r,\varepsilon_1) \subset C_n(r,\varepsilon),$$

$$\lim_{n\to\infty} \frac{\Lambda(\Theta_n)}{\Lambda(A_n(r,\varepsilon_1) \times B_n(r,\varepsilon_1))} = 1. \tag{7.3.15}$$

Note by (6.1.1) that the limit

$$\lim_{n\to\infty}\left[\frac{1}{n^2}\Lambda(B_n(r,\varepsilon_1))+\frac{N}{2}\log n\right]=\sum_{i=1}^{N}\chi_R(t_iS_i;r,\varepsilon_1) \tag{7.3.16}$$

exists.

First assume that $\chi(a_1,\dots,a_N)>-\infty$. Choose a subsequence $\{n_k\}\subset\{n\}$ such that

$$\lim_{k\to\infty}\left[\frac{1}{n_k^2}\log\Lambda(A_{n_k}(r,\varepsilon_1))+\frac{N}{2}\log n_k\right]=\chi_R(a_1,\dots,a_N;r,\varepsilon_1)\,. \tag{7.3.17}$$

Hence we get

$$\lim_{k\to\infty}\frac{1}{n_k^2}\log\frac{\Lambda(A_{n_k}(r,\varepsilon_1))}{\Lambda(B_{n_k}(r,\varepsilon_1))}=\chi_R(a_1,\dots,a_N;r,\varepsilon_1)-\sum_{i=1}^{N}\chi_R(t_iS_i;r,\varepsilon_1),$$

which is a finite real value (because $\chi_R(a_1,\dots,a_N;r,\varepsilon_1)\ge\chi(a_1,\dots,a_N)>-\infty$). So one can apply Lemma 7.3.6 to get

$$\Lambda(C_{n_k}(r,\varepsilon))^{2/Nn_k^2}\ge\Lambda(A_{n_k}(r,\varepsilon_1))^{2/Nn_k^2}+\Lambda(B_{n_k}(r,\varepsilon_1))^{2/Nn_k^2}$$

for large k. This implies that

$$\begin{aligned}
&\exp\Big(\frac{2}{N}\chi_{2R}(a_1+t_1S_1,\dots,a_N+t_NS_N;r,\varepsilon)\Big)\\
&\quad=\limsup_{n\to\infty}n\Lambda(C_n(r,\varepsilon))^{2/Nn^2}\\
&\quad\ge\limsup_{k\to\infty}n_k\Lambda(C_{n_k}(r,\varepsilon))^{2/Nn_k^2}\\
&\quad\ge\lim_{k\to\infty}\left[n_k\Lambda(A_{n_k}(r,\varepsilon_1))^{2/Nn_k^2}+n_k\Lambda(B_{n_k}(r,\varepsilon_1))^{2/Nn_k^2}\right]\\
&\quad=\exp\Big(\frac{2}{N}\chi_R(a_1,\dots,a_N;r,\varepsilon_1)\Big)+\exp\Big(\frac{2}{N}\sum_{i=1}^{N}\chi_R(t_iS_i;r,\varepsilon_1)\Big)\\
&\qquad\qquad\text{(by (7.3.16), (7.3.17))}\\
&\quad\ge\exp\Big(\frac{2}{N}\chi_R(a_1,\dots,a_N)\Big)+\exp\Big(\frac{2}{N}\sum_{i=1}^{N}\chi_R(t_iS_i)\Big)\,.
\end{aligned}$$

Since

$$\chi(t_1S_1,\dots,t_NS_N)=\sum_{i=1}^{N}\chi(t_iS_i)=\frac{1}{2}\sum_{i=1}^{N}\log(2\pi et_i^2)\,,$$

we have the result.

Next assume that $\chi(a_1,\dots,a_N) = -\infty$. For any $0 < \theta < 1$ fixed, by (7.3.15)

$$\Lambda(\Theta_n) \geq (1-\theta)\Lambda(A_n(r,\varepsilon_1) \times B_n(r,\varepsilon_1))$$

for large n. For such n there is $(A_1,\dots,A_N) \in A_n(r,\varepsilon_1)$ such that

$$\begin{aligned}
&(1-\theta)\Lambda(B_n(r,\varepsilon_1)) \\
&\quad \leq \Lambda(\{(B_1,\dots,B_N) : (A_1\dots,A_N,B_1\dots,B_N) \in \Theta_n\}) \\
&\quad \leq \Lambda(\{(B_1,\dots,B_N) : (A_1+B_1,\dots,A_N+B_N) \in C_n(r,\varepsilon)\}) \\
&\quad = \Lambda(C_n(r,\varepsilon))\,.
\end{aligned}$$

This implies that

$$\chi(a_1+t_1S_1,\dots,a_N+t_NS_n) \geq \chi(t_1S_1,\dots,t_NS_N)$$

which is the conclusion in this case.

□

Corollary 7.3.8 *If* $a_1,\dots,a_N \in \mathcal{M}^{sa}$ *and* $\chi(a_1,\dots,a_N) > -\infty$, *then* $\delta(a_1,\dots,a_N) = N$.

Proof: This is obvious because the previous proposition gives

$$\chi(a_1+\varepsilon S_1,\dots,a_N+\varepsilon S_N) \geq \chi(a_1,\dots,a_N) > -\infty$$

for every $\varepsilon > 0$.

□

Theorem 7.3.9 *Let* $a_1,\dots,a_N \in \mathcal{M}^{sa}$, *and let* $(S_1,\dots,S_N)$ *be a semicircular system free from* $\{a_1,\dots,a_N\}$. *Then the following are equivalent:*

(i) $(a_1,\dots,a_N)$ *has finite-dimensional approximants;*

(ii) $\chi(a_1+tS_1,\dots,a_N+tS_N) > -\infty$ *for all* $t > 0$*;*

(iii) $\delta(a_1,\dots,a_N) \geq 0$*;*

(iv) $\delta(a_1+tS_1,\dots,a_N+tS_N) = N$ *for all* $t > 0$.

Furthermore, if $(a_1,\dots,a_N)$ *does not have finite-dimensional approximants, then*

$$\chi(a_1+tS_1,\dots,a_N+tS_N) = -\infty \qquad (0 \leq t \leq \varepsilon)$$

for some $\varepsilon > 0$, *and* $\delta(a_1,\dots,a_N) = -\infty$.

Proof: (i) ⇒ (ii) is obvious from Proposition 7.3.7. Conversely, (ii) implies that $(a_1 + tS_1, \dots, a_N + tS_N)$ has f.d.a. for all $t > 0$, so (i) follows by approximation. Under the assumption (i), Proposition 7.3.7 yields

$$\limsup_{\varepsilon \to +0} \frac{\chi(a_1 + \varepsilon S_1, \dots, a_N + \varepsilon S_N)}{|\log \varepsilon|} \geq \limsup_{\varepsilon \to +0} \frac{\frac{N}{2}\log(2\pi e \varepsilon^2)}{|\log \varepsilon|} = -N\,.$$

Hence (i) ⇒ (iii). Conversely, (iii) implies that $\chi(a_1 + \varepsilon_k S_1, \dots, a_N + \varepsilon_k S_N) > -\infty$, and hence $(a_1 + \varepsilon_k S_1, \dots, a_N + \varepsilon_k S_N)$ has f.d.a. for some $\varepsilon_k \to +0$. This gives (i). (ii) ⇒ (iv) follows from the above corollary, and (iv) ⇒ (i) is shown similarly to (iii) ⇒ (i).

Next, assume that $(a_1, \dots, a_N)$ does not have f.d.a. It is clear from the above proof that $\chi(a_1 + tS_1, \dots, a_N + tS_N) = -\infty$ for all t near 0. This gives $\delta(a_1, \dots, a_N) = -\infty$.

□

According to Proposition 7.3.5 and Theorem 7.3.9, if $a_1, \dots, a_N$ are in a von Neumann algebra isomorphic to $\mathcal{L}(\mathbf{F}_M)$, then $(a_1, \dots, a_N)$ has f.d.a. But this is easy to check directly; in fact, if $(b_1, \dots, b_M)$ has f.d.a. and $a_1, \dots, a_N \in \{b_1, \dots, b_M\}''$, then $(a_1, \dots, a_N)$ has f.d.a. too.

The behavior of the free entropy dimension under linear transformations of the variables is similar to that of the free entropy in Corollary 6.3.2.

Proposition 7.3.10 *Let $a_1, \dots, a_N \in \mathcal{M}^{sa}$, let $A = [\alpha_{ij}]_{i,j=1}^N$ be an invertible real matrix, and let $\beta_1, \dots, \beta_N \in \mathbb{R}$. Then*

$$\delta(a_1, \dots, a_N) = \delta\Big(\sum_{j=1}^N \alpha_{1j} a_j + \beta_1 \mathbf{1}, \dots, \sum_{j=1}^N \alpha_{Nj} a_j + \beta_N \mathbf{1}\Big)\,.$$

Proof: It is obvious that $\delta(a_1 + \beta_1 \mathbf{1}, \dots, a_N + \beta_N \mathbf{1}) = \delta(a_1, \dots, a_N)$. So the case $\beta_i = 0$ $(1 \leq i \leq N)$ is enough. First assume that A is an orthogonal matrix. Let $(S_1, \dots, S_N)$ be as before and set $S_i' := \sum_{j=1}^N \alpha_{ij} S_j$. Then $(S_1', \dots, S_N')$ is a semicircular system again (cf. Proposition 2.6.6), and by Corollary 6.3.2

$$\chi\Big(\sum_{j=1}^N \alpha_{1j} a_j + \varepsilon S_1', \dots, \sum_{j=1}^N \alpha_{Nj} a_j + \varepsilon S_N'\Big) = \chi(a_1 + \varepsilon S_1, \dots, a_N + \varepsilon S_N)\,,$$

so we have $\delta\big(\sum_{j=1}^N \alpha_{1j} a_j, \dots, \sum_{j=1}^N \alpha_{Nj} a_j\big) = \delta(a_1, \dots, a_N)$. Now we may assume that $(a_1, \dots, a_N)$ has f.d.a. Let $\lambda_1 \geq \dots \geq \lambda_N\ (> 0)$ be the eigenvalues of $(A^t A)^{1/2}$. Since there are orthogonal matrices T_1, T_2 such that $T_1 A = \mathbf{Diag}(\lambda_1, \dots, \lambda_N) T_2$, it suffices to show the case $A = \mathbf{Diag}(\lambda_1, \dots, \lambda_N)$. For $\varepsilon > 0$ we have

$$\begin{aligned} &\chi(\lambda_1 a_1 + \varepsilon S_1, \dots, \lambda_N a_N + \varepsilon S_N) \\ &\quad = \chi\Big(a_1 + \frac{\varepsilon}{\lambda_1} S_1, \dots, a_N + \frac{\varepsilon}{\lambda_N} S_N\Big) + \log(\lambda_1 \cdots \lambda_N)\,. \end{aligned}$$

So it remains to check that

$$\delta(a_1,\dots,a_N) = N + \lim_{\varepsilon\to+0} \frac{\chi(a_1+\varepsilon t_1 S_1,\dots,a_N+\varepsilon t_N S_N)}{|\log\varepsilon|} \tag{7.3.18}$$

for any constants $t_1,\dots,t_N > 0$. For $0 < t < \min_i t_i$ we can write

$$\begin{aligned}&\chi(a_1+\varepsilon t_1 S_1,\dots,a_N+\varepsilon t_N S_N)\\ &\quad = \chi\Big(a_1+\varepsilon t S_1+\varepsilon\sqrt{t_1^2-t^2}\tilde{S}_1,\dots,a_N+\varepsilon t S_N+\varepsilon\sqrt{t_N^2-t^2}\tilde{S}_N\Big),\end{aligned}$$

where $(\tilde{S}_1,\dots,\tilde{S}_N)$ is a semicircular system free from $\{a_1,\dots,a_N\}\cup\{S_1,\dots,S_N\}$. Hence Proposition 7.3.7 yields

$$\begin{aligned}&\exp\Big(\frac{2}{N}\chi(a_1+\varepsilon t_1 S_1,\dots,a_N+\varepsilon t_N S_N)\Big)\\ &\quad\ge \exp\Big(\frac{2}{N}\chi(a_1+\varepsilon t S_1,\dots,a_N+\varepsilon t S_N)\Big)\\ &\qquad + \exp\Big(\frac{2}{N}\chi(\varepsilon\sqrt{t_1^2-t^2}\tilde{S}_1,\dots,\varepsilon\sqrt{t_N^2-t^2}\tilde{S}_N)\Big),\end{aligned}$$

so that

$$\chi(a_1+\varepsilon t_1 S_1,\dots,a_N+\varepsilon t_N S_N) \ge \chi(a_1+\varepsilon t S_1,\dots,a_N+\varepsilon t S_N).$$

Similarly, for $t' > \max_i t_i$ we have

$$\chi(a_1+\varepsilon t_1 S_1,\dots,a_N+\varepsilon t_N S_N) \le \chi(a_1+\varepsilon t' S_1,\dots,a_N+\varepsilon t' S_N).$$

Now (7.3.18) immediately follows from the above estimates.

□

It is worth noting that the free entropy dimension does not change when trivial elements (i.e. scalars) are added. Indeed, for $a_1,\dots,a_N \in \mathcal{M}^{sa}$ one gets

$$\begin{aligned}&\delta(a_1,\dots,a_N,0)\\ &\quad = N+1+\limsup_{\varepsilon\to+0}\frac{\chi(a_1+\varepsilon S_1,\dots,a_N+\varepsilon S_N,\varepsilon S_{N+1})}{|\log\varepsilon|}\\ &\quad = N+1+\limsup_{\varepsilon\to+0}\frac{\chi(a_1+\varepsilon S_1,\dots,a_N+\varepsilon S_N)+\chi(\varepsilon S_{N+1})}{|\log\varepsilon|}\\ &\quad = \delta(a_1,\dots,a_N).\end{aligned}$$

The above second equality is due to the strong additivity (6.4.3).

The propositions below are generalizations of the above with some additional assumptions.

Proposition 7.3.11 *Let $a_1,\dots,a_N \in \mathcal{M}^{sa}$ be such that $\chi(a_1,\dots,a_N) > -\infty$. If $b_j = b_j^* \in \{a_1,\dots,a_N\}''$ for $1 \le j \le M$, then*

$$\delta(a_1,\dots,a_N,b_1,\dots,b_M) \ge \delta(a_1,\dots,a_N) = N\,.$$

Proof: Let $(S_1,\dots,S_{N+M})$ be a semicircular system which is in free relation to $\{a_1,\dots,a_N\}$. By Propositions 6.1.3 and 6.1.6 we have

$$\begin{aligned}
&\chi(a_1+\varepsilon S_1,\dots,a_N+\varepsilon S_N,b_1+\varepsilon S_{N+1},\dots,b_M+\varepsilon S_{N+M})\\
&\quad\ge \chi(S_1,\dots,S_N,a_1+\varepsilon S_1,\dots,a_N+\varepsilon S_N,b_1+\varepsilon S_{N+1},\dots,b_M+\varepsilon S_{N+M})\\
&\qquad -\chi(S_1,\dots,S_N)\\
&\quad= \chi(S_1,\dots,S_N,a_1,\dots,a_N,b_1+\varepsilon S_{N+1},\cdots,b_M+\varepsilon S_{N+M}) - \sum_{i=1}^{N}\chi(S_i)\\
&\quad= \chi(S_1,\dots,S_N,a_1,\dots,a_N,\varepsilon S_{N+1},\dots,\varepsilon S_{N+M}) - \sum_{i=1}^{N}\chi(S_i)\\
&\quad= \chi(a_1,\dots,a_N) + \sum_{j=1}^{M}\chi(S_{N+j}) + M\log\varepsilon\,.
\end{aligned}$$

The last equality is due to repeated use of the strong additivity property (6.4.3). Therefore,

$$\delta(a_1,\dots,a_N,b_1,\dots,b_M) \ge (N+M) - M = N\,.$$

This, together with Corollary 7.3.8, gives the result.

□

Under a stronger assumption than the previous we have

Proposition 7.3.12 *Let $a_1,\dots,a_N \in \mathcal{M}^{sa}$. Let $F_j(X_1,\dots,X_N)$ $(1 \le j \le M)$ be noncommutative power series with a common multi-radius $(R_1,\dots,R_N)$ of convergence such that $R_i > \|a_i\|$ for $1 \le i \le N$. If $b_j := F_j(a_1,\dots,a_N)$ and $b_j = b_j^*$ for $1 \le j \le M$, then*

$$\delta(a_1,\dots,a_N,b_1,\dots,b_M) \le \delta(a_1,\dots,a_N)\,.$$

If $\chi(a_1,\dots,a_N) > -\infty$ in addition, then

$$\delta(a_1,\dots,a_N,b_1,\dots,b_M) = \delta(a_1,\dots,a_N) = N\,. \tag{7.3.19}$$

Proof: Let $(S_1,\dots,S_{N+M})$ be as in the previous proof. Since $\phi_j(z) := F_j(a_1+zS_1,\dots,a_N+zS_N)-b_j$ is analytic in a neighborhood of $0 \in \mathbb{C}$ and $\phi_j(0)=0$, we get

$$\inf\bigl\{\|b_j - x\| : x \in \{a_1+\varepsilon S_1,\dots,a_N+\varepsilon S_N\}''\bigr\} \le \|\phi_j(\varepsilon)\| = O(\varepsilon)\,. \tag{7.3.20}$$

If $x_j \in \{a_1 + \varepsilon S_1, \ldots, a_N + \varepsilon S_N\}''$ for $1 \le j \le M$, then

$$\begin{aligned}
&\chi(a_1 + \varepsilon S_1, \ldots, a_N + \varepsilon S_N, b_1 + \varepsilon S_{N+1}, \ldots, b_M + \varepsilon S_{N+M}) \\
&\quad = \chi(a_1 + \varepsilon S_1, \ldots, a_N + \varepsilon S_N, b_1 - x_1 + \varepsilon S_{N+1}, \ldots, b_M - x_M + \varepsilon S_{N+M}) \\
&\quad \le \chi(a_1 + \varepsilon S_1, \ldots, a_N + \varepsilon S_N) \\
&\qquad\quad + \chi(\varepsilon^{-1}(b_1 - x_1) + S_{N+1}, \ldots, \varepsilon^{-1}(b_M - x_M) + S_{N+M}) + M \log \varepsilon \\
&\quad \le \chi(a_1 + \varepsilon S_1, \ldots, a_N + \varepsilon S_N) \\
&\qquad\quad + \frac{M}{2} \log \left[\frac{2\pi e}{M} \sum_{j=1}^{M} \left(\frac{\|b_j - x_j\|}{\varepsilon} + 2 \right)^2 \right] + M \log \varepsilon
\end{aligned}$$

by Propositions 6.1.6, 6.1.3, Corollary 6.3.2 and Proposition 6.1.1. Hence, thanks to (7.3.20), we have

$$\begin{aligned}
&\chi(a_1 + \varepsilon S_1, \ldots, a_N + \varepsilon S_N, b_1 + \varepsilon S_{N+1}, \ldots, b_M + \varepsilon S_{N+M}) \\
&\quad \le \chi(a_1 + \varepsilon S_1, \ldots, a_N + \varepsilon S_N) + \text{const} + M \log \varepsilon
\end{aligned}$$

for sufficiently small $\varepsilon > 0$. This implies that

$$\delta(a_1, \ldots, a_N, b_1, \ldots, b_M) \le \delta(a_1, \ldots, a_N).$$

When $\chi(a_1, \ldots, a_N) > -\infty$, the reverse inequality is also valid by Proposition 7.3.11.

□

Here a quite interesting problem arises: is $\delta(a_1, \ldots, a_N)$ lower semicontinuous in the sense that if $a_i, a_{m,i} \in \mathcal{M}^{sa}$ and $a_{m,i} \to a_i$ $(m \to \infty)$ strongly for $1 \le i \le N$, then

$$\delta(a_1, \ldots, a_N) \le \liminf_{m \to \infty} \delta(a_{m,1}, \ldots, a_{m,N})\,?$$

For the single variable case this was proven in Corollary 7.3.3. When this happens to hold true in the multivariable case too, it is evident from Propositions 7.3.11 and 7.3.12 that we indeed have (7.3.19) under the assumption of Proposition 7.3.11. Then the isomrophism problem for free group factors would be solved. In fact, if $\mathcal{L}(\mathbf{F}_N) \cong \mathcal{L}(\mathbf{F}_M)$ for $N, M \in \mathbb{N}$, then there exist two semicircular systems $(a_1, \ldots, a_N)$ and $(b_1, \ldots, b_M)$ in $\mathcal{L}(\mathbf{F}_N)$ such that $\{a_1, \ldots, a_N\}'' = \{b_1, \ldots, b_M\}''$, and so (7.3.19) implies that

$$N = \delta(a_1, \ldots, a_N) = \delta(a_1, \ldots, a_N, b_1, \ldots, b_M) = \delta(b_1, \ldots, b_M) = M.$$

In this way, the lower semicontinuity of $\delta(a_1, \ldots, a_N)$ would imply that case (2) of Corollary 7.2.13 is indeed true. But the above lower semicontinuity problem may be extremely difficult to solve even though it is true.

Another notion named free dimension was introduced by K. Dykema in the course of his study on free products of von Neumann algebras. He defined the *free dimension* for a certain class of finite von Neumann algebras, including free group factors, finite-dimensional algebras and also hyperfinite finite von Neumann algebras. Von Neumann algebras treated here are of finite type, so they are tracial W^*-probability spaces equipped with a faithful normal tracial state. The following notation is convenient: When $\mathcal{M}_i$ $(i = 1, 2, \dots)$ are von Neumann algebras with associated trace τ_i and $\alpha_i \geq 0$ are such that $\sum_i \alpha_i = 1$, let

$$\underset{\alpha_1}{\mathcal{M}_1} \oplus \underset{\alpha_2}{\mathcal{M}_2} \oplus \cdots$$

denote the direct sum von Neumann algebra whose associated trace is

$$\tau(x_1 \oplus x_2 \oplus \cdots) = \sum_i \alpha_i \tau_i(x_i),$$

where the summands $\mathcal{M}_i$ with $\alpha_i = 0$ are considered to be removed. Dykema's definition of free dimension is given in the following way:

(1) The free dimension of

$$\mathcal{M} \cong \underset{\alpha_0}{\mathcal{L}(\mathbf{F}_r)} \oplus \underset{\alpha_1}{M_{n_1}(\mathbb{C})} \oplus \underset{\alpha_2}{M_{n_2}(\mathbb{C})} \oplus \cdots$$

is equal to

$$\mathbf{fdim}(\mathcal{M}) := \alpha_0^2 r + \sum_{i \geq 1} \alpha_i^2 (1 - n_i^{-2}) + \sum_{\substack{i,j \geq 0 \\ i \neq j}} \alpha_i \alpha_j .$$

In particular,

$$\mathbf{fdim}(\mathcal{L}(\mathbb{Z})) = 1, \quad \mathbf{fdim}(\mathcal{L}(\mathbf{F}_r)) = r \quad \text{for} \quad r > 1,$$

$$\mathbf{fdim}(M_n(\mathbb{C})) = 1 - n^{-2}.$$

(2) Let $\mathcal{M}$ be a hyperfinite von Neumann algebra with a trace specified. Decomposing it into its nonatomic and atomic parts, we may represent $\mathcal{M}$ as

$$\mathcal{M} \cong \underset{\alpha_0}{\mathcal{M}_0} \oplus \underset{\alpha_1}{M_{n_1}(\mathbb{C})} \oplus \underset{\alpha_2}{M_{n_2}(\mathbb{C})} \oplus \cdots,$$

where $\mathcal{M}_0$ is a nonatomic (or diffuse) finite von Neumann algebra. Then the free dimension of $\mathcal{M}$ is equal to

$$\mathbf{fdim}(\mathcal{M}) := \alpha_0^2 + \sum_{i \geq 1} \alpha_i^2 (1 - n_i^{-2}) + \sum_{\substack{i,j \geq 0 \\ i \neq j}} \alpha_i \alpha_j .$$

In particular, $\mathbf{fdim}(R) = 1$ for the hyperfinite II_1 factor R.

Of course, if the free group factors happen to be mutually isomorphic (see Corollary 7.2.13), then the above definition is not well-defined. However, in this case, it is needless to determine the free dimension itself (for instance, in the theorems below).

The free dimension defined above is of essential use in the next theorems, which were discovered by Dykema.

Theorem 7.3.13 *Let*

$$\mathcal{M} \cong \underset{\alpha_0}{\mathcal{L}(\mathbf{F}_r)} \oplus \underset{\alpha_1}{\mathbb{C}p_1} \oplus \underset{\alpha_2}{\mathbb{C}p_2} \oplus \cdots \qquad (r \ge 1,\ \alpha_i \ge 0),$$

$$\mathcal{N} \cong \underset{\beta_0}{\mathcal{L}(\mathbf{F}_s)} \oplus \underset{\beta_1}{\mathbb{C}q_1} \oplus \underset{\beta_2}{\mathbb{C}q_2} \oplus \cdots \qquad (s \ge 1,\ \beta_i \ge 0),$$

where $\alpha_0 + \beta_0 > 0$. *Then*

$$\mathcal{M} \star \mathcal{N} \cong \underset{\gamma}{\mathcal{L}(\mathbf{F}_t)} \oplus \bigoplus_{i,j} \underset{\gamma_{ij}}{\mathbb{C}(p_i \wedge q_j)}\,,$$

where $\gamma_{ij} := \max\{\alpha_i + \beta_j - 1, 0\}$ *(and* $\gamma = 1 - \sum_{i,j} \gamma_{ij}$*) and* t *is determined so that* $\mathbf{fdim}(\mathcal{M} \star \mathcal{N}) = \mathbf{fdim}(\mathcal{M}) + \mathbf{fdim}(\mathcal{N})$.

Theorem 7.3.14 *Let* $\mathcal{M}$ *and* $\mathcal{N}$ *be hyperfinte (possibly finite-dimensional) von Neumann algebras having the linear dimensions* $\dim(\mathcal{M}) \ge 2$ *and* $\dim(\mathcal{N}) \ge 3$. *Represent*

$$\mathcal{M} \cong \underset{\alpha_0}{\mathcal{M}_0} \oplus \bigoplus_{i \in I} \underset{\alpha_i}{M_{m_i}(\mathbb{C})}\,,$$

$$\mathcal{N} \cong \underset{\beta_0}{\mathcal{N}_0} \oplus \bigoplus_{j \in J} \underset{\beta_j}{M_{n_j}(\mathbb{C})}\,,$$

where $\mathcal{M}_0$ *and* $\mathcal{N}_0$ *are diffuse von Neumann algebras. Then*

$$\mathcal{M} \star \mathcal{N} \cong \underset{\gamma}{\mathcal{L}(\mathbf{F}_r)} \oplus \bigoplus_{(i,j) \in I \times J} \underset{\gamma_{ij}}{M_{N(i,j)}(\mathbb{C})}\,,$$

where $N(i,j) := \max\{m_i, n_j\}$, $\gamma_{ij} := N(i,j)^2 \max\{\alpha_i m_i^{-2} + \beta_j n_j^{-2} - 1, 0\}$, *and* r *is determined so that* $\mathbf{fdim}(\mathcal{M} \star \mathcal{N}) = \mathbf{fdim}(\mathcal{M}) + \mathbf{fdim}(\mathcal{N})$. *(Note that* $\gamma_{ij} > 0$ *implies* $m_i = 1$ *or* $n_j = 1$, *and that* $\gamma_{ij} > 0$ *only for finitely many pairs* (i,j).*)*

Let $\mathcal{M}$ and $\mathcal{N}$ be von Neumann algebras as in Theorem 7.3.14. Then the theorem implies that $\mathcal{M} \star \mathcal{N}$ is a factor if and only if

$$\max_{i \in I} \frac{\alpha_i}{m_i^2} + \max_{j \in J} \frac{\beta_j}{n_j^2} \le 1\,.$$

This tells us that the free product of finite (particularly finite-dimensional) von Neumann algebras mostly becomes a type II_1 factor.

One may expect that the two notions of free entropy dimension and free dimension are naturally related to each other. The next theorem shows that this is indeed so when $a_1, \dots, a_N$ are free.

Theorem 7.3.15 *Let $a_1, \dots, a_N \in \mathcal{M}^{sa}$ be in free relation. Then*

$$\delta(a_1, \dots, a_N) = \mathbf{fdim}(\{a_1, \dots, a_N\}'').$$

Moreover, if $\{a_1, \dots, a_N\}''$ is a factor, then

$$\{a_1, \dots, a_N\}'' \cong \mathcal{L}(\mathbf{F}_{\delta(a_1, \dots, a_N)}).$$

Proof: First we show that $\delta(a) = \mathbf{fdim}(\{a\}'')$ for any $a \in \mathcal{M}^{sa}$. If $p_1, p_2, \dots$ are the atoms of $\{a\}''$ and $\alpha_i := \tau(p_i)$, then

$$\{a\}'' \cong \underset{\alpha_0}{\mathcal{L}(\mathbb{Z})} \oplus \underset{\alpha_1}{\mathbb{C}p_1} \oplus \underset{\alpha_2}{\mathbb{C}p_2} \oplus \cdots,$$

and Theorem 7.3.2 implies that

$$\delta(a) = 1 - \sum_{i \geq 1} \alpha_i^2 = \alpha_0^2 + \sum_{\substack{i,j \geq 0 \\ i \neq j}} \alpha_i \alpha_j = \mathbf{fdim}(\{a\}'').$$

Since $a_1, \dots, a_N$ are free, we can write

$$\{a_1, \dots, a_N\}'' = \star_{i=1}^{n} \left(\{a_i\}'', \tau|_{\{a_i\}''}\right).$$

Hence Corollary 7.3.4 (3) and Theorem 7.3.13 imply that

$$\delta(a_1, \dots, a_N) = \sum_{i=1}^{N} \delta(a_i) = \sum_{i=1}^{N} \mathbf{fdim}(\{a_i\}'') = \mathbf{fdim}(\{a_1, \dots, a_N\}'').$$

Moreover, assume that $\{a_1, \dots, a_N\}''$ is a factor. Repeated application of Theorem 7.3.13 implies that $\{a_1, \dots, a_N\}'' \cong \mathcal{L}(\mathbf{F}_t)$, where

$$t := \sum_{i=1}^{N} \mathbf{fdim}(\{a_i\}'') = \delta(a_1, \dots, a_N).$$

□

7.4 Applications of free entropy

The subject of this section is the study of applications of Voiculescu's free entropy to von Neumann algebras, in particular, to the structure theory on free group factors. The first success in this direction was obtained by Voiculescu himself. Let $\mathcal{M}$ be a general von Neumann algebra (always acting on a separable Hilbert space). A von Neumann subalgebra $\mathcal{A}$ of $\mathcal{M}$ is said to be *regular* (in $\mathcal{M}$) if the *normalizer*

$$\mathcal{N}(\mathcal{A}) := \{u \in \mathcal{M} : \text{unitary}, \ u\mathcal{A}u^* = \mathcal{A}\}$$

generates $\mathcal{M}$. When $\mathcal{A}$ is a *maximal abelian subalgebra* (abbreviated as *MASA*) of $\mathcal{M}$, $\mathcal{A}$ is called a *Cartan subalgebra* if it is regular and there is a faithful normal conditional expectation from $\mathcal{M}$ onto $\mathcal{A}$ (this latter condition is automatically satisfied whenever $\mathcal{M}$ is of type II_1). Note that in a type II_1 factor a MASA is automatically diffuse (or nonatomic), i.e. it has no nonzero minimal projections.

Since a Cartan subalgebra of a type II_1 factor is a regular and diffuse hyperfinite von Neumann subalgebra, the next theorem, due to Voiculescu, implies the absence of Cartan subalgebras in free group factors, providing a negative answer to the long-standing open question of whether every type II_1 factor has a Cartan subalgebra. This means that free group factors cannot be realized by the measure space construction via a measurable equivalence relation (see [75] for this construction and its relation with Cartan subalgebras).

Theorem 7.4.1 *Let $(\mathcal{M}, \tau)$ be a tracial W^*-probability space and $a_1, \dots, a_N \in \mathcal{M}^{sa}$ with $N \geq 2$. If $\chi(a_1, \dots, a_N) > -\infty$, then $\{a_1, \dots, a_N\}''$ does not have a regular and diffuse hyperfinite von Neumann subalgebra. In particular, the free group factor $\mathcal{L}(\mathbf{F}_N)$ $(2 \leq N < \infty)$ does not have a regular and diffuse hyperfinite von Neumann subalgebra.*

For the proof of this it is convenient to modify the free entropy of noncommutative selfadjoint multivariables. Let $a_1, \dots, a_N, b_1, \dots, b_L \in \mathcal{M}^{sa}$. For $n, r \in \mathbb{N}$, $\varepsilon > 0$ and $R > 0$ define

$$\begin{aligned} \Gamma_R(a_1, \dots, a_N : b_1, \dots, b_L; n, r, \varepsilon) := \big\{ &(A_1, \dots, A_N) \in (M_n(\mathbb{C})^{sa})^N : \\ &(A_1, \dots, A_N, B_1, \dots, B_L) \in \Gamma_R(a_1, \dots, a_N, b_1, \dots, b_L; n, r, \varepsilon) \\ &\text{for some } (B_1, \dots, B_L) \in (M_n(\mathbb{C})^{sa})^L \big\} \end{aligned}$$

and

$$\begin{aligned} &\chi_R(a_1, \dots, a_N : b_1, \dots, b_L) \\ &:= \lim_{\substack{r \to \infty \\ \varepsilon \to +0}} \limsup_{n \to \infty} \left[\frac{1}{n^2} \log \Lambda(\Gamma_R(a_1, \dots, a_N : b_1, \dots, b_L; n, r, \varepsilon)) + \frac{N}{2} \log n \right]. \end{aligned}$$

Then

$$\chi(a_1, \dots, a_N : b_1, \dots, b_L) := \sup_{R > 0} \chi_R(a_1, \dots, a_N : b_1, \dots, b_L) \tag{7.4.1}$$

is called the *modified* (or *conditional*) *free entropy* of $(a_1, \dots, a_N)$ in the presence of $(b_1, \dots, b_L)$. The modified free entropy can be similarly defined when (some of) $b_1, \dots, b_L$ are non-selfadjoint. In fact, when $b_1, \dots, b_K$ are selfadjoint and $b_{K+1}, \dots, b_L$ are non-selfadjoint with $b_j = c_j + \mathrm{i}\, d_j$, $c_j, d_j \in \mathcal{M}^{sa}$ $(K+1 \le j \le L)$, we may define $\chi(a_1, \dots, a_N : b_1, \dots, b_L)$ as

$$\chi(a_1, \dots, a_N : b_1, \dots, b_K, c_{K+1}, \dots, c_L, d_{K+1}, \dots, d_L).$$

The basic properties of $\chi(a_1, \dots, a_N : b_1, \dots, b_L)$ are similar to those of $\chi(a_1, \dots, a_N)$ given in Sec. 6.1. Here we state some of them which will be used below. The next proposition can be shown in a way similar to Proposition 6.1.4.

Proposition 7.4.2 $\chi(a_1, \dots, a_N : b_1, \dots, b_L) = \chi_R(a_1, \dots, a_N : b_1, \dots, b_L)$ *whenever* $R > \|a_i\|, \|b_j\|$.

Proposition 7.4.3 *If* $c_1, \dots, c_K \in \{a_1, \dots, a_N, b_1, \dots, b_L\}''$, *then*

$$\chi(a_1, \dots, a_N : b_1, \dots, b_L) = \chi(a_1, \dots, a_N : b_1, \dots, b_L, c_1, \dots, c_K).$$

In particular, if $c_1, \dots, c_K \in \{a_1, \dots, a_N\}''$, *then*

$$\chi(a_1, \dots, a_N) = \chi(a_1, \dots, a_N : c_1, \dots, c_K).$$

Proof: We need to show that

$$\chi(a_1, \dots, a_N : b_1, \dots, b_L) \le \chi(a_1, \dots, a_N : b_1, \dots, b_L, c_1, \dots, c_K),$$

because the reverse inequality is obvious. By assumption one can choose selfadjoint noncommutative polynomials $P_{m,k}(X_1, \dots, X_N, Y_1, \dots, Y_L)$ $(m \in \mathbb{N},\ 1 \le k \le K)$ such that $c_{m.k} := P_{m,k}(a_1, \dots, a_N, b_1, \dots, b_L) \to c_k$ strongly as $m \to \infty$. Choose $R > \sup\{\|a_i\|, \|b_j\|, \|c_{m,k}\| : i, j, m, k\}$. Since an argument similar to the proof of Proposition 6.1.5 yields the upper semicontinuity

$$\begin{aligned}\limsup_{m \to \infty} &\chi_R(a_1, \dots, a_N : b_1, \dots, b_L, c_{m,1}, \dots, c_{m,K}) \\ &\le \chi_R(a_1, \dots, a_N : b_1, \dots, b_L, c_1, \dots, c_K),\end{aligned}$$

it suffices by Proposition 7.4.2 to show that

$$\chi_R(a_1, \dots, a_N : b_1, \dots, b_L) \le \chi(a_1, \dots, a_N : b_1, \dots, b_L, c_{m,1}, \dots, c_{m,K}) \quad (7.4.2)$$

for every $m \in \mathbb{N}$. For each $m, r \in \mathbb{N}$, $\varepsilon > 0$ and $R > 0$ there exist $r_1 \in \mathbb{N}$, $\varepsilon_1 > 0$ and $R' \ge R$ such that, for every $n \in \mathbb{N}$, if $(A_1, \dots, A_N, B_1, \dots, B_L) \in \Gamma_R(a_1, \dots, a_N, b_1, \dots, b_L; n, r_1, \varepsilon_1)$, then $\|P_{m,k}(A_1, \dots, A_N, B_1, \dots, B_L)\| \le R'$

$(1 \le k \le K)$ and

$$\begin{aligned}&(A_1,\dots,A_N,B_1,\dots,B_L,P_{m,1}(A_1,\dots,A_N,B_1,\dots,B_L),\dots,\\&\qquad\qquad\qquad\qquad P_{m,K}(A_1,\dots,A_N,B_1,\dots,B_L))\\&\quad\in\Gamma_{R'}(a_1,\dots,a_N,b_1,\dots,b_L,c_{m,1},\dots,c_{m,K};n,r,\varepsilon)\,.\end{aligned}$$

This implies that

$$\begin{aligned}&\Gamma_R(a_1,\dots,a_N:b_1,\dots,b_L;n,r_1,\varepsilon_1)\\&\quad\subset\Gamma_{R'}(a_1,\dots,a_N:b_1,\dots,b_L,c_{m,1},\dots,c_{m,K};n,r,\varepsilon)\,.\end{aligned}$$

Hence (7.4.2) is obtained.

□

Now let $\mathcal{B}$ be a finite-dimensional *-subalgebra of $\mathcal{M}$, and corresponding to the decomposition $\mathcal{B}\cong\bigoplus_{k=1}^K M_{d_k}(\mathbb{C})$ let us have a system of matrix units $\bigcup_{k=1}^K(e_{rs}^{(k)})_{1\le r,s\le d_k}$ in $\mathcal{B}$. Let Ω denote the union of the mixed strings of selfadjoint $(e_{rr}^{(k)})_{1\le r\le d_k}$ and non-selfadjoint $(e_{rs}^{(k)})_{1\le r<s\le d_k}$ for $1\le k\le K$. Then the subset $\hat\Gamma_R(\Omega;n,r,\varepsilon)$ of $\prod_{k=1}^K\big((M_n(\mathbb{C})^{sa})^{d_k}\times M_n(\mathbb{C})^{d_k(d_k-1)/2}\big)$ is defined as in Sec. 6.5. We state the following facts.

(I) For every $\rho>0$ there exists $\delta>0$ such that, for each $n\in\mathbb{N}$, if two systems of matrix units $\bigcup_{k=1}^K(E_{rs}^{(k)})_{1\le r,s\le d_k}$ and $\bigcup_{k=1}^K(F_{rs}^{(k)})_{1\le r,s\le d_k}$ in $M_n(\mathbb{C})$ satisfy

$$|\mathrm{tr}_n(E_{rr}^{(k)})-\mathrm{tr}_n(F_{rr}^{(k)})|\le\delta\qquad(1\le r\le d_k,\ 1\le k\le K),$$

then

$$\|UE_{rs}^{(k)}U^*-F_{rs}^{(k)}\|_2\le\rho\qquad(1\le r,s\le d_k,\ 1\le k\le K)$$

for some $U\in\mathcal{U}(n)$.

(II) Let Ω be as above. For every $\delta>0$ and $R\ge1$ there exist $r\in\mathbb{N}$ and $\varepsilon>0$ such that, for sufficiently large $n\in\mathbb{N}$, if

$$\big((B_{rr}^{(k)})_{1\le r\le d_k},(B_{rs}^{(k)})_{1\le r<s\le d_k}\big)_{1\le k\le K}\in\hat\Gamma_R(\Omega;n,r,\varepsilon)\,,$$

then there is a system of matrix units $\bigcup_{k=1}^K(E_{rs}^{(k)})_{1\le r,s\le d_k}$ in $M_n(\mathbb{C})$ such that

$$\|E_{rs}^{(k)}-B_{rs}^{(k)}\|_2\le\delta\qquad(1\le r\le s\le d_k,\ 1\le k\le K).$$

Above, $\|\cdot\|_2$ denotes the 2-norm with respect to tr_n, i.e. $\|X\|:=\mathrm{tr}_n(X^*X)^{1/2}$ (also with respect to τ below). Note that any system of matrix units

$\bigcup_{k=1}^{K}(F_{rs}^{(k)})_{1\le r,s\le d_k}$ in $M_n(\mathbb{C})$ is unitarily conjugate to $\bigcup_{k=1}^{K}(I_{m_k}\otimes E_{rs})_{1\le r,s\le d_k}$ corresponding to the embedding

$$\bigoplus_{k=1}^{K}(M_{m_k}(\mathbb{C})\otimes M_{d_k}(\mathbb{C}))\oplus 0_l\subset M_n(\mathbb{C}),$$

where $\sum_{k=1}^{K} m_k d_k + l = n$ and $(E_{rs})_{1\le r,s\le d_k}$ is a system of usual matrix units of $M_{d_k}(\mathbb{C})$. The proof of (I) is elementary. On the other hand, (II) can be shown, for instance, by induction on $d := \sum_{k=1}^{K} d_k$ and using Lemma 4.3.4. The details are left to the reader.

Here is one more result which will play an essential role in the proof of Lemma 7.4.5. This is due to Szarek ([184], [185]).

Lemma 7.4.4 *For every $\delta > 0$ and $n\in\mathbb{N}$ there is a δ-net $(U_t)_{t\in T(n)}$ in $\mathcal{U}(n)$ in the operator norm such that $\#T(n)\le (C/\delta)^{n^2}$, where C is a universal constant.*

The main technical result to prove Theorem 7.4.1 is the following estimate of free entropy.

Lemma 7.4.5 *Let $a_1,\dots,a_N\in\mathcal{M}^{sa}$, and let $\mathcal{B}$ be a finite-dimensional *-subalgebra of $\{a_1,\dots,a_N\}''$. Assume that there are $0<\theta<1$, $a_{ij}\in\mathcal{M}$ and projections $p_{ij},q_{ij}\in\mathcal{B}$ ($1\le i\le N$, $1\le j\le K(i)$) such that*

$$a_{ij} = p_{ij}a_{ij}q_{ij} \qquad (1\le i\le N,\ 1\le j\le K(i)), \tag{7.4.3}$$

$$\Big\|a_i - \sum_{j=1}^{K(i)}(a_{ij}+a_{ij}^*)\Big\|_2 < \theta \qquad (1\le i\le N), \tag{7.4.4}$$

$$2\sum_{i=1}^{N}\sum_{j=1}^{K(i)}\tau(p_{ij})\tau(q_{ij}) < \theta\,. \tag{7.4.5}$$

Then

$$\chi(a_1,\dots,a_N)\le N\log(CR)+(N-1-\theta)\log\theta\,,$$

where C is a universal constant and $R := 1+\max_{1\le i\le N}\|a_i\|_2$.

Proof: Each a_{ij} can be replaced by its conditional expectation onto $\{a_1,\dots,a_N\}''$, so we can assume $a_{ij}\in\{a_1,\dots,a_N\}''$. Furthermore, we may assume by approximation that each a_{ij} is a noncommutative polynomial $P_{ij}(a_1,\dots,a_N)$ of $a_1,\dots,a_N$.

Let Ω be given as above for $\mathcal{B}$. The above (I) implies that for every $\rho>0$ there exists $\delta>0$ such that if $\phi,\phi':\mathcal{B}\to M_n(\mathbb{C})$ are *-homomorphisms (not necessarily unital) and $|\mathrm{tr}_n(\phi(e))-\mathrm{tr}_n(\phi'(e))|\le 3\delta$ for any $e\in\Omega$, then one has $U\in\mathcal{U}(n)$ satisfying

$$\|U(\phi(b))U^*-\phi'(b)\|_2\le\rho \qquad (b\in\mathcal{B},\ \|b\|\le 1).$$

For this $\delta > 0$ and the $R \geq 1$ given in the lemma, choose $r \in \mathbb{N}$ and $0 < \varepsilon < \delta$ for which the assertion of the above fact (II) holds.

For any large n one can choose a fixed *-homomorphism $\phi_n : \mathcal{B} \to M_n(\mathbb{C})$ such that

$$|\mathrm{tr}_n(\phi_n(e)) - \tau(e)| \leq \delta \qquad (e \in \Omega). \tag{7.4.6}$$

Assume that

$$(A_1, \ldots, A_N, (B(e))_{e\in\Omega}) \in \Gamma_R(a_1, \ldots, a_N, \Omega; n, r, \varepsilon). \tag{7.4.7}$$

Then, due to the assertion of (II), there is a *-homomorphism $\phi : \mathcal{B} \to M_n(\mathbb{C})$ such that

$$\|\phi(e) - B(e)\|_2 \leq \delta \qquad (e \in \Omega). \tag{7.4.8}$$

Combining (7.4.6)–(7.4.8), we get $|\mathrm{tr}_n(\phi(e)) - \mathrm{tr}_n(\phi_n(e))| \leq 3\delta$ for any $e \in \Omega$, so there exists $U \in \mathcal{U}(n)$ such that

$$\|U(\phi(b))U^* - \phi_n(b)\|_2 \leq \rho \qquad (b \in \mathcal{B},\ \|b\| \leq 1). \tag{7.4.9}$$

When ρ (hence δ, ε) is small enough and r is large enough, we estimate, for $1 \leq i \leq N$,

$$\left\|A_i - \sum_{j=1}^{K(i)} (P_{ij}(A_1, \ldots, A_N) + P_{ij}(A_1, \ldots, A_N)^*)\right\|_2 < \theta \tag{7.4.10}$$

from (7.4.4) with $a_{ij} = P_{ij}(a_1, \ldots, a_N)$. On the other hand, since p_{ij} and q_{ij} are written as linear combinations of e, e^* ($e \in \Omega$), (7.4.7) and (7.4.8) provide an approximation of (a_{ij}, p_{ij}, q_{ij}) by $(P_{ij}(A_1, \ldots, A_N), \phi(p_{ij}), \phi(q_{ij}))$ in their joint moments. So, according to (7.4.3), we estimate

$$\|P_{ij}(A_1, \ldots, A_N) - \phi(p_{ij})P_{ij}(A_1, \ldots, A_N)\phi(q_{ij})\|_2 < \frac{\theta}{2K(i)} \tag{7.4.11}$$

for all $1 \leq i \leq N$, $1 \leq j \leq K(i)$. Combining the estimates (7.4.10), (7.4.11) and (7.4.9) (for ρ sufficiently small), we infer that

$$\left\|UA_iU^* - \sum_{j=1}^{K(i)} (X_{ij} + X_{ij}^*)\right\|_2 < 3\theta \qquad (1 \leq i \leq N)$$

for some $U \in \mathcal{U}(n)$ and $X_{ij} \in \phi_n(p_{ij})M_n(\mathbb{C})\phi_n(q_{ij})$. Set

$$\mathcal{W}_n := \left\{ \left(\sum_{j=1}^{K(i)} (X_{ij} + X_{ij}^*) \right)_{1\leq i\leq N} : X_{ij} \in \phi_n(p_{ij})M_n(\mathbb{C})\phi_n(q_{ij}) \right\},$$

which is a subspace of the Nn^2-dimensional Euclidean space $(M_n(\mathbb{C})^{sa})^N$. The dimension of $\mathcal{W}_n$ is not greater than

$$2n^2 \sum_{i=1}^{N} \sum_{j=1}^{K(i)} (\tau(p_{ij}) + \delta)(\tau(q_{ij}) + \delta) < n^2\theta$$

(for δ small enough) according to (7.4.5). In this way, we have shown that if n, r are large and ε is small, then for every $(A_1, \dots, A_N) \in \Gamma_R(a_1, \dots, a_N : \Omega; n, r, \varepsilon)$ there exist $U \in \mathcal{U}(n)$ and $(X_1, \dots, X_N) \in \mathcal{W}_n$ such that $\|UA_iU^* - X_i\|_2 < 3\theta$ $(1 \le i \le N)$.

Now apply Lemma 7.4.4 to get a $\theta/2R$-net $(U_t)_{t \in T(n)}$ in $\mathcal{U}(n)$ so that $\#T(n) \le (CR/\theta)^{n^2}$. Then for every $(A_1, \dots, A_N) \in \Gamma_R(a_1, \dots, a_N : \Omega; n, r, \varepsilon)$ we have

$$\mathrm{dist}((A_1, \dots, A_N), U_{t,N}^* \mathcal{W}_n U_{t,N}) < (Nn)^{1/2} \cdot 4\theta \tag{7.4.12}$$

for some $t \in T(n)$, where dist is the distance in $(M_n(\mathbb{C})^{sa})^N$ with respect to the Euclidean norm and $U_{t,N} := \bigoplus_1^N U_t$. Enlarge $\mathcal{W}_n$ to a subspace $\tilde{\mathcal{W}}_n$ of dimension $[n^2\theta]$ in $(M_n(\mathbb{C})^{sa})^N$. Note that the Euclidean norm of the above $(A_1, \dots, A_N)$ is $\left(\sum_{i=1}^N \mathrm{Tr}(A_i^* A_i)\right)^{1/2} \le (Nn)^{1/2} R$. Let $\mathbf{B}_n$ be the ball of radius $(Nn)^{1/2}R$ and center 0 in $\tilde{\mathcal{W}}_n$, and $\mathbf{B}'_n$ the ball of radius $(Nn)^{1/2} \cdot 4\theta$ and center 0 in the orthogonal complement of $\tilde{\mathcal{W}}_n$ in $(M_n(\mathbb{C})^{sa})^N$. The estimate (7.4.12) means that

$$\Gamma_R(a_1, \dots, a_N : \Omega; n, r, \varepsilon) \subset \bigcup_{t \in T(n)} U_{t,N}^* (\mathbf{B}_n \oplus \mathbf{B}'_n) U_{t,N}$$

whenever n, r are large enough and ε is small enough. Therefore,

$$\begin{aligned} &\Lambda(\Gamma_R(a_1, \dots, a_N : \Omega; n, r, \varepsilon)) \\ &\quad \le \#T(n) \cdot \mathrm{vol}_{[n^2\theta]}(\mathbf{B}_n)\, \mathrm{vol}_{Nn^2 - [n^2\theta]}(\mathbf{B}'_n) \\ &\quad \le \left(\frac{CR}{\theta}\right)^{n^2} \frac{\pi^{Nn^2/2} (Nn)^{Nn^2/2} R^{[n^2\theta]} (4\theta)^{Nn^2 - [n^2\theta]}}{\Gamma\left(1 + \frac{[n^2\theta]}{2}\right) \Gamma\left(1 + \frac{Nn^2 - [n^2\theta]}{2}\right)}. \end{aligned}$$

Using Propositions 7.4.3, 7.4.2 and the Stirling formula, we have

$$\begin{aligned} &\chi(a_1, \dots, a_N) = \chi_R(a_1, \dots, a_N : \Omega) \\ &\quad \le \limsup_{n \to \infty} \left[\frac{1}{n^2} \log \Lambda(\Gamma_R(a_1, \dots, a_N : \Omega; n, r, \varepsilon)) + \frac{N}{2} \log n \right] \\ &\quad = \limsup_{n \to \infty} \bigg[\log\left(\frac{CR}{\theta}\right) + \frac{N}{2} \log \pi + \frac{N}{2} \log(Nn) \\ &\qquad\qquad + \theta \log R + (N - \theta) \log(4\theta) - \frac{\theta}{2} \log \frac{n^2\theta}{2} + \frac{\theta}{2} \\ &\qquad\qquad - \frac{N - \theta}{2} \log \frac{n^2(N - \theta)}{2} + \frac{N - \theta}{2} + \frac{N}{2} \log n \bigg] \end{aligned}$$

$$
\begin{aligned}
&= \log(CR) + \theta \log R + (N-\theta)\log 4 + \frac{N}{2}\log(2\pi e) \\
&\qquad + (N-1-\theta)\log\theta - \frac{N}{2}\Big(\frac{\theta}{N}\log\frac{\theta}{N} + \frac{N-\theta}{N}\log\frac{N-\theta}{N}\Big) \\
&\le \log(CR) + N\log(R+4) + \frac{N}{2}\log(4\pi e) + (N-1-\theta)\log\theta \\
&\le N\log(C_1 R) + (N-1-\theta)\log\theta\,,
\end{aligned}
$$

where C_1 is another universal constant. Here the inequality

$$
-t\log t - (1-t)\log(1-t) \le \log 2
$$

$(0 < t < 1)$ was used.

□

Now we are in a position to complete the proof of the theorem.

Proof of Theorem 7.4.1: Assume that $\tilde{\mathcal{M}} := \{a_1, \dots, a_N\}''$ has a regular and diffuse hyperfinite von Neumann subalgebra $\mathcal{A}$. For any $\theta > 0$ let us show that one can choose a finite-dimensional *-subalgebra $\mathcal{B} \subset \mathcal{A}$, $a_{ij} \in \tilde{\mathcal{M}}$ and projections $p_{ij}, q_{ij} \in \mathcal{B}$ with $a_{ij} = p_{ij}a_{ij}q_{ij}$ $(1 \le i \le N,\ 1 \le j \le K(i))$ such that (7.4.4) and (7.4.5) of Lemma 7.4.5 are satisfied. When this is shown, Lemma 7.4.5 implies that

$$
\chi(a_1, \dots, a_N) \le N\log(CR) + (N-1-\theta)\log\theta\,,
$$

so we obtain $\chi(a_1, \dots, a_N) = -\infty$, a contradiction.

Since the linear span of the normalizer $\mathcal{N}(\mathcal{A})$ (in $\tilde{\mathcal{M}}$) is dense in $\tilde{\mathcal{M}}$ in the strong operator topology (hence in the 2-norm), it suffices to prove the similar assertion for any $u_1, \dots, u_M \in \mathcal{N}(\mathcal{A})$ instead of $a_1, \dots, a_N$. Given $\theta > 0$, since $\mathcal{A}$ is diffuse, choose $L \in \mathbb{N}$ with $L > M/\theta$ and projections $\tilde{p}_1, \dots, \tilde{p}_L \in \mathcal{A}$ such that $\tilde{p}_1 + \cdots + \tilde{p}_L = \mathbf{1}$ and $\tau(\tilde{p}_j) = 1/L$ $(1 \le j \le L)$. If we set $\tilde{q}_{ij} := u_i^*\tilde{p}_j u_i \in \mathcal{A}$, then $\sum_{j=1}^{L} \tilde{p}_j u_i \tilde{q}_{ij} = u_i$ and

$$
\sum_{i=1}^{M}\sum_{j=1}^{L} \tau(\tilde{p}_j)\tau(\tilde{q}_{ij}) = \frac{M}{L} < \theta\,.
$$

Since $\mathcal{A}$ is hyperfinite, one can choose a finite-dimensional *-subalgebra $\mathcal{B} \subset \mathcal{A}$ and projections $p_j, q_{ij} \in \mathcal{B}$ $(1 \le i \le M,\ 1 \le j \le L)$ such that $\|p_j - \tilde{p}_j\|_2$ and $\|q_{ij} - \tilde{q}_{ij}\|_2$ are sufficiently small so that

$$
\Big\| u_i - \sum_{j=1}^{L} p_j u_i q_{ij} \Big\|_2 < \theta\,, \qquad \sum_{i=1}^{M}\sum_{j=1}^{L} \tau(p_j)\tau(q_{ij}) < \theta\,.
$$

Now the assertion follows with $a_{ij} := p_j u_j q_{ij}$ and $p_{ij} := p_j$.

□

It is known ([205]) that if $a_1, \ldots, a_N \in \mathcal{M}^{sa}$ ($N \geq 2$) and $\chi(a_1, \ldots, a_N) > -\infty$, then $\{a_1, \ldots, a_N\}''$ is a factor and moreover it is not hyperfinite. In fact, by estimating $\chi(a_1, \ldots, a_N : p, \mathbf{1}-p)$ for a projection $p \in \mathcal{M}$ in terms of the quantities $\max_i \|pa_i - a_i p\|_2$ and $\tau(p)$, Voiculescu proved a stronger result: $\{a_1, \ldots, a_N\}''$ is a *non-*Γ *II_1 factor* (i.e. it has no non-trivial central sequences) in the above case. This fact tells us that the free entropy method could be useful in a really non-hyperfinite situation.

There are some other applications of the free entropy method to free group factors. For instance, the next theorem, due to L. Ge, shows another remarkable property of free group factors.

Theorem 7.4.6 *Let $\mathcal{M}$ be a type II_1 factor generated by $a_1, \ldots, a_N \in \mathcal{M}^{sa}$ with $N \geq 2$. If $\chi(a_1, \ldots, a_N) > -\infty$, then $\mathcal{M}$ is prime, that is, $\mathcal{M}$ is not isomorphic to the tensor product of any two factors of type II_1. In particular, the free group factor $\mathcal{L}(\mathbf{F}_N)$ ($2 \leq N < \infty$) is prime.*

In particular, the theorem says that $\mathcal{L}(\mathbf{F}_N)$ is not isomorphic to $\mathcal{L}(\mathbf{F}_N) \otimes \mathcal{L}(\mathbf{F}_N)$, and it answers Sakai's question of whether $\mathcal{M} \cong \mathcal{M} \otimes \mathcal{M}$ for any type II_1 factor $\mathcal{M}$.

It is not so difficult to see that any type II_1 factor is generated by a sequence of projections with trace $1/2$. So Theorem 7.4.6 is a consequence of the following estimate of free entropy. The proof of this follows a pattern similar to that of Lemma 7.4.5, although it is more complicated. The details are omitted here.

Lemma 7.4.7 *Let $\mathcal{M}$ be a type II_1 factor, and let $\mathcal{R}_1, \mathcal{R}_2$ be mutually commuting hyperfinite subfactors of $\mathcal{M}$. Let $a_1, \ldots, a_N \in \mathcal{M}^{sa}$ ($N \geq 2$) be such that $\mathcal{M} = \{a_1, \ldots, a_N\}''$. Assume that there are $\theta > 0$, projections $p_k \in \mathcal{R}_1'$ ($1 \leq k \leq K$), $q_l \in \mathcal{R}_2'$ ($1 \leq l \leq L$), and selfadjoint noncommutative polynomials $P_i(X_1, \ldots, X_K, Y_1, \ldots, Y_L)$ ($1 \leq i \leq N$) such that*

$$\tau(p_k) = \tau(q_l) = \frac{1}{2} \qquad (1 \leq k \leq K,\ 1 \leq l \leq L),$$
$$\|a_i - P_i(p_1, \ldots, p_K, q_1, \ldots, q_L)\|_2 < \theta \qquad (1 \leq i \leq N).$$

Then

$$\chi(a_1, \ldots, a_N) \leq N \log(CR) + (n - 1 - \theta) \log \theta\,,$$

where C is a universal constant and $R := 1 + \max_{1 \leq i \leq N} \|a_i\|$.

Notes and Remarks. When G is a discrete countable group, it is known that the group von Neumann algebra $\mathcal{L}(G)$ is hyperfinite if and only if G is amenable. For instance, the group $\mathbf{S}_\infty$ of all finite permutations on $\mathbb{N}$ is an amenable ICC group, so $\mathcal{L}(\mathbf{S}_\infty)$ is another realization of R. Free group factors $\mathbf{F}_n$ ($2 \leq n \leq \infty$) and $SL(n, \mathbb{Z})$ are among typical examples of non-amenable ICC groups. The isomorphism question of whether $\mathcal{L}(\mathbf{F}_n) \not\cong \mathcal{L}(\mathbf{F}_m)$ if $n \neq m$ has been one of the most famous problems in theory of operator algebras since the early 1950's. Besides the free group factors, the reduced free group C^*-algebra $C_r^*(\mathbf{F}_n)$ is defined as the norm closure of the linear span of the left regular representation $\{L_g : g \in \mathbf{F}_n\}$. Non-isomorphism $C_r^*(\mathbf{F}_n) \not\cong C_r^*(\mathbf{F}_m)$ $(n \neq m)$ is seen from the difference of the K_1-groups ([150]). The Fock space representation in Theorem 7.1.4 may be also used to construct the Cuntz algebras $\mathcal{O}_n$. In fact, if $\dim \mathcal{H} = n < \infty$, then the C^*-algebra $C^*(\ell(\mathcal{H}))$ generated by $\{\ell(f) : f \in \mathcal{H}\}$ includes $K(\mathcal{F}(\mathcal{H}))$, the compact operators on $\mathcal{F}(\mathcal{H})$, and $C^*(\ell(\mathcal{H}))/K(\mathcal{F}(\mathcal{H})) \cong \mathcal{O}_n$, and if $\dim \mathcal{H} = \infty$ then $C^*(\ell(\mathcal{H})) \cong \mathcal{O}_\infty$ (cf. [74]).

The formulas (7.2.1) and (7.2.2) were obtained in [199], the paper which opened the promising study on free group factors, by using machinery of (semi)circular systems. Theorem 7.2.4, along with (7.2.3), was given in [61], while Theorem 7.2.6 is taken from [63]. The one-parameter family $\mathcal{L}(\mathbf{F}_r)$ $(1 < r \leq \infty)$ of interpolated free group factors was independently constructed in [158] and [66]. Corollaries 7.2.10 and 7.2.11 were first proven in [155] and [157], respectively, before construction of interpolated free group factors. Moreover, it is remarkable that a trace-scaling continuous one-parameter automorphism group of $\mathcal{L}(\mathbf{F}_\infty) \otimes B(\mathcal{H})$ was constructed in [156] by using semicircular systems; thus another proof of Corollary 7.2.10 is provided.

The distribution of pqp in Lemma 7.2.5 can be more systematically computed by using the $\mathcal{S}$-transform formula for the multiplicative free convolution. In fact, it was shown in [198] that if $\{p, q\}$ is a free pair of projections such that $\tau(p) = \alpha$ and $\tau(q) = \beta$, where $0 < \alpha, \beta < 1$, then the distribution of pqp in $(\mathcal{M}, \tau)$ is

$$c_0\delta(0) + c_1\delta(1) + \chi_{(r,s)}(t)\frac{\sqrt{(t-r)(s-t)}}{2\pi t(1-t)}\,dt\,,$$

where $c_0 := 1 - \min\{\alpha, \beta\}$, $c_1 := \max\{\alpha + \beta - 1, 0\}$ and

$$s, r := \alpha + \beta - 2\alpha\beta \pm \sqrt{4\alpha\beta(1-\alpha)(1-\beta)}\,.$$

Moreover, let $\{x, y\}$ be a free pair of positive elements in $(\mathcal{M}, \tau)$ and μ, ν the distributions of x, y. Then the following is not difficult to show:

$$\begin{cases} \|x^{1/2}yx^{1/2}\| = \|x\|\|y\| & \text{if } \mu(\{\|x\|\}) + \nu(\{\|y\|\}) \geq 1, \\ \|x^{1/2}yx^{1/2}\| < \|x\|\|y\| & \text{otherwise.} \end{cases}$$

Compare this with the remark after Example 3.2.3.

The free entropy dimension was introduced in [203], and the subordination inequalities (7.3.1) and (7.3.2) were also obtained there. Lemma 7.3.6 on restricted Minkowski sums is from [186], where the free entropy power inequality (for the single variable case) was proven. The contents of Propositions 7.3.7–7.3.10 are new, but the ideas are from [208]. Propositions 7.3.11 and 7.3.12 are from [203], with improvements.

The terminology of f.d.a. is due to Voiculescu [208]. A fundamental open problem is whether $(a_1, \dots, a_N)$ has f.d.a. for every tracial W^*-probability space $(\mathcal{M}, \tau)$ and every $a_1, \dots, a_N \in \mathcal{M}^{sa}$. As was pointed out in [203], this problem is equivalent to Connes' approximate embedding problem of whether any countably generated type II_1 factor can be approximately embedded into the hyperfinite II_1 factor ([52], p. 105). Rădulescu's work [160] is toward the affirmative solution to this problem. Theorem 7.3.9 says that if the solution is affirmative, then the free entropy dimension $\delta(a_1, \dots, a_N)$ always admits a nonnegative value (i.e. the exceptional value $-\infty$ is excluded).

In [205], along with the modified free entropy (7.4.1), the *modified free entropy dimension* was introduced as

$$\delta_0(a_1, \dots, a_N) := N + \limsup_{\varepsilon \to +0} \frac{\chi(a_1 + \varepsilon S_1, \dots, a_N + \varepsilon S_N : S_1, \dots, S_N)}{|\log \varepsilon|}.$$

The inequality $\delta_0(a_1, \dots, a_N) \le \delta(a_1, \dots, a_N)$ is clear. The two concepts δ_0 and δ have similar properties and coincide in some cases (in particular, $\delta_0(a) = \delta(a)$); however, $\delta_0 = \delta$ in general is not known. It seems that δ_0 is technically more convenient than δ.

The free dimension of certain von Neumann algebras was introduced in [62], and the proofs of Theorems 7.3.13 and 7.3.14 are found there.

When $\chi(a_1, \dots, a_N) > -\infty$, the free information $\Phi(a_1, \dots, a_N)$ in the microstates approach may be defined in the following way:

$$\Phi(a_1, \dots, a_N) := 2 \limsup_{\varepsilon \to +0} \frac{\chi(a_1 + \sqrt{\varepsilon} S_1, \dots, a_N + \sqrt{\varepsilon} S_N) - \chi(a_1, \dots, a_N)}{\varepsilon}$$

with a semicircular system $(S_1, \dots, S_N)$ free from $\{a_1, \dots, a_N\}$. For the single variable case this definition coincides with the free information in [202]. The function $t \mapsto \chi(a_1 + \sqrt{t} S_1, \dots, a_N + \sqrt{t} S_N)$ is increasing and right-continuous on $[0, \infty)$. An interesting question is whether this function is concave on $[0, \infty)$. If it is, then one can express $\chi(a_1, \dots, a_N)$ in terms of the above defined free information:

$$\begin{aligned} &\chi(a_1, \dots, a_N) \\ &= \frac{1}{2} \int_0^\infty \Big(\frac{N}{1+t} - \Phi(a_1 + \sqrt{t} S_1, \dots, a_N + \sqrt{t} S_N) \Big)\, dt + \frac{N}{2} \log(2\pi e), \end{aligned}$$

and the expression has completely the same form as in the microstates-free approach (see the Notes and Remarks section of the previous chapter).

The first application of the (modified) free entropy method to von Neumann algebra theory was given in [205], where Theorem 7.4.1 was proven. Moreover, the same assertion of Theorem 7.4.1 for the interpolated free group factors $\mathcal{L}(\mathbf{F}_r)$ $(1 < r \le \infty)$ was proven there. Theorem 7.4.6, due to L. Ge [81], solved the longstanding question about the existence of prime factors. In [80] he proved that the free group factors (and the interpolated free group factors) have no simple MASA.

A slightly stronger result was given in Dykema [66]. Furthermore, Theorem 7.4.6 was strengthened by M.B. Ştefan [182]; he proved that any subfactor of $\mathcal{L}(\mathbf{F}_r)$ with finite Jones index (in particular, $\mathcal{L}(\mathbf{F}_r)$ itself) is prime. A further application is found in [82].

Finally, some recent developments on (amalgamated) free products of operator algebras are briefly surveyed. Free product von Neumann algeras with respect to non-tracial states were studied in [13], [64], [65], [194], and it turned out that when φ_1 or φ_2 is not tracial, the free product von Neumann algebra $(\mathcal{M}_1, \varphi_1) \star (\mathcal{M}_2, \varphi_2)$ is mostly a factor of type III_λ $(0 < \lambda \leq 1)$ and it is usually *full* (in the type II_1 case, the fullness is equivalent to non-Γ). Amalgamated free products of von Neumann algebras were first treated in [152] with respect to tracial states, and the precise definition of those with respect to general normal states is found in [193]. The technique of (amalgamated) free products is sometimes useful in constructing (sub)factors or actions on them with specified properties. The first success in this direction was made by S. Popa [152], who constructed an irreducible type II_1 subfactor with Jones index s for any $s \in \{4\cos^2(\pi/n) : n \geq 3\} \cup [4, \infty)$. (See also [34], [153].) This method was further exemplified in [158], where subfactors of $\mathcal{L}(\mathbf{F}_\infty)$ with finite Jones index were constructed. The free product construction was adopted in [192] to show the existence of a minimal coaction of the compact quantum group $SU_q(n)$ on a full factor of type III_{q^2}.

Much work has been done on (amalgamated) free products of C^*-algebras. Only a few of those results are mentioneded here. In [69] it was shown that the reduced C^*-free product $(\mathcal{A}_1, \varphi_1) \star (\mathcal{A}_2, \varphi_2)$ has stable rank 1 whenever the $\mathcal{A}_i$ $(i = 1, 2)$ are C^*-algebras with faithful tracial states φ_i and satisfy the Avitzour condition. [67] is related. Dykema [68] proved that the reduced amalgamated free product of exact C^*-algebras is exact. In [169] and [170], Shlyakhtenko studied the C^*-probability space $(\Gamma(\mathcal{H}_\mathbb{R}, U_t), \varphi_U)$ associated with a one-parameter orthogonal transformation group U_t on a real Hilbert space $\mathcal{H}_\mathbb{R}$. The construction of φ_U is a free analogue of the construction of quasi-free states on the CAR and CCR algebras, and the corresponding von Neumann algebra $\Gamma(\mathcal{H}_\mathbb{R}, U_t)''$ via the GNS representation is a free analogue of the Araki-Woods factors.

The compression formula $(\mathcal{M} \star \mathcal{L}(\mathbf{F}_r))_t \cong \mathcal{M}_t \star \mathcal{L}(\mathbf{F}_{r/t^2})$ for any II_1 factor $\mathcal{M}$, $1 < r \leq \infty$ and $0 < t < 1$ was shown in [170]. This contains Corollary 7.2.10. A more extended compression formula for free products of von Neumann algebras can be found in [70].

Bibliography

[1] L. Accardi, Y. Hashimoto and N. Obata, Notions of independence related to the free group, *Infin. Dimens. Anal. Quantum Probab. Relat. Top.* **1** (1998), 201–220.

[2] N.I. Akhiezer, *The Classical Moment Problems*, Oliver & Boyd, Edinburgh-London, 1965.

[3] M. Akiyama and H. Yoshida, The distributions for linear combinations of a free family of projections and their orthogonal polynomials, preprint.

[4] T.W. Anderson, *An Introduction to Multivariate Statistical Analysis*, Second edition, John Wiley, New York, 1971.

[5] G.E. Andrews, *The Theory of Partitions*, Addison-Wesley, Reading, 1976.

[6] L. Arnold, On the asymptotic distribution of the eigenvalues of random matrices, *J. Math. Anal. Appl.* **20** (1967), 262–268.

[7] R. Askey and M. Ismail, *Recurrence Relations, Continued Fractions and Ortogonal Polynomials*, Mem. Amer. Math. Soc. **49**, 1984.

[8] D. Avitzour, Free products of C*-algebras, *Trans. Amer. Math. Soc.* **271** (1982), 423–435.

[9] Z.D. Bai, Convergence rate of expected spectral distribution of large random matrices. Part I. Wigner matrices and Part II. Sample covariance matrices, *Ann. Prob.* **21** (1993), 625–672.

[10] Z.D. Bai and Y.Q. Yin, Necessary and sufficient conditions for almost sure convergence of the largest eigenvalue of a Wigner matrix, *Ann. Probab.* **16** (1988), 1729–1741.

[11] R. Balian, Random matrices and information theory, *Nuovo Cimento B* **57** (1968), 183–193.

[12] T. Banica, On the polar decomposition of cricular variables, *Integral Equations Operator Theory* **24** (1996), 372–377.

[13] L. Barnett, Free product von Neumann algebras of type III, *Proc. Amer. Math. Soc.* **123** (1995), 543–553.

[14] A.R. Barron, Entropy and the central limit theorem, *Ann. Probab.* **14** (1986), 336–342.

[15] G. Ben Arous and A. Guionnet, Large deviation for Wigner's law and Voiculescu's noncommutative entropy, *Probab. Theory Related Fields* **108** (1997), 517–542.

[16] G. Ben Arous and O. Zeitouni, Large deviations from the circular law, *ESAIM: Probability and Statistics* **2** (1998), 123–134.

[17] H. Bercovici and D. Voiculescu, Lévy-Hinčin type theorems for multiplicative and additive free convolution, *Pacific J. Math.* **153** (1992), 217–248.

[18] H. Bercovici and D. Voiculescu, Free convolution of measures with unbounded support, *Indiana Univ. Math. J.* **42** (1993), 733–773.

[19] H. Bercovici and D. Voiculescu, Superconvergence to the central limit and failure of the Cramér theorem for free random variables, *Probab. Theory Related Fields* **102** (1995), 215–222.

[20] C. Berg, J.P.R. Christensen and P. Ressel, *Harmonic Analysis on Semigroups. Theory of Positive Definite and Related Functions*, Springer, New York, 1984.

[21] N. Berline, E. Getzler and M. Vergne, *Heat Kernels and Dirac Operators*, Springer, Berlin-Heidelberg-New York, 1992.

[22] R. Bhatia, *Matrix Analysis*, Springer, 1997.

[23] P. Biane, Permutation model for semicircular systems and quantum random walks, *Pacific J. Math.* **171** (1995), 373–387.

[24] P. Biane, Representations of unitary groups and free convolutions, *Publ. Res. Inst. Math. Sci.* **31** (1995), 63–79.

[25] P. Biane, Free brownian motion, free stochastic calculus and random matrice, in *Free Probability Theory*, D.V. Voiculescu (ed.), Fields Inst. Commun. **12**, Amer. Math. Soc., 1997, pp. 1–19.

[26] P. Biane, Free hypercontractivity, *Comm. Math. Phys.* **184** (1997), 457–474.

[27] P. Biane, On the free convolution with semi-circular distribution, *Indiana Univ. Math. J.* **46** (1997), 705–718.

[28] P. Biane, Segal-Bargmann transform, functional calculus on matrix spaces and the theory of semi-circular and circular systems, *J. Funct. Anal.* **144** (1997), 232–286.

[29] P. Biane, Some propreties of crossings and partitions, *Discrete Math.* **175** (1997), 41–53.

[30] P. Biane, Processes with free increments, *Math. Z.* **227** (1998), 143–174.

[31] P. Biane, Representations of symmetric groups and free probability, *Adv. Math.* **138** (1998), 126-181.

[32] P. Biane and R. Speicher, Stochastic calculus with respect to free brownian motion, and analysis on Wigner space, *Probab. Theory Related Fields*, **112** (1998), 373–410.

[33] P. Billingsley, *Probability and Measure*, Second edition, John Wiley, New York, 1986.

[34] F. Boca, On the method of constructing irreducible finite index subfactors of Popa, *Pacific J. Math.* **161** (1993), 201-231.

[35] A. Boutet de Monvel, L. Pastur and M. Shcherbina, On the statistical mechanics approach in random matrix theory: Integrated density of states, *J. Stat. Phys.* **79** (1995), 585–611.

[36] M. Bożejko, On $\Lambda(p)$ sets with minimal constant in discrete noncommutative groups, *Proc. Amer. Math. Soc.* **51** (1975), 407–412.

[37] M. Bożejko, B. Kümmerer and R. Speicher, q-Gaussian processes: non-commutative and classical aspects, *Comm. Math. Phys.* **185** (1997), 129–154.

[38] M. Bożejko, M. Leinert and R. Speicher, Convolution and limit theorems for conditionally free random variables, *Pacific J. Math.* **175** (1996), 357–388.

[39] M. Bożejko and R. Speicher, An example of a generalized Brownian motion I, *Comm. Math. Phys.* **137** (1991), 519–531.

[40] M. Bożejko and R. Speicher, ψ-independent and symmetrized white noises, in *Quantum Probability and Related Topics VII*, L. Accardi (ed.), World Scientific, Singapore, 1992, pp. 219–236.

[41] M. Bożejko and R. Speicher, An example of a generalized Brownian motion II, in *Quantum Probability and Related Topics VII*, L. Accardi (ed.), World Scientific, Singapore, 1992, pp. 67–77.

[42] M. Bożejko and R. Speicher, Interpolation between bosonic and fermionic relations given by generalized Brownian motions, *Math. Z.* **222** (1996), 135–160.

[43] H.J. Brascamp, E.H. Lieb and J.M. Luttinger, A general rearrangement inequality for multiple integrals, *J. Funct. Anal.* **17** (1974), 227–237.

[44] O. Bratteli and D.W. Robinson, *Operator Algebras and Quantum Statistical Mechanics I, II*, Springer, New York, 1979, 1981.

[45] E. Brézin, Dyson's universality in generalized ensembles of random matrices, in *The Mathematical Beauty of Physics*, J.M. Drouffe and J.B. Zuber (eds.), World Scientific, 1997, pp. 1–11.

[46] E. Brézin, C. Itzykson, G. Parisi and J.B. Zuber, Planar diagrams, *Comm. Math. Phys.* **59** (1978), 35–51.

[47] M.T. Cabanal-Duvillard, *Probabilités libres et calcul stochastique. Application aux grandes matrices aléatoires*, Ph.D. Thesis, Université Paris VI, 1999.

[48] D.I. Cartwright and P.M. Soardi, Random walks on free products, quotients and amalgams, *Nagoya Math. J.* **102** (1986), 163–180.

[49] T.S. Chihara, *An Introduction to Orthogonal Polynomials*, Gordon and Breach, 1978.

[50] W.-M. Ching, Free products of von Neumann algebras, *Trans. Amer. Math. Soc.* **178** (1973), 147–163.

[51] M. Choda, Reduced free products of completely positive maps and entropy for free products of automorphisms, *Publ. Res. Inst. Math. Sci.* **32** (1996), 371–382.

[52] A. Connes, Classification of injective factors, *Ann. of Math.* **104** (1976), 73–115.

[53] T.M. Cover and J.A. Thomas, *Elements of Information Theory*, John Wiley, New York, 1991.

[54] I. Cuculescu and A.G. Oprea, *Noncommutative Probability*, Kluwer, Dordrecht, 1994.

[55] A. Dembo and O. Zeitouni, *Large Deviation Techniques and Applications*, Second edition, Springer, New York, 1998.

[56] J.D. Deuschel and D.W. Stroock, *Large Deviations*, Academic Press, Boston, 1989.

[57] W. Donoghue, *Monotone Matrix Functions and Analytic Continuation*, Springer, New York, 1974.

[58] J.L. Doob, *Stochastic Processes*, John Wiley, New York, 1953.

[59] R.G. Douglas, *Banach Algebra Techniques in Operator Theory*, Academic Press, New York, 1972.

[60] P.L. Duren, *Univalent Functions*, Springer, New York, 1983.

[61] K. Dykema, On certain free product factors via an extended matrix model, *J. Funct. Anal.* **112** (1993), 31–60.

[62] K. Dykema, Free products of hyperfinite von Neumann algebras and free dimension, *Duke Math. J.* **69** (1993), 97–119.

[63] K. Dykema, Interpolated free group factors, *Pacific J. Math.* **163** (1994), 123–135.

[64] K.J. Dykema, Factoriality and Connes' invariant $T(\mathcal{M})$ for free products of von Neumann algebras, *J. Reine Angew. Math.* **450** (1994), 159-180.

[65] K.J. Dykema, Free products of finite dimensional and other von Neumann algebras with respect to non-tracial states, in *Free Probability Theory*, D.V. Voiculescu (ed.), Fields Inst. Commun. **12**, Amer. Math. Soc., 1997, pp. 41-88.

[66] K. Dykema, Two applications of free entropy, *Math. Ann.* **163** (1997), 547–558.

[67] K.J. Dykema, Simplicity and the stable rank of some free product C^*-algebras, *Trans. Amer. Math. Soc.* **351** (1999), 1–40.

[68] K.J. Dykema, Exactness of reduced amalgamated free product C^*-algebras, preprint.

[69] K. Dykema, U. Haagerup and M. Rørdam, The stable rank of some free product C^*-algebras, *Duke Math. J.* **90** (1997), 95–121; Correction, *ibid.* **94** (1998), 213.

[70] K.J. Dykema and F. Rădulescu, Compressions of free products of von Neumann algebras, preprint.

[71] A. Edelman, The probability that a random real Gaussian matrix has k real eigenvalues, related distributions, and the circular law, *J. Multivariate Anal.* **60** (1997), 203–232.

[72] P.H. Edelman, Chain enumeration and non-crossing partitions, *Discrete Math.* **31** (1980), 171–180.

[73] R.S. Ellis, *Entropy, Large Deviations and Statistical Mechanics*, Springer, New York-Berlin, 1985.

[74] D.E. Evans, On $\mathcal{O}_n$, *Publ. Res. Inst. Math. Sci.* **16** (1980), 915–927.

[75] J. Feldman and C.C. Moore, Ergodic equivalence relations, cohomology, and von Neumann algebras. I, II, *Trans. Amer. Math. Soc.* **234** (1977), 289–324, 325–359.

[76] W. Feller, *An Introduction to Probability Theory and Its Applications I*, Third edition, John Wiley, New York-London-Sydney, 1968.

[77] W. Feller, *An Introduction to Probability and Its Applications II*, Second edition, John Wiley, New York-London-Sydney, 1971.

[78] B. Fuglede and R.V. Kadison, Determinant theory in finite factors, *Ann. of Math.* **55** (1952), 520–530.

[79] Z. Füredi and J. Komlós, The eigenvalues of random symmetric matrices, *Combinatorics* **1** (1981), 233-241.

[80] L. Ge, Applications of free entropy to finite von Neumann algebras, *Amer. J. Math.* **119** (1997), 467–485.

[81] L. Ge, Applications of free entropy to finite von Neumann algebras, II, *Ann. of Math. (2)* **147** (1998), 143–157.

[82] L. Ge and S. Popa, On some decomposition properties for factors of type II_1, *Duke Math. J.* **94** (1998), 79–101.

[83] S. Geman, A limit theorem for the norm of random matrices, *Ann. Probab.* **8** (1980), 252–261.

[84] S. Geman, The spectral radius of large random matrices, *Ann. Probab.* **14** (1986), 1318–1328.

[85] J. Ginibre, Statistical ensembles of complex, quaternion and real matrices, *J. Math. Phys.* **6** (1965), 440–449.

[86] V.L. Girko, Elliptic law, *Theory Probab. Appl.* **30** (1986), 677–690.

[87] V.L. Girko, *Spectral Theory of Random Matrices* (Russian), Nauka, Moscow, 1988.

[88] V.L. Girko, *Theory of Random Determinants*, Kluwer, Dordrecht 1990.

[89] V.L. Girko, The circular law: ten years later, *Random Oper. and Stoch. Equ.* **2** (1994), 235–276, 377–398.

[90] V.L. Girko, Elliptic law: ten years later I, II, *Random Oper. and Stoch. Equ.* **3** (1995), 257–302, 377–398.

[91] P. Glockner, M. Schürmann and R. Speicher, Realization of free white noises, *Arch. Math.* **58** (1992), 407–416.

[92] B. Gnedenko, *The Theory of Probability*, Mir Publishers, Moscow, 1976.

[93] O.W. Greenberg, Particles with small violations of Fermi and Bose statistics, *Phys. Rev.* **D43** (1991), 4111–4120.

[94] D.J. Gross and E. Witten, Possible third-order phase transition in the large-N lattice gauge theory, *Phys. Rev.* **D 21** (1980), 446–453.

[95] T. Guhr, A. Müller-Groeling and H.A. Weidenmüller, Random matrix theories in quantum physics: Common concepts, *Phys. Rep.* **299** (1998), 190–425.

[96] U. Haagerup, On Voiculescu's R- and S-transforms for free non-commutative random variables, in *Free Probability Theory*, D.V. Voiculescu (ed.), Fields Inst. Commun. **12**, Amer. Math. Soc., 1997, pp. 127–148.

[97] U. Haagerup and F. Larsen, Brown's spectral distribution measure for R-diagonal elements in finite von Neumann algebras, Preprint 1999/12, Institut for Matematik, Odense Universitet.

[98] U. Haagerup and S. Thorbjørnsen, Random matrices with complex Gaussian entries, Preprint 1998/7, Institut for Matematik, Odense Universitet.

[99] U. Haagerup and S. Thorbjørnsen, Random matrices and K-theory for exact C^*-algebras, *Doc. Math.* **4** (1999), 341–450

[100] F. Hiai and D. Petz, Maximizing free entropy, *Acta Math. Hungar.* **80** (1998), 325–346.

[101] F. Hiai and D. Petz, A large deviation theorem for the empirical eigenvalue distribution of random unitary matrices, Preprint No. 17/1997, Math. Inst. HAS, Budapest, to appear in *Ann. Inst. Henri Poincaré, Probabilités et Statistiques.*

[102] F. Hiai and D. Petz, Eigenvalue density of the Wishart matrix and large deviations, *Infin. Dimens. Anal. Quantum Probab. Relat. Top.* **1** (1998), 633–646.

[103] F. Hiai and D. Petz, Properties of free entropy related to polar decomposition, *Comm. Math. Phys.* **202** (1999), 421-444.

[104] O. Hiwatashi, T. Kuroda, N. Nagisa and H. Yoshida, The free analogue of noncentral chi-square distributions and symmetric quadratic forms in free random variables, *Math. Z.* **230** (1999), 63–77.

[105] O. Hiwatashi, M. Nagisa and H. Yoshida, The characterizations of a semicircle law by the certain freeness in a C^*-probability space, *Probab. Theory Related Fields* **113** (1999), 115–133.

[106] M.E. Ismail, D. Stanton and G. Viennot, The combinatorics of q-Hermite polynomials and the Askey-Wilson integral, *Europ. J. Combinatorics* **8** (1987), 379–392.

[107] K. Johansson, On fluctuations of eigenvalues of random hermitian matrices, *Duke Math. J.* **91** (1998), 151–204

[108] R.V. Kadison and J.R. Ringrose, *Fundamentals of the Theory of Operator Algebras I, II*, Providence, Amer. Math. Soc., 1986.

[109] A.M. Khorunzhy and L.A. Pastur, On the eigenvalue distribution of the deformed Wigner ensemble of random matrices, in *Advances in Soviet Math.* **19**, V.A. Marchenko (ed.), Amer. Math. Soc., 1994, pp. 97–127.

[110] P. Koosis, *Introduction to H_p Spaces*, Cambridge Univ. Press, Cambridge, 1980.

[111] B. Krawczyk and R. Speicher, Combinatorics of free cumulants, preprint.

[112] G. Kreweras, Sur le partitions noncroissées d'un cycle, *Discrete Math.* **1** (1972), 333–350.

[113] B. Kümmerer and R. Speicher, Stochastic integration on the Cuntz algebra O_∞, *J. Funct. Anal.* **103** (1992), 372-408.

[114] N.S. Landkof, *Foundations of Modern Potential Theory*, Springer, Berlin-Heidelberg-New York, 1972.

[115] F. Larsen, Brown measures and R-diagonal elements in finite von Neumann algebras, Ph.D. Thesis, Odense Universitet, 1999.

[116] F. Larsen, Powers of R-diagonal elements, Preprint 1999/13, Institut for Matematik, Odense Universitet.

[117] H. van Leeuwen and H. Maassen, A q-deformation of the Gauss distribution, *J. Math. Phys.* **36** (1995), 4743–4756.

[118] Y.G. Lu, On the interacting free Fock space and the deformed Wigner law, *Nagoya Math. J.* **145** (1997), 1–28.

[119] H. Maassen, Addition of freely independent random variables, *J. Funct. Anal.* **106** (1992), 409–438.

[120] V.A. Marchenko and L.A. Pastur, The distribution of eigenvalues in certain sets of random matrices, *Mat. Sb.* **72** (1967), 507–536; English transl., *Math. USSR Sb.* **1** (1967), 457–483.

[121] M.L. Mehta, *Random Matrices*, Second edition, Academic Press, Boston, 1991.

[122] F.J. Murray and J. von Neumann, On rings of operators, *Ann. of Math.* **37** (1936), 116–229.

[123] F.J. Murray and J. von Neumann, On rings of operators IV, *Ann. of Math.* **44** (1943), 716–808.

[124] N.I. Muskhelishvili, *Singular Integral Equations*, Noordhoff, Groningen, 1953.

[125] N.I. Muskhelishvili, *Some Basic Problems of the Mathematical Theory of Elasticity*, Noordhoff, Groningen, 1963.

[126] M. Nagisa, Stable rank of some full group C^*-algebras of groups obtained by the free product, *Internat. J. Math.* **8** (1997), 375–382.

[127] P. Neu and R. Speicher, A self-consistent master equation and a new kind of cumulants, *Z. Phys. B* **92** (1993), 399-407.

[128] P. Neu and R. Speicher, Non-linear master equation and non-crossing cumulants. In *Quantum Probability and Related Topics IX*, L. Accardi (ed.), World Scientific, Singapore, 1994, pp. 311-326.

[129] P. Neu and R. Speicher, Rigorous mean-field theory for coherent-potential approximation: Anderson model with free random variables, *J. Stat. Phys.* **80** (1995), 1279-1308.

[130] P. Neu and R. Speicher, Random matrix theory for CPA: Generalization of Wegner's n-orbital model, *J. Phys.* **A 28** (1995), L79-L83.

[131] A. Nica, A one-parameter family of transforms, linearizing convolution laws for probability distributions, *Comm. Math. Phys.* **168** (1995), 187–207.

[132] A. Nica, R-transform of free joint distributions and non-crossing partitions, *J. Funct. Anal.* **135** (1996), 271–296.

[133] A. Nica, R-diagonal pairs arising as free off-diagonal compressions, *Indiana Univ. Math. J.* **45** (1996), 529–544

[134] A. Nica, D. Shlyakhtenko and R. Speicher, Some minimization problems for the free analogue of the Fisher information, *Adv. Math.* **141** (1999), 282–321.

[135] A. Nica, D. Shlyakhtenko and R. Speicher, Maximality of the microstates free entropy for R-diagonal elements, *Pacific J. Math.* **187** (1999), 333–347

[136] A. Nica and R. Speicher, On the multiplication of free N-tuples of noncommutative random variables, *Amer. J. Math.* **118** (1996), 799–837.

[137] A. Nica and R. Speicher, R-diagonal pairs—a common approach to Haar unitaries and circular elements, in *Free Probability Theory*, D.V. Voiculescu (ed.), Fields Inst. Commun. **12**, Amer. Math. Soc, 1997, pp. 149–188.

[138] A. Nica and R. Speicher, A "Fourier transform" for multiplicative functions on non-crossing partitions, *J. Alg. Comb.* **6** (1997), 141–160.

[139] A. Nica and R. Speicher, Commutators of free random variables, *Duke Math. J.* **92** (1998), 553–592

[140] M. Ohya and D. Petz, *Quantum Entropy and Its Use*, Springer-Verlag, Heidelberg, 1993

[141] F. Oravecz, Powers of Voiculescu's circular element, Preprint No. 15/1998, Math. Inst. HAS, Budapest.

[142] F. Oravecz and D. Petz, On the eigenvalue distribution of some symmetric random matrices, *Acta Sci. Math.* **63** (1997), 483–495.

[143] K.R. Parthasarathy, *An Introduction to Quantum Stochastic Calculus*, Birkhäuser, Basel, 1992.

[144] L. Pastur, On the universality of the level spacing distribution for some ensembles of random matrices, *Lett. Math. Phys.* **25** (1992), 259–265.

[145] L. Pastur, A simple approach to global regime of the random matrix theory, preprint.

[146] L. Pastur and A. Fitogin, *Spectra of Random and Almost-Periodic Operators*, Springer, Berlin, 1992.

[147] L. Pastur and M. Sherbina, Universality of the local eigenvalue statistics for a class of unitarily invariant random matrix ensembles, *J. Stat. Phys.* **86** (1997), 109–147.

[148] D. Petz, *An Invitation to the Algebra of the Canonical Commutation Relation*, Leuven University Press, 1990.

[149] D. Petz and F. Hiai, Logarithmic energy as entropy functional, in *Advances in Differential Equations and Mathematical Physics*, E. Carlen et al. (eds.), Contemp. Math. **217**, Amer. Math. Soc., 1998, pp. 205–221.

[150] M. Pimsner and D. Voiculescu, K-groups of reduced crossed products by free groups, *J. Operator Theory*, **8** (1982), 131–156.

[151] G. Pisier, *The Volume of Convex Bodies and Banach Space Geometry*, Cambridge Univ. Press, Cambridge, 1989.

[152] S. Popa, Markov traces on universal Jones algebras and subfactors of finite index, *Invent. Math.* **111** (1993), 375-405.

[153] S. Popa, Free-independent sequences in type II_1 factors and related problems, in *Recent Advances in Operator Algebras*, *Astérique* **232** (1995), 187-202.

[154] C.E. Porter and N. Rosenzweig, Statistical properties of atomic and nuclear spectra, *Ann. Acad. Sci. Fennicae A, VI Physica* **44** (1960), 1–66.

[155] F. Rădulescu, The fundamental group of the von Neumann algebra of a free group with infinitely many generators is $\mathbb{R}_+ \setminus \{0\}$, *J. Amer. Math. Soc.* **5** (1992), 517–532.

[156] F. Rădulescu, A one-parameter group of automorphisms of $L(\mathbf{F}_\infty) \otimes B(H)$ scaling the trace, *C. R. Acad. Sci. Paris Sér. I Math.* **314** (1992), 1027-1032.

[157] F. Rădulescu, Stable equivalence of the weak closures of free groups convolution algebras, *Comm. Math. Phys.* **156** (1993), 17–36.

[158] F. Rădulescu, Random matrices, amalgamated free product and subfactors of the von Neumann algebra of a free group, of noninteger index, *Invent. Math.* **115** (1994), 347–389.

[159] F. Rădulescu, A type III_λ factor with core isomorphic to the von Neumann algebra of a free group, tensor $B(H)$, in *Recent Advances in Operator Algebras, Astérique* **232** (1995), 203-209.

[160] F. Rădulescu, Convex sets associated with von Neumann algebras and Connes' approximate embedding problem, *Math. Res. Lett.* **6** (1999), 229–236.

[161] C.R. Rao, *Linear Statistical Inference and Its Applications*, Second edition, John Wiley, New York-London-Sidney, 1973.

[162] M. Reed and B. Simon, *Methods of Modern Mathematical Physics I. Functinal Analysis*, Second edition, Academic Press, New York-London, 1975.

[163] F. Riesz and B. Sz.-Nagy, *Leçons d'analyse fonctionelle*, Akadémiai Kiadó, Budapest, 1952, 1953, 1955, 1965.

[164] Ø. Ryan, On the limit distribution of random matrices with independent or free entries, *Comm. Math. Phys.* **193** (1998), 631–650.

[165] E.B. Saff and V. Totik, *Logarithmic Potentials with External Fields*, Springer, Berlin-Heidelberg-New York, 1997.

[166] A.N. Shiryayev, *Probability*, Springer, New York-Berlin, 1984.

[167] D. Shlyakhtenko, Limit distributions of matrices with bosonic and fermionic entries, in *Free Probability Theory*, D.V. Voiculescu (ed.), Fields Inst. Commun. **12**, Amer. Math. Soc., 1997, pp. 241–252.

[168] D. Shlyakhtenko, R-transform of certain joint distributions, in *Free Probability Theory*, D.V. Voiculescu (ed.), Fields Inst. Commun. **12**, Amer. Math. Soc., 1997, pp. 253–256.

[169] D. Shlyakhtenko, Free quasi-free states, *Pacific J. Math.* **177** (1997), 329-368.

[170] D. Shlyakhtenko, Some applications of freeness with amalgamation, *J. Reine Angew. Math.* **500** (1998), 191-212.

[171] R. Simon, Combinatorial statistics on non-crossing partitions, *J. Combinatorial Th. A* **66** (1994), 270–301.

[172] R. Speicher, A new axample of independence and white noise, *Probab. Theory Related Fields* **84** (1990), 141–159.

[173] R. Speicher, A non-commutative central limit theorem, *Math. Z.* **209** (1992), 55–66.

[174] R. Speicher, Free convolution and the random sum of matrices, *Publ. Res. Inst. Math. Sci.* **29** (1993), 731-744.

[175] R. Speicher, The lattice of admissible partitions, in *Quantum Probability and Related Topics VIII*, L. Accardi (ed.), World Scientific, Singapore, 1993, pp. 347-352.

[176] R. Speicher, Multiplicative functions on the lattice of non-crossing partitions and free convolution, *Math. Ann.* **298** (1994), 611-628.

[177] R. Speicher, On universal products, in *Free Probability Theory*, D.V. Voiculescu (ed.), Fields Inst. Commun. **12**, Amer. Math. Soc., 1997, pp. 257–266..

[178] R. Speicher, *Combinatorial Theory of the Free Product with Amalgamation and Operator-Valued Free Probability Theory*, Mem. Amer. Math. Soc. **627**, 1998.

[179] R. Speicher and R. Woroudi, Boolean Convolution, in *Free Probability Theory*, D.V. Voiculescu (ed.), Fields Inst. Commun. **12**, Amer. Math. Soc., 1997, pp. 267–280.

[180] E. Spiegel and C.J. O'Donnell, *Incidence Algebras*, Marcel Dekker, New York, 1997.

[181] H. Stahl and V. Totik, *General Orthogonal Polynomials*, Cambridge Univ. Press, Cambridge, 1992.

[182] M.B. Ştefan, The primality of subfactors of finite index in the interpolated free group factors, *Proc. Amer. Math. Soc.* **126** (1998), 2299–2307.

[183] E.M. Stein, *Singular Integrals and Differentiability Properties of Functions*, Princeton Univ. Press, Princeton, 1970.

[184] S.J. Szarek, Nets of Grassmann manifold and orthogonal group, in *Proceedings of Research Workshop on Banach Space Theory*, Univ. Iowa, Iowa City, Iowa, 1982, pp. 169–185.

[185] S.J. Szarek, Metric entropy of homogeneous spaces, in *Quantum Probability*, Banach Center Publ. **43**, Polish Acad. Sci., 1998, pp. 395–410.

[186] S.J. Szarek and D. Voiculescu, Volumes of restricted Minkowski sums and the free analogue of the entropy power inequality, *Comm. Math. Phys.* **178** (1996), 563–670.

[187] G. Szegő, *Orthogonal Polynomials*, Fourth edition, Amer. Math. Soc., Providence, 1975.

[188] M. Takesaki, *Theory of Operator Algebras I*, Springer, New York-Heidelberg-Berlin, 1979.

[189] S. Thorbjørnsen, Mixed moments of Voiculescu's gaussian random matrix, Preprint 1999/6, Institut for Matematik, Odense Universitet.

[190] V. Totik, *Weighted Approximation with Varying Weight*, Lecture Notes in Math. **1569**, Springer, 1994.

[191] C.A. Tracy and H. Widom, Introduction to random matrices, in *Geometric and Quantum Aspects of Integrable Systems (Schweningen, 1992)*, Lecture Notes in Physics **424**, Springer, 1993, pp. 103–130.

[192] Y. Ueda, A minimal action of the compact quantum group $SU_q(n)$ on a full factor, *J. Math. Soc. Japan* **51** (1999), 449–461.

[193] Y. Ueda, Amalgamated free product over Cartan subalgebra, *Pacific J. Math.*, to appear.

[194] Y. Ueda, Remarks on free products with respect to non-tracial states, *Math. Scand.*, to appear.

[195] W. Van Assche, *Asymptotics for orthogonal polynomials*, Lecture Notes in Math. **1265**, Springer, 1987.

[196] D. Voiculescu, Symmetries of some reduced free product C*-algebras, in *Operator Algebras and Their Connection with Topology and Ergodic Theory*, Lecture Notes in Math. **1132**, Springer, 1985, pp. 556–588.

[197] D. Voiculescu, Addition of certain non-commuting random variables, *J. Funct. Anal.* **66** (1986), 323–346.

[198] D. Voiculescu, Multiplication of certain non-commuting random variables, *J. Operator Theory* **18** (1987), 223–235.

[199] D. Voiculescu, Circular and semicircular systems and free product factors, in *Operator Algebras, Unitary Representations, Enveloping Algebras, and Invariant Theory*, A. Connes el al. (eds.), Birkhäuser, 1990, pp. 45–60.

[200] D. Voiculescu, Noncommutative random variables and spectral problems in free product C*-algebras, *Rocky Mountain J. Math.* **20** (1990), 263–283.

[201] D. Voiculescu, Limit laws for random matrices and free products, *Invent. Math.* **104** (1991), 201–220.

[202] D. Voiculescu, The analogues of entropy and of Fisher's information measure in free probability theory, I, *Comm. Math. Phys.* **155** (1993), 71–92.

[203] D. Voiculescu, The analogues of entropy and of Fisher's information measure in free probability theory, II, *Invent. Math.* **118** (1994), 411–440.

[204] D. Voiculescu, Operations on certain non-commutative operator-valued random variables, in *Recent Advances in Operator Algebras*, *Astérisque* **232** (1995), 243–275.

[205] D. Voiculescu, The analogues of entropy and of Fisher's information measure in free probability theory III: The absence of Cartan subalgebras, *Geom. Funct. Anal.* **6** (1996), 172–199.

[206] D. Voiculescu, The analogues of entropy and of Fisher's information measure in free probability theory, IV: Maximum entropy and freeness, in *Free Probability Theory*, D.V. Voiculescu (ed.), Fields Inst. Commun. **12**, Amer. Math. Soc., 1997, pp. 293–302.

[207] D. Voiculescu, The analogues of entropy and of Fisher's information measure in free probability theory, V: Noncommutative Hilbert transforms, *Invent. Math.* **132** (1998), 189–227.

[208] D. Voiculescu, A strengthened asymptotic freeness result for random matrices with applications to free entropy, *Internat. Math. Res. Notices* **1998**, 41–63.

[209] D. Voiculescu, The analogues of entropy and of Fisher's information measure in free probability theory VI: Liberation and mutual free information, *Adv. Math.* **146** (1999), 101–166.

[210] D.V. Voiculescu, K.J. Dykema and A. Nica, *Free Random Variables*, CRM Monograph Ser., Vol. 1, Amer. Math. Soc., 1992.

[211] K.W. Wachter, The strong limits of random matrix spectra for sample matrices of independent elements, *Ann. Probab.* **6** (1978), 1–18.

[212] E.P. Wigner, Characteristic vectors of bordered matrices with infinite dimensions, *Ann. of Math.* **62** (1955), 548–564.

[213] E.P. Wigner, On the distribution of the roots of certain symmetric matrices, *Ann. of Math.* **67** (1958), 325–327.

[214] E.P. Wigner, Random matrices in physics, *SIAM Review* **9** (1967), 1–23.

[215] W. Woess, Nearest neighbour random walks on free products of discrete groups, *Boll. Un. Mat. Ital. B* **5** (1986), 961–982.

[216] F. Xu, A random matrix model from two dimensional Yang-Mills theory, *Comm. Math. Phys.* **190** (1997), 287–307.

Index

Selected Titles in This Series

(Continued from the front of this publication)

For a complete list of titles in this series, visit the AMS Bookstore at **www.ams.org/bookstore/**.